Springer Proceedings in Mathematics & Statistics

Volume 93

More information about this series at http://www.springer.com/series/10533

Springer Proceedings in Mathematics & Statistics

This book series features volumes composed of select contributions from workshops and conferences in all areas of current research in mathematics and statistics, including OR and optimization. In addition to an overall evaluation of the interest, scientific quality, and timeliness of each proposal at the hands of the publisher, individual contributions are all refereed to the high quality standards of leading journals in the field. Thus, this series provides the research community with well-edited, authoritative reports on developments in the most exciting areas of mathematical and statistical research today.

Jan Awrejcewicz
Editor

Applied Non-Linear Dynamical Systems

Springer

Editor
Jan Awrejcewicz
Department of Automation, Biomechanics
and Mechatronics
Łódź University of Technology
Łódź, Poland

ISSN 2194-1009 ISSN 2194-1017 (electronic)
ISBN 978-3-319-08265-3 ISBN 978-3-319-08266-0 (eBook)
DOI 10.1007/978-3-319-08266-0
Springer Cham Heidelberg New York Dordrecht London

Library of Congress Control Number: 2014951679

Mathematics Subject Classification (2010): 34-xx, 34A12, 34A26, 34A34, 34A35, 34A36, 34B15, 34C15, 34C23, 34C25, 34C29, 34Dxx, 34D06, 34D08, 34D10, 34D20, 24D23, 34E13, 34F10, 34F15, 34H10,37-xx, 37Bxx, 37B55, 37C27, 37C70, 37C75, 37D45, 37Gxx, 37G15, 37H20, 37Jxx, 37J20, 37J25, 37J55,70-xx, 70Exx, 70Hxx, 70Jxx, 70Kxx, 74-xx, 74Gxx, 74Jxx

Printed on acid-free paper

Springer is part of Springer Science+Business Media (www.springer.com)

Preface

The 12th International Conference "Dynamical Systems-Theory and Applications" (DSTA) was held on December 2–5, 2013, in Łódź, Poland. Organized by the staff of the Department of Automation, Biomechanics and Mechatronics of the Lodz University of Technology, the aim of the conference was to discuss and illustrate the present state and perspective for modeling, simulation, and control of nonlinear dynamical systems—a rapidly developing research front that includes various disciplines in science, bioscience, and high technology.

The Scientific Committee of the conference include the following researchers: Marcilio Alves (Brazil), Igor V. Andrianov (Ukraine), José M. Balthazar (Brazil), Wojciech Blajer (Poland), Tadeusz Burczyński (Poland), Czesław Cempel (Poland), Simona-Mariana Cretu (Romania), Virgil-Florin Duma (Romania), Horst Ecker (Austria), Michal Fečkan (Slovakia), Barry Gallacher (UK), Józef Giergiel (Poland), Peter Hagedorn (Germany), Katica Hedrih (Serbia), Ivana Kovacić (Serbia), Janusz Kowal (Poland), Jan Kozanek (Czech Republic), Vadim A. Krysko (Russia), Lidiya V. Kurpa (Ukraine), Claude-Henri Lamarque (France), Gennady A. Leonov (Russia), Nuno M.M. Maia (Portugal), Leonid I. Manevitch (Russia), Yuriy Mikhlin (Ukraine), Gerard Olivar (Colombia), Carla M.A. Pinto (Portugal), Bogdan Posiadała (Poland), Stanisław Radkowski (Poland), Bogusław Radziszewski (Poland), Giuseppe Rega (Italy), Christos H. Skiadas (Greece), Alexander Seyranian (Russia), Gábor Stépán (Hungary), Jerzy Świder (Poland), Andrzej Tylikowski (Poland), Tadeusz Uhl (Poland), Ferdinand Verhulst (The Netherlands), Jerzy Warmiński (Poland), Edmund Wittbrodt (Poland), Józef Wojnarowski (Poland), Ludmila V. Yakushevich (Russia), Hamad M. Yehia (Egypt), Mikhail V. Zakrzhevsky, and Klaus Zimmermann (Germany).

In the following 38 chapters we present only a small sample of different approaches and understandings of dynamical systems and their applications in physics, mechanics, automation, biomechanics, and applied mathematics.

In Chap. 1 Kovacic and Rand consider nonlinear oscillators with period independent of amplitude and with Duffing-type restoring force. They present Lagrangians, conservation laws, equations of motions, and solutions of the governing equations for oscillators with hardening and softening cubic-type nonlinearities.

Pilipchuk in Chap. 2 presents physical insight and methodologies of asymptotics of "rigid-body" motions for nonlinear dynamics. He shows that by tracking rigid Euclidean transformation nonlinear models can be revealed.

In Chap. 3 Krysko et al. use the Bubnov–Galerkin method and the finite difference method for studies of vibrations of flexible cylindrical and sector shells subjected to the action of uniformly distributed loads. A few novel nonlinear phenomena have been reported.

Bhem and Schwebke analyze in Chap. 4 the wormlike locomotion system. They focused on gear shift patterns and presented procedure for their adjustments to optimize speed and gait/crawling to predefined limits of actuator or spike force load.

Awrejcewicz et al. illustrate and discuss (in Chap. 5) periodic and chaotic dynamics of plates and shells as well as a weak turbulent behavior exhibited by these solid structures, while modeling them as 2D infinite objects. They present also novel approaches to obtain reliable results of nonlinear differential equations, as well as new methods of chaos monitoring.

In Chap. 6 Morcillo et al. propose a methodology of using bifurcation diagrams for computation chaos controllers. They apply this to PWM-controlled power converters method based on an adaptive control, where the sawtooth signal is redefined as a function of the output and reference voltages.

In Chap. 7 Andrianov et al. present studies of anti-plane shear waves. They focus on wave propagation through a cylindrical structured cancellous bone, applying model of a two-dimensional mesh of elastic trabeculae filled by a viscous marrow.

Kizilova et al. focused in Chap. 8 on dynamics of postural sway in human. Analyzing body sway patterns for the group of young healthy individuals and two groups of patients with pathologies of spine and joints, they observed quasi-regular and chaotic dynamics with certain asymmetry of the body acceleration, respectively.

In Chap. 9 Adamiec-Wójcik et al. present modeling of the slender system by means of the rigid finite element method. They considered large deformations and friction using rope as the model. Presented and analyzed are also dynamics of an offshore crane lifting a load from a vessel.

Syta and Litak in Chap. 10 presented results of the investigations of Van der Pol–Duffing system. They focused on the dynamical response of the system with an external harmonic excitation and a memory of a fractional characteristic. The obtained results indicate occurrence of bifurcations of quasiperiodic solutions with short intervals of chaotic solutions.

Analysis of fluid flow around the cylinder is described by Akhmetov and Kutluev in Chap. 11. They apply asymptotic methods for problem of the vortex structure appearance in a stationary viscous incompressible fluid and investigate properties of the flow function in boundary layer.

In Chap. 12 Awrejcewicz et al. present results of application of the asymptotical approach in the form of limiting phase trajectories and multiple timescale methods for analysis of dynamical problem of a two-degree-of-freedom mechanical system. Application of those approaches in the case of spring pendulum allows for determination of critical values of the parameters responsible for change of the character of vibrations.

Kaliński et al. present (Chap. 13) an idea of the workpiece holder with adjustable stiffness to be applied in milling of flexible details. This holder, together with vibration surveillance system, allows for more efficient milling using slender ball-end tools without interference of vibrations.

In Chap. 14 Blajer presents problem of servo-constraint in the inverse simulation problem for underactuated mechanical system. Simulation results are discussed and confronted with computational issues of the governing equations in the form of ordinary differential equations and differential algebraic equations.

Global analysis of nonlinear dynamics of simple hybrid electronic systems through application of the method of complete bifurcation groups is presented by Pikulin in Chap. 15. He also proves that it is possible to design reliable switching power convertor. The applied method of complete bifurcation groups allows to predict and avoid occurrence of undesirable regimes in operation.

Udwadia and Mylapilli in Chap. 16 discuss interrelations and connections between tracking control of nonlinear systems and constrained motion of mechanical systems. By providing diverse examples, they illustrate how ease, simplicity, and efficiency in control can be achieved. They also show that closed-form forces obtained are optimal and minimize the control cost at each instant of time still providing exact trajectory tracking.

In Chap. 17 Ruchkin presents new method of the intellectual investigations of the nonlinear dynamical system. By application of this method the system with a special Hamiltonian structure is studied and its regular and chaotic behavior is analyzed.

Lateral dynamics behavior of the two-axle freight wagon with the UIC double-link suspension in dependence on chosen parameters is shown in Chap. 18 by Matej et al. Based on the Coulomb law regarding friction, applied are the non-smooth mechanics, which allows for derivation of the mathematical models with and without lateral bump-stop.

Zimmerman et al. focus on kinematics and dynamics of a mechanical system with mecanum wheels (Chap. 19). They proceed with comparison of nonholonomic model and approximate model used in robotics, obtaining similar results, which resulted in the production of prototype of a mobile robot with four mecanum wheels.

Chapter 20 by Ritto and Sampaio is devoted to reliability analysis of horizontal drill-string dynamics. It focuses on reliability of the operation, defined as the probability of not achieving a target efficiency and measured by the mean input/output ratio.

Hedrih presents in Chap. 21 changes of elasticity 3D matrix surrounding mammalian oocyte. She considers changes occurring in zona pellucida during maturation and fertilization processes. Applying an oscillatory spherical net model of mouse, eigen circular frequencies of mouse oocyte and mouse embryo have been estimated.

Problem of mass points interacting gravitationally is presented by Szumiński and Przybylska in Chap. 22. They illustrate complicated behavior of trajectories of system with applied certain holonomic constraints using Poincarè cross sections.

Verhulst applies in Chap. 23 slow–fast timescales in the framework of Fenichel geometric singular perturbation theory for analysis of equations with periodic coefficients for singularly perturbed growth.

In Chap. 24 basing on fractional calculus and the concept of fractals Abramov proposes a model of nonlinear fractal oscillator. Investigated are stress field for different structural states of structural dislocation in a nanosystem, as well as its deformation and behavior of energy spectrum.

Dynamics in diagnostic expert systems are addressed by Cholewa in Chap. 25. He introduces dynamic statement network, where statements consist of contents and values. After detailed comparison of networks, he proposes manner of transformation from a static network into dynamic one.

In Chap. 26 Palej proposes method for solving the nonstandard two-point boundary value problem. He considers the case, where the number of boundary conditions is higher than the number of first-order ordinary differential equations containing certain number of unknown parameters and where difference between standard and nonstandard boundary value problem consists in the size of the initial value problem that needs to be solved.

Lainscsek et al. in Chap. 27 propose a differential equation with time delay for automatic sleep scoring from single electrode. Assuming that there is a parameter, where at least one of coefficients varies depending on sleep stages, it is possible to construct hypnogram. In this case brain activity is considered as resulting from a dynamical system.

Chapter 28 by Gyebrószki and Csernák is focused on numerical methods for quick analysis of micro-chaos. They describe and compare different methods used for characterizing chaotic behavior and compare them with simple cell mapping for investigations of chaotic behavior for the case of digitally controlled inverted pendulum.

Studies of a bouncing ball impacting with a periodically moving limiter are presented in Chap. 29 by Okniński and Radziszewski. They analyze this problem within two defined frameworks of the table motion for four cubic polynomial and sinusoidal motion models.

In Chap. 30 Pascal and Stepanov present results of investigations of the behavior of the strongly nonlinear vibrating system excited by dry friction and harmonic force. They consider as the model a system composed of two masses connected by linear springs, where one of the masses is in contact with a driving belt moving at a constant velocity and friction force with Coulomb's characteristics acting between mass and belt.

Selyutskiy in Chap. 31 focuses on studies of aerodynamic pendulum dynamics in low-speed airflow. He applies as an investigation model a phenomenological mode with the internal dynamics of the flow simulated by an oscillator attached to the pendulum and obtained results of simulations are in good correlation with experimental results.

Studies of nonlinear interactions of two coupled oscillators at different timescales are presented by Ture Savadkoohi and Lamarque in Chap. 32. Analytical developments of those investigations are compared with numerical results and discussed is the possibility of the passive control of the main system by means of the time-dependent nonlinear energy sink.

In Chap. 33 Szklarz and Jarzębowska discuss a systematic coordinate-free approach for the formulation of the no-slip condition for wheeled robot models that can be used for derivation of the nonholonomic constraint equations. This approach yields models that are unified, verifiable, comparable, and repeatable.

Original unitary theory of the optical choppers with rotating wheels working with top-hat laser beams is discussed in Chap. 34 by Circa and Duma. They present program developed for designing of the optical choppers with rotating wheels and verified it by application to classical and eclipse choppers.

In Chap. 35 modification of the mathematical model of the HIV dynamics in HIV-specific helper cells is presented by Pinto and Carvalho. Considered are two types of the delay—a latent period for the interval for cells, first one with contact with the virus, to be infected by the virons and released by them and the second one is a viron production period for the virons to be infected by virons and released to the blood stream from the infected cells.

Gallacher et al. discuss in Chap. 36 application of parametric excitation and amplification for MEMS ring gyroscope. Theoretically proved is that the parametric excitation offers suitable excitation method for the rate integrating gyroscope.

In Chap. 37 Piccirillo et al. present analysis of the influence of external parameters on the oscillator dynamics. They investigate two smart materials, i.e., shape memory alloy and magneto rheological damper as well as effect of system dissipation energy related to their hysteretic behavior.

Generalized method of studying plane topographical Poincarè systems to higher dimensions is presented by Shamolin in Chap. 38. He shows also elaborated methods for qualitative study of dissipative systems and systems with anti-dissipation yielding a possibility of obtaining conditions for bifurcation of birth of stable and unstable auto-oscillations.

As can be noticed by the variability of topics of the chapters, dynamical systems analysis can be found in very wide range of scientific disciplines and their constant development is unavoidable and necessary from both theoretical and practical point of view.

I do hope that the readers of this book will be attracted by the chosen topics. I greatly appreciate help of Springer Editor Dr. Eve Mayer in publishing the presented chapters recommended by the Scientific Committee of the DSTA 2013 after the standard review procedure. Finally, I would like to thank all referees for their help in reviewing the manuscripts.

Łódź and Warsaw, Poland Jan Awrejcewicz

Contents

Duffing-Type Oscillators with Amplitude-Independent Period

Ivana N. Kovacic and Richard H. Rand

Abstract Nonlinear oscillators with hardening and softening cubic Duffing nonlinearity are considered. Such classical conservative oscillators are known to have an amplitude-dependent period. In this work, we design oscillators with the Duffing-type restoring force but an amplitude-independent period. We present their Lagrangians, equations of motion, conservation laws, as well as solutions for motion.

1 Introduction

Classical Duffing oscillators are governed by

$$\ddot{x} + x \pm x^3 = 0. \tag{1}$$

Their restoring force $F = x \pm x^3$ includes a linear geometric term as well as a cubic geometric term: a positive sign in front of the cubic term corresponds to a hardening Duffing oscillator (HDO) and the negative one to a softening Duffing oscillator (SDO) [4, 6]. Unlike the majority of nonlinear oscillators, both of them have a closed-form exact solution, which is expressed in terms of Jacobi cn or sn elliptic functions. These solutions corresponding to the following initial conditions

$$x(0) = A, \quad \dot{x}(0) = 0, \tag{2}$$

I.N. Kovacic (✉)
Department of Mechanics, Faculty of Technical Sciences, University of Novi Sad, 21125 Novi Sad, Serbia
e-mail: ivanakov@uns.ac.rs

R.H. Rand
Department of Mathematics, Cornell University, Ithaca, NY 14853, USA

Department of Mechanical and Aerospace Engineering, Cornell University, Ithaca, NY 14853, USA
e-mail: rhr2@cornell.edu

J. Awrejcewicz (ed.), *Applied Non-Linear Dynamical Systems*, Springer Proceedings in Mathematics & Statistics 93, DOI 10.1007/978-3-319-08266-0_1

Table 1 Solutions for motion of the HDO and SDO (*in case of the SDO one requires $|A| < 1$ for the closed orbits around the stable origin)

	Solution for motion	Frequency	Elliptic modulus
HDO:	$x_{\text{HDO}} = A\,\text{cn}\,(\omega_{\text{HDO}}t, k_{\text{HDO}})$	$\omega_{\text{HDO}}^2 = 1 + A^2$	$k_{\text{HDO}}^2 = \frac{A^2}{2\omega_{\text{HDO}}^2} = \frac{A^2}{2(1+A^2)}$
SDO*:	$x_{\text{SDO}} = A\,\text{sn}\,(\omega_{\text{SDO}}t, k_{\text{SDO}})$	$\omega_{\text{SDO}}^2 = 1 - \frac{A^2}{2}$	$k_{\text{SDO}}^2 = \frac{A^2}{2\omega_{\text{SDO}}^2} = \frac{A^2}{2\left(1-\frac{A^2}{2}\right)}$

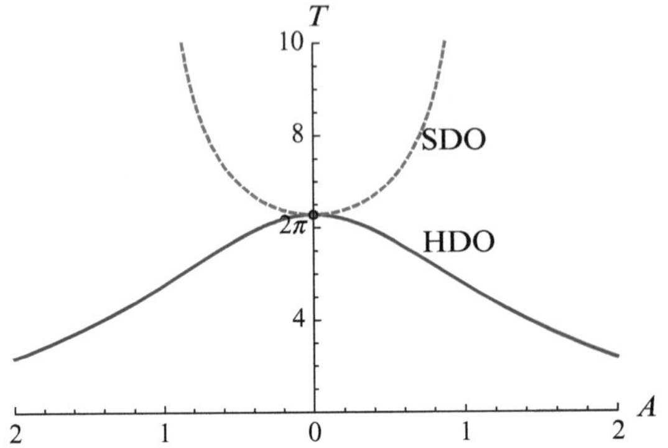

Fig. 1 Period of the HDO (*blue solid line*) and SDO (*red dashed line*) as a function of amplitude A

are given in Table 1 together with their frequencies ω and elliptic moduli k (see, for example, [4, 6] for more detail).

Given the fact that the period of cn and sn is $T = 4K(k)/\omega$, where $K(k)$ stands for the complete elliptic integral of the first kind, the following expressions are obtained for the period of the HDO and SDO:

$$T_{\text{HDO}} = \frac{4K\left(\sqrt{\frac{A^2}{2(1+A^2)}}\right)}{\sqrt{1+A^2}}, \quad T_{\text{SDO}} = \frac{4K\left(\sqrt{\frac{A^2}{2\left(1-\frac{A^2}{2}\right)}}\right)}{\sqrt{1-\frac{A^2}{2}}}. \tag{3}$$

As seen from these expressions and Fig. 1, the period of the HDO and SDO depends on the amplitude A. This leads us to the question of designing an oscillator having the Duffing-type restoring force $F = x \pm x^3$, but a constant, amplitude-independent period, corresponding to the so-called isochronous oscillators [1, 2]. In what follows we present such Duffing-type oscillators, both hardening and softening, modelled by

$$\ddot{x} + G(x, \dot{x}) + x \pm x^3 = 0, \tag{4}$$

finding also the corresponding Lagrangians, conservation laws, as well as solutions for motion.

2 Derivation

The derivation of the mathematical model (4) is based on the transformation approach [5], in which the kinetic energy E_k and potential energy E_p of nonlinear oscillators are made equal to the one of the simple harmonic oscillator (SHO)

$$E_{k\text{SHO}} = \frac{1}{2}\dot{X}^2, \quad E_{p\text{SHO}} = \frac{1}{2}X^2, \tag{5}$$

which is known to have a constant, amplitude-independent period (note that its generalized coordinate is labelled here by X).

2.1 *Case I*

To find the first mathematical model of the form (4), we assume that $E_p = E_p(x)$ and let $E_p \equiv E_{p\text{SHO}} = X^2/2$, obtaining

$$X = \sqrt{2E_p}. \tag{6}$$

Then, we also make the kinetic energy E_k of nonlinear oscillators equal to the one of the SHO and use Eq. (6) to derive

$$E_k = \frac{\dot{X}^2}{2} = \frac{\left(E_p'\right)^2}{4E_p}\dot{x}^2, \tag{7}$$

where $E_p' = dE_p/dx$. The differential equation of motion stemming from the Lagrangian $L = E_k - E_p$ has a general form

$$\ddot{x} + \left(\frac{E_p''}{E_p'} - \frac{E_p'}{2E_p}\right)\dot{x}^2 + \frac{2E_p}{E_p'} = 0. \tag{8}$$

The last term on the left-hand side $2E_p/E_p'$ is required to correspond to the Duffing restoring force F=$x \pm x^3$, which gives the potential energy

$$E_p = \frac{x^2}{2\left(1 \pm x^2\right)}. \tag{9}$$

Equation (7) now yields the kinetic energy

$$E_k = \frac{1}{2\left(1 \pm x^2\right)^3}\dot{x}^2, \tag{10}$$

so that Eq. (8) becomes

$$\ddot{x} \mp \frac{3x}{1 \pm x^2}\dot{x}^2 + x \pm x^3 = 0. \tag{11}$$

Based on $X^2/2 + \dot{X}^2/2 = const.$, the system exhibits the first integral

$$\frac{1}{(1 \pm x^2)^3}\dot{x}^2 + \frac{x^2}{1 \pm x^2} = \frac{A^2}{1 \pm A^2}. \tag{12}$$

Taking the solution for motion of the SHO in the form $X = a\cos(t + \alpha)$, where a and α are constants, and using Eqs. (6), (9) and (2), the solution for motion of Eq. (11) is obtained

$$\frac{A}{\sqrt{1 \pm A^2}}\cos t = \frac{x}{\sqrt{1 \pm x^2}}. \tag{13}$$

Numerical verifications of the analytical results for the motion (13) and phase trajectories (12) are shown in Figs. 2 and 3 for the HDO and SDO, respectively. These figures confirm that the analytical results coincide with the solutions obtained by solving the equation of motion (11) numerically. In addition, they illustrate the fact that the period stays constant despite the fact that the amplitude A changes, i.e., that the systems perform isochronous oscillations.

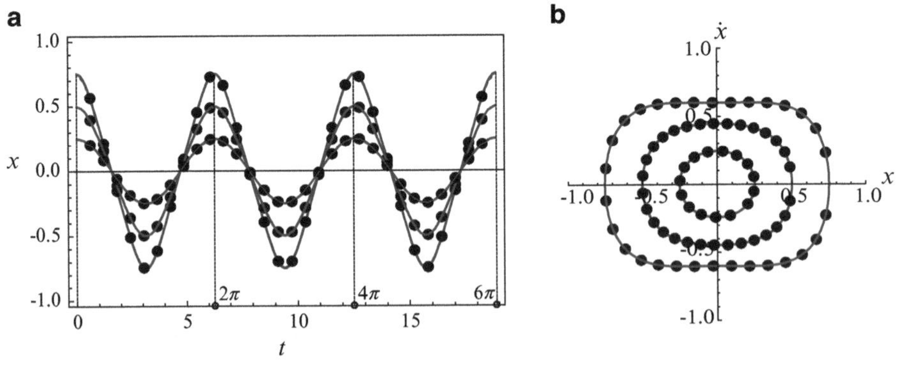

Fig. 2 Isochronous oscillations of the HDO, Eq. (11) for $A = 0.25; 0.5; 0.75$: (**a**) time histories obtained numerically from Eq. (11) (*black dots*) and from Eq. (13) (*blue solid line*); (**b**) phase trajectories obtained numerically from Eq. (11) (*black dots*) and from Eq. (12) (*blue solid line*) (upper signs are used in all these equations)

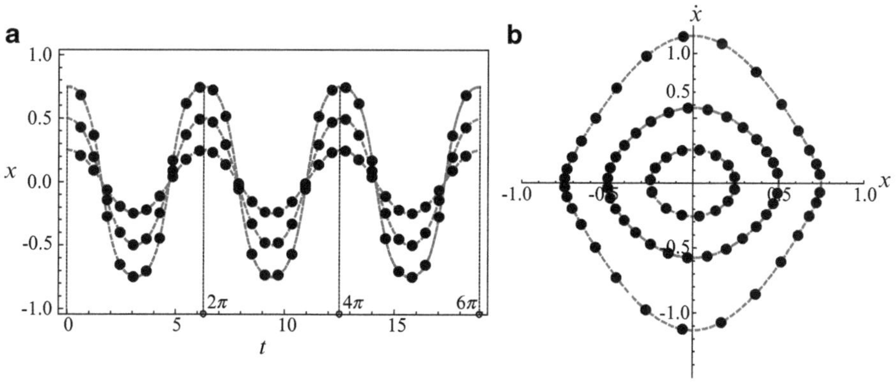

Fig. 3 Isochronous oscillations of the SDO, Eq. (11) for $A = 0.25; 0.5; 0.75$: (**a**) time histories obtained numerically from Eq. (11) (*black dots*) and from Eq. (13) (*red dashed line*); (**b**) phase trajectories obtained numerically from Eq. (11) (*black dots*) and from Eq. (12) (*red dashed line*) (lower signs are used in all these equations)

2.2 *Case II*

In this case we consider the system whose potential and kinetic energies are

$$E_p = \frac{1}{2}X^2 = \frac{1}{2}(x\,f)^2, \quad E_k = \frac{1}{2}\dot{X}^2 = \frac{1}{2}\left(\dot{x}\,f + x^2 f\,'\right)^2, \tag{14}$$

where $f \equiv f(I)$, $I = \int_0^t x(t)\,dt$, and $f\,' = df/dI$.

The corresponding Lagrange's equation is

$$\ddot{x} + 3x\dot{x}\frac{f\,'}{f} + x + \frac{f\,''}{f}x^3 = 0. \tag{15}$$

This system has two independent first integrals. The first one is the energy conservation law stemming from $\dot{X}^2/2 + X^2/2 = const$:

$$\left(\dot{x}\,f + x^2 f\,'\right)^2 + (x\,f)^2 = h_1, \quad h_1 = const. \tag{16}$$

The other first integral is related to the principle of conservation of momentum for the SHO $\dot{X} + \int_0^t X dt = \dot{X}(0) = const$. By using X and $\dot{X}$ from (14) and knowing that $dI/dt = x$, we obtain

$$\dot{x}\,f + x^2 f\,' + \int f(I)\,dI = h_2, \quad h_2 = const. \tag{17}$$

In addition, as the solution for motion for the SHO can be written down as $X = a\sin(t+\alpha)$, the following should be satisfied:

$$a\sin(t+\alpha) = x\,f, \quad a\cos(t+\alpha) = \dot{x}\,f + x^2 f'. \tag{18}$$

For the hardening-type nonlinearity in Eq. (15), one requires $f''/f = 1$, which leads to

$$f_{\mathrm{HDO}} = \exp\left(\int_0^t x(t)\,dt\right), \tag{19}$$

and the equation of motion takes the form

$$\ddot{x} + 3x\dot{x} + x + x^3 = 0. \tag{20}$$

Two first integrals (16) and (17) are

$$\exp(2x_1)\left[\left(x_3 + x_2^2\right)^2 + x_2^2\right] = h_1, \tag{21}$$

and

$$\exp(x_1)\left(x_3 + x_2^2 + 1\right) = h_2, \tag{22}$$

where

$$x_1 = \int_0^t x(t)\,dt, \quad x_2 = \dot{x}_1 = x, \quad x_3 = \dot{x}_2 = \dot{x}, \tag{23}$$

with initial conditions being [see Eq. (2)]

$$x_1(0) = 0, \quad x_2(0) = \dot{x}_1(0) = A, \quad x_3(0) = \ddot{x}_1(0) = 0. \tag{24}$$

Equation (21) is plotted in Fig. 4a for $h_1 = 1$. To analyze phase trajectories in more detail, Eq. (22) is squared and divided by Eq. (21) to obtain

$$\frac{\left(x_3 + x_2^2 + 1\right)^2}{\left(x_3 + x_2^2\right)^2 + x_2^2} = B, \quad B = const. \tag{25}$$

This expression agrees with the first integral obtained and studied in [3] and is plotted in Fig. 4b, where periodic solutions correspond to the case $B > 1$. Note that for the initial conditions (24), one has $B = 1 + 1/A^2$, which implies that B is always higher than unity.

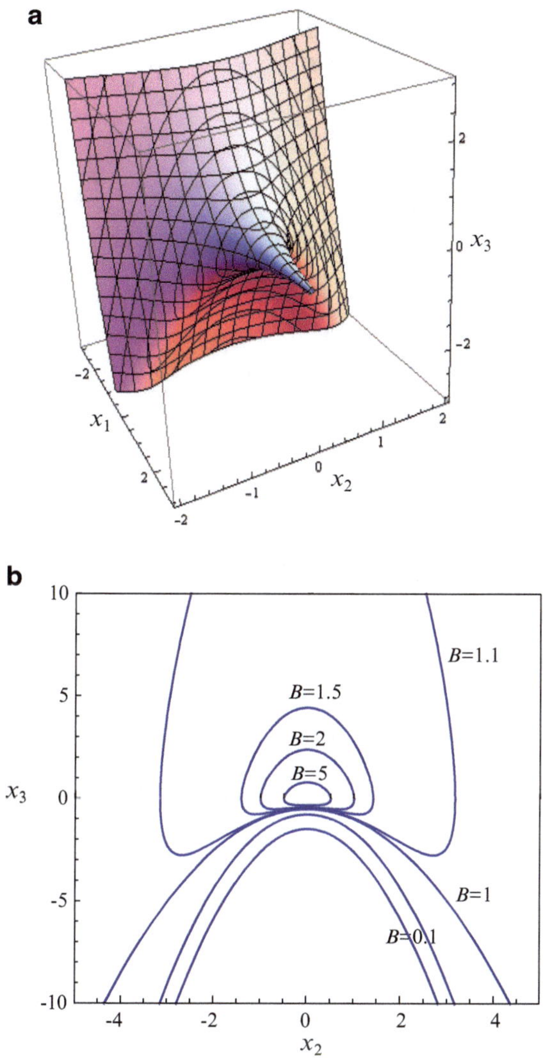

Fig. 4 (**a**) 3D plot of Eq. (21); (**b**) phase trajectories obtained from Eq. (25) for different values of B

By using (18) and (19) one can derive

$$\dot{x}\,\sin(t+\alpha)+x^2\sin(t+\alpha)-x\cos(t+\alpha)=0. \tag{26}$$

Its solution satisfying Eq. (2) is

$$x=\frac{\sin\left(t+\arctan\frac{1}{A}\right)}{\sqrt{1+\frac{1}{A^2}}-\cos\left(t+\arctan\frac{1}{A}\right)}. \tag{27}$$

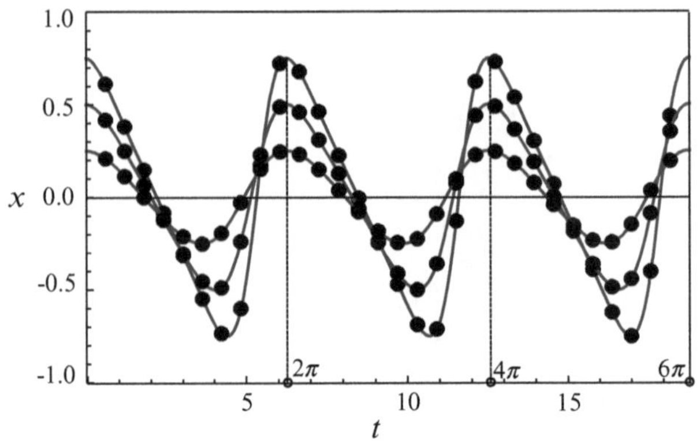

Fig. 5 Time response of the HDO, Eq. (20) for $A = 0.25; 0.5; 0.75$: numerically obtained solution from Eq. (28) (*black dots*) and from Eq. (27) (*blue solid line*)

To compare the analytical solution for motion (27) with a numerically obtained solution of the equation of motion (20), the latter is written down in the form

$$\dddot{x}_1 + 3\dot{x}_1\ddot{x}_1 + \dot{x}_1 + \dot{x}_1^3 = 0, \tag{28}$$

and numerically integrated by using the initial conditions (24). This comparison, plotted in Fig. 5, shows that these two types of solution are in full agreement as well as that the period is amplitude-independent.

The equation of motion (15) corresponds to the SDO if $f\,''/f = -1$, which is satisfied for

$$f_{\mathrm{SDO}} = \cos\left(\int_0^t x\,(t)\,dt\right). \tag{29}$$

This equation of motion is now given by

$$\ddot{x} - 3x\dot{x}\tan\left(\int_0^t x\,(t)\,dt\right) + x - x^3 = 0. \tag{30}$$

By using the notation given in Eq. (23), the equation of motion (30) transforms to

$$\dddot{x}_1 - 3\dot{x}_1\ddot{x}_1\tan x_1 + \dot{x}_1 - \dot{x}_1^3 = 0, \tag{31}$$

with the initial conditions given in Eq. (24). Two first integrals (16) and (17) become

$$\left(x_3\,\cos x_1 - x_2^2\,\sin x_1\right)^2 + x_2^2\,\cos^2 x_1 = h_1, \tag{32}$$

and

$$x_3 \cos x_1 - x_2^2 \sin x_1 + \sin x_1 = h_2. \tag{33}$$

These two integrals can be manipulated to exclude x_1 and to derive

$$\begin{aligned}(x_2^2 - 1)^2(h_1 + (-1 - h_1 + h_2^2)x_2^2 + x_2^4)^2 + 2(1 - h_1 - h_2^2 \\ +(-1 + h_1 - h_2^2)x_2^2)(h_1(x_2^2 - 1) - x_2^2(h_2^2 + x_2^2 - 1))x_3^2 \\ +((-1 + h_1)^2 - 2(1 + h_1)h_2^2 + h_2^4)x_3^4 = 0.\end{aligned} \tag{34}$$

For the initial conditions (24) one has $h_1 = A^2$ and $h_2 = 0$. Introducing these values into Eq. (34) and solving it with respect to x_3, the following explicit solution for phase trajectories is obtained:

$$x_3 = \pm(x_2^2 - 1)\sqrt{\frac{A^2 - x_2^2}{1 - A^2}}. \tag{35}$$

Combining equations in (18) and using $a = A$ and $\alpha = \pi/2$, we derive

$$\dot{x}\, A \cos t - x^2\sqrt{x^2 - A^2 \cos^2 t} + xA \sin t = 0. \tag{36}$$

Its solution satisfying Eq. (2) is

$$x = \frac{A \cos t}{\sqrt{1 - A^2 \sin^2 t}}. \tag{37}$$

This solution is plotted in Fig. 6 together with the numerical solution of Eq. (31) with Eq. (24) for different values of A. These solutions coincide and confirm isochronicity.

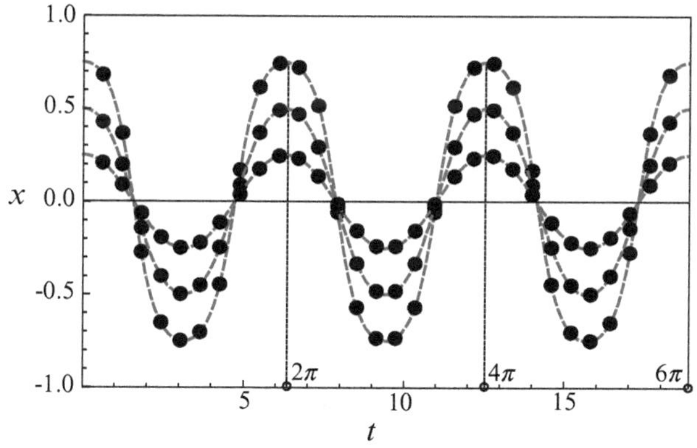

Fig. 6 Time response of the SDO, Eq. (30) with $A = 0.25$; 0.5; 0.75: numerically obtained solution from Eq. (31) (*black dots*) and from Eq. (37) (*red dashed line*)

3 Conclusions

In this work we have considered nonlinear oscillators with a hardening and softening Duffing restoring force. Unlike classical conservative Duffing oscillators, which have an amplitude-dependent period, the designed Duffing-type oscillators have the period that does not change with their amplitude and are, thus, isochronous. Two separate cases are considered with respect to the form of their potential and kinetic energy, which are made equal to the corresponding energies of the SHO, which is known to be isochronous. Corresponding equations of motions are derived, as well as their solutions for motion. Numerical verifications of these isochronous solutions are provided. In addition, two independent first integrals are presented: the energy-conservation law and the principle of conservation of momentum.

Acknowledgements Ivana Kovacic acknowledges support received from the Provincial Secretariat for Science and Technological Development, Autonomous Province of Vojvodina (Project No. 114-451-2094).

References

1. Calogero, F.: Isochronous Systems. Oxford University Press, Oxford (2008)
2. Calogero, F.: Isochronous dynamical systems. Phil. Trans. R. Soc. A **369**, 1118–1136 (2011)
3. Iacono,R., Russo, F.: Class of solvable nonlinear oscillators with isochronous orbits. Phys. Rev. E **83**, 027601 (4 pages) (2011)
4. Kovacic, I., Brennan, M.J.: The Duffing Equation: Nonlinear Oscillators and their Behaviour. Wiley, Chichester (2011)
5. Kovacic, I., Rand, R.: About a class of nonlinear oscillators with amplitude-independent frequency. Nonlinear Dyn. **74**, 455–465 (2013)
6. Rand, R.H.: Lecture notes on nonlinear vibrations (version 53). http://dspace.library.cornell.edu/handle/1813/28989

Asymptotic of "Rigid-Body" Motions for Nonlinear Dynamics: Physical Insight and Methodologies

Valery Pilipchuk

Abstract The purpose of the present work is to show that an adequate basis for understanding the essentially nonlinear phenomena must also be essentially nonlinear but still simple enough to play the role of a basis. It is shown that such types of "elementary" nonlinear models can be revealed by tracking the hidden links between analytical tools of analyses and subgroups of the rigid-body motions or, in other terms, rigid Euclidean transformation. While the subgroup of rotations is linked with linear and weakly nonlinear vibrations, the translations with reflections can be viewed as a geometrical core of the strongly nonlinear dynamics associated with the so-called vibro-impact behaviors. It is shown that the corresponding analytical approach develops through non-smooth temporal substitutions generated by the impact models.

1 Introduction

This work is motivated by the intent to introduce a unified physical basis for analyzing vibrations of essentially unharmonic, non-smooth, or maybe discontinuous time shapes. Transitions to non-smooth limits can make investigations especially difficult due to the fact that the dynamic methods were originally developed within the paradigm of smooth motions based on the classical theory of differential equations. From the physical standpoint, such way is natural for modeling the low-energy motions. Although the impact dynamics has also quite a long prehistory, non-smooth behaviors are often viewed as an exemption rather than a rule. Notice that the classical theory of differential equations usually avoids non-differentiable and discontinuous functions. Presently, however, many theoretical and applied areas are dealing with the high-energy phenomena accompanied by strongly nonlinear spatiotemporal behaviors making the classical smooth methods difficult to apply. For instance, such phenomena are considered in engineering analyses of dynamical systems under constraint conditions, friction-induced vibrations, structural damages

V. Pilipchuk (✉)
Wayne State, Detroit, MI, USA
e-mail: pilipchuk@wayne.edu

J. Awrejcewicz (ed.), *Applied Non-Linear Dynamical Systems*, Springer Proceedings in Mathematics & Statistics 93, DOI 10.1007/978-3-319-08266-0_2

due to cracks, liquid sloshing impacts, etc. Similarly to the well-known analogy between mechanical and electrical harmonic oscillators, the so-called Schmitt trigger circuits generate non-smooth signals whose temporal shapes resemble the mechanical vibro-impact processes. In many such cases, it is still possible to adapt different smooth methods of the dynamic analyses through strongly nonlinear algebraic manipulations with state vectors or by splitting the phase space into multiple "smooth domains." As a result, formulations are often reduced to the discrete mappings in a wide range of the dynamics from periodic to stochastic. It will be shown that a complementary analytical tool can be built on generating models developing essentially unharmonic behaviors as their inherent properties. For instance, the methodology presented in this work employs elementary impact systems as a physical basis for describing different types of unharmonic processes. This is implemented through the non-smooth time substitutions introduced originally for strongly nonlinear but smooth models [5]. Besides, as was shown in [4], such methodology reveals explicit links between the impact dynamics and the algebra of hyperbolic numbers analogously to the link between harmonic vibrations and conventional (elliptic) complex analyses.

2 Physical and Mathematical Principles

2.1 Linear and Elementary Nonlinear Phenomena

Although the notion of linearity is quite clear in terms of mathematical formulations, the attempt to directly associate the mathematical definition with physical phenomena faces ambiguities due to the fact that the differential equations of motion for the same model may appear to be either linear or nonlinear if switching between different types of coordinates. For instance, a mechanical system, which is linear in Cartesian coordinates, becomes nonlinear in polar coordinates. Nevertheless, recognizing the unique role of Cartesian coordinates, it was suggested to define a mechanical system as linear if the corresponding differential equations of motion are linear in Cartesian coordinates.[1] The purpose of this subsection is to determine the most elementary dynamic *phenomena* that can already be qualified as essentially nonlinear. For that reason, it is convenient to consider the linear situation first, for instance, on a typical mass-spring model; see Fig. 1a, where all the springs are assumed to be linearly elastic. The corresponding position vector is represented in the following complex form:

$$\overline{q} = \mathbf{A}_1 \exp\left(i\omega_1 t + \varphi_1\right) + \mathbf{A}_2 \exp\left(i\omega_2 t + \varphi_2\right) \tag{1}$$

where $\mathbf{A}_k$ are constant complex vectors and ω_k and φ_k ($k = 1, 2$) are the modal frequencies and initial phases.

---

[1]This definition was suggested by V. Zhuravlev (private communication, Moscow, 1989).

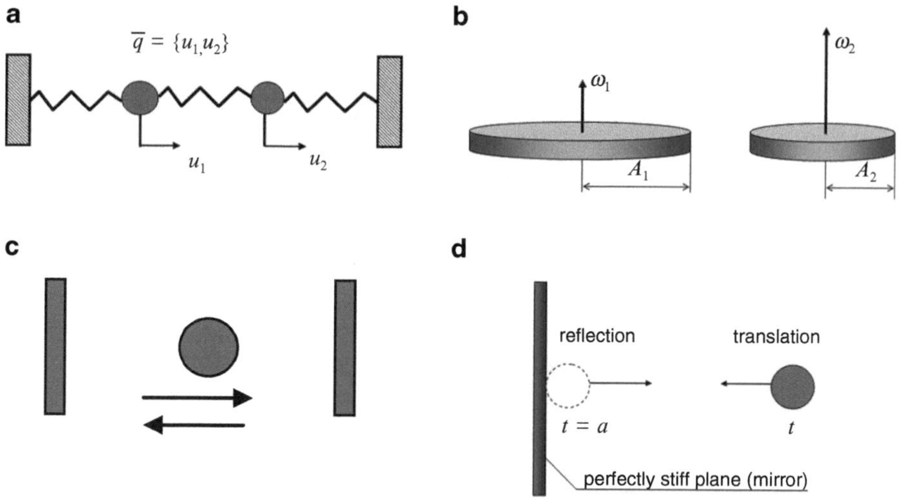

Fig. 1 Vibration of the linearly elastic system (**a**) is represented by the rigid-body rotations (**b**); the impact oscillator; (**c**) is associated with the rigid-body translation and reflection (**d**)

The fragment (b) in Fig. 1 shows another mechanical system whose dynamics is described by the same expression (1), if the system's position is identified by the two points fixed at the edges of the discs. Therefore, the one-dimensional dynamics generated by linearly elastic restoring forces can be represented by the rigid-body rotations. In other words, the linearly elastic forces are effectively eliminated by increasing the dimension of space. To some extent, such simple observation provides an explanation why the sine and cosine waves possess so convenient mathematical properties. Namely these harmonic functions represent the subgroup of elementary rigid-body motions, such as rotation. Furthermore, the link between the two models, (a) and (b), in Fig. 1, enables one to associate the linearity with the subgroup of rotations.

Apparently, the rotations do not cover all the rigid-body motions – rigid Euclidean transformations. These include also translations and reflections. The translation itself seems to have a little physical content. However, combinations of translation and reflection appear to be more interesting as shown in Fig. 1c, d through the corresponding mechanical representations. In particular, the fragment (c) shows a typical impact oscillator, whereas the fragment (d) illustrates a single reflection case, which can be viewed as the most elementary nonlinear phenomenon.

2.2 *Impact Models and Non-smooth Temporal Substitutions*

Based on the class of linear generating models, whose typical representative is shown in Fig. 1a, the quasi-harmonic methods for nonlinear vibration theory are

	Mechanical model	Basic function	Generated Temporal substitution
1	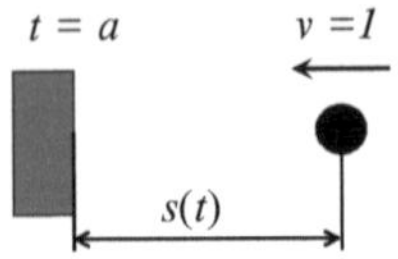	$s = \lvert t-a \rvert$	$t = a + s\dot{s}$ $\dot{s}^2 = 1$
2	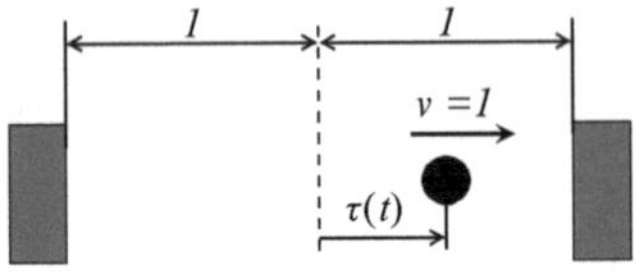	$\tau = \frac{2}{\pi}\arcsin\sin\left(\frac{\pi t}{2}\right)$	$t = 1 + (\tau - 1)\dot{\tau}$ $\dot{\tau}^2 = 1$
3	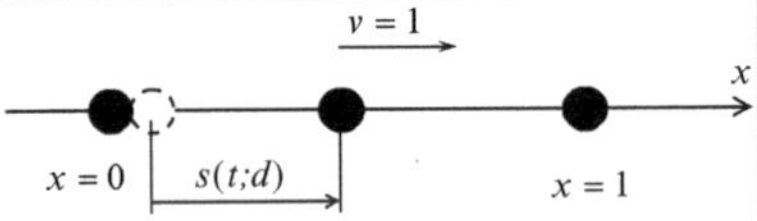	$s(t;d)$ $= \frac{1}{2}(d + \lvert t \rvert - \lvert t-d \rvert)$	$t = \sum_{i=0}^{\infty}(t_i + s_i)\dot{s}_i$ $\dot{s}_i^2 = 1$

Fig. 2 Different versions of non-smooth temporal substitutions generated by the impact models

well developed and widely used. In contrast, using the models associated with the translation-reflection, as those shown in Fig. 1c, d, is less common. Despite of being geometrically simple, such models are not described within the classical theory of differential equations due to the presence of discontinuities in the dynamic states. This fact essentially complicates any direct use of the impact models as a basis for building asymptotic or iterative procedures. It is shown, however, that appropriate preliminary adaptations of the differential equations of motion can be conducted through the non-smooth time substitutions listed in the third column of Fig. 2. In particular, the first row explains how such type of substitutions is introduced. Namely, the basic function $s = |t - a|$ is the model's coordinate, which can be interpreted as an *eigentime* of the system provided that nothing else happens except the translation and reflection under consideration. The goal is to introduce a new temporal argument, say s, in order to obtain the differential equation of motion for the model, which is shown on the left of first row in Fig. 2. Apparently, the substitution $t \to s$ cannot work directly since no inversion $t = t(s)$ does exist on the entire time domain. Nevertheless, the following generalization holds $t = t\,(s, \dot{s})$; see the first row on its right in Fig. 2 for details. In particular, this inversion appears to have the specific algebraic structure of hyperbolic numbers with the basis $\{1, \dot{s}\}$. In contrast to the conventional elliptic complex numbers, the *unipotent* $\dot{s}$, if squared, gives the positive sign, namely, $\dot{s}^2 = 1$. Interestingly enough, the hyperbolic numbers have been known for about one and a half century as abstract algebraic elements with no relation to the nonlinear dynamics or non-

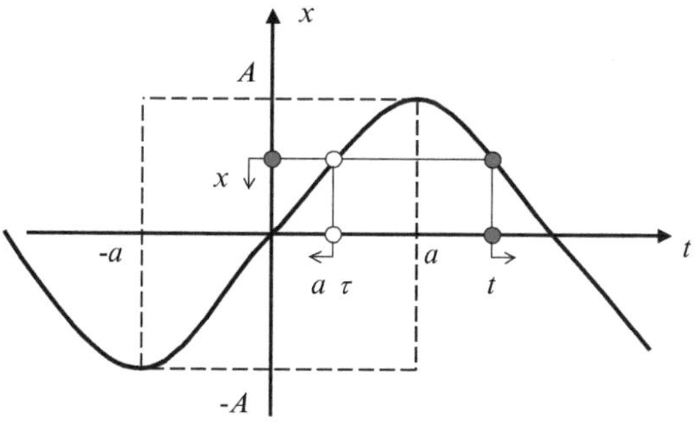

Fig. 3 The "oscillating time" τ changes its direction whenever a system makes a U-turn, while the original physical time t runs to infinity

smooth functions (http://en.wikipedia.org/wiki/Split-complex_number). Moreover, such numbers have been reintroduced multiple times under different names; see [1] for an explanation of this fact and further details. In the present case, the hyperbolic structures are generated by the non-smooth temporal arguments of impact systems, while the unipotent $\dot{s}$ possesses a clear physical interpretation as the normalized velocity of impact system. Different applications to the vibration problems are based on the periodic version illustrated by the second row of Fig. 2. In particular, the equation on the right of the row means that *during one period, the time argument is expressed through the dynamic states of impact oscillator in the form of hyperbolic number with the unipotent* $\dot{\tau}$ [4, 5]. This statement therefore applies to any periodic process. Figure 3 provides a geometrical interpretation of the "oscillating time" τ for the particular case of periodic process described by the even function $x(t)$ with respect to the quarter of period, $t = a$.

Finally, in the third row of Fig. 2, the original time argument is structurised to match the one-dimensional dynamics of rigid-body chain of identical particles. The continuous "global" time is associated with the propagation of linear momentum, whereas a sequence of non-smooth "local" time arguments describes the behavior of individual physical particles. Such an idea helps to incorporate the temporal symmetries of the dynamics into the differential equations of motion in many other cases of regular or irregular sequences of internal impacts or external pulses. Since the local times are bounded, a wider range of analytical tools becomes applicable.

2.3 *Different Asymptotic Approaches to the Vibration Theory*

Figure 4 provides further illustration for logical links between two alternative approaches to the vibration problems. Interestingly enough, the illustration is possible within the same one-degree-of-freedom model, which is shown at the first row of Fig. 4. Note that the oscillators with power-form characteristics were considered for quite a long time. For instance, Lyapunov obtained such oscillators while investigating degenerated cases of the dynamic stability problems [2]. Besides, he introduced a couple of special functions, *cs* and *sn*, in order to invert the

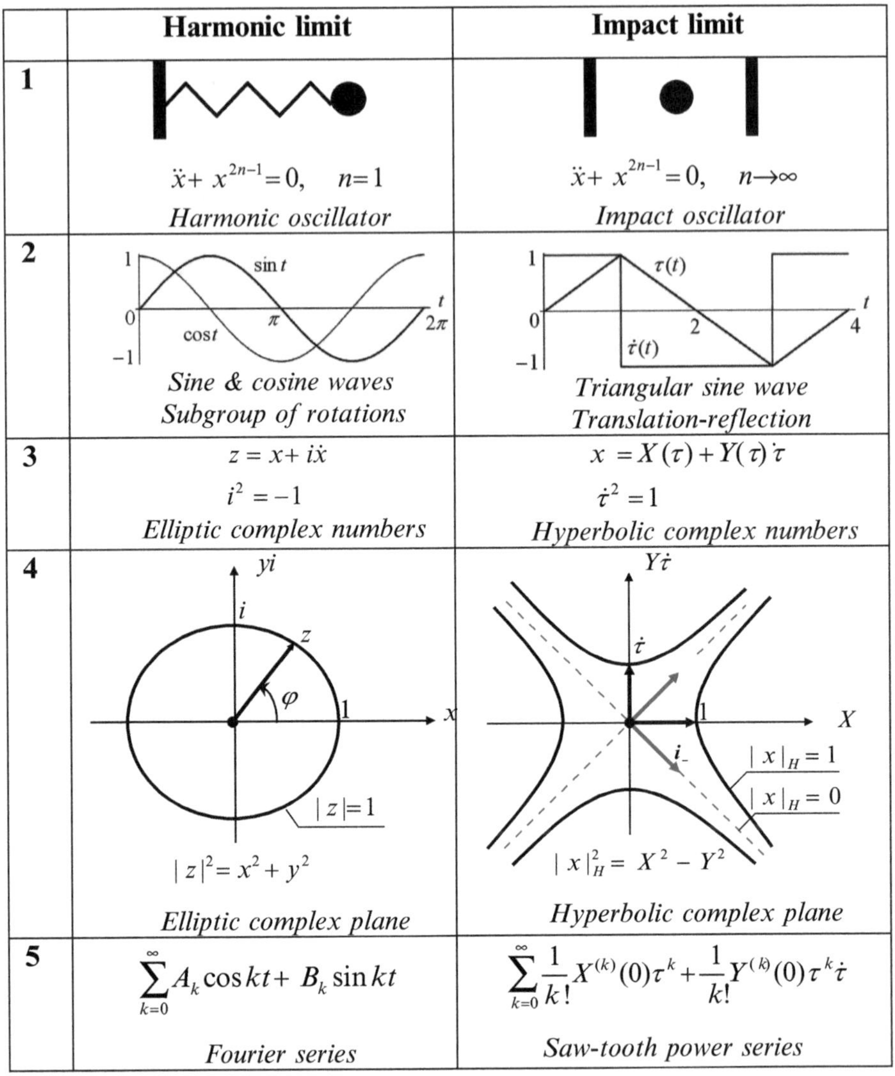

Fig. 4 Two alternative approaches to the vibration theory based on the harmonic and impact limits

corresponding quadratures. This relatively simple model nevertheless depicts the gradual transition from linear to strongly nonlinear dynamics as the exponent n runs from unity to infinity. Notably, all the temporal mode shapes of the oscillator are described by the special functions *cs* and *sn*, except the two boundaries of the interval $1 \leq n < \infty$. Both boundaries represent simple asymptotic limits described within the class of elementary functions.

Consider first the limit of harmonic oscillator ($n = 1$), generating the *sine* and *cosine* waves; see the left of the second row. The widely known convenience of using this couple of functions can be explained by their links to the elementary rigid-body motions, namely the subgroup of rotations. The algebra of conventional (elliptic) complex numbers with the corresponding complex plane representation can be viewed as a next hierarchic step here due to the well-known Euler formula. Finally, taking the linear combination of harmonic waves with different frequencies and keeping in mind the idea of parameter variations lead the area of harmonic and quasi-harmonic analyses of vibrating systems. Such tools therefore represent the dynamic processes as a combination of the elementary rigid-body rotations with different angular speeds.

Let us consider now the limit $n \to \infty$, when the restoring force vanishes inside the interval $-1 < x < 1$ but becomes infinitely large as the system reaches the potential barriers at $x = \pm 1$. The physical meaning of this limit is introduced at the top of the right column in Fig. 4. Despite of the strong (impact) nonlinearity, the limiting oscillator is also described by quite simple elementary functions such as the triangular sine and rectangular cosine, say τ and $\dot{\tau}$. These two non-smooth functions are associated with another subgroup of the rigid-body motions, namely translation and reflection. Therefore, analogously to the case $n = 1$, the upper limit $n = \infty$ can play the same fundamental role by generating a hierarchy of tools as listed in the right column of Fig. 4.

3 Further Mathematical Properties and Examples

This section describes the basic mathematical properties of the non-smooth temporal substitutions introduced in Fig. 2. These properties are used then for derivations in different illustrating examples.

3.1 Mathematical Properties

Consider first the single reflection case; see the first row of Fig. 2. Algebraic, differential, and integral properties are as follows:

- Isomorphism with 2×2 symmetric matrixes:

$$t^2 = (a + s\dot{s})^2 = a^2 + s^2 + 2as\dot{s}$$

$$\hat{t}^2 = \begin{pmatrix} a & s \\ s & a \end{pmatrix}^2 = \begin{pmatrix} a^2 + s^2 & 2as \\ 2as & a^2 + s^2 \end{pmatrix}$$

- Functional linearity holds for any function *x(t)*:

$$x(t) = x\,(a + s\dot{s}) = X(s) + Y(s)\dot{s} \tag{2}$$

$$X(s) = \tfrac{1}{2}\left[x(a+s)+x(a-s)\right]$$
$$Y(s) = \tfrac{1}{2}\left[x(a+s)-x(a-s)\right]$$

For instance, $\exp(t) = \exp(a)(\cosh s + \dot{s}\sinh s)$

- Division is conditioned to exclude the possibility of zero denominators:

$$t^{-1} = \frac{(a - s\dot{s})}{(a + s\dot{s})\,(a - s\dot{s})} = \frac{a - s\dot{s}}{a^2 - s^2} = \frac{a}{a^2 - s^2} - \frac{s}{a^2 - s^2}\dot{s}$$
$$(s \neq |a|)$$

- Sequential differentiation remains in the algebra of hyperbolic numbers under the smoothness conditions at $s = 0$:

$$\begin{aligned}&\frac{d}{dt}\left[X(s) + Y(s)\dot{s}\right] = Y'(s) + X'(s)\dot{s}\\&if \quad Y(0) = 0\\[6pt]&\frac{d^2}{dt^2}\left[X(s) + Y(s)\dot{s}\right] = X''(s) + Y''(s)\dot{s}\\&if \quad Y(0) = 0, \quad X'(0) = 0\end{aligned} \tag{3}$$

- Integration remains in the algebra of hyperbolic numbers:

$$\begin{aligned}&\int \left[X\left(s(t)\right) + Y\left(s(t)\right)\dot{s}(t)\right]dt\\&= \left[\int_0^s Y(z)dz + C\right] + \left[\int_0^s X(z)dz\right]\dot{s}\end{aligned}$$

3.2 Sample Solution Procedure for the s-Case

Let us consider the following initial value problem:

$$\begin{aligned}&\dot{x} + \lambda x = 2p\delta\,(t - a) \equiv p\ddot{s}\\&x(0) = 0\\&s = |t - a|\end{aligned} \tag{4}$$

Introducing the new temporal argument, $t \to s$, generates the following substitution for the unknown function: $x(t) = X(s) + Y(s)\dot{s}$. As a result, the differential equation takes the form

$$\underbrace{Y' + \lambda X}_{\text{regular}} + \underbrace{\left(X' + \lambda Y\right)\dot{s}}_{\text{step-wise discont.}} + \underbrace{(Y - p)\ddot{s}}_{\text{singular term}} = 0 \tag{5}$$

Equating separately the terms of different levels of singularity in (5) to zero leads to the following boundary value problem with *no singular terms*:

$$\begin{cases} Y' + \lambda X = 0 \\ X' + \lambda Y = 0 \\ p - Y(0) = 0 \end{cases} \tag{6}$$

The initial condition in (4) yields

$$X(a) - Y(a) = 0 \tag{7}$$

The boundary value problem (6) and (7) is easy to solve in few steps. Then the corresponding solution of the original initial value problem (4) is obtained in the closed-form $x(t) = p \exp(-\lambda s)(1 + \dot{s})$; see Fig. 5 for illustration.

3.3 *Properties and Sample Solution for the τ-Case*

Algebraic and differential properties of the periodic τ-version, which is shown in the second row of Fig. 2, are similar to those listed in Sect. 3.1. Note that, although the analytical definitions for the basic functions τ and $\dot{\tau}$ look more complicated in this case, there is no need for memorizing them. What is necessary for solving problems is the following properties:

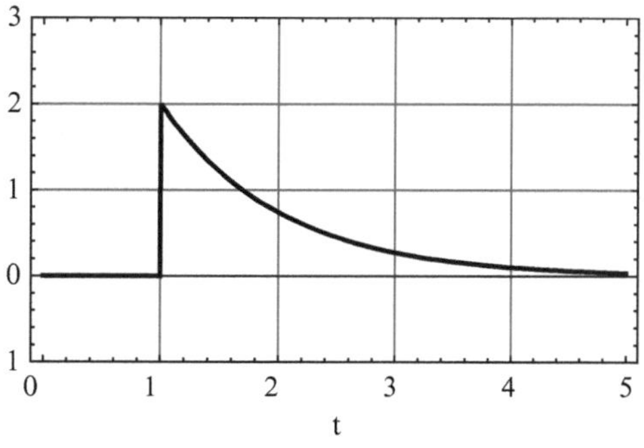

Fig. 5 Solution of problem (4) under the following parameters: $P = 1.0$; $a = 1.0$; $\lambda = 1.0$

$$\dot{\tau} = e, \quad e^2 = 1, \quad \dot{e} \neq 0 \Longleftrightarrow \tau = \pm 1 \tag{8}$$

The third relationship in (8) means that whenever τ reaches its amplitude values, a δ-spike occurs from the following series:

$$\dot{e}(t) = 2 \sum_{k=-\infty}^{\infty} [\delta(t + 1 - 4k) - \delta(t - 1 - 4k)]$$

In other words, the present situation is quite similar to analytical manipulations with the conventional trigonometric functions using only the function properties with no involvement of their definitions. In order to illustrate the manipulations, let us consider the overdamped oscillator under the rectangular cosine loading:

$$\dot{x} + \lambda x = pe(t) \tag{9}$$

The unknown periodic solution is represented in the form [4]

$$x = X(\tau) + Y(\tau)\dot{\tau} \tag{10}$$

Substituting (10) in (9) and taking into account properties (8) gives

$$\underbrace{Y' + \lambda X}_{\text{regular}} + \underbrace{(X' + \lambda Y - p)e}_{\text{step-wise discont.}} + \underbrace{Y\dot{e}}_{\text{singular}} = 0 \tag{11}$$

Equating separately to zero the terms of different levels of singularity gives the autonomous boundary value problem with no discontinuities:

$$\begin{cases} X' + \lambda Y = p \\ Y' + \lambda X = 0 \\ Y(\pm 1) = 0 \end{cases} \tag{12}$$

Substituting solution of the boundary value problem (12) in (10) gives finally the periodic closed-form solution of the original equation (9); see Fig. 6 for illustration:

$$x = \frac{p}{\lambda} \left\{ \frac{\sinh(\lambda\tau)}{\cosh\lambda} + \left[1 - \frac{\cosh(\lambda\tau)}{\cosh\lambda} \right] e \right\} \tag{13}$$

Obviously, (9) can be also solved by means of either Fourier series or Laplace transforms or directly by matching different pieces of solution under periodicity conditions. However, using the Fourier series, for instance, requires a very large number of terms near the non-smoothness points as shown in Fig. 6.

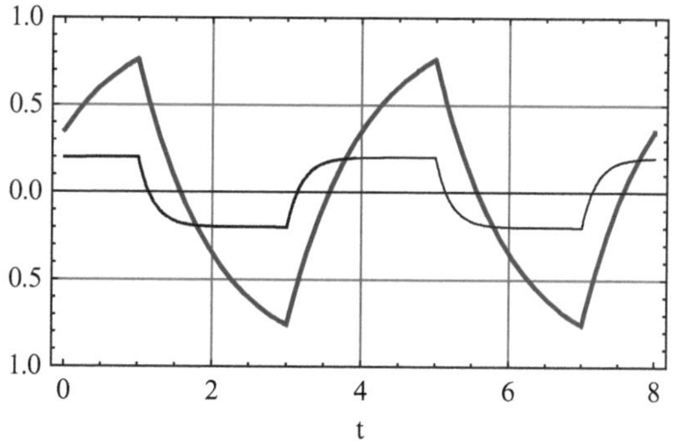

Fig. 6 Solution (13) for two different magnitudes of the parameter: $\lambda = 1.0$ – *solid line*, larger amplitude; $\lambda = 5.0$ – *thin line*, smaller amplitude

4 Conclusions

This paper outlines the basic principles of non-smooth temporal substitutions and gives exactly solvable illustrating examples. Note that the transition to non-smooth temporal argument should be viewed as a preliminary stage of analyses. Such stage of transformation incorporates the specifics of external loading or/and inherent properties of a physical system into the new equations. As a result, a much wider range of analytical and numerical methods becomes possible to apply. This is due to the fact that the new temporal arguments vary within narrower domains and automatically capture the major temporal symmetries of the dynamics, such as periodicity and reflections. The corresponding analytical algorithms and solutions for different strongly nonlinear oscillators can be found in the reference [4]. The typical form of such solutions is power series with respect to the triangular sine τ; see row 5 in Fig. 4 for explanation. Note that direct power series expansions with respect to the original time t usually make little sense for vibration problems due to the loss of periodicity. The amplitude and frequency modulated motions can be analyzed by adding a slow time argument to representation (10) and then using the idea of two variables or multiple-scale expansions. Let us mention also a new area of applications, which is being developed due to an interesting observation that the temporal mode shapes of the phase variable, describing the 1:1 resonance energy exchange between weakly nonlinear oscillators, resemble the dynamic states of impact oscillator [3]. In particular, it was found that such "impacts" take place when the entire energy is involved into the exchange process.

Finally, let us summarize the revealed links between specific cases of the Euclidean rigid transformations, the induced algebraic structures, and linear and nonlinear dynamics. The class of rigid transformations T of an arbitrary vector $\boldsymbol{r}$ is described by the expression $T(\mathbf{r}) = \mathbf{Ar} + \mathbf{b}$, where $\boldsymbol{A}$ and $\boldsymbol{b}$ are the orthogonal matrix and a constant translation vector, respectively. Then the abovementioned logical links are illustrated by the table, which can be viewed as an extension of Fig. 4:

$$\det(\mathbf{A}) = \begin{cases} +1 & \Longleftrightarrow \text{Rotation} & \Longleftrightarrow i^2 = -1 \ \text{Elliptic numbers} & \Longleftrightarrow \text{Linear dynamics} \\ 0 & \Longleftrightarrow \text{Singular case} & \Longleftrightarrow i^2 = 0 \ \text{Parabolic numbers} & \Longleftrightarrow \ ? \\ -1 & \Longleftrightarrow \text{Reflection} & \Longleftrightarrow i^2 = +1 \ \text{Hyperbolic numbers} & \Longleftrightarrow \text{Nonlinear dynamics} \end{cases}$$

where the singular case generates an open question.

References

1. Kisil, V.V.: Induced representations and hypercomplex numbers. Adv. Appl. Clifford Algebras **23**(2), 417–440 (2013)
2. Lyapunov, A.M.: Investigation of a singular case of the problem of stability of motion. Math. Sbornik **17**, 252–333 (1893)
3. Manevitch, L.I.: New approach to beating phenomenon in coupled nonlinear oscillatory chains. Arch. Appl. Mech. **77**, 301–312 (2007)
4. Pilipchuk, V.N.: Nonlinear Dynamics: Between Linear and Impact Limits. Springer, Berlin (2010)
5. Pilipchuk, V.N.: Transformation of oscillating systems by means of a pair of nonsmooth periodic functions. Dokl. Akad. Nauk Ukrain. SSR, Ser. A (4), 37–40 (1988)

Non-linear Phenomena Exhibited by Flexible Cylindrical and Sector Shells

V.A. Krysko, J. Awrejcewicz, I.V. Papkova, V.B. Baiburin, and T.V. Yakovleva

Abstract Vibrations of flexible cylindrical and sector shells subjected to the action of uniformly distributed static loads are studied. The analyzed problems are solved using two methods: the Bubnov–Galerkin method (BGM) and the finite difference method (FDM). Validity and reliability of the results is verified through a comparison to the results obtained by Andreev et al. (Stability of Shells Under Non-Symmetric Deformation. Nauka, Moscow, 1988) in the case of a nonlinear static problem.

1 Introduction

The variety of loading applied plays a crucial role while estimating the strength of materials in numerous constructions working in high-temperature fields. Proper estimation of the construction strength requires the detailed analysis of the elastic-plasticmaterial behavior, initial deflections, interaction of the construction elements, or interaction of those elements with the surrounding medium. The proper estimation of the construction stability requires development of suitable computational algorithms [12].

V.A. Krysko (✉) • I.V. Papkova • T.V. Yakovleva
Department of Mathematics and Modeling, Saratov State Technical University, Politehnicheskaya 77, 410054 Saratov, Russian Federation
e-mail: tak@san.ru; ikravzova@mail.ru; Yan-tan1987@mail.ru

J. Awrejcewicz
Department of Automation, Biomechanics and Mechatronics, Lodz University of Technology, 1/15 Stefanowski Str., 90-924 Lodz, Poland

Department of Vehicles, Warsaw University of Technology, 84 Narbutta Str., 02-524 Warsaw, Poland
e-mail: jan.awrejcewicz@p.lodz.pl

V.B. Baiburin
Department of Information Security of Automated Systems, Saratov State Technical University, Politehnicheskaya 77, 410054 Saratov, Russian Federation
e-mail: tak@san.ru

J. Awrejcewicz (ed.), *Applied Non-Linear Dynamical Systems*, Springer Proceedings in Mathematics & Statistics 93, DOI 10.1007/978-3-319-08266-0_3

Many researchers have applied the Bubnov–Galerkin methods (BGMs) in the Vlasov form as well as Ritz and FDM methods to solve problems of the stability of beams, plates, and shells subjected to the action of a constant transversal load and taking into account the geometric nonlinearities. The mentioned numerical approaches yield reliable and validated results regarding a wide class of both stationary and nonstationary problems of mathematical physics. In the case of periodic loading, chaotic vibrations of the mentioned structural members may appear [3, 13, 14]. The so far mentioned computational approaches reduce the continuous problems to those of finite degrees of freedom [9, 18].

In order to investigate stability loss one may apply a few different criteria. Since stability loss of an arbitrary deformed object takes place in time, therefore it should be studied using various approaches to dynamics. However, a majority of the stability problems of construction can be studied within static approaches, where the equilibrium states are formulated without the inclusion of inertial forces.

Investigation of the stability loss is carried out using a dynamic criterion. Namely, we define it through a buckling of an equilibrium form. Those loads being responsible for the buckling occurrence are further named the critical loads.

We omit here an overview of the fundamental works dealing with the mentioned problems, but we mention the method developed by Feodosev [10] regarding nonlinear problems of shells, which is rather omitted in English-language literature.

In the latter one being originally named by Feodosev as the variational-step method, the system deformation is considered as a process independent of either fast or slow changes of the external load. For this purpose, time is introduced artificially and equations of motion are derived. Nowadays this method refers to the iteration process of finding solutions to nonlinear algebraic equations, where results obtained in each computational step are improved, finally approaching the desired exact solution of the problem. In this method of relaxation a solution to PDEs is reduced to the Cauchy problem of ODEs.

The proposed algorithm is used to solve a wide class of static and dynamic problems. We show a few possibilities of this approach to solve geometrically nonlinear static and dynamic problems. We consider a mechanical system subjected to the action of the transversal uniformly distributed constant load over the shell surface and we consider the load in the form of the impulse with infinite action. Since the problem of a critical static load plays here a crucial role, we briefly describe the known criteria of stability proposed by numerous researchers.

Already Volmir [19] proposed the following criterion: either a fast increase of deflection corresponding to a small decrease of load appears or an inflexion point of the relation $q(w)$ $\left(\frac{\partial^2 q}{\partial w^2} = 0\right)$ occurs. On the other hand a load, where the increased process of time is responsible for the achievement of the first maximum in the load–time characteristic, is treated as the critical one. Kantor, who solved numerous problems of axially symmetric spherical shells using the Ritz method, proposed the following dynamic criterion responsible for the beam buckling [11]. The buckling occurs if in the shell center its deflection achieves $K \cong 2\overline{f}$, where $\overline{f} = f/h$ and f denotes deflection, whereas h is the shell thickness.

In references [7, 16] different criteria are proposed. Namely, the system transits into a new dynamic state with the corresponding zero velocity. It can be explained in the following manner. In the beginning the inertial forces act against the external load, and after transition through zero they change sign and support the action of the external load. It means that in a certain time instant the beam center velocity achieves zero and then a sudden change of deflection occurs. In reference [8] the time instant is taken as the stability loss criterion, where the displacement of an elastic body changes without a change of the associated accelerations and velocities. In some works the problem of dynamical stability loss is reduced to a quasi-dynamical problem. Owing to this approach, the precritical stress of the middle shell process is analyzed via static approaches. There are also works where a dynamic criterion of the stability loss is matched with the occurrence of plastic deformations of shell structures.

In reference [17], arcs were investigated and their buckling process was characterized by two different mechanisms. In the case of the direct buckling mechanism an unstable construction state was realized via symmetric forms. In the case of indirect buckling, the system lost its stability via nonsymmetric forms. Since the system stability loss via symmetric and nonsymmetric forms is qualitatively different, one may expect two different dynamic criteria of the stability loss.

In this work by a critical load we mean limiting load values or the point of inflexion of the relation $w_{max}(q)$. Further on, we will investigate critical loads acting on axially symmetric spherical and conical shells, on a closed cylindrical shell as well as on a spherical sector shell.

2 Shallow Closed Cylindrical Shells

We study shallow shells, i.e., objects in R^3 with the associated curvilinear coordinates $\overline{x}, \overline{y}, \overline{z}$, introduced in the following manner. In the shell body the middle surface $\overline{z} = 0$ is fixed; axes ox and oy overlap the main shell curvatures, whereas axis oz shows curvature surface origin (Fig. 1). In the given coordinates the shell is defined as follows: $\Omega = \left\{\overline{x}, \overline{y}, \overline{z}/(\overline{x}, \overline{y}, \overline{z}) \in [0, \overline{a}] \times [0, \overline{b}] \times \left[-\overline{h}/2, \overline{h}/2\right]\right\}$, where dimensional quantities are denoted by bars.

The governing nonlinear dynamics of the shell shown in Fig. 1 is obtained assuming that the shell material is isotropic, homogeneous, and elastic and it satisfies the Kirchhoff–Love hypotheses. Furthermore, we assume that the length of the shell fiber along shell thickness remains unchanged [19].

Therefore, in the nondimensional form, the equation of motion of the shell element as well as the deformation compatibility equations have the following nondimensional forms:

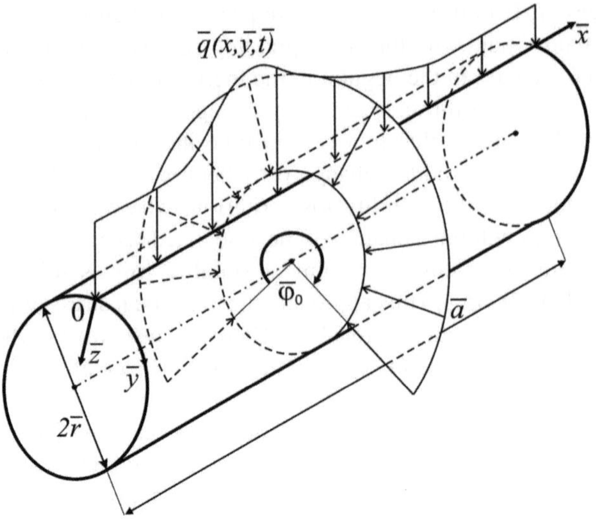

Fig. 1 Computational scheme of a cylindrical shell

$$\left[\frac{1}{\lambda^2}\frac{\partial^2 w}{\partial x^2}\frac{\partial^2(\cdot)}{\partial x^2}+\lambda^2\frac{\partial^2 w}{\partial y^2}\frac{\partial^2(\cdot)}{\partial y^2}+2(1-\mu)\frac{\partial^2 w}{\partial x\partial y}\frac{\partial^2(\cdot)}{\partial x\partial y}+\mu\left(\frac{\partial^2 w}{\partial x^2}\frac{\partial^2(\cdot)}{\partial y^2}+\right.\right.$$

$$\left.\left.+\frac{\partial^2 w}{\partial y^2}\frac{\partial^2(\cdot)}{\partial x^2}\right)\right]-\nabla_k^2 F-L(w,F)+Mq(t)-\left(\frac{\partial^2 w}{\partial t^2}+\varepsilon\frac{\partial w}{\partial t}\right)=0,$$

$$\left[\left(\lambda^2\frac{\partial^2 F}{\partial y^2}-\mu\frac{\partial^2 F}{\partial x^2}\right)\frac{\partial^2(\cdot)}{\partial y^2}+\left(\frac{1}{\lambda^2}\frac{\partial^2 F}{\partial x^2}-\mu\frac{\partial^2 F}{\partial y^2}\right)\frac{\partial^2(\cdot)}{\partial x^2}+\right. \tag{1}$$

$$\left.+2(1+\mu)\frac{\partial^2 F}{\partial x\partial y}\frac{\partial^2(\cdot)}{\partial x\partial y}\right]+\nabla_k^2 w+\frac{1}{2}L(w,w)=0.$$

The following relations hold between dimensional and nondimensional quantities:

$$w=h\overline{w},\quad F=Eh^2\overline{F},\quad t=t_0\overline{t},\quad \varepsilon=\overline{\varepsilon}/\tau,\quad x=L\overline{x},\quad y=R\overline{y},$$

$$k_y=\overline{k}_y\frac{h}{R^2}\quad (k_x=0),\quad q=\overline{q}\frac{Eh^4}{L^2R^2},\quad \tau=\frac{LR}{h}\sqrt{\frac{\rho}{Eg}}, \tag{2}$$

$$M=k_y^2,\quad \lambda=\frac{L}{R},$$

where L and $R = R_y$ correspond to the shell length and radius, respectively. In addition, we have t, time; ε, damping coefficient; $\mu = 0.3$; and $q(x, y, t)$, transversal load. One of the following boundary conditions is taken:

1. Moving clamping

$$w = 0; \quad \frac{\partial w}{\partial x} = 0; \quad F = 0; \quad \frac{\partial F}{\partial x} = 0 \quad \text{for} \quad x = 0; 1, \tag{3}$$

$$w = g(x, y, t); \ \frac{\partial w}{\partial y} = p(x, y, t); \ F = u(x, y, t); \ \frac{\partial F}{\partial y} = v(x, y, t) \text{ for } y = 0; \xi.$$

2. Pinned support

$$w = 0; \ \frac{\partial w}{\partial x} = 0; \ F = 0; \ \frac{\partial^2 F}{\partial x^2} = 0 \text{ for } x = 0; 1, \tag{4}$$

$$w = g(x, y, t); \ \frac{\partial w}{\partial y} = p(x, y, t); \ F = u(x, y, t); \ \frac{\partial F}{\partial y} = v(x, y, t) \text{ for } y = 0; \xi.$$

3. Moving clamping with ribs

$$w = 0; \quad \frac{\partial^2 w}{\partial x^2} = 0; \quad F = 0; \quad \frac{\partial F}{\partial x} = 0 \quad \text{for} \quad x = 0; 1, \tag{5}$$

$$w = g(x, y, t); \ \frac{\partial w}{\partial y} = p(x, y, t); \ F = u(x, y, t); \ \frac{\partial F}{\partial y} = v(x, y, t) \text{ for } y = 0; \xi.$$

4. Pinned support with flexible ribs

$$w = 0; \quad \frac{\partial^2 w}{\partial x^2} = 0; \quad F = 0; \quad \frac{\partial^2 F}{\partial x^2} = 0 \quad \text{for} \quad x = 0; 1, \tag{6}$$

$$w = g(x, y, t); \ \frac{\partial^2 w}{\partial y^2} = r(x, y, t); \ F = u(x, y, t); \ \frac{\partial^2 F}{\partial y^2} = z(x, y, t) \text{ for } y = 0; \xi.$$

Here we take $\xi = 2\pi$ for a closed cylindrical shell. In addition, the following initial conditions are applied:

$$w|_{t=0} = w_0, \qquad \dot{w}|_{t=0} = w_0. \tag{7}$$

2.1 The Bubnov–Galerkin Method (BGM)

After application of the BGM the following system of algebraic-differential equations is obtained:

$$\begin{aligned} \mathbf{G}(\ddot{\mathbf{A}} + \varepsilon\dot{\mathbf{A}}) + \mathbf{HA} + \mathbf{C}_1\mathbf{B} + \mathbf{D}_1\mathbf{AB} &= \mathbf{Q}q(t), \\ \mathbf{C}_2\mathbf{A} + \mathbf{PB} + \mathbf{D}_2\mathbf{AA} &= 0, \end{aligned} \tag{8}$$

where $\mathbf{H} = \|H_{ijrs}\|$, $\mathbf{G} = \|G_{ijrs}\|$, $\mathbf{C}_1 = \|C_{1ijrs}\|$, $\mathbf{C}_2 = \|C_{2ijrs}\|$, $\mathbf{D}_1 = \|D_{1ijklrs}\|$, $\mathbf{D}_2 = \|D_{2ijklrs}\|$, $\mathbf{P} = \|P_{ijrs}\|$ - square matrices of dimensions $2 \cdot N_1 \cdot N_2 \times 2 \cdot N_1 \cdot N_2$, $\mathbf{A} = \|A_{ij}\|$, $\mathbf{B} = \|B_{ij}\|$, $\mathbf{Q} = \|Q_{ij}\|$ matrices of dimension $2 \cdot N_1 \cdot N_2 \times 1$.

The second equation of system (8) is solved regarding $\mathbf{B}$ on each of the computational steps:

$$\mathbf{B} = [-\mathbf{P}^{-1}\mathbf{D}_2\mathbf{A} - \mathbf{P}^{-1}\mathbf{C}_2]\mathbf{A}. \tag{9}$$

Multiplying by $\mathbf{G}^{-1}$ and taking $\dot{\mathbf{A}} = \mathbf{R}$, the following Cauchy problem is formulated for nonlinear ODEs:

$$\begin{aligned} \dot{\mathbf{R}} &= -\overline{\varepsilon}\mathbf{R} - [\mathbf{G}^{-1}\mathbf{C}_1 + \mathbf{G}^{-1}\mathbf{D}_1\mathbf{A}] \cdot \mathbf{B} - \mathbf{G}^{-1}\mathbf{HA} + \mathbf{G}^{-1}\mathbf{Q}q(\overline{t}), \\ \dot{\mathbf{A}} &= \mathbf{R}. \end{aligned} \tag{10}$$

It is solved via the fourth-order Runge–Kutta method, and the computational step in time is chosen using the Runge rule. We apply the method of relaxation for the closed cylindrical shells with $\lambda = 2$ and we compare our results with the solution obtained by Andreev et al. [1] for the corresponding static problem. We consider the case of transversal external load whose location is defined by the central angle φ_0. In order to get $q_{cr}(\varphi_0)$ we need to construct a set $\{q_i, w_i\}$ for $\forall\varphi_0 \in [0; 2\pi]$, which yields the critical load q_{cr}. As it has been shown in [15], an increase in the number of approximations yields a remarkable improvement of the obtained results.

The following conclusion can be formulated: in the case of nonhomogeneous load, the use of a small number of the series terms yields large computational errors and the obtained results depend essentially on the number of introduced approximations. However, the situation changes qualitatively beginning from $N = 13$. Namely, the dynamical properties of the cylindrical shell are stabilized and a further increase of N does not improve the obtained results either qualitatively or quantitatively. Therefore, beginning from $N = 13$, a convergent series is obtained and all further computations are carried out for $N = 13$. Consequently, we constructed the relation of the critical loads versus width of the pressure zone $\overline{q}_{cr}(\varphi_0)$ for $N = 13$ reported in [15].

Dependencies $\overline{q}_{cr}(\varphi_0)$ reported by Andreev et al. [1] for the closed cylindrical shell for $\lambda = 2$ are in agreement with the results obtained by our method. Hence, the obtained results indicate high efficiency of the proposed method for solving static problems.

3 Shallow Sector Shells

Consider now a non-axially symmetric spherical shell in R^2 in the polar coordinates bounded by a contour Γ, introduced in the following way: $\overline{\Omega} = \Omega + \Gamma = \{(r,\theta,z)|r \in [0,b], \theta \in [0,\theta_k], z \in [-h/2, h/2]$. Equations governing the dynamics of shallow shells are obtained from a system of equations of the rectangular spherical shell via transition to the polar coordinates:

$$\begin{gathered} w'' + \varepsilon w' = -\nabla^2\nabla^2 w + N(w,F) + \nabla^2 F + 4q, \\ \nabla^2\nabla^2 F = -\nabla^2 w - N(w,w), \end{gathered} \tag{11}$$

where

$$\nabla^2(\cdot) = \frac{\partial^2(\cdot)}{\partial r^2} + \frac{1}{r}\frac{\partial(\cdot)}{\partial r} + \frac{1}{r^2}\frac{\partial^2(\cdot)}{\partial \theta^2},$$

$$\nabla^2\nabla^2(\cdot) = \frac{\partial^4(\cdot)}{\partial r^4} + \frac{2}{r}\frac{\partial^3(\cdot)}{\partial r^3} - \frac{1}{r^2}\frac{\partial(\cdot)}{\partial r^2} + \frac{1}{r^3}\frac{\partial(\cdot)}{\partial r} +$$

$$+\frac{2}{r^2}\frac{\partial^4(\cdot)}{\partial\theta^2\partial r^2} - \frac{2}{r^3}\frac{\partial^3(\cdot)}{\partial\theta^2\partial r} + \frac{4}{r^4}\frac{\partial^2(\cdot)}{\partial\theta^2} + \frac{1}{r^4}\frac{\partial^4(\cdot)}{\partial\theta^4},$$

$$\begin{aligned} N(w,F) = {} & \frac{\partial^2 w}{\partial r^2}\left(\frac{1}{r}\frac{\partial F}{\partial r} + \frac{1}{r^2}\frac{\partial^2 F}{\partial\theta^2}\right) + \frac{\partial^2 F}{\partial r^2}\left(\frac{1}{r}\frac{\partial w}{\partial r} + \frac{1}{r^2}\frac{\partial^2 w}{\partial\theta^2}\right) - \\ & -2\cdot\frac{\partial}{\partial r}\left(\frac{1}{r}\frac{\partial w}{\partial\theta}\right)\frac{\partial}{\partial r}\left(\frac{1}{r}\frac{\partial F}{\partial\theta}\right), \end{aligned} \tag{12}$$

$$N(w,w) = 2\cdot\frac{\partial^2 w}{\partial r^2}\left(\frac{1}{r}\frac{\partial w}{\partial r} + \frac{1}{r^2}\frac{\partial^2 w}{\partial\theta^2}\right) - 2\cdot\left[\frac{\partial}{\partial r}\left(\frac{1}{r}\frac{\partial w}{\partial\theta}\right)\right]^2.$$

Boundary conditions follow:

1. Pinned support of arc slices

$$w = 0, \quad \frac{\partial^2 w}{\partial r^2} + \frac{\nu}{r}\frac{\partial w}{\partial r} = 0, \quad F = 0, \quad \frac{\partial F}{\partial r} = 0. \tag{13}$$

2. Pinned support of radial slices

$$w = 0, \quad \frac{\partial^2 w}{\partial\theta^2} = 0, \quad F = 0, \quad \frac{\partial^2 F}{\partial\theta^2} = 0. \tag{14}$$

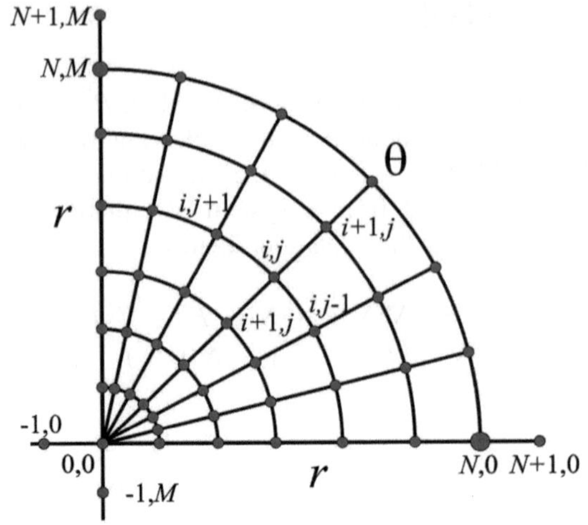

Fig. 2 Mesh of a sector shell

3. Sliding clamping of arc slices

$$w = 0, \quad \frac{\partial w}{\partial r} = 0, \quad F = 0, \quad \frac{\partial F}{\partial r} = 0. \tag{15}$$

4. Sliding clamping of radial slices

$$w = 0, \quad \frac{\partial w}{\partial \theta} = 0, \quad F = 0, \quad \frac{\partial^2 F}{\partial \theta^2} = 0. \tag{16}$$

Initial conditions are as follows:

$$w = f_1(r,\theta) = 0, \quad w' = f_2(r,\theta) = 0 \quad \text{for} \quad t = 0. \tag{17}$$

3.1 Finite Difference Method

In order to reduce the continuous system governed by (12)–(17) to a lumped system by the FDM with the approximation $O(\Delta^2)$ versus spatial coordinates r and θ (Fig. 2), the following difference operators are applied:

$$-\Lambda(\Lambda w) + \Lambda_{rr} w(\Lambda F + \Lambda_{rr} F) + \Lambda_{rr} F(\Lambda w + \Lambda_{rr} w) -$$

$$-2 \cdot \Lambda_{r\theta} w \Lambda_{r\theta} F + \Lambda F + 4q_i = (w_{tt} + \varepsilon w_t)_{i,j}, \tag{18}$$

$$\Lambda(\Lambda F) = -\Lambda_{rr} w(\Lambda w + \Lambda_{rr} w) + (\Lambda_{r\theta} w)^2 - \Lambda w,$$

where

$$\Lambda(\cdot) = \Lambda_{rr}(\cdot)+\Lambda_r(\cdot),\ \Lambda_r(\cdot) = \frac{1}{r_i^2}(\cdot)_r,\ \Lambda_{rr}(\cdot) = (\cdot)_{rr},\ \Lambda_{r\theta}(\cdot) = -\frac{1}{r_i^2}(\cdot)_\theta+\frac{1}{r_i}(\cdot)_{r\theta},$$

$$\Lambda_{rr}(\cdot) = \frac{1}{\Delta^2}[(\cdot)_{i+1} - 2(\cdot)_i + (\cdot)_{i-1}],\quad \Lambda_r(\cdot) = \frac{1}{2\cdot\Delta\cdot r_i^2}[(\cdot)_{i+1} - (\cdot)_{i-1}].$$

Boundary conditions:

1. Pinned support of arc slices

$$w_{N,j} = 0,\quad \Lambda_{rr}w-\frac{v}{b}\Lambda_r w = 0,\quad F_{N,j} = 0,\quad \Lambda_r w = 0,\quad j = 1,\ldots,M-1. \tag{19}$$

2. Pinned support of radial slices

$$w_{i,j} = 0,\quad \Lambda_{\theta\theta}w = 0,\quad F_{i,j} = 0,\quad \Lambda_{\theta\theta}F = 0,\quad j = 0, M,\quad i = 0,\ldots,N. \tag{20}$$

3. Sliding clamping of arc slices

$$w_{N,j} = 0,\quad \Lambda_r w = 0,\quad F_{N,j} = 0,\quad \Lambda_r F = 0,\quad j = 1,\ldots,M-1. \tag{21}$$

4. Sliding clamping of radial slices

$$w_{i,j} = 0,\quad \Lambda_{\theta\theta}w = 0,\quad F_{i,j} = 0,\quad \Lambda_{\theta\theta}F = 0,\quad j = 0, M,\quad i = 0,\ldots,N. \tag{22}$$

The system of (18)–(22) should be supplemented by conditions to be satisfied in the shell cusp and the matching conditions. In the majority of cases it is assumed that a shell has a circular hole of small dimension in its cusp, and this assumption does not influence computational results essentially. In this work, while solving nonsymmetric problems for $\theta = 2\cdot\pi$, the approximating functions in the point $r = 0$ are interpolated by the Lagrange formula of the second order. We have

$$f_{0,j} = 3\cdot f_{1,j} - 3\cdot f_{2,j} + f_{3,j}, \tag{23}$$

where $f_{i,j} = f(r_i)_j$, $r_i = i\cdot h$ $(i = 0, 1, 2, 3)$, $0 \le j \le M-1$, and h is the distance between the nodes of interpolation. In the case of a point lying out of the contour the following symmetry condition holds:

$$f_{-1,j} = f_{1,j} \quad \text{for} \quad 0 \le j \le M-1. \tag{24}$$

Matching conditions for non-axially symmetric problems $\theta = 2\cdot\pi$ follow:

$$w_{i,j} = w_{i,M+j},\quad F_{i,j} = F_{i,M+j} \quad \text{for} \quad j = 0; -1,\quad 0 \le i \le N-1. \tag{25}$$

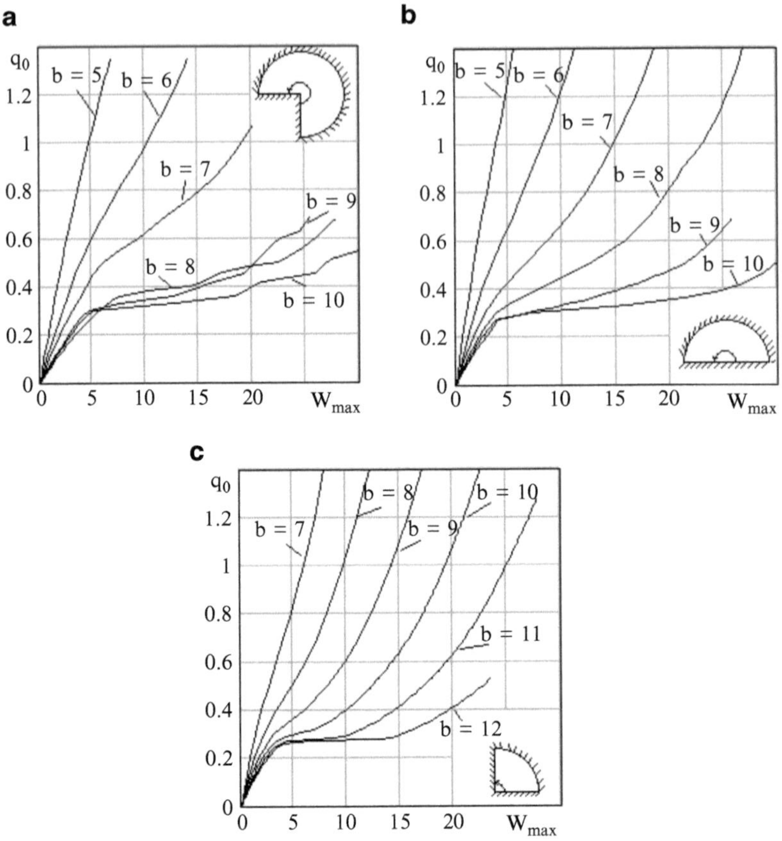

Fig. 3 Computational results of sector shells analysis

The Cauchy problem (18)–(25) is solved via the fourth-order Runge–Kutta method, where the computational step has been chosen due to the Runge rule [2, 4–6].

Figure 3 shows the dependencies $q(w)$ for sector angles $\theta_k = \frac{3}{2} \cdot \pi$ (a), for $\theta_k = \pi$ (b), and for $\theta_k = \frac{\pi}{2}$ (c) and for various shallow parameters for $\theta_k = \frac{3}{2} \cdot \pi$ and $\theta_k = \pi$ ($b = 5, 6, 7, 8, 9, 10$), and for $\theta_k = \frac{\pi}{2}$ for the shallow parameter $b = 7, 8, 9, 10, 11, 12$ (lower values of b are not used, since the sector shell behaves like a plate).

The obtained graphs imply the following conclusions. Beginning from a certain value of the parameter b the critical points appear. For a shell with $\theta_k = \frac{3}{2} \cdot \pi; \pi; \frac{\pi}{2}$ and $b = 8, 9$ and 11, respectively, the jump-type buckling occurs. Figure 4 illustrates the curves of equal deflections (isoclines) for all studied angles $\theta = \frac{\pi}{2}; \pi; \frac{3 \cdot \pi}{2}$. In what follows we compare the behavior of curves for the critical and post-critical load q_0 for different θ and b. For $\theta = \frac{\pi}{2}$ and $\theta = \pi$ and for an arbitrary shallow parameter the obtained results coincide and the maximum deflection is achieved on the intersection of the bisectrix of angle θ and the central shell radius. For $\theta = \frac{3 \cdot \pi}{2}$ and for $b = 7$ the jump-type buckling is not observed, and the system dynamics

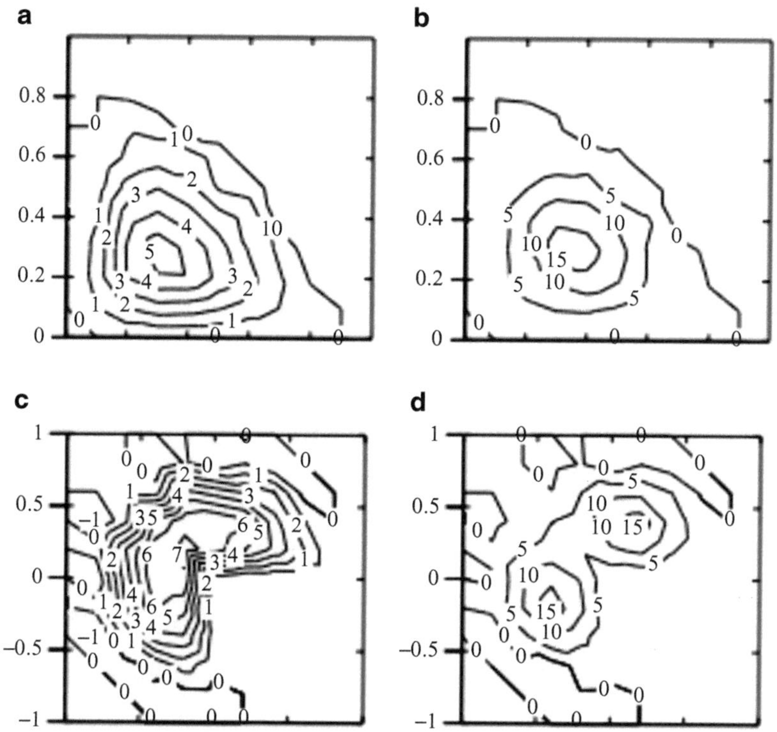

Fig. 4 Computational results of sector shells (isoclines) for $\theta = \frac{\pi}{2}, b = 12, q_0 = 0.245$ (**a**), $\theta = \frac{\pi}{2}, b = 12, q_0 = 0.246$ (**b**), $\theta = \frac{3 \cdot \pi}{2}, b = 10, q_0 = 0.27$ (**c**) and $\theta = \frac{3 \cdot \pi}{2}, b = 10, q_0 = 0.28$ (**d**)

is not changed qualitatively. In Fig. 3a the same sector angle and the first-order discontinuity imply qualitative changes in the location of isoclines for both critical and post-critical load (see Fig. 4). For $\theta = \frac{\pi}{2}$, $b = 12$ a fast shell deflection is observed but without a change of the shell form. For $\theta = \frac{3 \cdot \pi}{2}$, $b = 10$ for the critical load, the maximum shell deflection occurs in the middle of the bisectrix, whereas in the case of post-critical load two zones of maximum deflection appear, being symmetric with respect to the angle bisectrix.

4 Concluding Remarks

In this work the relaxation method has been applied to study flexible cylindrical and sector shells. The Bubnov–Galerkin and the finite difference approaches allowed us to reduce the problems to the algebraic-ordinary differential equations. A few interesting nonlinear phenomena exhibited by the analyzed continuous systems have been reported.

Acknowledgements This work has been supported by the grant RFBR No. 12-01-31204 and the National Science Centre of Poland under the grant MAESTRO 2, No. 2012/04/A/ST8/00738, for the years 2013–2016.

References

1. Andreev, L.V., Obodan, N.I., Lebedev, A.G.: Stability of Shells Under Non-Symmetric Deformation. Nauka, Moscow (1988) (in Russian)
2. Awrejcewicz, J., Krysko, V.A., Krysko, A.V.: Thermodynamics of Plates and Shells. Springer, Berlin (2007)
3. Awrejcewicz, J., Krylova, E.Yu., Papkova, I.V., Yakovleva, T.V., Krysko, V.A.: On the memory of nonlinear differential equations in theory of plates. Vestnik Niznegorodskogo Universiteta im. Lobachevskogo **4**(2), 21–23 (2011) (in Russian)
4. Awrejcewicz, J., Krysko, A.V., Papkova, I.V., Krysko, V.A.: Routes to chaos in continuous mechanical systems. Part 1: mathematical models and solution methods. Chaos Solitons Fractals **45**, 687–708 (2012)
5. Awrejcewicz, J., Krysko, A.V., Papkova, I.V., Krysko, V.A.: Routes to chaos in continuous mechanical systems: Part 2. Modelling transitions from regular to chaotic dynamics. Chaos Solitons Fractals **45**, 709–720 (2012)
6. Awrejcewicz, J., Krysko, A.V., Papkova, I.V., Krysko, V.A.: Routes to chaos in continuous mechanical systems. Part 3: the Lyapunov exponents, hyper, hyper-hyper and spatial-temporal chaos. Chaos, Solitons Fractals **45**, 721–736 (2012)
7. Bakhtieva, L.U.: Investigation of stability of thin shells and plates subjected to dynamic loads. Ph.D. Thesis, Kazan (1981) (in Russian)
8. Darevskiy, V.M.: Nonlinear equations in theory of shells and their linearizations in problems of stability. In: Proceedings of VI Russian Federation Conference on Theory of Plates and Shells, pp. 355–368. Nauka, Moscow (1969) (in Russian)
9. Deresiewicz, H.: Solution of the equations of thermoelasticity. In: Proceedings of 3rd National Congress of Applied Mechanics ASME, pp. 685–690, 1958
10. Feodosev, V.I.: On the solution of non-linear stability problems of deformable systems. PMM **27**(2), 265–274 (1963) (in Russian)
11. Kantor, B.Ya.: On the Non-linear Theory of Shells. Dynamics and Strength of Machines. Kharkiv National University, Kharkiv (1967)
12. Krysko, V.A., Awrejcewicz, J., Kuznetsova, E.S.: Chaotic vibrations of shells in a temperature field. In: Proceedings of the International Conference on Engineering Dynamics, Carvoeiro, Algarve, Portugal, , pp. 21–28, 2007
13. Krysko, V.A., Saltykova, O.A., Yakovleva, T.V.: Nonlinear dynamics of antennas in cosmonautic coupling tools. Izvestia VUZ. Aviation Tech. **2**, 60–62 (2011) (in Russian)
14. Krysko, A.V., Koch, M.I., Yakovleva, T.V., Nackenhorst, U., Krysko, V.A.: Chaotic nonlinear dynamics of cantilever beams under the action of signs-variables loads. In: PAMM, Special Issue: 82nd Annual Meeting of the International Association of Applied Mathematics and Mechanics (GAMM), vol. 11(1), pp. 327–328, Graz, 2011
15. Krysko, V.A., Awrejcewicz, J., Papkova, I.V., Baiburin, V.B., Yakovleva, T.V.: Method of relaxation in theory of flexible shells. In: Dynamical Systems Theory. pp. 477–490. TU of Lodz Press, Lodz (2013)
16. Kuznetsov, E.B.: On action of dynamic load on systems with buckling. Probl. Appl. Mech. 12–35 (1974) (in Russian)

17. Lock, M.H.: Snapping of a shallow sinusoidal arch under a step pressure load. AIAA J. **4**(7), 31–41 (1966)
18. Takezono, S., Tao, K., Inamura, E., Inoue, M.: Thermal stress and deformation in functionally graded material shells of revolution under thermal loading due to fluid. JSME Int. J. Ser. A Solid Mech. Mater. Eng. **39**(4), 573–581 (1996)
19. Volmir, A.S.: Stability of Elastic Systems. Fizmatgiz, Moscow (1963) (in Russian)

Gear Shift Patterns in Uncertain Terrestrial Locomotion Systems

Carsten Behn and Silvan Schwebke

Abstract The paper focusses on various gear shift patterns in an uncertain terrestrial locomotion system, i.e., in a wormlike locomotion system (WLLS). A WLLS in this theory is understood as a system living in a straight line (one dominant linear dimension) with no active (i.e., driving) legs nor wheels. A mechanical model comprises a chain of discrete mass points (1) connected by viscoelastic force actuators and (2) having ground interaction via spikes which make the velocities unidirectional. A spike means any device which realizes this restriction. The distances between two consecutive mass points are changed by an adaptive controller. In combination with the ground contact spikes, this results in an undulatory locomotion of the system. Optimal gaits which achieve a defined number and succession of resting mass points as well as the resulting velocity are developed in numerical investigations. We present a gait shifting procedure incorporating a combination of speed adjustment and gait change that enables optimal crawling for predefined limits of actuator or spike force load.

1 Introduction: Model and Control Approach

A mechanical model of a WLLS comprises of a chain of interconnected, discrete mass points. Akin to the earthworm, the worm is equipped with *spikes* that inhibit backward motion. Actuators are assumed to be positioned between two adjacent mass points to change the distance between them (change of shape of the worm). This results in a global movement (undulatory locomotion [5]).

Regarding the movement pattern of WLLS, current literature (see [3, 7, 8]) often takes the basic idea from the movement of biological worms. However, the mathematical description of these patterns is based on a purely kinematic view, at times just using two states for the system's actuating elements—elongated and contracted. The generation of specific gaits is discussed, but they are not analyzed and compared in terms of dynamics. Such analyses are the focus of this work.

C. Behn (✉) • S. Schwebke
Department of Technical Mechanics, Ilmenau University of Technology,
Max-Planck-Ring 12, 98693 Ilmenau, Germany
e-mail: carsten.behn@tu-ilmenau.de; silvan.schwebke@tu-ilmenau.de

J. Awrejcewicz (ed.), *Applied Non-Linear Dynamical Systems*, Springer Proceedings in Mathematics & Statistics 93, DOI 10.1007/978-3-319-08266-0_4

For that, the existing theory for gait generation as described in [9] is used. Since all system parameters are assumed to be unknown or, more precisely, uncertain, an adaptive controller is used to achieve the tracking of certain gaits. After finding optimal gaits, a procedure for automatic gait shift based on the measurement of certain restricted system variables is implemented.

1.1 Modeling

The starting point is the chain of mass points in a common straight line as shown in Fig. 1. The variables $x_i(t)$ $(i = 0, \ldots, n)$ are the current coordinates of the segments. The distance between m_i and m_{i-1} are labeled as $l_j(t) := x_{i-1}(t) - x_i(t)$. Each mass point is equipped with *spikes*, which are assumed to be ideal, i.e., they restrict velocities from being negative [10].

To achieve the desired movement due to the paradigm earthworm, the distances $l_j(t)$ have to be changed in an appropriate way. In this model, viscoelastic actuators are used to apply forces between mass points. The acting forces on mass points are $(j = 1, \ldots, n)$ (see Fig. 2):

- spring forces $F_{c,j} = c_j(x_{j-1} - x_j - l_{0,j})$, where $l_{0,j}$ is the initial length of the spring;
- damping forces $F_{d,j} = d_j(\dot{x}_{j-1} - \dot{x}_j)$;

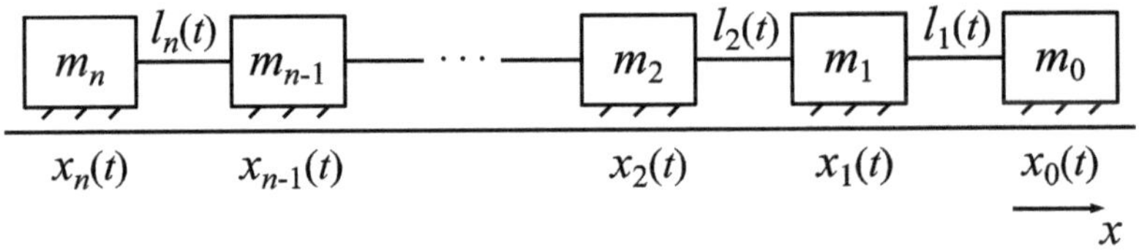

Fig. 1 Chain of mass points with spikes

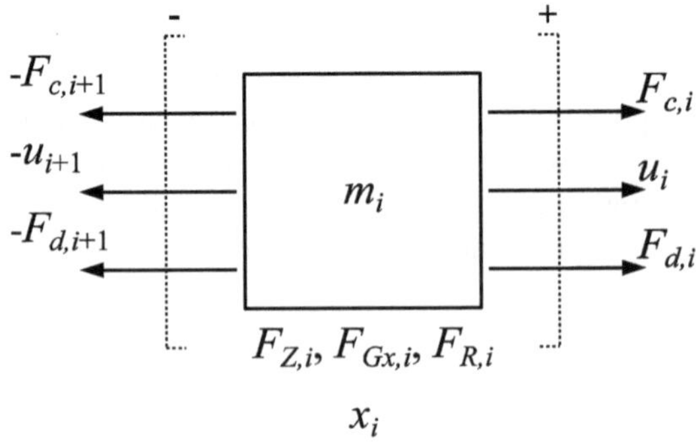

Fig. 2 Mass point with forces

- actuator forces u_j and $u_0 = u_{n+1} = 0$;
- spike forces $F_{Z,i}$, used to model the ideal spikes, therefore fulfilling three conditions according to [9]:

$$\dot{x}_i \geq 0, \quad F_{Z,i} \geq 0, \quad \dot{x}_i \cdot F_{Z,i} = 0, \tag{1}$$

with $i = 0, \ldots, n$. These conditions are fulfilled by the following explicit equation given in [9]:

$$F_{Z,i} = -\frac{1}{2}\big(1 - \text{sign}(\dot{x}_i)\big)\big(1 - \text{sign}(F_i)\big) F_i, \tag{2}$$

with $i = 0, \ldots, n$, where F_i is the sum of all other forces acting upon m_i;
- the weight of a mass point is considered in the x direction only and thus depends on the ground inclination α: $F_{Gx,i}(\alpha) = -m_i g \sin(\alpha)$.

The equations of motion are (using Newton's second law):

$$\begin{aligned} m_i \ddot{x}_i = &- c_{i+1} \cdot (x_i - x_{i+1} - l_{0,i+1}) + c_i \cdot (x_{i-1} - x_i - l_{0,i}) \\ &- d_{i+1} \cdot (\dot{x}_i - \dot{x}_{i+1}) - u_{i+1} + u_i + d_i \cdot (\dot{x}_{i-1} - \dot{x}_i) + F_{Z,i} + F_{Gx,i} \end{aligned} \tag{3}$$

with $i = 0, \ldots, n$ and $c_0 = c_{n+1} = d_0 = d_{n+1} = u_0 = u_{n+1} = 0$. This system can only be influenced using the actuator forces $u_1(t)$ through $u_n(t)$. They shall serve as control inputs to achieve global movement: the distances between mass points shall follow prescribed time-variant functions.

1.2 Control

To achieve the mentioned objective, the following variables are considered:

- $l_j(t) = x_{j-1}(t) - x_j(t)$, the distances between adjacent mass points, which represent the system outputs;
- $l_{ref,j}(t)$, the reference functions of these distances;
- $e_j(t) := l_j(t) - l_{ref,j}(t)$, the error of the output variables.

The task of the controller is to find appropriate actuator forces, i.e., find a mapping $\mathbf{e} \mapsto \mathbf{u}$. The controller shall stabilize the WLLS with unknown or uncertain (known within an interval) parameters. Therefore, *adaptive control* as described in [1] is used. The controller consists of a PD-structure and a gain adaption law to achieve stability. The control goal is to track a reference function of the output variables within a given tolerance λ (accuracy of tracking), which is called λ-tracking [4]. The following controller was proposed in [2]:

$$\left.\begin{aligned} \mathbf{e}(t) &:= \mathbf{l}(t) - \mathbf{l}_{ref}(t) \\ \mathbf{u}(t) &= k(t)\,\mathbf{e}(t) + k(t)\,\kappa\,\dot{\mathbf{e}}(t) \\ &= k(t)\cdot(\mathbf{e}(t) + \kappa\,\dot{\mathbf{e}}(t)) \\ \dot{k}(t) &= \begin{cases} \gamma\cdot(\|\mathbf{e}(t)\| - \lambda)^2, & \|\mathbf{e}(t)\| \geq \lambda + 1 \\ \gamma\cdot(\|\mathbf{e}(t)\| - \lambda)^{0.5}, & \|\mathbf{e}(t)\| \in [\lambda;\lambda{+}1) \\ 0, & (\|\mathbf{e}(t)\| < \lambda) \wedge (t - t_E < t_d) \\ -\sigma k(t), & (\|\mathbf{e}(t)\| < \lambda) \wedge (t - t_E \geq t_d) \end{cases} \end{aligned}\right\} \tag{4}$$

with $\gamma > 1,\ \kappa > 0,\ \sigma > 0,\ t_d \geq 0,\ \lambda > 0,\ k(0) > 0$.

The adaption law results in an increase of the gain $k(\cdot)$ if the error norm exceeds λ. The quadratic increase in case of $\|\mathbf{e}(t)\| \geq \lambda + 1$ forces the error signal to the tube in generating a quick rise of the gain for these very large error values. To make sure the gain increase is not too small for error norms that are smaller, but exceed λ, the square-root function is used to determine $\dot{k}(\cdot)$—now the error signal is forced inside the tube. The variable t_E denotes the time of the last entrance of the error norm into the λ tube. When $\|\mathbf{e}(t)\|$ is smaller than λ, the gain remains at a constant value for a predefined time of stay t_d and then decreases exponentially.

2 Gaits: Construction and Analysis

The *construction of reference trajectories* $l_{ref,j}(t)$ (i.e., gaits) that enable the desired global movement of the WLLS consists of two steps, as described in [9]:

First, it is chosen which mass points should have active spikes (i.e., are resting) at which periods of time. The number of active spikes shall be identical for every time and is denoted by $a \in \{1,\dots,n\}$. All sequences $\mathbf{A}(t)$ (e.g., $\mathbf{A}(t) = \{0\} \to \{1\} \to \{2\} \to \{3\}$), which describe the set of active spikes as a function of time, shall be periodic: mode-functions. Once the active spikes for specific time intervals, i.e., equal parts of one period, are determined, it can be deduced which $l_j(t)$ have to contract, which have to elongate, and which have to keep their length in each time interval.

A short notation of the gait shall consist of the set of active spikes at $t = 0$, $\mathbf{A}_0$, and the direction in which the traveling wave of active spikes moves, dir. The direction from head to rear through the WLLS is called ℓ (left).

In addition to the previous definitions, the detailed $l_{ref,j}(t)$ functions are yet to be chosen. They should be of class C^2 (continuous up to the second derivative) to ensure there are no discontinuous steps in the acceleration terms, as this is undesirable in applications.

The reference functions are built as described in [9] on time intervals

$$t \in \left[p\frac{T}{N}, (p+1)\frac{T}{N}\right], \quad p \in \mathbb{N}_0\,.$$

Table 1 Optimal gaits for WLLS with $N = 4$

$a = 1$
$dir = \ell, \mathbf{A}_0 = \{1\}$
$dir = r, \mathbf{A}_0 = \{2\}$
$a = 2$
$dir = \ell, \mathbf{A}_0 = \{1, 2\}$
$dir = r, \mathbf{A}_0 = \{1, 2\}$
$a = 3$
$dir = \ell, \mathbf{A}_0 = \{2, 3, 0\}$
$dir = r, \mathbf{A}_0 = \{3, 0, 1\}$

using a $\sin^2(\cdot)$-approximation. Using $\tau = t - p\frac{T}{N}$ we exemplarily get as a part of the reference gait

$$\left.\begin{aligned} \dot{l}(\tau) &= \varepsilon l_0 2Nf \sin^2(\pi f N\tau) \\ l(\tau) &= l_{0*} + \varepsilon l_0 N f\tau - \frac{1}{2\pi}\varepsilon l_0 \sin(2\pi f N\tau) \ , \end{aligned}\right\} \tag{5}$$

- $|\varepsilon| \in (0; 1)$ is the relative factor of the maximum contraction/elongation;
- f is the frequency of the $\mathbf{A}(t)$ sequence with periodic time $T = \frac{1}{f}$;
- $l_0 > 0$ is the initial (non-contracted, non-elongated) length;
- l_{0*} is the length at the beginning of a time interval ($\tau = 0$), either l_0 or $l_0 \cdot (1 + \varepsilon)$, depending on the previous movement.

Numerical simulations are carried out to compare all gaits for the WLLS with $N = 4$ that have a set of consecutive active spikes that moves in a traveling wave through the system, like the paradigm earthworm does it. With spike forces, actuator forces, and actuator powers—all of which can be relevant for applications—as primary criteria, a set of six gaits is found, two for each $a \in \{1; 2; 3\}$ [6]. This selection of *optimal gaits* is shown in Table 1.

3 Shift of Gaits

The change of gaits can adjust the whole system to a changing environment, e.g., a varying slope. It is desired to implement an automatic, internal gait shifting relying solely on measured system variables. The options for varying the movement pattern are twofold:

1. changing the *number of active spikes* (value a) by switching the $\mathbf{A}(t)$ sequence; and/or
2. changing the *gait frequency* (value f) while keeping the general $\mathbf{A}(t)$ pattern.

This is possible at any time.

Isolated implementations of either of these principles limit the gait adjustment—especially the mere change of the number of active spikes while keeping a constant frequency, which only allows for n different reference speeds. The reference speed can be obtained from kinematical considerations of the reference functions; it is

$$\bar{v}_{ref}(a, f) = (N - a)\,\varepsilon\, l_0\, f. \tag{6}$$

This reference speed can be altered by changing a and f. To be able to adjust the speed of the WLLS in an optimal way, a combined switching of the two variables is desired.

Possible *criteria for gait shift* can include certain limits of *actuator or spike forces*, to avoid reaching prohibited values; *actuator power*, as a maximum or (moving) average value; and *gain*, which is influenced by external conditions such as friction or slope of the terrain.

In simulations with a number of slopes, gaits, and speeds, shown in [6], it was found that—if the maximum allowed spike forces or actuator forces are chosen as criteria—the choice of the optimal gait depends on the current slope and speed, i.e., there is a certain gait that allows for the largest speed for given force limits. The implementation of an automatic gait shift requires the determination of $n - 1$ speed thresholds, at which the number of active spikes should be switched. Between the thresholds, the gait frequency should be adjusted. The exact values of these thresholds can vary during locomotion, as they are not only dependent on internal system parameters but also on external variables such as the slope.

Investigating the spike forces will enable the choice of *speed thresholds*. In case of $\dot{x}_i = 0$, the spike force $F_{Z,i}$ equals the sum of all other forces acting upon mass point m_i, with opposite sign:

$$F_{Z,i}(0) = u_{i+1} - u_i - F_{Gx,i} + F_{c,i+1} - F_{c,i} + F_{d,i-1} - F_{d,i} - F_{R,i} \tag{7}$$

For the following estimation, all spring, damping, and friction forces are neglected. As shown in parameter studies in [6] (p. 29 ff.), larger damping has little effect on the spike forces, since it is compensated by higher actuator forces. The estimation of the expected maximum value of $u_{i+1} - u_i =: u_{max}$ can be found with knowledge of the reference functions $\mathbf{l}_{ref}(\cdot)$ and the masses. For equal masses $m_i = m$, the actuator force maximum according to the reference functions (5) is

$$u_{max} = m(N - a)\left[2N^2\pi f^2 \varepsilon l_0 + g\sin(\alpha)\right] \tag{8}$$

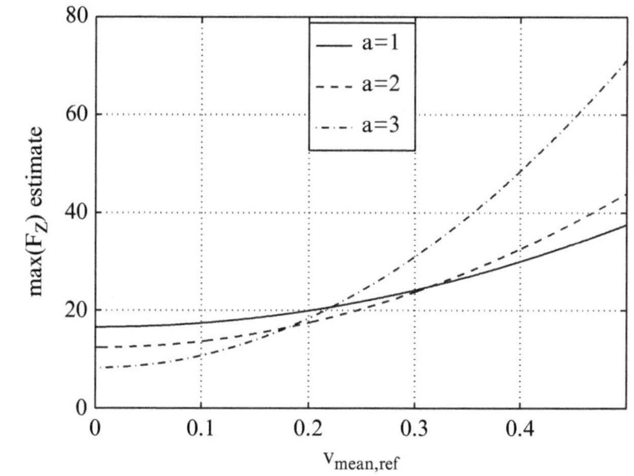

Fig. 3 System with $N = 4$; estimate of the maximum spike force according to (9) vs. $\bar{v}_{ref}$ according to (6), shown for different numbers of active spikes; slope $\alpha = 25°$

The spike force estimate is completed by adding the x-component of the weight of mass m_i:

$$F_{Z,i,max}(f, a) = u_{max} + m_i\, g \sin(\alpha)\,, \tag{9}$$

which makes $F_{Z,i,max}$ a function of frequency (or reference speed) and number of active spikes a. The estimated spike force maxima are shown in Fig. 3. For the WLLS with $N = 4$ and given parameters, the estimates for different numbers of active spikes a, $a = 1, 2, 3$, are plotted. As observed in previous simulations, the actual speed resulting from a certain gait frequency can slightly vary, due to the allowed error of λ.

The combination of the plotted functions exhibits certain speed intervals:

- for speeds of up to $0.18\,\text{ms}^{-1}$, the gait with $a = 3$ requires the lowest spike force maxima;
- between 0.18 and $0.31\,\text{ms}^{-1}$, the gait with $a = 2$ is advantageous;
- for speeds higher than $0.31\,\text{ms}^{-1}$, the gait with $a = 1$ is superior.

Remark 1. Consider driving a car as an analogy. A large value of a is equivalent to a low gear of the car. To perform a gear shift in order to increase the speed, the driver has to accelerate (open the throttle) first, i.e., increase f of the gait for the worm system.

This corresponds qualitatively to the results of the simulations of the worm with $N = 4$. The exact values of the speed thresholds differ slightly, possibly due to neglecting spring and damping forces.

The relevant parts of Fig. 3 are the intersections of the curves. They represent the speed thresholds at which the number of active spikes should be changed to achieve the lowest possible spike forces. The speed thresholds are noted as $\bar{v}_{ref,q\leftrightarrow q+1}$ (with $q = 1,\dots,n-1$), meaning the gait reference speed at which the number of active spikes should be switched from q to $q+1$ or $q+1$ to q.

The switching rule is the following:

- for speeds smaller than $\bar{v}_{ref,q\leftrightarrow q+1}$, switch to $a = q+1$ (larger number of active spikes);
- for speeds larger than $\bar{v}_{ref,q\leftrightarrow q+1}$, switch to $a = q$ (smaller number of active spikes).

Using (9) and (6), the general form of these speed thresholds is, for arbitrary N (number of mass points),

$$\bar{v}_{ref,q\rightarrow q+1} = \left(1-\frac{q}{N}\right)\cdot\sqrt{\frac{1}{2\pi}\left(1-\frac{1}{N-q}\right)\varepsilon l_0 g\sin(\alpha)}, \tag{10}$$

whereby these speed thresholds only depend on N and α.

Remark 2. Note that we only assumed that all masses are equal, i.e., $m_i = m$; their actual value is not required to find these speed thresholds for switching the number of active spikes.

Now, if the slope is known (measured), it is then possible to choose adequate gaits, with respect to low spike forces.

Remark 3. The estimate of the actuator forces (see (8)) gives identical speed thresholds, as it differs from the spike force estimate only by the addition of a weight term of one mass; this term does not influence the intersection of curves in Fig. 3 for different a.

It should therefore be possible to use the speed thresholds above to achieve optimal locomotion, i.e., to use the gait that yields the lowest actuator/spike forces (see Remark 3).

To adjust the reference speed in variable steps, it appears useful to apply closed-loop control between the input variable *reference frequency* and an output variable, which may be the measured (maximum) value of actuator forces, spike forces, or actuator powers. This nice idea is based on [6]. The implementation is based on simple P-feedback. This forms a *proportional frequency control*. First, a reference value X_{ref} for the output criterion is chosen. The actual output is built by measurements of the chosen criterion over given intervals of time, in this case $\frac{T}{N}$. Measurement and frequency adjustment according to the control are evaluated after each of these $\frac{T}{N}$-time intervals, so that the result is a time-discrete controller which provides discrete frequency values $f_k = f\left(k\frac{T}{N}\right)$.

The deviation between the measurement variable and its reference is evaluated and fed back into the loop to form the new frequency. The measurement shall be performed continuously over given time intervals, in order to find the maximum value, i.e.,

$$X_{err,k} = X_{ref} - X_{meas,k}\,, \quad \text{with } X_{meas,k} = \max_{\tau \in \left[\frac{T}{4}(k-1), \frac{T}{4}k\right]} \left(|X_0(\tau)|, \ldots, |X_n(\tau)|\right). \tag{11}$$

This error shall contribute proportionally to the factor of frequency adjustment. For the frequency adjustment, a factor is used instead of absolute values, because a feasible range of frequencies might not be known in advance and a relative factor should enable proper adjustment for any range. The new frequency is calculated according to

$$f_{k+1} = f_k + f_k\, X_{err,k}\, k_P = f_k \cdot (1 + X_{err,k}\, k_P)\,, \tag{12}$$

with a gain $k_P \in \mathbb{R}_+$, $f_0 \in \mathbb{R}_+$. As described before, this implementation entails relative frequency changes, where a certain deviation of the measured variable from its reference always results in the same percental frequency change.

For practical implementations, the variable X may represent *actuator forces*, *actuator powers*, or *spike forces*.

4 Simulation

In the following simulation of an automatic gait shift for the WLLS with $N = 4$, the parameter set shown in Table 2 is used.

The slope is defined as a function of time:

$$\alpha(t) = \begin{cases} 15^\circ, & t \leq 10\,\text{s} \\ 60^\circ, & 10\,\text{s} < t < 20\,\text{s} \\ 25^\circ, & t \geq 20\,\text{s}. \end{cases} \tag{13}$$

Three optimal gaits are chosen from Table 1:

- $\mathbf{A}_0 = \{1\}, dir = \ell;$
- $\mathbf{A}_0 = \{1, 2\}, dir = \ell;$

Table 2 Parameter set

$t = [0, 20\,\text{s}]$	$l_0 = 1\,\text{m}$	$\lvert\varepsilon\rvert = 0.4$
$\alpha = 25^\circ$	$f = 0.5\,\text{Hz}$	$\sigma = 0.2\text{s}^{-1}$
$m_i = 1\,\text{kg}$	$\lambda = 0.05\,\text{m}$	$t_d = 2\,\text{s}$
$c_j = 10\,^{\text{N}/\text{m}}$	$\gamma = 500$	$k(0) = 10^{\text{N}/\text{m}}$
$d_j = 5\,\text{kg}\,\text{s}^{-1}$	$\kappa = 1\,\text{s}$	

- $\mathbf{A}_0 = \{2, 3, 0\}, dir = \ell$.

As an initial value, the gait with $a = 1$ and a frequency of $f|_{t=0} = 0.5\,\text{Hz}$ are selected. The frequency is adjusted by the frequency P-feedback (12) presented before.

The frequency thresholds for the shift of the number of active spikes are derived according to (10); they are

$$f_1(\alpha) = \sqrt{\frac{g \sin(\alpha)}{3N^2\pi\varepsilon l_0}} \approx 0.4032 \cdot \sqrt{\sin(\alpha)}\,\text{Hz and}$$

$$f_2(\alpha) = \sqrt{\frac{g \sin(\alpha)}{4N^2\pi\varepsilon l_0}} \approx 0.3492 \cdot \sqrt{\sin(\alpha)}\,\text{Hz.}$$

With these thresholds, the rule for switching the number of active spikes is

$$\left\{\begin{array}{ll} \text{if} a = 2 \wedge f \geq \frac{3}{2} f_1 : & \text{switch to } a = 1 \\ \text{if}(a = 1 \wedge f < f_1) \vee (a = 3 \wedge f > 2f_2) : & \text{switch to } a = 2 \\ \text{if} a = 2 \wedge f < f_2 : & \text{switch to } a = 3 \\ \text{else} : & \text{no switch} \end{array}\right\}$$

When switching a, a frequency change is applied to keep the same reference speed, according to (6), as before. The reference speed shall only be changed by the frequency P-feedback. For this feedback, the following parameters are chosen according to (11) and (12): $F_{Z,ref} = 20\,\text{N}$, $k_P = 0.03\,\text{N}^{-1}$. In the following plots, the points of slope change (at 10 s and at 20 s) are marked vertical lines; additional vertical lines denote points of gait shift. In the plot of the *worm motion* [Fig. 4(left)], the different movement patterns of the gaits are visible.

For the larger slope, the *gait* is almost instantly shifted to $a = 2$ [see Fig. 4(right)]. This happens even before the reference speed is reduced, because the

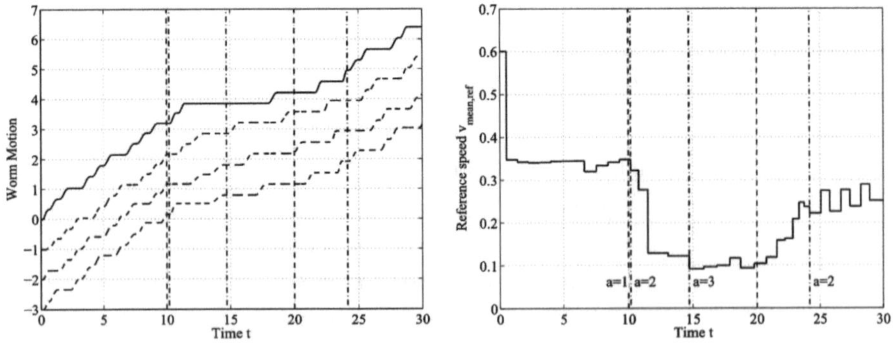

Fig. 4 *Left*, worm movement; *right*, reference speed $\bar{v}_{ref}$

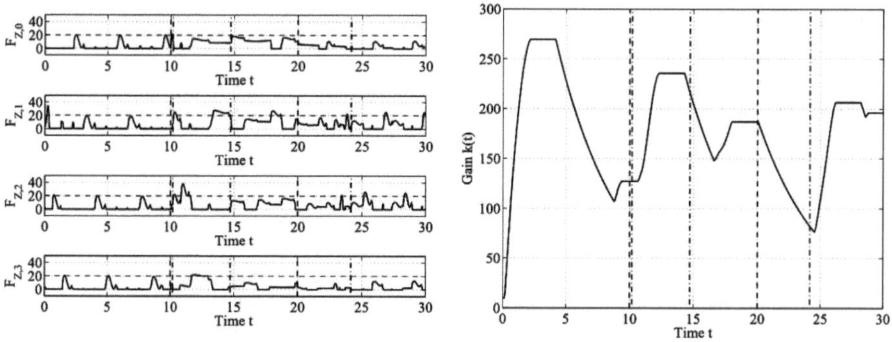

Fig. 5 *Left*, spike forces; *right*, gain $k(t)$

(measured) larger slope influences $\bar{v}_{1\to 2,N=4}$, as seen in (10), so that the gait with $a = 2$ is considered optimal. The speed is then reduced to a value that requires the gait with $a = 3$ to be used. Later, when the slope is smaller, the gait is shifted back to $a = 2$.

Within $t = [0, 10\,\text{s}]$ the maxima of the *spike forces* [see Fig. 5(left)] adhere to the reference value very closely. Later, when different gaits are used, the maxima of some of the spike forces clearly exceed the reference value once per period by up to 30%. The gait and slope switches require different gain levels [Fig. 5(right)]. The necessary adjustments are performed quickly, within fractions of a period.

5 Conclusion

Overall, the adaptive controller is capable of quickly adjusting the gain when frequency and gait pattern are changed, so that the control goal can be accomplished. The combined switching of frequency and number of active spikes enables quick speed adjustments in order to adhere to given limits of a system variable such as spike or actuator force. The gait shift, i.e., the change of the number of active spikes, helps to choose the gait that is optimal for a certain speed and slope. With this, predefined restrictions of forces or powers of the system are utilized in such a way that the maximum speed is reached.

Future work shall be directed to *improvements* of the concepts for automatic gait shift (in order to achieve a faster and more robust shifting), *analysis* of shifting concepts for WLLS with $N > 4$, and *experimental verification*.

References

1. Behn, C., Zimmermann, K.: Robotics Auton. Syst. **54**, 529–545 (2006)
2. Behn, C., Steigenberger, J.: Int. Rev. Mech. Eng. (IREME) **4**(7), 886–898 (2010)
3. Chen, I.M., Yeo, S.H., Gao, Y.: Robotica **19**, 535–542 (2001)
4. Ilchmann, A.: Non-Identifier-Based High-Gain Adaptive Control. Springer, London (1993)
5. Ostrowski, J.P., Burdick, J.W., Lewis, A.D., Murray, R.M.: The mechanics of undulatory locomotion: the mixed kinematic and dynamic case. In: Proceedings IEEE International Conference on Robotics and Automation, Nagoya, Japan, 1995
6. Schwebke, S.: Contributions to adaptive control strategies of biomimetic, worm-like locomotion systems and their use for gait shifting. Bachelor Thesis, TU Ilmenau (2012)
7. Seok, S., Onal, C.D., Wood, R., Rus, D., Kim, S.: Peristaltic locomotion with antagonistic actuators in soft robotics. In: Proceedings 2010 IEEE International Conference on Robotics and Automation, 2010
8. Slatkin, A.B., Burdick, J., Grundfest, W.: The development of a robotic endoscope. In: Proceedings 1995 IEEE/RSJ International Conference on Intelligent Robots and Systems: Human Robot Interaction and Cooperative Robots, vol. 2, pp.162–171, 1995
9. Steigenberger, J., Behn, C.: Robotics Auton. Syst. **5**, 555–562 (2011)
10. Steigenberger, J., Behn, C.: Worm-like locomotion systems: an intermediate theoretical approach. Oldenbourg, Munich (2012)

Turbulent Phenomena in Flexible Plates and Shells

J. Awrejcewicz, A.V. Krysko, V.A. Krysko, E.Yu. Krylova, S.A. Mitskievich, I.V. Papkova, T.V. Yakovleva, V.M. Zakharov, and V. Dobriyan

Abstract The aim of this chapter is to present a study of periodic and chaotic dynamics of plates and shells and weak turbulent behavior exhibited by these solid structural members modeled as 2D infinite objects. Besides the new results obtained with respect to the transition from a regular to weak turbulent and weak hyper turbulent behavior, we also present novel methods and approaches to get reliable and validated results of numerical analysis of nonlinear partial differential equations. In particular, besides the standard numerical techniques for chaos monitoring, new effective approaches are presented and applied including the wavelet-based analysis, charts of vibration regimes, computation of the spectra of Lyapunov exponents via generalization of the classical Benettin's approach, and application of the neural network technique. This common strategy aimed at numerical computations through various types of robust discretization allowed us to obtain novel scenarios of transition from regular/laminar (periodic, quasiperiodic) to spatiotemporal chaotic (weak turbulent) dynamics of flexible shells either parametrically excited (Sect. 2) or through the periodic shear load action in the shell volume unit (Sect. 3), flexible multilayer rectangular (Sect. 4) and cylindrical shells with gaps (Sect. 5), as well as a flexible plate of infinite length (Sect. 6).

J. Awrejcewicz (✉)
Department of Automation, Biomechanics and Mechatronics,
Lodz University of Technology, 1/15 Stefanowski Str., 90-924 Lodz, Poland

Department of Vehicles, Warsaw University of Technology, 84 Narbutta Str.,
02-524 Warsaw, Poland
e-mail: jan.awrejcewicz@p.lodz.pl

A.V. Krysko • V.A. Krysko • E.Yu. Krylova • S.A. Mitskievich • I.V. Papkova
T.V. Yakovleva • V.M. Zakharov • V. Dobriyan
Saratov State Technical University, Applied Mathematics and Systems Analysis,
Politekhnicheskaya 77, 410054 Saratov, Russian Federation
e-mail: anton.krysko@gmail.com; tak@san.ru

J. Awrejcewicz (ed.), *Applied Non-Linear Dynamical Systems*, Springer Proceedings
in Mathematics & Statistics 93, DOI 10.1007/978-3-319-08266-0_5

1 Introduction

This chapter addresses an important problem of turbulent wave dynamics exhibited by a solid which has not been satisfactorily explained and clarified yet through the investigations of vibrations of flexible plates and shells. It extends our earlier studies on beams/plates/shells spatiotemporal chaotic dynamics. In particular, the wavelet approach is applied, which contrary to the Fourier analysis (FFT) makes it possible to follow the frequency spectrum time evolution and hence reveal the so far hidden nonlinear phenomena that have not been described in the existing literature.

Recently, the attention of numerous researchers has been focused on the application of the wavelet analysis to study nonlinear dynamics of continuous structural members (beams, plates and shells) governed by partial difference equations (PDEs) using the finite elements method (FEM), finite difference method (FDM), higher-order Bubnov–Galerkin approach, or semi-discrete techniques.

It has been shown that wavelets allow for a solution decomposition into a set depending on the scale and location, and the wavelet bases in multi-scales combined either with FEM or FDM yield sometimes novel results, which have been omitted using the standard techniques. The applied scaling functions provide a multi-resolution analysis and a robust strategy for mesh refinement. For example, the feasibility of a hybrid scheme of the combination of Daubechies wavelet functions and FEM is outlined in [1], and a few results regarding the analysis of a Euler–Bernoulli beam element and a Mindlin–Reissner plate element are given. The theory of wavelets is widely used in identification problems, where a studied system evolution is deduced from the wavelet transform response. Wavelets of the Daubechies, Morlet, and Gauss families are used to detect faults in gears and gearboxes as well as to estimate the location of cracks [21, 25, 30]. The wavelet-type analysis is applied to lumped mechanical systems governed by ordinary differential equations [12, 17]. The proposed procedure can be viewed as a logarithmic transform applied to a fitted component of the original response. Damping ratios of a multi-degrees-of-freedom systems are estimated via the wavelet-based formula allowing for the separation of purely random modal contributions from the fundamental mode. Linear time-varying dynamic systems governed by ODEs were studied in [2, 10] using the wavelet-based Galerkin procedure. In particular, in the latter paper [10], the damping and stiffness parameters were identified using the robust wavelet-based algorithm. On the other hand, the Daubechies wavelets of order 1 (Haar wavelets) are simple in applications due to their orthonormality and compact support, and they are used while studying the dynamics of elastic–plastic beams [13, 18–20].

So far, we have briefly overviewed the application of wavelet concepts to studies on nonlinear dynamics of the systems governed by ODEs and PDEs. Below, we describe the state of part of the research on turbulent behavior exhibited by continuous solid dynamic systems.

Mordant [24] applied an experimental method to monitor both temporal and spatial evolutions of wave turbulence in a thin elastic plate. Various Fourier spectra of the wave deformations have been analyzed. His research follows a series of earlier works devoted to a study of the Föppl–von Kármán model [8, 9, 23]. Despite qualitative good agreement with the kinetic weak turbulence theory, the obtained energy spectrum has not been confirmed by a theoretical prediction. It has been shown that the wave turbulence can also be exhibited by vibrating structural members including beams, plates, and shells, where in spite of the standard Fourier components various wavelet-type methods have been applied allowing for a deeper insight into the wave turbulence in a solid [3–6]. Recently, the transition scenario from periodic to wave turbulence regime in the forced vibrations of thin free-edge circular and simply supported rectangular plates has been reported in [26, 27]. It has been shown that only two bifurcations are required to achieve a spatiotemporal chaotic dynamics. The first bifurcation is associated with a stability loss of the directly excited plate mode in favor of the quasiperiodic dynamics. The second bifurcation implies the stability loss and the wave turbulence appearance. It has been confirmed numerically that the obtained power spectra are in good agreement with the theoretical prediction given by the wave turbulence theory with respect to the von Kármán equations for perfect undamped plates.

The paper is organized in the following manner. Section 2 is dedicated to the analysis of chaotic/turbulent dynamics of rectangular periodically forced flexible shells. Section 3 illustrates and discusses the application of various types of wavelets to study nonlinear dynamics of flexible rectangular isotropic shells subjected to the periodic shear load action. Section 4 presents turbulent dynamics of multilayer flexible spherical panels putting emphasis on the simultaneous occurrence of both temporal and spatial chaos. In Sect. 5 a turbulent behavior of two-layer cylindrical shells coupled via boundary conditions and subjected to local periodic transversal load action is reported. In Sect. 6 a weak hyper turbulent plate behavior is observed while monitoring numerically the spectrum of Lyapunov exponents.

2 Chaotic Parametric Vibrations of Flexible Plates

2.1 Mathematical Model

A mathematical model of the flexible rectangular plate with constant stiffness and density under the action of a periodic load (Fig. 1) is constructed using the Kirchhoff–Love hypotheses and taking into account the nonlinear relations between deformations and displacements in the von Kármán form. In the rectangular coordinates the 3D plate space is defined as $\Omega = \{x_1, x_2, x_3 | (x_1, x_2) \in [0; a] \times [0; b], x_3 \in [-h; h]\}$, $0 \leq t < \infty$. In initial time interval $t \in [0; 1]$ we introduce a small static load (its lack defines the governing differential equations as homogenous ones).

We study the following non-dimensional PDEs governing dynamics of the shallow shells [29]:

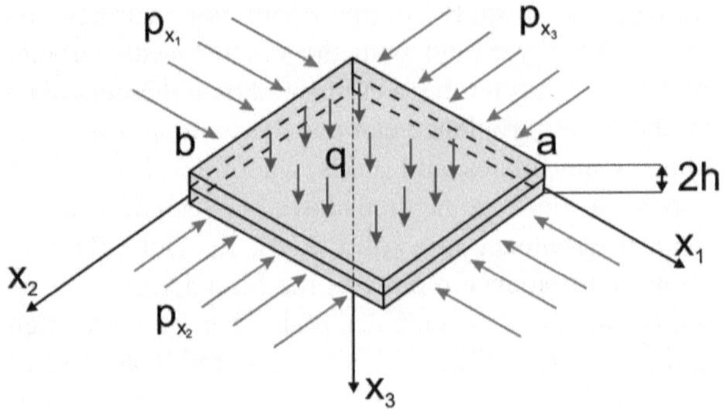

Fig. 1 Plate computational scheme

$$\begin{gathered}\frac{1}{12(1-\mu^2)}\left(\nabla_\lambda^4 w\right)-\nabla_k^2 F-L(w,F)-\frac{\partial^2 w}{\partial t^2}-\varepsilon\frac{\partial w}{\partial t}-q(x_1,x_2,t)=0,\\ \nabla_\lambda^4 F+\nabla_k^2 w+\frac{1}{2}L(w,w)=0,\end{gathered} \tag{1}$$

where ∇_λ^4, $L(w,F)$ and ∇_k^2 are the known nonlinear operators, whereas w and F are the functions of deflection and stress, respectively.

The following non-dimensional parameters are introduced: $\lambda = a/b$, $x_1 = a\overline{x}_1$, $x_2 = b\overline{x}_2$, $k_{x_1} = a^2/R_{x_1}(2h)$, $k_{x_2} = b^2/R_{x_2}(2h)$ are the non-dimensional shell curvature parameters regarding x_1 and x_2, respectively; $w = 2h\overline{w}$ is the deflection; $F = E(2h)^3\overline{F}$ is the stress function; $t = t_0\overline{t}$ is the time; $q = \frac{E(2h)^4}{a^2b^2}\overline{q}$ is the external load; $\varepsilon = (2h)\overline{\varepsilon}$ is the damping coefficient; and $P = E(2h)^3\overline{P}$ is the external longitudinal load. Bars over non-dimensional quantities in the governing equations have already been omitted. The following notation is introduced: a, b are the plate plane dimensions in x_1 and x_2 directions, respectively; μ is the Poisson's coefficient. Equation (1) is supplemented by the following boundary conditions [16]:

$$\begin{aligned} &w=0;\quad \frac{\partial^2 w}{\partial x_1^2}=0;\quad F=0;\quad \frac{\partial^2 F}{\partial x_1^2}=p_{x_2}\quad \text{for}\quad x_1=0;1,\\ &w=0;\quad \frac{\partial^2 w}{\partial x_2^2}=0;\quad F=0;\quad \frac{\partial^2 F}{\partial x_2^2}=p_{x_1}\quad \text{for}\quad x_2=0;1,\end{aligned} \tag{2}$$

and the following initial conditions:

$$w(x_1,x_2)|_{t=0}=0,\quad \frac{\partial w}{\partial t}=0. \tag{3}$$

2.2 Method of Solution and Results

Our mechanical object is studied keeping fixed the following parameters: $\lambda = 1$, Poisson's coefficient $\mu = 0.3$. We apply the following longitudinal load $p_{x_1} = p_{x_2} = p_0 \sin(\omega_p t)$. After the application of FDM with approximation (h^2) regarding spatial coordinates the differential problem (1)–(3) is solved by the Runge–Kutta method of the fourth order. In addition, on each time step we need to solve a large system of linear algebraic equations (AEs) regarding the stress function. Time step is yielded through the Runge principle. The number of partitions of spatial coordinates is $n = 14$ (the number of partitions choice and convergence of the obtained numerical results were discussed by Awrejcewicz et al. [6]).

In this work we take the amplitude and frequency of exciting longitudinal load acting on the shell perimeter. Our aim was to construct charts displaying the system vibration regimes in the plane of parameters $[\omega_p, p_0]$ with the resolution of 300×300. In order to construct each of the charts 90,000 differential problems were solved. Each of the mentioned problems required the analysis of signals (time histories), phase and modal portraits, Poincaré cross-sections and maps, Fourier and wavelet frequency power spectra, autocorrelation functions, and signs of the Lyapunov exponents. Figure 2 gives charts of the shell vibration regimes depending on the geometric shell parameters. In the first subharmonic regime zones of Hopf bifurcations are wide. It is seen that an increase of the geometric parameters implies the increase of bifurcation zones and chaotic zones and a decrease of the zones of periodic vibrations. Besides, one may observe that in the case of shell curvatures $k_{x_1} = k_{x_2} = 24$ a transition from periodic into chaotic/turbulent vibrations appears suddenly without any other transitional zones. The following general remarks are based on the computational results. Small exciting load amplitudes generate damped vibrations. Low values of the applied frequencies $\omega \leq 2$ of subharmonic zones of vibrations (excitation frequency is doubled in comparison to the shell vibration frequency) are mixed with the zones of periodic vibrations. An increase of the excitation frequency implies an extension of these zones into higher frequencies, and they are interlaced with rather large chaotic/turbulent zones.

While constructing the shell vibration charts chaotic dynamics occurred already after the second or even first Hopf bifurcation. Additionally, the wavelet spectra imply that these bifurcations appear even at fixed amplitude and frequency of the excitation for $t \geq t_0$ (they depend only on time giving rise to intermittency). In the numerical example with parameters $k_{x_1} = 12$, $k_{x_2} = 0$ (cylindrical panel), $\omega_p = 8.4$, $p_0 = 8.5$, $t \in [0; 300]$, using both Fourier and wavelets spectra, it is evidently demonstrated how power spectrum essentially changes in time. For instance, at $t < 150$, the first Hopf bifurcation takes place ($\omega_1 = \omega_p/2$), and at $t > 150$ the second Hopf bifurcation appears ($\omega_2 = \omega_p/4$) (Fig. 3). A more detailed analysis is provided by the wavelet analysis which allows us to monitor local phenomena of the studied signal.

It should be emphasized that the charts reported in this section (Fig. 2) allow us to control the investigated continuous mechanical systems. The choice of the control

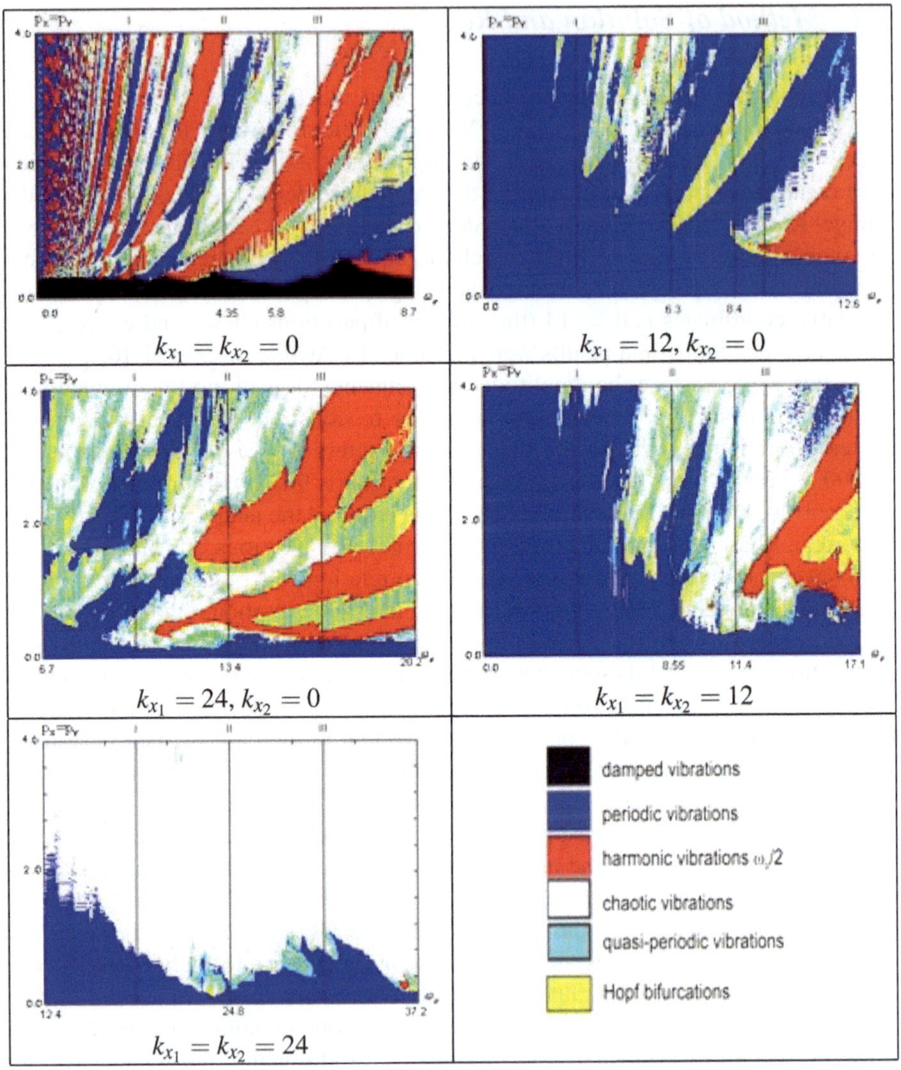

Fig. 2 Charts of the shell vibration regimes versus geometric (shell curvature) parameters

parameters should be made in a way to keep the system within a safe (periodic) zone. Otherwise, transition into the chaotic zone implies the loss of system stability and its catastrophe.

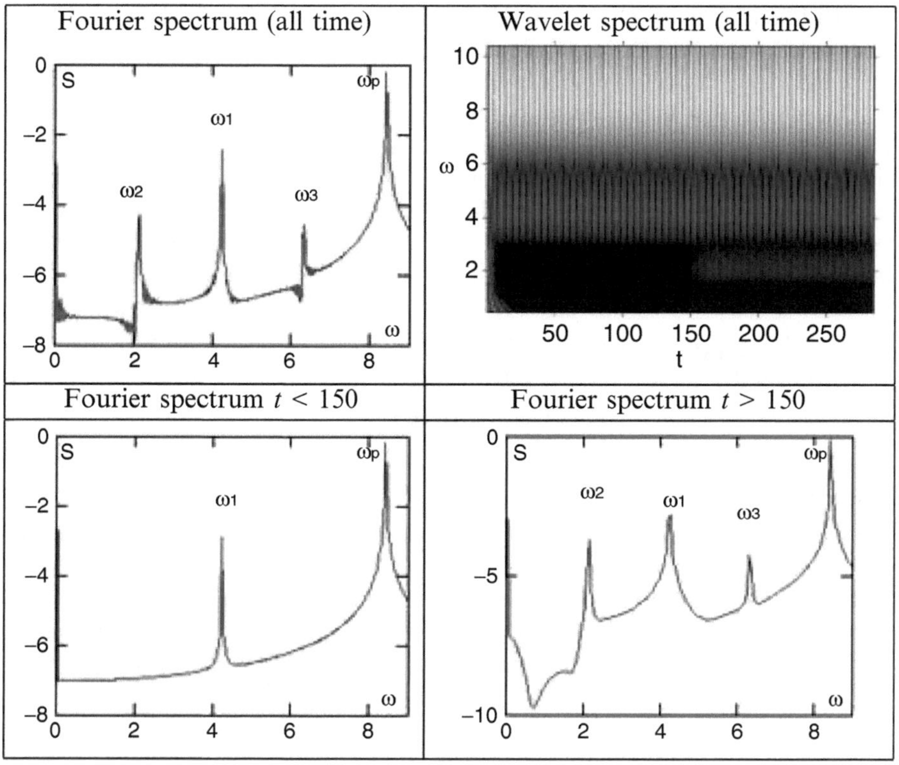

Fig. 3 Fourier and wavelet spectra $k_{x_1} = 12$, $k_{x_2} = 0$, $\omega_p = 8.4$, $p_0 = 8.5$

3 Wavelet Versus Fourier Analysis

3.1 Governing Equations and Results

It is well known that the data provided by numerical experiments are presented in the time domain. In other words, we take time as an independent coordinate and amplitude as a dependent coordinate, and the studied signal is analyzed through its amplitude–time representation. However, in order to understand profoundly nonlinear continuous systems subjected to various types of load actions and in order to fully understand the occurring dynamics, we have to apply the information hidden in the spectral signal characteristics. Predominantly, the Fourier transformation has been applied for a long time. However, it has been demonstrated that the Fourier analysis (FFT) is reliable only for the study of frequency components of stationary processes, i.e., the processes which through the whole period of investigation keep constant frequency components in time. It happens that, in particular, the dynamics of continuous mechanical systems may exhibit quite complicated output and their frequency characteristics may change strongly in time. This is why in spite of the standard Fourier approach the wavelet analysis is applied, allowing us to detect and understand many interesting nonlinear phenomena of the mentioned mechanical systems.

One of the first important tasks to be solved is associated with the choice of a suitable wavelet, which fits well with the stated problem. In order to solve this query we have studied the non-stationary signal (Fig. 4) obtained via the numerical experiment as an output of the mathematical model of the rectangular flexible isotropic shell subjected to the periodic shear load acting in the shell volume unit. The mathematical model is as follows [29]:

$$\begin{gathered}\frac{1}{12(1-\mu^2)}\left(\nabla_\lambda^4 w\right)-L(w,F)+\frac{\partial^2 w}{\partial t^2}+\varepsilon\frac{\partial w}{\partial t}-q(x_1,x_2,t)+2S\frac{\partial^2 w}{\partial x_1\partial x_2}=0,\\ \nabla_\lambda^4 F+\frac{1}{2}L(w,w)=0,\end{gathered} \tag{4}$$

where $\nabla_\lambda^4=\frac{1}{\lambda^2}\frac{\partial^4}{\partial x_1^4}+\lambda^2\frac{\partial^4}{\partial x_2^4}+2\frac{\partial^4}{\partial x_1^2\partial x_2^2}$, $L(w,F)=\frac{\partial^2 w}{\partial x_1^2}\frac{\partial^2 F}{\partial x_2^2}+\frac{\partial^2 w}{\partial x_2^2}\frac{\partial^2 F}{\partial x_1^2}-2\frac{\partial^2 w}{\partial x_1\partial x_2}\frac{\partial^2 F}{\partial x_1\partial x_2}$ are the known nonlinear operators, whereas w and F stand for the plate deflection and Airy's function, respectively.

System (4) is reduced to the non-dimensional form using the following non-dimensional parameters: $\lambda = a/b$; $x_1 = a\overline{x}_1$, $x_2 = b\overline{x}_2$ are the non-dimensional parameters regarding x_1 and x_2, respectively; $w = 2h\overline{w}$ is the deflection; $F = E(2h)^3\overline{F}$ is Airy's function; $t = t_0\overline{t}$ is the time; $q = \frac{E(2h)^4}{a^2b^2}\overline{q}$ is the external load; $\varepsilon = (2h)\overline{\varepsilon}$ is the damping coefficient of the surrounding medium; and $S = \frac{E(2h)^3}{ab}\overline{S}$ is the external shear load. In the equations bars over the non-dimensional quantities have already been omitted. The following notation is introduced: a, b are the shell dimensions regarding x_1 and x_2, respectively; μ is Poisson's coefficient. Zero-value initial conditions and the following boundary value conditions are attached to system (4):

$$\begin{aligned} &w=0; \quad \frac{\partial^2 w}{\partial x_1^2}=0; \quad F=0; \quad \frac{\partial^2 F}{\partial x_1^2}=0 \quad \text{for} \quad x_1=0;1,\\ &w=0; \quad \frac{\partial^2 w}{\partial x_2^2}=0; \quad F=0; \quad \frac{\partial^2 F}{\partial x_2^2}=0 \quad \text{for} \quad x_2=0;1.\end{aligned} \tag{5}$$

The external harmonic shear load has the form $S = s_0 \sin\omega_p t$. PDEs governing dynamics of our investigated shell are reduced to the ODEs via the FDM with the approximation $O(h^2)$ regarding spatial coordinates. Next, ODEs are solved via the fourth-order Runge–Kutta method, and additionally on each of the iterations a large system of linear algebraic equations should be solved with respect to the stress (Airy's) function. Time integration step has been chosen using the Runge rule. The partition number of spatial coordinates is $n = 14$ while applying FDM. Validity and reliability of the obtained results regarding the number of partitions have been discussed by Awrejcewicz et al. [6]. The term $2S\frac{\partial^2 w}{\partial x_1 \partial x_2}$ introduced in the governing equations exhibits the action of shear stresses located in shell middle plane and it influences essentially nonlinear dynamics of the investigated shell. The numerical simulation indicates that the output signal (shell vibrations) may change in time rapidly. We apply this signal to choose a methodology suitable for the investigation of non-stationary processes, and in addition we illustrate advantages and disadvantages of the standard Fourier approach versus the wavelet transform procedure. The studied signal has been obtained, analyzing the system with the following fixed parameters: $s_0 = 8.4$ and $\omega_p = 26$. We show that frequency characteristics taken in different time intervals essentially differ from each other. It should be emphasized that the system stability loss is due not only to the change of chosen control parameters but is caused even by keeping all of them fixed. The system time evolution may lead to shell instability. In the first time interval $t \in [50; 56]$ the shell exhibits two quasiperiodic frequency vibrations. Instead of the vanished excitation frequency, two independent frequencies appear. A further long-time evolution of chaotic vibrations with the exhibition of a few dependent frequencies is observed. In the Fourier spectrum the excitation frequency is not visible. The last studied time interval corresponds to periodic vibrations, which is also in agreement with the Fourier power spectrum (Fig. 4).

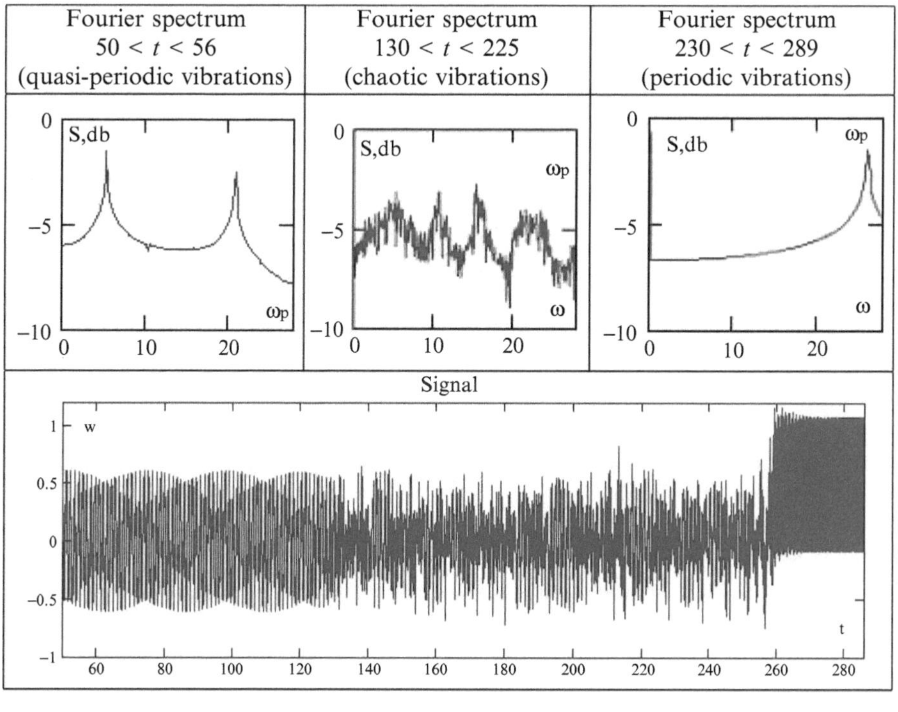

Fig. 4 Fourier spectra and signal

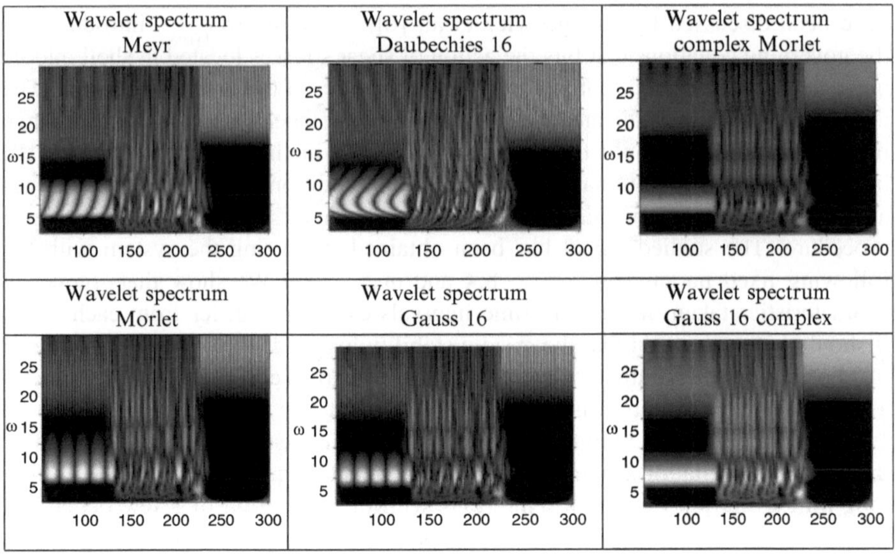

Fig. 5 Wavelets spectra

We constructed wavelet spectra to study the mentioned signal. We applied the following wavelets: Haar, Shannon–Kotelnikov, Meyer, Daubechies from db2 up to db 16, Coiflets and Symlet, as well as the Morlet and Gauss (real and imaginary), on the basis of the derivatives from 2 to 16. The Haar and Shannon–Kotelnikov wavelets appeared to be unsuitable for the analysis of shell structures. The first one is badly localized regarding frequency, whereas the second one, contrary to the previous wavelets, is badly localized in time. On the other hand, the analysis based on the Daubechies wavelets, as well as Coiflets and Symlet wavelets, implies an increase of the frequency resolution, assuming that the filter properties are increased. Neglecting differences in the wavelet forms and the associated filters, the wavelet spectra obtained through the Daubechies wavelets as well as Coiflets and Symlet wavelets are practically identical. However, their localization with respect to frequency is not suitable for the analysis of continuous systems dynamics. In the case of the Gauss functions, an increase of their derivative order implies an increase of the frequency resolution.

Figure 5 gives results associated with the application of different wavelets (Meyer, Morlet, complex Morlet, real and complex Gauss with 16 derivative orders, Daubechies) to analyze nonlinear shell vibrations.

One may conclude from Fig. 5 that the localization regarding frequency increases with an increase of the number of the wavelet zero moments.

This section has shown that the complex Morlet and Gauss wavelets have better localization with respect to frequency than their real analogues, but time localization is better exhibited for the real wavelets. Therefore, one may apply either real or complex Morlet and Gauss wavelets of the order bigger than 16 while studying plates/shells dynamics.

4 Turbulence of Flexible Multilayer Shells with Gaps Between Layers

4.1 Problem Statement

Significant results have been achieved recently during the analysis of chaotic dynamics of simple dynamic systems yielded by truncated original ODEs or the recurrence equations (maps). It should be emphasized, however, that in these simple systems we may observe only temporal chaos. This may be helpful while analyzing the beginning of turbulence, when velocity field starts with the evolution of chaotic velocities, but keeping all the time the well self-organized space. In this work we illustrate that both chaotic phenomena, i.e., temporal and spatial chaos, exist simultaneously. Spatiotemporal chaos (weak turbulence) has been detected while analyzing the Ginzburg–Landau equation [11], but as far as we know this phenomenon has not been detected so far with respect to the shell structural members. In this section nonlinear multilayer shells are studied, where in spite of the geometric nonlincaritywe also take into account a design-type nonlinearity. It means that on each time step we need to solve a contact problem yielded by the interaction of shell members.

We apply a classical nonlinear theory to study the two-layer spherical flexible isotropic elastic rectangular shell with constant stiffness and density and subjected to the action of longitudinal time periodic load. The load is applied only to the upper shell layer. The layers exhibit hybrid dynamics: they are either contacting between each other without friction or they vibrate without a contact. The occurrence of stick zones is rather unlikely because the contact pressure is small. Contact conditions between the layers may depend on the coordinates and may include all kinds of the one-sided contact. Welding conditions regarding normal and tangential directions are not considered in this work. The behavior of layers agrees with the Kármán-Vlasov theory and is the same for all layers. The function describing the contact pressure is removed from the number of unknowns. Since the order of the governing differential equations is a product of the number of layers and the system equations order, we deal with the Bolotin–Novikov model (Fig. 6). In this work we analyze two-layer panels:

$$\Omega_1 = \{x_1, y_1, z_1 | (x_1, y_1) \in [0; a] \times [0; b], z_1 \in [-h_1, h_1]\},$$

$$\Omega_2 = \{x_2, y_2, z_2 | (x_2, y_2) \in [0; a] \times [0; b], z_2 \in [-h_2, h_2]\}, \quad 0 \leq t < \infty.$$

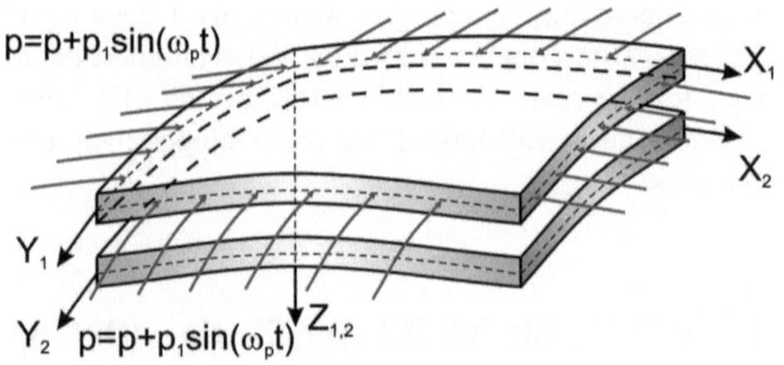

Fig. 6 Computational scheme of a two-layer panel

We study the following equations of the theory of shallow shells presented in the non-dimensional form [15, 29]:

$$\begin{gathered}\nabla^4 w_m + L(w_m, F_m) + \nabla^2 F_m + q_1 + p_{x_m}\frac{\partial^2 w_m}{\partial y_m^2} + p_{y_m}\frac{\partial^2 w_m}{\partial x_m^2} \pm \\ \pm K(w_1 - h_k - w_2)\psi = \frac{\partial^2 w_m}{\partial t^2} + \varepsilon_1\frac{\partial w_m}{\partial t}, \\ \nabla^4 F_m = -\tfrac{1}{2}L(w_m, w_m) - \nabla^2 w_m,\end{gathered} \tag{6}$$

where

$$\begin{aligned} L(w_m, F_m) &= \frac{\partial^2 w_m}{\partial x_m^2}\frac{\partial^2 F_m}{\partial y_m^2} + \frac{\partial^2 w_m}{\partial x_m^2}\frac{\partial^2 F_m}{\partial y_m^2} - 2\frac{\partial^2 w_m}{\partial x_m \partial y_m}\frac{\partial^2 F_m}{\partial x_m \partial y_m}, \\ \nabla^4 &= \frac{1}{\lambda^2}\frac{\partial^4}{\partial x_m^4} + \lambda^2\frac{\partial^4}{\partial y_m^4} + 2\frac{\partial^4}{\partial x_m^2 \partial y_m^2}, \\ L(w_m, w_m) &= \frac{\partial^2 w_m}{\partial x_m^2}\frac{\partial^2 w_m}{\partial y_m^2} - \left(\frac{\partial^2 w_m}{\partial x_m \partial y_m}\right)^2, \quad i = 1, 2. \end{aligned} \tag{7}$$

Here $\psi = \frac{1}{2}[1 + \mathrm{sign}(w_1 - h_k - w_2)]$, w_m and F_m are the functions of deflection and stresses, respectively, where the layers $m = 1, 2$ and $K = 17,500$ are the stiffness coefficients of the transversal contact zone. If $w_1 > w_2 + h_k$, the contact between layers occurs, which means that $\psi = 1$ and otherwise $\psi = 0$, w_1, w_2 deflection of the upper and lower panel, respectively.

The system of (6) is transformed to the non-dimensional form through the following relations:

$$\begin{gathered} x_m = a\overline{x}_m, \quad y_m = b\overline{y}_m, \quad \overline{k}_{x_m} = k_{x_m}\frac{a^2}{h_m}, \quad \overline{k}_{y_m} = k_{y_m}\frac{b^2}{h_m}, \\ k_{x_m} = \frac{1}{r_{x_m}}, \quad k_{y_m} = \frac{1}{r_{y_m}}, \quad q_m = \overline{q}_m\frac{E_m h_m^4}{a^2 b^2}, \quad p_{x_m} = \overline{p}_{x_m}\frac{E_m h_m^3}{b^2}, \\ p_{y_m} = \overline{p}_{y_m}\frac{E_m h_m^3}{a^2}, \quad \tau_m = \frac{ab}{h_m}\sqrt{\frac{\rho_m}{E_m g_m}}, \quad \lambda_1 = \frac{a}{b}, \end{gathered} \tag{8}$$

where a, b are the sides of a cylindrical panel regarding x_m and y_m, respectively; h_m is the shell thickness; g_m is the acceleration due to gravity; $\rho_m = \gamma_m h_m$, where γ_m is the specific material density; r_{x_m}, r_{y_m} is the radius of the middle surface curvature regarding x_m and y_m, respectively. In addition, t is the time; ε_m is the damping coefficient of the surrounding medium; ν is Poisson's coefficient (for an isotropic material $\nu = 0.3$); E_m is the elasticity modulus; $p_{x_m}(t)$, $p_{y_m}(t)$ are the longitudinal loads; and $q_m(x, y, t)$ is the transversal load. Further, bars over non-dimensional quantities are omitted.

Equations (6) are supplemented with the following boundary conditions:

$$\begin{aligned} w_m = 0, \quad \frac{\partial^2 w_m}{\partial x_m^2} = 0, \quad F_m = 0, \quad \frac{\partial^2 F_m}{\partial x_m^2} = 0 \quad \text{for} \quad x_m = 0; 1, \\ w_m = 0, \quad \frac{\partial^2 w_m}{\partial y_m^2} = 0, \quad F_m = 0, \quad \frac{\partial^2 F_m}{\partial y_m^2} = 0 \quad \text{for} \quad y_m = 0; 1, \end{aligned} \tag{9}$$

and the initial conditions:

$$w_m(x_m, y_m)|_{t=0} = \varphi_1(x_m, y_m), \quad \frac{\partial w_m}{\partial t} = \varphi_2(x_m, y_m). \tag{10}$$

In order to reduce PDEs (6) to a system with lumped parameters, the FDM is applied with the approximation $O(h^2)$ with respect to spatial variables x_m and y_m.

The Cauchy problem is solved by the fourth-order Runge–Kutta method. On each of the time steps we need to solve a large system of linear algebraic equations (yielded by the second equation of the governing ones with respect to the stress function) using the inverse matrix method. Time step follows from the Runge principle.

4.2 Numerical Experiment

We consider nonlinear dynamics of a two-layer cylindrical panel with curvatures $k_{x_m} = 12$, $k_{y_m} = 0$. First, the panel is subjected to the action of time periodic load $p_{x_1}(t) = p_{y_1}(t) = p_1 \sin(\omega_p t)$ with frequency $\omega_p = 8.4$, excitation amplitude $p_1 = 7.74$, and shell clearance $h_k = 0.5$. The numerically obtained time characteristics include the signal $w_m(t)$, wavelet spectrum, and Fourier spectrum, whereas the spatial characteristics include the curves of equal deflections and contact pressure (see Fig. 7). While analyzing signal (Fig. 7m) and applying the wavelet analysis (Fig. 7b) three phases are clearly distinguished. After a transitional state the signal becomes stationary, i.e., periodic (Fig. 7c, d). Owing to the Fourier analysis carried out in time interval $t \in (30; 60)$ (Fig. 7f), first periodic vibrations are observed, and then the first Hopf bifurcation occurs for $t \in (100; 250)$, which is reported in Fig. 7g. As it can be seen in the drawings, in the case of stationary vibrations (periodic vibrations and vibrations with the frequency after the Hopf bifurcation), the contact pressure (Fig. 7i, k) and isoclines regarding deflections (Fig. 7j, l) are symmetric with respect to the axes. Next, the system jumps into a turbulent regime.

Stiff stability loss is exhibited by the system, and the deflection amplitude suddenly increases (Fig. 7e); both contact pressure (Fig. 7m) and deflection isoclines (Fig. 7n) (spatial characteristics) become nonsymmetric. Power spectrum (Fig. 7h), which exhibits time properties of vibrations, is a broad band implying the occurrence of chaos for $t \in (250; 500)$.

The so far described scenario from regularity to chaos for our investigated structural member has been reported for the first time. It allows us to conclude that both temporal and spatial chaos appears simultaneously, yielding the shell turbulent behavior.

5 Turbulent Behavior of Two-Layer Cylindrical Shells Coupled via Boundary Conditions and Subjected to Local Periodic Transversal Load Action

5.1 *Problem Formulation*

In this section we study the mathematical model of a two-layer cylindrical shell (with clearance between both shells) having constant stiffness and density and subjected to a periodic transversal load action. The derived mathematical model allows us to study nonlinear dynamics of both structural members with regard to external load and internal interaction between the two layers, as well as the force action coming from the inside of the second cylinder. The developed mathematical model includes geometric nonlinearity of both cylinders and their contact interactions. Differential equations governing the dynamics of both shells are solved by the Bubnov–Galerkin higher-order approximation method, whereas the obtained Cauchy problem is solved using the fourth-order Runge–Kutta method. Convergence of the Bubnov–Galerkin method versus the number of approximating series terms is considered. Namely, we solve the problem of two embedded cylindrical shells including their interaction as systems with an infinite number of degrees of freedom.

We consider cylindrical shells as closed 3D objects embedded into space R^3 with the curvilinear coordinates $\overline{x}_1, \overline{y}_1, \overline{z}_1$ and $\overline{x}_2, \overline{y}_2, \overline{z}_2$ introduced in the following way: in the shell body we fix the middle surface for $\overline{z}_1 = 0$ and $\overline{z}_2 = 0$; axes $0x_1, 0x_2$ and $0y_1, 0y_2$ are directed along the main curvatures of this middle surface, whereas axes $0z_1, 0z_2$ go into the curvature centers (Fig. 8). In the given system of coordinates the following 3D subspaces Ω_1 and Ω_2 are defined:

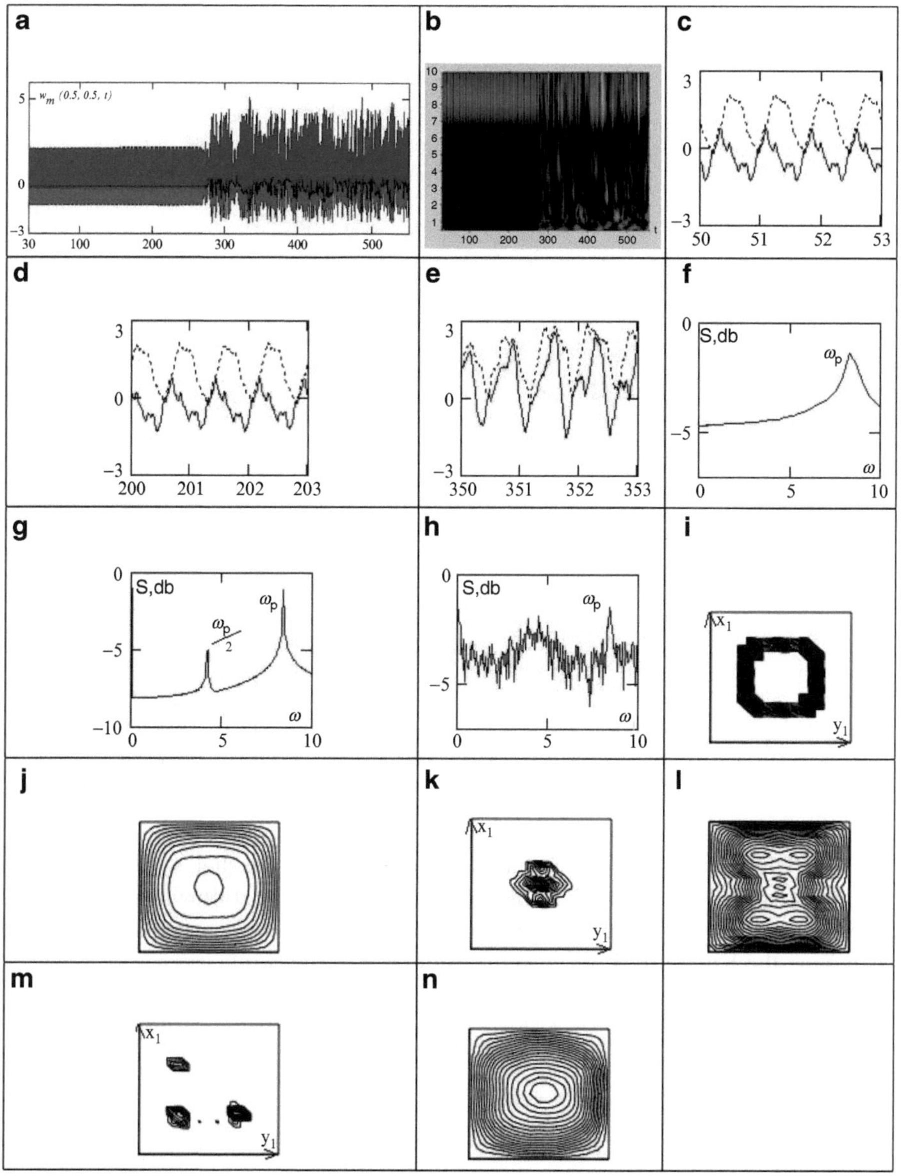

Fig. 7 System characteristics. (**a**) Signal spectrum. (**b**) Wavelet. (**c**) Periodic vibrations $w_m(0.5; 0.5; t)$. (**d**) Hopf bifurcation $w_m(0.5; 0.5; t)$. (**e**) Chaotic vibrations $w_m(0.5; 0.5; t)$. (**f**) Power spectrum $t \in (30; 60)$. (**g**) Power spectrum $t \in (100; 250)$. (**h**) Power spectrum $t \in (250; 500)$. (**i**) Contact pressure. (**j**) Deflection isoclines $w_1(x_1; y_1; 35)$. (**k**) Contact pressure. (**l**) Deflection isoclines $w_1(x_1; y_1; 150)$. (**m**) Contact pressure. (**n**) Deflection isoclines $w_1(x_1; y_1; 500)$

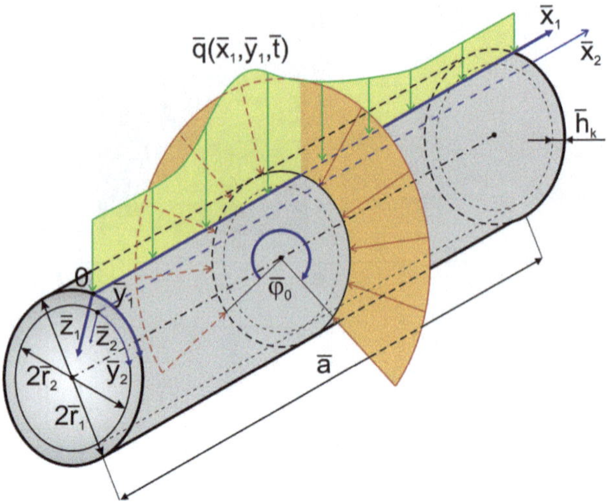

Fig. 8 Computational scheme

$$\Omega_1 = \left\{ \overline{x}_1, \overline{y}_1, \overline{z}_1 / (\overline{x}_1, \overline{y}_1, \overline{z}_1) \in [0; \overline{a}] \times [0; \overline{b}] \times \left[-\frac{\overline{h}_1}{2}, \frac{\overline{h}_1}{2} \right] \right\},$$

$$\Omega_2 = \left\{ \overline{x}_2, \overline{y}_2, \overline{z}_2 / (\overline{x}_2, \overline{y}_2, \overline{z}_2) \in [0; \overline{a}] \times [0; \overline{b}] \times \left[-\frac{\overline{h}_2}{2}, \frac{\overline{h}_2}{2} \right] \right\},$$

and bars refer to dimensional quantities here.

The PDEs given below govern nonlinear dynamics of the shell system presented in Fig. 8, and they are derived taking into account the following hypotheses: two-layer shells are made from an isotropic, homogeneous, and elastic material satisfying the Kirchhoff–Love hypothesis of each of the two layers:

$$\begin{gathered}
\frac{1}{12(1-\nu^2)}\Delta^2 w_1 - L(F_1, w_1) - k_{y_1}\frac{\partial^2 F_1}{\partial x_1^2} + \frac{\partial^2 w_1}{\partial t^2} + \varepsilon_1 \frac{\partial w_1}{\partial t} - \\
-q_1(t) + K(w_1 - h_k - w_2)\psi = 0, \\
\Delta^2 F_1 + L(w_1, w_1) + k_{y_1}\frac{\partial^2 w_1}{\partial x_1^2} = 0, \\
\frac{1}{12(1-\nu^2)}\Delta^2 w_2 - L(F_2, w_2) - k_{y_2}\frac{\partial^2 F_2}{\partial x_2^2} + \frac{\partial^2 w_2}{\partial t^2} + \varepsilon_2 \frac{\partial w_2}{\partial t} - \\
-q_2(t) + K(w_1 - h_k - w_2)\psi = 0, \\
\Delta^2 F_2 + L(w_2, w_2) + k_{y_2}\frac{\partial^2 w_2}{\partial x_2^2} = 0,
\end{gathered} \tag{11}$$

where

$$\Delta^2 = \frac{1}{\lambda^2}\frac{\partial^4}{\partial x_i^4} + \lambda^2\frac{\partial^4}{\partial y_i^4} + 2\frac{\partial^4}{\partial x_i^2 \partial y_i^2},$$

$$L(w_i, F_i) = \frac{\partial^2 w_i}{\partial x_i^2}\frac{\partial^2 F_i}{\partial y_i^2} + \frac{\partial^2 w_i}{\partial y_i^2}\frac{\partial^2 F_i}{\partial x_i^2} - 2\frac{\partial^2 w_i}{\partial x_i \partial y_i}\frac{\partial^2 F_i}{\partial x_i \partial y_i},$$

$$L(w_i, w_i) = \frac{\partial^2 w_i}{\partial x_i^2}\frac{\partial^2 w_i}{\partial y_i^2} - \left(\frac{\partial^2 w_i}{\partial x_i \partial y_i}\right)^2, \quad i = 1, 2.$$

Non-dimensional quantities are as follows:

$$\begin{gathered} \overline{w}_i = \overline{h}_i w_i, \quad \overline{F}_i = \overline{E}_i \overline{h}_i^{-2} F_i, \quad \overline{t} = t\overline{\tau}, \quad \overline{\varepsilon}_i = \varepsilon_i/\overline{\tau}, \\ \overline{x}_i = \overline{a} x_i, \quad \overline{y}_i = \overline{r}_i y_i, \quad \overline{k}_y = \frac{1}{\overline{r}_{y_i}}, \quad \overline{q} = q\frac{\overline{E}\overline{h}^4}{\overline{a}^2 \overline{b}^2}, \\ \overline{\tau} = \frac{\overline{a}\overline{r}_i}{\overline{h}_i}\sqrt{\frac{\overline{\gamma}_i}{\overline{E}_i \overline{g}_i}}, \quad \lambda = \frac{\overline{a}}{\overline{r}_i}, \end{gathered} \tag{12}$$

where w_i stands for deflection; F_i is the stress function; t is the time; ε_i is the damping coefficient; a and $r_i = r_{y_i}$ are the length and radius of the closed cylindrical shell regarding x_i and y_i, respectively; h_i is the shell thickness; g_i is the Earth acceleration; γ_i denotes the specific material density; and r_{y_i} is the radius of the middle surface. In addition, ν is Poisson's coefficient of the isotropic material; E_i is the Young modulus; and $q(x_i, y_i, t)$ is the external transversal load.

The following boundary and initial conditions are attached to (11)

$$\begin{gathered} w_i = 0, \quad \frac{\partial^2 w_i}{\partial x_i^2} = 0, \quad F_i = 0, \quad \frac{\partial^2 F_i}{\partial x_i^2} = 0 \quad \text{for} \quad x = 0; 1, \\ w_i = g_i(x_i, y_i, t), \quad \frac{\partial^2 w_i}{\partial y_i^2} = r_i(x_i, y_i, t), \quad F_i = u_i(x_i, y_i, t), \\ \frac{\partial^2 F_i}{\partial y_i^2} = z_i(x_i, y_i, t) \quad \text{for} \quad y_i = 0; 2\pi, \end{gathered} \tag{13}$$

$$w_i(x_i, y_i)|_{t=0} = 0, \quad \frac{\partial w_i}{\partial t} = 0, \quad i = 1, 2. \tag{14}$$

We apply the Bubnov–Galerkin method to reduce PDEs to ODEs, and the latter ones are solved via the fourth-order Runge–Kutta method. The numerical computation step is given by the Runge rule.

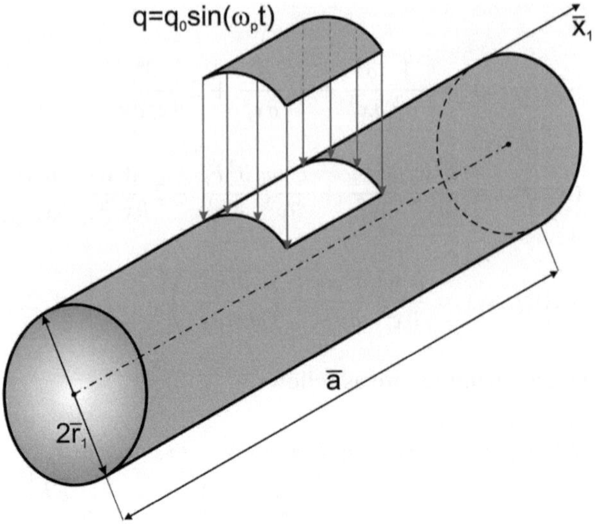

Fig. 9 Computational scheme

5.2 *Computational Results*

Reliability and validity of the obtained results are illustrated and discussed elsewhere (Fig. 12); we consider vibrations of one-layer cylindrical shell under the action of the local transversal load regarding x_1 and y_1 (Fig. 9).

The maximum shell deflection amplitude versus load amplitude $w_{max}(q_0)$ is shown in Fig. 8, where the so-called vibration scales are also included. Fourier spectra for each of the control parameters have been computed in order to construct the scales. Below, we analyze the influence of both the loading surface and coordinates of the load action on the vibration regimes of the studied cylindrical shells.

In Fig. 10 functions $w_{max}(q_0)$ and scales of dynamic regimes for various cases of loading are applied. We apply the square loading surface, excluding the case reported in Fig. 10e, where the load along y_1 is distributed around a disc. Analysis of the obtained results implies that an increase of the loading surface yields an increase of the chaotic vibrations. In addition, location of the load has also an essential influence on the properties of chaotic vibrations. Longitudinal shift of the concentrated load into the shell center is accompanied by an increase of chaotic vibration zones. However, if the external load is applied to the shell edges (Fig. 10a), there are chaotic vibrations, since the so-called boundary layer effect plays a crucial role here.

We study the spatiotemporal (turbulence) characteristics of the shell behavior in two-frequency quasiperiodic and chaotic regimes in the case of the action of transversal local load $x_1 \in (0.2; 0.4)$. Figures 11 and 12 show wave forms versus

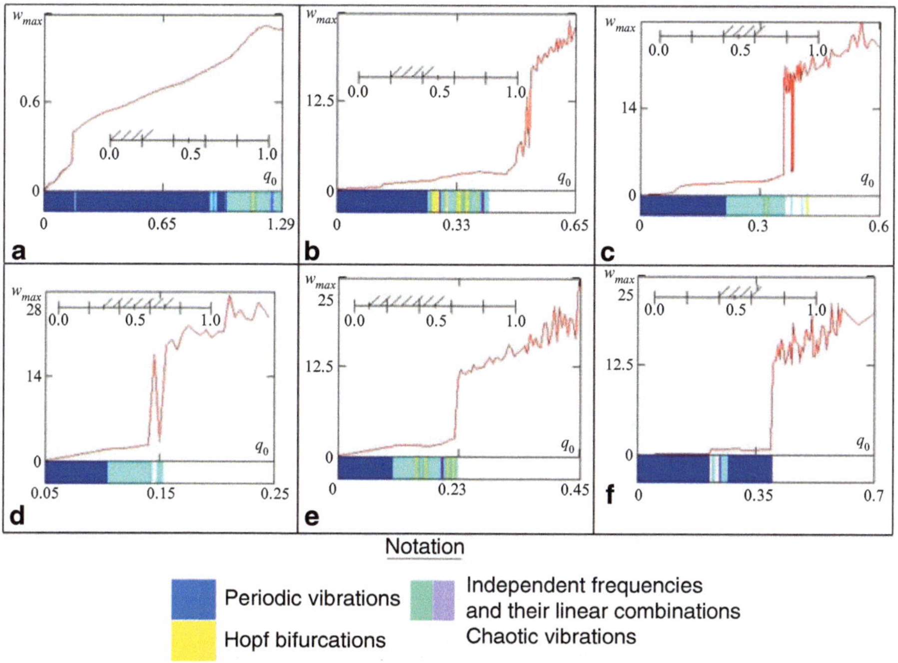

Fig. 10 Functions $w_{max}(q_0)$, scales of dynamic regimes and load action schemes. (**a**) $x_1 \in (0.0; 0.2)$. (**b**) $x_1 \in (0.2; 0.4)$. (**c**) $x_1 \in (0.4; 0.6)$. (**d**) $x_1 \in (0.3; 0.7)$. (**e**) $x_1 \in (0.1; 0.5)$. (**f**) $x_1 \in (0.4; 0.6)$, $y_1 \in (0; 2\pi)$

time $w_1(x_1; y_1; t)$, $0 \le x_1 \le 1$, $0 \le y_1 \le 2\pi$, signal $w(0.3; 0.0; t)$ in the loading center, phase portrait $w_1(\dot{w}_1)$, Poincaré map $w_{1t}(w_{1t+T})$, and the frequency power spectrum $S(\omega)$. In the case of quasiperiodic vibrations with two frequencies, a_1 and ω_p, all shell points move in a regular way, and the points under the action of the load move inside the shell curvature, whereas free points regarding the load action move up and vice versa. Maximum deflections along the circle direction are located in the external loading zones. In the case of turbulent vibrations a complex movement of all shell points is observed. Maximum deflections are distributed onto the whole shell surface, whereas shell points under the load and free of the load action may move in an unpredictable manner (Fig. 12e, $t = 3, 7, 10$, and $t = 2, 4$). Step changes of the deflection signs are exhibited, which has not been observed in the case of quasiperiodic vibrations. In the latter case the sign changes appeared in a smooth way.

The so far reported study allows for a conclusion that the spatial chaos and temporal chaos appear simultaneously, which validates our earlier observations on the shell transition into turbulent shell vibrations (see Sect. 4).

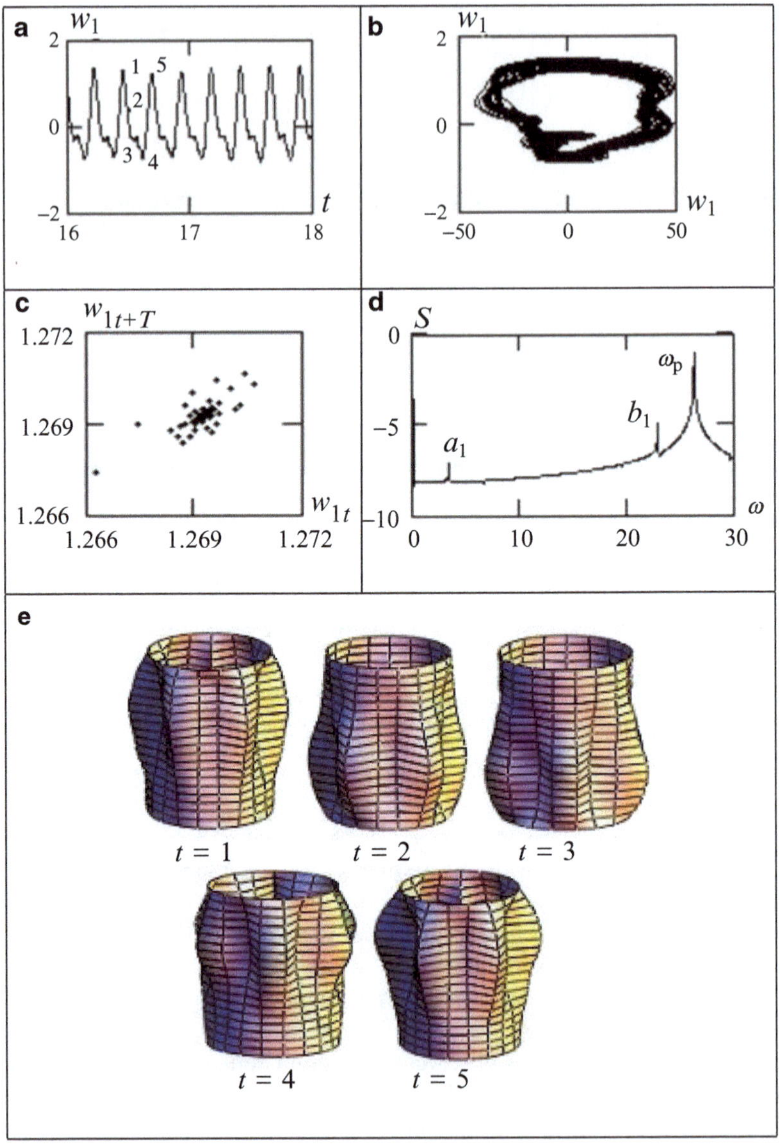

Fig. 11 Spatiotemporal characteristics of the shell quasiperiodic dynamics. (**a**) $w_1(0.3; 0.0; t)$. (**b**) $w_1(\dot{w}_1)$. (**c**) $w_{1t}(w_{1t+T})$. (**d**) $S(\omega)$. (**e**) $w_1(x_1; y_1; t), 0 \leq x_1 \leq 1, 0 \leq y_1 \leq 2\pi$

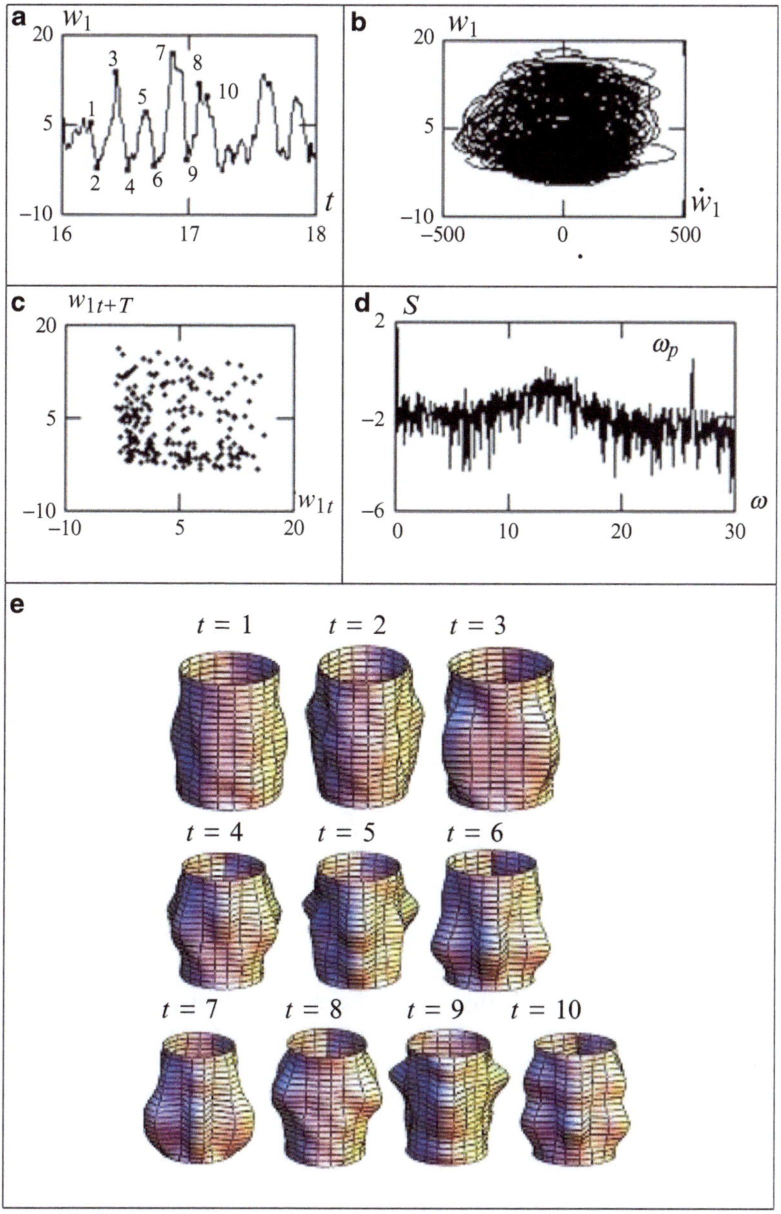

Fig. 12 Spatiotemporal characteristics of the shell turbulent dynamics. (**a**) $w_1(0.3; 0.0; t)$. (**b**) $w_1(\dot{w}_1)$. (**c**) $w_{1t}(w_{1t+T})$. (**d**) $S(\omega)$. (**e**) $w_1(x_1; y_1; t), 0 \le x_1 \le 1, 0 \le y_1 \le 2\pi$

6 Analysis of the Lyapunov Exponents for Nonlinear Dynamics of Plates and Shells

6.1 Governing Equations

This section aims at a description of a novel method for the estimation of Lyapunov exponents using neural networks. In the attempts described earlier, many approaches and algorithms have been proposed to compute the Lyapunov exponents, but in general almost all of them apply the algorithm proposed already by Benettin [7]. However, while studying nonlinear dynamics of structural members (beams, plates, and shells) it is of great importance to estimate the spectrum of Lyapunov exponents.

We also illustrate and demonstrate the efficiency of the neural network approach to study a flexible plate with an infinite length. The governing equations are within the Kirchhoff hypothesis and they have the following non-dimensional form [29]:

$$\begin{gathered}\frac{\partial^2 u}{\partial x^2} + L_3(w,w) - \frac{\partial^2 u}{\partial t^2} = 0\\ \frac{1}{\lambda^2}\left\{-\frac{1}{12}\frac{\partial^4 w}{\partial x^4} + L_1(u,w) + L_2(w,w)\right\} + q - \frac{\partial^2 w}{\partial t^2} - \varepsilon\frac{\partial w}{\partial t} = 0,\end{gathered} \tag{15}$$

$$L_1(u,w) = \frac{\partial^2 u}{\partial x^2}\frac{\partial w}{\partial x} + \frac{\partial u}{\partial x}\frac{\partial^2 w}{\partial x^2},$$

$$L_2(w,w) = \frac{3}{2}\left(\frac{\partial w}{\partial x}\right)^2\frac{\partial^2 w}{\partial x^2},$$

$$L_3(w,w) = \frac{\partial w}{\partial x}\frac{\partial^2 w}{\partial x^2},$$

where $L_1(u,w)$, $L_2(w,w)$, $L_3(w,w)$ are the nonlinear operators; $w(x,t)$ is the plate element bending in normal direction; $u(x,t)$ is the plate element longitudinal displacement; $\varepsilon$is the dissipation coefficient; E is the Young modulus; h is the height of the transversal panel cross section; $\gamma$is the specific plate material gravity; g is the acceleration due to gravity; t is the time; and $q = q_0\sin(\omega_p t)$ is the external load.

The non-dimensional parameters are as follows:

$$\begin{gathered}\lambda = \frac{a}{h},\quad \overline{w} = \frac{w}{h},\quad \overline{u} = \frac{ua}{h^2},\quad \overline{x} = \tfrac{x}{a},\quad \overline{t} = \frac{t}{\tau},\\ \tau = \frac{a}{p},\quad p = \sqrt{\frac{Eg}{\gamma}},\quad \overline{\varepsilon} = \frac{a}{p},\quad \overline{q} = \frac{qa^4}{h^4 E},\end{gathered} \tag{16}$$

and bars over the non-dimensional quantities have already been omitted in (15). We demonstrate how to determine four first Lyapunov exponents applying pinned boundary conditions:

$$\begin{aligned} w(0,t) = w(a,t) = u(0,t) = u(a,t) = w''_{xx}(0,t) = w''_{xx}(a,t) = 0, \\ w(0,t) = w(1,t) = u(0,t) = u(1,t) = w''_{xx}(0,t) = w''_{xx}(1,t) = 0, \end{aligned} \tag{17}$$

and the following initial conditions

$$w(x,0) = \dot{w}(x,0) = u(x,0) = \dot{u}(x,0). \tag{18}$$

The boundary value problem (15), (17), (18) is reduced to the Cauchy problem via FDM of the second-order accuracy. The obtained ODEs are solved by the Runge–Kutta method of the fourth and sixth orders. Validity and reliability of the obtained numerical results are confirmed by the FEM results. The initial problem of infinite dimension is substituted by that of finite dimension via partition of the interval $x \in [0, 1]$ into 120 parts.

One of the ways to compute the spectrum of Lyapunov exponents is the neural network approach based on the generalized Benettin algorithm. It includes the following steps: (i) choice of the appropriate time delay via tests; (ii) computation of an embedding space dimension; (iii) reconstruction of pseudo-phase trajectories using the method of time delays; (iv) neural network approximation; (v) teaching of neural networks to compute successive iteration vectors; (vi) computation of the spectrum through the generalized Benettin algorithm with the help of the neural networks approach.

We apply the neural network with the following properties: it is an analogue network regarding the input data (information is delivered through real numbers); it is self-organized with respect to its teaching aspects (output space of solutions is defined only through the input data); it belongs to the neural networks of straight signal distributions (all neural network couplings come from the input neurons and go to the output neurons); the neural network has dynamic couplings (control and improvement of synaptic couplings is carried out during the neural network learning process ($dW/dt \neq 0$), where W stands for the net weight coefficients).

6.2 *Generalization of the Benettin Algorithm*

Let point x_0 belong to attractor A of a dynamic system. The trajectory of evolution of point x_0 is called the unperturbed trajectory. We choose the positive quantity ε essentially lesser than the attractor dimension. Next, we choose an arbitrary (perturbed) point $\tilde{x}_0$ in a way to satisfy the relation $\|\tilde{x}_0 - x_0\| = \varepsilon$. Then, we monitor the evolution of the chosen x_0 and $\tilde{x}_0$ in time interval T, and the corresponding new points obtained after that time interval are denoted as x_1 and $\tilde{x}_1$, respectively. Vector $\Delta x_1 = \tilde{x}_1 - x_1$ is called the perturbation vector. We are ready to estimate λ

$$\tilde{\lambda}_1 = \frac{1}{T} \ln \frac{\|\Delta x_1\|}{\varepsilon}. \tag{19}$$

Time interval T is chosen in a way to keep the perturbation amplitude lesser than the linear dimensions of the space nonhomogeneity as well as lesser than the attractor dimension. We consider the unit normalized perturbation vector $\Delta x_1' = \varepsilon \Delta x_1 / \|\Delta x_1\|$ and the corresponding new perturbation point $\tilde{x'}_1 = x_1 + \Delta x_1'$. We extend the so far described approach by using points x_1 and $\tilde{x}_1$ instead of x_0 and $\tilde{x}_0$, respectively. We repeat the described procedure M times, and we may estimate λ as an average arithmetic quantity $\tilde{\lambda}_i$ of those obtained on each computation step. The proposed approach has been tested using the standard classical examples including that of the Henon map [14], the Lorenz system [22], and the logistic map [28].

We consider vibrations of our mechanical object with the following fixed parameters: $\lambda = 50$, $\varepsilon = 1$, $\omega_p = 7$, $q = q_0 \sin(\omega_p t)$, and for the following amplitudes of the harmonic excitation: $q_0 = 0.125 \times 10^3$; 5×10^3; 7×10^3. In order to study chaotic dynamics of flexible plates, we need to monitor and analyze the following output characteristics: time histories (a), phase (c), and modal portraits; phase portraits yielded by the neural networks approach (d); Fourier power spectra (b); wavelet spectra, Poincaré sections (e); spectra of Lyapunov exponents, where d stands for the fractional part of dimension and h is the Kolmogorov–Sinai entropy (f); and autocorrelation functions (some of them are reported in Fig. 13). Analysis of the obtained results implies that for $q_0 = 0.125 \times 10^3$ periodic vibrations appear, whereas for 5×10^3 a spatiotemporal chaos is exhibited, and for 7×10^3 the hyper-spatio-temporal chaos occurs, since two of the Lyapunov exponents spectrum are positive.

The obtained validated spectra of the Lyapunov exponents allow for the estimation of Kaplan–Yorke dimension, Sinai–Kolmogorov entropy, and velocity of the phase-space compression.

7 Concluding Remarks

We studied numerically two-dimensional solid objects represented by shells and plates putting emphasis on high-dimensional spatiotemporal chaos versus the weak wave turbulence framework. Although all analyzed nonlinear PDEs were reduced to ordinary differential equations and possibly algebraic equations, all of the studied mechanical objects were modeled as infinite dimensional systems using the robust discretization methods and then by a careful checking of validation and reliability of the results. Next, we briefly summarized the obtained novel results.

In Sect. 2 spatiotemporal chaotic parametric vibrations of the flexible rectangular plate with constant stiffness and density periodically loaded were investigated.

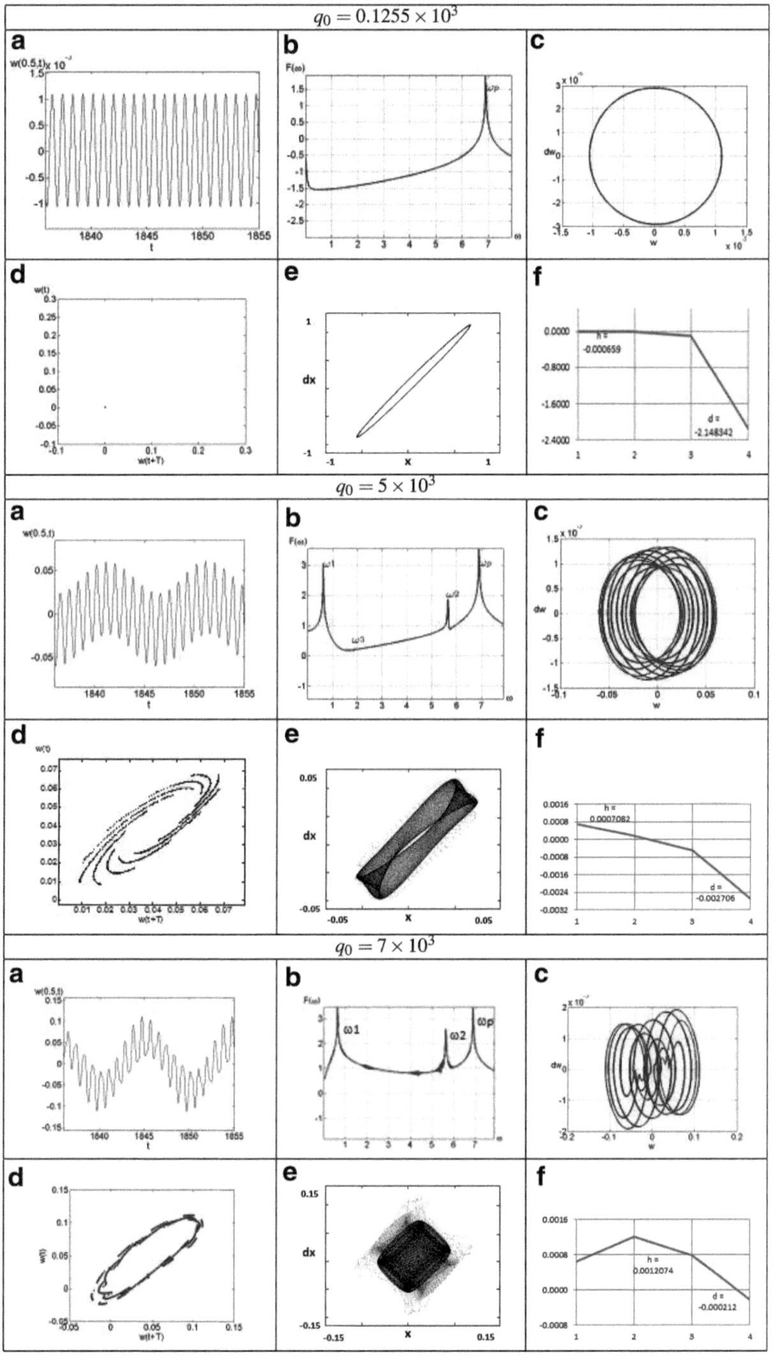

Fig. 13 Plate characteristics

Its 2D model was constructed using the Kirchhoff–Love hypotheses and taking into account nonlinear relations between deformations and displacements in the form proposed by von Kármán. The governing PDEs were reduced to ODEs and AEs via application of the FDM with approximation $O(h^2)$ regarding spatial coordinates. Among others a sudden transition from periodic (laminar) into chaotic (turbulent) shell dynamics was reported without any other transitional zones. It was also shown how low values of the applied frequencies ($\omega \leq 2$) of subharmonic zones were interlaced with the zones of periodic vibrations. The excitation frequency increase implies extension of these zones into higher frequency spectrum and finally they started mixing with large chaotic zones strengthening the shell spatiotemporal chaotic dynamics. We reported also another transition when chaotic dynamics occurred already after the second or even first Hopf bifurcation.

Section 3 deals with the irregular dynamics of the rectangular flexible isotropic shell subjected to the periodic shear load. The governing PDEs were reduced to ODEs via the FDM with approximation $O(h^2)$ regarding spatial coordinates and then they were solved using the fourth-order Runge–Kutta method. The reported result had two important aspects. First, the standard Fourier approach (being often used) might lead even to erroneous results. The reason was that the independent variable (time) might play a role as a parameter. Namely, it was shown that in the first time interval the shell exhibited two-frequency quasiperiodic dynamics, then a few linearly dependent frequencies appeared, making the signal more complicated, then chaotic dynamics was exhibited, where finally ($230 < t < 286$) periodic/laminar shell dynamics appeared. It is clear that the different Fourier frequency spectra were obtained depending on the applied time intervals. Furthermore, assuming that one takes the FFT regarding the whole studied time interval ($0 < t < 286$), the obtained results will have no physical interpretation. This is why another supplemented technique should be applied, namely the wavelet-based approach. Another important point yielded by our numerical simulation is a question if the so far reported scenario violates the theoretical prediction of the so-called wave (or weak) turbulence relying on the relaxation of strong assumptions exhibited by the fully developed turbulence and, in particular, where the presence of the intermittency is removed [31]. Second, different wavelets (Meyer, Morlet, complex Morlet, real and complex Gauss, Daubechies) were analyzed with respect to their application to study nonlinear dynamics of shells.

In Sect. 4 we apply the Bolotin–Novikov model to study two-layer panels with a gap. A sudden jump into a turbulent regime is reported accompanied by a stiff stability loss. The previous symmetry of a contact pressure and deflection isoclines is also suddenly broken. It means that this sudden scenario of transition into a spatiotemporal chaos is associated with a catastrophe and symmetry breaking. Furthermore, we observed that spatial and temporal chaos appeared simultaneously, yielding shell turbulent dynamics.

Next, Sect. 5 focuses on a study of contact interaction of two-layer cylindrical shells coupled via boundary conditions and subjected to local periodic transversal load action. The governing PDEs are solved via the Bubnov–Galerkin higher-order approximation method, whereas the obtained ODEs are solved through the fourth-order Runge–Kutta method. Again a weak shell turbulent behavior is observed, and again the spatial and temporal chaos appears simultaneously.

Finally, Sect. 6 deals with nonlinear dynamics of plates/shells governed by the Kirchhoff–Love PDEs. The stated boundary value problem was reduced to ODEs through two different approaches (FDM and FEM), which allowed us to achieve validity and reliability of the results. From the theoretical/numerical point of view we propose here a novel method for computation of the Lyapunov exponents within the neural network framework. From a physical point of view, a weak turbulent plate behavior was observed.

Acknowledgements This work has been supported by the grant RFBR No 12-01-31204 and the National Science Centre of Poland under the grant MAESTRO 2, No. 2012/04/A/ST8/00738, for years 2012–2015.

References

1. Alvarez Diaz, L., Martin, M.T., Vampa, V.: Daubechies wavelet beam and plate finite elements. Finite Elem. Anal. Des. **45**, 206–209 (2009)
2. Amaratunga, K., Williams, J.R., Quian, S., Weiss, J.: Wavelet-Galerkin solutions for one dimensional partial differential equations. Int. J. Natl. Meth. Eng. **37**, 2703–2716 (1994)
3. Awrejcewicz, J., Krysko, A.V., Papkova, I.V., Krysko, V.A.: Routes to chaos in continuous mechanical systems. Part 1: mathematical models and solution methods. Chaos Solitons Fractals **45**, 687–708 (2012)
4. Awrejcewicz, J., Krysko, A.V., Papkova, I.V., Krysko, V.A.: Routes to chaos in continuous mechanical systems. Part 2: modelling transitions from regular to chaotic dynamics. Chaos Solitons Fractals **45**, 709–720 (2012)
5. Awrejcewicz, J., Krysko, A.V., Papkova, I.V., Krysko, V.A.: Routes to chaos in continuous mechanical systems. Part 3: the Lyapunov exponents, hyper, hyper-hyper and spatial-temporal chaos. Chaos Solitons Fractals **45**, 721–736 (2012)
6. Awrejcewicz, J., Krylova, E.Y., Papkova, I.V., Krysko, V.A.: Wavelet-based analysis for the regular and chaotic dynamics of rectangular flexible plates subjected to shear-harmonic loading. Shock Vib. **19**, 979–994 (2012)
7. Benettin, G., Galgani L., Giorgilli, A., Strelcyn, J.M.: Lyapunov characteristic exponents for smooth dynamical systems and for Hamiltonian systems: a method for computing all of them. Part I: theory. Part II: numerical application. Meccanica **15**, 21–30 (1980)
8. Boudaut, A., Cadot, O., Odille, B., Touzé, C.: Observation of wave turbulence in vibrating plates. Phys. Rev. Let. **100**, 234504 (2008)
9. Connaughton, C.: Numerical solutions of the isotropic 3-wave kinetic equation. Physica D **238**, 2282–2297 (2009)
10. Ghanem, R., Romeo, F.: A wavelet-based approach for the identification of linear time-varying dynamical systems. J. Sound Vib. **234**, 555–576 (2000)
11. Ginzburg, V.L., Landau, L.D.: On the theory of superconductivity. ZhETF **20**, 1064–1081 (1950) (in Russian)
12. Hans, S., Ibraim, E., Pernot, S., Boutin, C., Lamarque, C.-H.: Damping identification in multi-degree-of-freedom system via a wavelet-logarithmic decrement-Part 2: Study of a civil engineering building. J. Sound Vib. **235**, 375–403 (2000)
13. Hariharan, G.: Solving finite length beam equation by the Haar wavelet method. Int. J. Comput. Appl. **9**, 27–24 (2010)

14. Henon, M.: A two-dimensional mapping with a strange attractor. Comm. Math. Phys. **50**, 69–77 (1976)
15. Kantor, B.Ya.: Problems of Nonlinear Theory of Rotational Shells. Naukova Dumka, Kiev (1990) (in Russian)
16. Kornichin, M.S.: Non-Linear Problems of the Theory Plates and Shallow Shells and Methods of Their Solutions. Nauka, Moscow (1964) (in Russian)
17. Lamarque, C.-H., Pernot, S., Cuer, A.: Damping identification in multi-degree-of-freedom system via a wavelet-logarithmic decrement-Part I. Theory. J. Sound Vib. **235**, 361–374 (2000)
18. Lepik, U.: Impulsively loaded fully clamped elastic-plastic beams by Galerkin's method. Int. J. Impact Eng. **15**, 17–23 (1994)
19. Lepik, U.: Vibrations of elastic-plastic fully clamped beams and flat arches under impulsive loading. Int. J. Non-Linear Mech. **29**, 613–623 (1994)
20. Lepik, U.: Dynamic response of elastic-plastic beams with axial constraints. Int. J. Impact Eng. **15**, 3–16 (1994)
21. Liew, K.M., Wang, Q.: Application of wavelet theory for crack identification in structures. J. Eng. Mech. **124**, 152–157 (1998)
22. Lorenz, E.N.: Deterministic non-periodic flow. J. Atm. Sci. **20**, 130–141 (1963)
23. Mordant, N.: Are there waves in elastic wave turbulence? Phys. Rev. Let. **100**, 124505 (2008)
24. Mordant, N.: Fourier analysis of wave turbulence in a thin elastic plate. Eur. Phys. J. B **76**, 537–545 (2010)
25. Staszewski, W.J., Tomlinson, R.G.: Application of the wavelet transform to fault detection in a spur gear. Mech. Syst. Signal Process **8**, 289–307 (1994)
26. Touze, C., Thomas, O., Amabili, M.: Transition to chaotic vibrations for harmonically forced perfect and imperfect circular plates. Int. J. Non-linear Mech. **46**, 234–246 (2011)
27. Touze, C., Bilbao, S., Cadot, O.: Transition scenario to turbulence in thin vibrating plates. J. Sound Vib. **331**, 412–433 (2012)
28. Verhulst, P.-F.: Recherches mathématiques sur la loi d'accroissement de la population. Nouv. Mém. Acad. Roy. Sci. Belles-Lett. Bruxelles **18**, 1–41 (1845) (in French)
29. Volmir, A.S.: Nonlinear Dynamics of Plates and Shells. Nauka, Moscow (1972) (in Russian)
30. Wang, W.J., McFadden, P.D.: Application of wavelets to gearbox vibration signals for fault detection. J. Sound Vib. **192**, 927–939 (1996)
31. Zakharov, V.E., Lvov, V.S., Falkovisch, G.: Kolmogorov Spectra of Turbulence I: Wave Turbulence. Springer, Berlin (1992)

Using Bifurcation Diagrams for Controlling Chaos

Daniel Morcillo, Daniel Burbano, Fabiola Angulo, and Gerard Olivar

Abstract In this chapter, a methodology to compute chaos controllers based on bifurcation diagrams is proposed and applied to PWM-controlled power converters. The technique is based on an adaptive control where the offset of the T-periodic sawtooth signal is modified. Basically, the sawtooth is redefined as a function of the output and reference voltages. This control technique reduces the percentage of regulation error as well as eliminates orbits of period greater than one and the chaotic behavior when the voltage source is varied. Finally, numerical results are obtained to validate the performance of the proposed scheme.

1 Introduction

Power converters are widely used in industrial and household appliances. Many researchers have focused their attention to the study of control techniques for switching power supplies, specially the buck converter which is one of the most common converters used in a variety of applications [10]. Due to thc presence of the PWM, the power converters exhibit a plethora of complex behaviors. These behaviors strongly affect the performance of the converter. Nevertheless, many techniques have been developed to counteract chaotic regimes and another undesired behaviors [1, 4, 9]. In order to eliminate high-period orbits and chaos two techniques have been mainly proposed: OGY [7] and $TDAS$ [8]. Both methods work properly when they are used in the buck converter and an example can be found in [3]. However, these techniques require complex schemes to be implemented as well as they need digital devices such as FPGAs, DSPs, microcontrollers, and so on.

D. Morcillo (✉) • F. Angulo • G. Olivar
Department of Ingeniería Eléctrica, Electrónica y Computación, Universidad Nacional de Colombia, Campus La Nubia, Manizales, Colombia
e-mail: jdmorcillob@unal.edu.co; fangulog@unal.edu.co; golivart@unal.edu.co

D. Burbano
Department of Systems and Computer Sciences, University of Naples Federico II, Via Claudio 21, 80125 Napoli, Italy
e-mail: danielalberto.burbanolombana@unina.it

J. Awrejcewicz (ed.), *Applied Non-Linear Dynamical Systems*, Springer Proceedings in Mathematics & Statistics 93, DOI 10.1007/978-3-319-08266-0_6

Deane et al. [2] proposed a basic voltage-mode control applied to a DC–DC buck converter based on a PWM, using a simple ramp as a T-periodic signal. The circuit is depicted in Fig. 1a. This simple technique makes that the system regulates for some V_{in} values, but orbits of period greater than one and chaotic bands appear when V_{in} varies. After the knowledge of these phenomena, some authors have developed controllers for this power converter. In [5] a method which is called "V_{in} feed-forward" is discussed; this method changes the amplitude of the ramp in such a way that the amplitude is proportional to V_{in} values. On the other hand, in [6] are presented different ways to automatically adjust the slope of the ramp to avoid output variations or subharmonic oscillations. However, these papers do not provide bifurcation analysis to observe clearly the performance improvement in a wider operation range of V_{in}, and they do not consider the offset voltage of the ramp as a variable.

With the aim to develop a control technique simple to implement, in this paper it is proposed a novel control scheme which suppresses chaotic bands and orbits of period greater than one in a wide range of power source, reference voltage, and load value for the buck power converter controlled by ramp. Essentially, this method, which is named "adaptive ramp control," consists in adapting the ramp waveform $(Vr(t))$ to the control signal $(Vco(t))$ in such a way $Vr(t)$ becomes similar to $Vco(t)$. In this way the ramp signal can change its offset voltage over the time. In addition, with this control technique, the frequency of the ramp signal is not altered; thus the buck converter continues as a periodically forced system. Besides, after applying the proposed technique, the performance of the system has been proven numerically. It was proven that $1T$-periodic orbits remain stable within the V_{in} range $[13, 40]$ V.

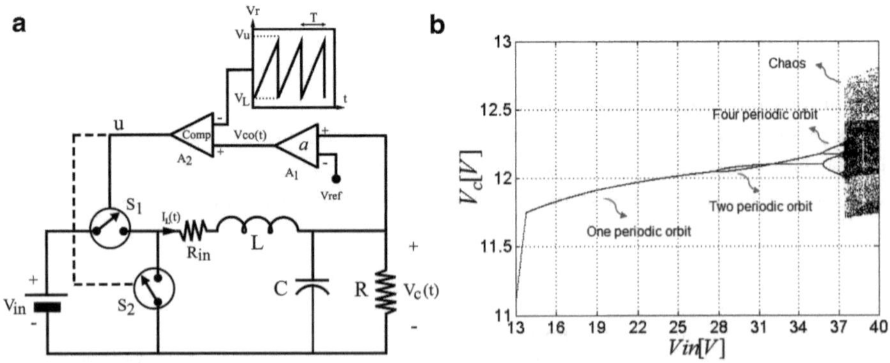

Fig. 1 (**a**) Simplified circuit of DC–DC buck converter. (**b**) Bifurcation diagram. V_{in} is used as bifurcation parameter

2 DC–DC Buck Converter

A simplified schematic diagram of the PWM voltage-controlled buck converter is depicted in Fig. 1a and is described by Eq. (1):

$$\begin{bmatrix} \dot{V}_C \\ \dot{I}_L \end{bmatrix} = \begin{bmatrix} -1/RC & 1/C \\ -1/L & -R_{in}/L \end{bmatrix} \begin{bmatrix} V_C \\ I_L \end{bmatrix} + \begin{bmatrix} 0 \\ V_{in}/L \end{bmatrix} u \tag{1}$$

V_C and I_L are the state variables and correspond to capacitor voltage and inductor current. Switches S_1 and S_2 are operating in a complementary way. R, L, and C are the load resistance, the inductor, and the capacitor, respectively. R_{in} is the equivalent resistance of the current sensor and inductor internal resistance. V_{in} is the value of the power source. u takes values into the discrete set $\{0, 1\}$ and its value is given according to the controller.
The buck converter controlled by ramp consists in a control signal $Vco(t)$ which is compared with a T-periodic sawtooth waveform $Vr(t)$ which is given by Eq. (2):

$$Vr(t) = V_L + (V_u - V_L)\frac{t}{T} \tag{2}$$

where V_L and V_u are constant voltages (lower and upper voltage values) and T is the switching period. The control signal $Vco(t)$ is proportional to the output error, which is the difference between the reference voltage $Vref$ and the output voltage V_C. The amplifier A_1 with gain a is used to obtain $Vco(t)$, and A_2 is as comparator which has infinite gain, and it is used to obtain the control action u. $Vco(t)$ and u are computed according to the following equations:

$$Vco(t) = a\,(V_C(t) - Vref) \tag{3}$$

$$u = \begin{cases} 1 \; if \; Vco(t) < Vr(t) \\ 0 \; if \; Vco(t) > Vr(t) \end{cases} \tag{4}$$

To perform numerical analysis SimPowerSystems tool has been used. Table 1 contains all the parameter values used in this work. In Fig. 1b a bifurcation diagram as V_{in} varies is shown. This bifurcation scenario was firstly reported in 1990 by Deane and Hamill in [2]. Bifurcation diagram observed in Fig. 1b exhibits the same behaviors as it has been described in other works. Using parameters of Table 1, close to $V_{in} = 28$ V a period doubling orbit is detected and after $V_{in} = 37.5$ V chaos is presented. With the aim to eliminate the chaotic behavior and orbits with period greater than one, a control technique based on adaptive ramp is proposed.

Table 1 Parameter values used in simulations

Parameter	Value
Input voltage (V_{in})	(20–40) V
Reference voltage ($Vref$)	11.3 V
Inductance (L)	20 mH
Inductor resistance (R_{in})	3.8 Ω
Capacitance (C)	47 μF
Load (R)	22 Ω
Lower voltage (V_u)	3.5 V
Upper voltage (V_L)	8.1 V
Ramp period (T)	400 μs
A_1 gain (a)	7.8
A_2 gain ($Comp$)	∞

3 Methodology to Compute the Adaptive Ramp

Analysis of the signal $Vco(t)$ shows that as V_{in} increases $Vco(t)$ also increases and $Vco(t)$ is unable to interact with the voltage of ramp $Vr(t)$, inducing the bifurcation scenario presented in Fig. 1b. Then, the main idea is to change $Vr(t)$ value depending on $Vco(t)$ value without changing the slope of the ramp which makes easy to implement the controller. Remaining fixed the slope of the ramp and avoiding undesired behaviors it is possible to obtain fixed frequency switching and a simple circuit design. However, in order to find a suitable constant slope, the design of the controller is based on the bifurcation diagram when the slope changes. In this way, the basic "*adaptive ramp* ($Var(t)$)" in which k and V_{in} vary is defined as

$$Var(t) = \frac{Vco(t)}{k} + \frac{V_{in}}{k}\frac{t}{T} \tag{5}$$

where $Var(t)$ corresponds to the signal that will replace $Vr(t)$, k is a constant to be computed, and V_{in} will be fixed to a constant value. Values of V_{in} and k are to be calculated in such a way that undesired behaviors are avoided. From Eq. (5) can be observed that a change in $Vco(t)$ immediately updates $Var(t)$. Notice that the term $Vco(t)$ makes the adaptive-ramp waveform ($Var(t)$) change its voltage offset. Now, the control signal is described by

$$u = \begin{cases} 1\ if\ \ Vco(t) < Var(t) \\ 0\ if\ \ Vco(t) > Var(t) \end{cases} \tag{6}$$

In what follows the methodology to compute k and to fix the sawtooth slope value is explained.

The objective now is to find the values for expression V_{in} of Eq. (5) and constant k which guarantee $1T$-periodic solution with good regulation performance. Bifurcation diagrams have been used to find suitable values for V_{in} and k. Initially,

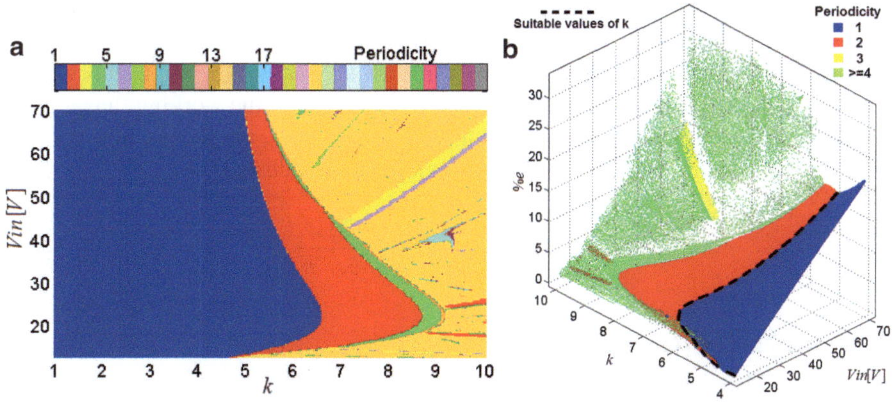

Fig. 2 (**a**) Two-dimensional bifurcation diagram. Bifurcation parameters are V_{in} and k. (**b**) Three-dimensional diagram. Bifurcation parameters are V_{in} and k, showing the percentage of regulation error ($\%e$)

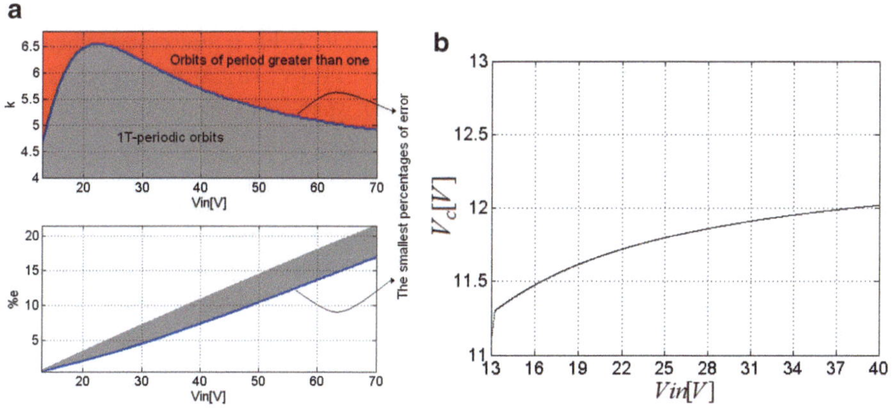

Fig. 3 (**a**) Function that provides suitable values of k according to V_{in}. (**b**) One-dimensional bifurcation diagram of the buck converter controlled by adaptive ramp taking $k = 5.7$ for all V_{in} values. $V_C(t)$ versus V_{in}

a two-dimensional bifurcation diagram (depicted in Fig. 2a) is computed using V_{in} and k as bifurcation parameters. The aim of this diagram is to find a region where the system presents a $1T$-periodic solution, which is obtained if k is properly chosen. Now, the values for which the system exhibits the lowest regulation error are found. Figure 2b shows the steady-state regulation error (in percentage) as V_{in} and k vary. In Fig. 3a the lowest percentage error and the stability limit of $1T$-periodic orbit for each combination of V_{in} and k are presented; in fact, the limit of the stability coincides with the lowest error. Curves were smoothed by cubic interpolation. $1T$-Periodic orbits taken from Fig. 3a upper are mapped to Fig. 3a lower. Hence, from Fig. 3a can be observed that it is possible to obtain a $1T$-periodic orbit for

$V_{in} \in [13, 70]$ V by selecting a proper value of k. According to this figure, the procedure to compute V_{in} and k is explained.

The first step is to define the working range of V_{in}, $V_{in} \in [13, 40]$ is chosen, and then V_{in} in Eq. (5) is replaced by 40 as it is the last value of the working range. Now, taking into account stability range and lowest error curves, the second step consists of choosing k value. In this case $k = 5.7$; then the final design is

$$Var(t) = \frac{Vco(t)}{5.7} + \frac{40}{5.7}\frac{t}{T} \tag{7}$$

4 Numerical Results

In the following, the designed controller will be proven when the input voltage, reference voltage, and load resistance change. The control design was successful and bifurcations and chaotic attractor were suppressed. Figures 3b and 4a, b are bifurcation diagrams numerically computed, using the voltage source, load resistance, and reference voltage as bifurcation parameters, respectively. In these bifurcation diagrams it is observed that $1T$-periodic orbit is preserved when: (i) $V_{in} \in [13, 40]$ V and $R = 22\,\Omega$ and $Vref = 11.3$ V, (ii) $R \in [2, 500]\,\Omega$ and $V_{in} = 40$ V and $Vref = 11.3$ V, and (iii) $Vref \in [8, 26]$ V and $V_{in} = 40$ V and $R = 22\,\Omega$.

Finally, three-dimensional bifurcation diagrams (taking Vin, $Vref$, and R as bifurcation parameters) have been computed in order to observe the broader range of $1T$-periodic orbits achieved applying the "adaptive-ramp control" to the buck converter. In Fig. 5a, b are shown the bifurcation diagrams computed from the system controlled by ramp and for the system controlled by adaptive ramp, respectively.

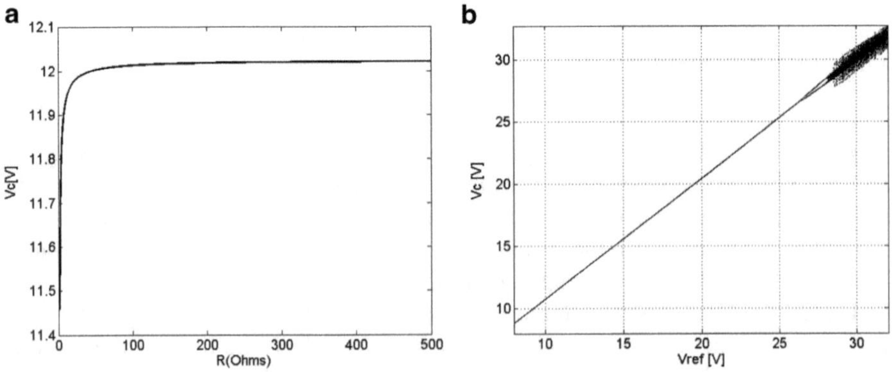

Fig. 4 One-dimensional bifurcation diagram. $k = 5.7$ and $V_{in} = 40$ V. (**a**) Load resistance R varies. (**b**) Reference voltage $Vref$ varies

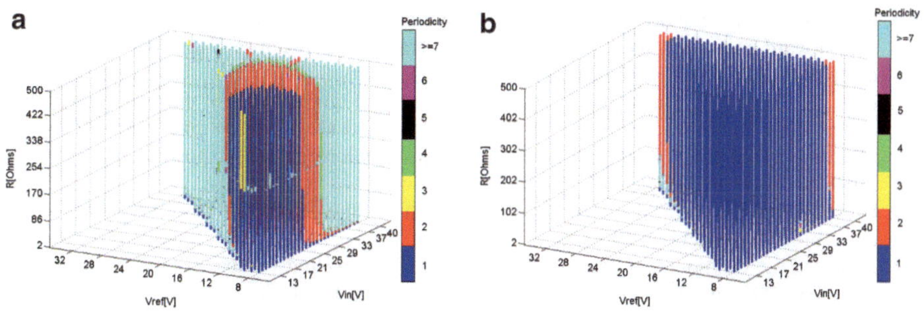

Fig. 5 Three-dimensional bifurcation diagram of the buck converter (**a**) controlled by ramp, (**b**) controlled by adaptive ramp

5 Conclusions

A new technique for chaos control applied to a PWM voltage-controlled buck converter has been proposed. The design approach is based on the analysis of bifurcation diagrams to compute constant parameters which define the controller. The methodology provides an adaptive ramp whose offset voltage changes as Vco changes, which prevents complex behaviors within the range $V_{in} \in [13, 40]$ V. The proposed control is easy to implement, does not need many components, and yields excellent results. Simulation results were used to show the feasibility of the proposed control method.

Acknowledgements This work was partially supported by Universidad Nacional de Colombia - Manizales, project 12475, Vicerrectoría de Investigaci6n, DIMA, and Programa J6venes Investigadores e Innovadores 2011 (COLCIENCIAS) convenio No. 0043 de 2012.

References

1. Bouzahir, H., El Guezar, F., El Aroudi, A., Ueta, T.: Partial time delayed feedback control of chaos in a hybrid model of a DC–DC converter. In: Proceedings of 3rd International Symposium on Communications, Control and Signal Processing, 2008. doi:10.1109/ISCCSP.2008.4537211
2. Deane, J.H.B., Hamill, D.C.: Analysis, simulation and experimental study of chaos in the buck converter. In: Proceedings of 21st Annual IEEE Power Electronics Specialists Conference, PESC'90, 1990. doi:10.1109/PESC.1990.131228
3. El Aroudi, A., Debbat, M., Giral, R., Olivar, G., Benadero, L., Toribio, E.: Bifurcations in DC-DC switching converters: review of methods and applications. Int. J. Bifurcat. Chaos Appl. Sci. Eng. **15**, 1549–1578 (2005). doi:10.1142/S0218127405012946
4. Giaouris, D., Banerjee, S., Zahawi, B., Pickert, V.: Stability analysis of the continuous-conduction-mode buck converter Via Filippov's method. IEEE Trans. Circ. Syst. I Regular Papers **55**(4), 1084–1096 (2008). doi:10.1109/TCSI.2008.916443

5. Karppanen, M., Suntio, T., Sippola, M.: Dynamical characterization of input-voltage-feedforward-controlled buck converter. IEEE Trans. Ind. Electron **54**, 1005–1013 (2007). doi:10.1109/TIE.2007.892732
6. Liyu, Y., Jinseok, P., Huang, A.Q.: An adaptive external ramp control of the peak current controlled Buck converters for high control bandwidth and wide operation range. In: Proceedings of 25th Annual IEEE Applied Power Electronics Conference and Exposition, APEC'2010, 2010. doi:10.1109/APEC.2010.5433539
7. Ott, E., Grebogi, C., Yorke, J.A.: Controlling chaos. Phys. Rev. Lett **64**, 1196–1199 (1990). doi:10.1103/PhysRevLett.64.1196
8. Pyragas, K.: Continuous control of chaos by self-controlling feedback. Phys. Lett. A **170**, 421–428(1992). http://dx.doi.org/10.1016/0375-9601(92)90745-8
9. Samosir, A.S., Yatim, A.H.M.: Dynamic evolution control for synchronous buck DC-DC converter: theory, model and simulation. Simul. Model. Pract. Theory **18**(5), 663–676 (2010). http://dx.doi.org/10.1016/j.simpat.2010.01.010
10. Xiaohui, Q., Siu-Chung, W., Tse, C.K.: Resonant assisted buck converter for offline driving of high brightness LED replacement lamps. In: Proceedings of 2010 IEEE Energy Conversion Congress and Exposition (ECCE), 2010. doi:10.1109/ECCE.2010.5618054

Shear Waves Dispersion in Cylindrically Structured Cancellous Viscoelastic Bones

I.V. Andrianov, V.V. Danishevs'kyy, and J. Awrejcewicz

Abstract In this chapter we study anti-plane shear waves propagating through a cylindrically structured cancellous bone represented by a two-dimensional mesh of elastic trabeculae filled by a viscous marrow. In the long-wave limit, the original heterogeneous medium can be approximately substituted by a homogeneous one characterized by an effective complex shear modulus. The effect of dispersion is caused by the transmission of mechanical energy to heat due to the viscosity of the marrow (viscoelastic damping). We derive an approximate analytical solution using the asymptotic homogenization method; the cell problem is solved by means of a boundary shape perturbation and a lubrication theory approaches. For short waves, when the wavelength is comparable to the trabeculae size, the effect of dispersion is caused by successive reflections and refractions of local waves at the trabecula-marrow interfaces (Bloch dispersion). Decrease in the wavelength reveals a sequence of pass and stop frequency bands, so the heterogeneous bone can act like a discrete wave filter.

I.V. Andrianov (✉)
Department of General Mechanics, RWTH Aachen University, Templergraben 64, 52056 Aachen, Germany
e-mail: andrianov@gmail.com

V.V. Danishevs'kyy
Department of Structural Mechanics and Strength of Materials, Prydniprovs'ka State Academy of Civil Engineering and Architecture, Chernyshevs'kogo 24a, Dnipropetrovs'k 49600, Ukraine
e-mail: vdanish@ukr.net

J. Awrejcewicz
Department of Automation, Lodz University of Technology, Biomechanics and Mechatronics, 1/15 Stefanowski St., 90-924 Lodz, Poland

Department of Vehicles, Warsaw University of Technology, 84 Narbutta Str., 02-524 Warsaw, Poland
e-mail: jan.awrejcewicz@p.lodz.pl

J. Awrejcewicz (ed.), *Applied Non-Linear Dynamical Systems*, Springer Proceedings in Mathematics & Statistics 93, DOI 10.1007/978-3-319-08266-0_7

1 Introduction

Animal and human bones are heterogeneous materials with a complicated hierarchical structure. Bone tissues occur in the two main forms: as a dense solid (cortical or compact bone) and as a porous medium filled by a viscous marrow (trabecular or cancellous bone) [17]. The basic mechanical discrepancy between these two types consists in their relative densities measured by a volume fraction of solids. Both types can be found in the most bones of the body. A classical example of the macroscopic bone structure can be given by the long bones (e.g., humerus, femur, and tibia). They include an outer shell of a dense cortical tissue surrounding an inner core of a porous cancellous tissue.

The microstructure of cancellous bones is often described by two- or three-dimensional mesh of interconnected rods and plates [12, 17, 21]. Despite the obvious simplicity, such idealized models can provide a satisfactory agreement between theoretical predictions and experimental results for the mechanical properties of real bones [10, 11, 15, 19, 25]. In the present paper, we shall deal with a two-dimensional model of cylindrically structured cancellous bones.

Williams and Lewis [27] considered a real 2D section of a trabecular bone and evaluated its elastic constants using the plane-strain finite elements method. The developed approach enables one to predict mechanical properties of cylindrically structured cancellous bones basing on the morphological measurements in the transverse plane.

A challenging problem consists in the detection of the bone structure using non-invasive measurements. The inverse homogenization approach ("dehomogenization procedure") [7] can help to derive information about the microgeometry of the bone tissue from the magnitudes of static effective moduli. However, the static moduli supply only limited morphological data. Much more information can be obtained studying dynamic response of the bones. Measuring velocities and attenuation of acoustic waves at different frequencies provide us with additional information about the microstructure. Generally, frequency-dependent dynamic properties of the bone may be considered as a kind of "identification portrait", which is unique for every sample. The larger is the explored frequency range, the more accurate is the "portrait" that can be compiled. This should give a possibility to detect even very small variations of the internal bone texture.

Acoustic waves propagating through cancellous bones undergo dispersion and damping. There are two different physical effects influencing on the dynamic properties of the bone: (i) transmission of mechanical energy to heat due to the viscosity of the marrow (viscoelastic damping and dispersion) [16] and (ii) successive reflections and refractions of local waves at the trabecula-marrow interfaces (Bloch dispersion) [2, 20]. From the theoretical standpoint, both effects are realized simultaneously. However, their intensities are very frequency dependent. For many real materials the effects of viscoelastic damping and Bloch dispersion are observed in rather distant frequency ranges. In such a case they can be analyzed separately.

We study propagation of anti-plane shear waves through a 2D cylindrically structured cancellous bones. In our work, we have shown what kind of dispersion mechanism is dominant in the propagation of viscoelastic waves in the bone. This is an important theoretical result, because it allows choosing an adequate simulation model for the solution of this class of problems. Namely, it is necessary to consider Bloch dispersion and at the same time viscoelastic effects can be neglected, i.e. use an elastic model of the media.

The input dynamic problem is formulated in Sect. 2. In Sect. 3, the long-wave limit is considered. The solution is evaluated by means of the asymptotic homogenization procedure, and the effect of viscoelastic damping is predicted. In Sect. 4, the short-wave case is studied and the effect of Bloch dispersion is analyzed by means of the plane-wave expansion method. Conclusive remarks are presented in Sect. 5.

2 Input Dynamic Problem

We study transverse anti-plane shear waves propagating in the x_1x_2 plane through a regular cancellous structure consisting of a spatially infinite elastic matrix (trabeculae) $\Omega^{(1)}$ and viscous inclusions (marrow) $\Omega^{(2)}$ (Fig. 1). The governing two-dimensional wave equation is

$$\nabla_x (G\nabla_x u) = \rho\frac{\partial^2 u}{\partial t^2}, \tag{1}$$

where G is the complex shear modulus, ρ is the mass density, u is the longitudinal displacement (in the x_3 direction), and $\nabla_x = \mathbf{e}_1\partial/\partial x_1 + \mathbf{e}_2\partial/\partial x_2$, $\mathbf{e}_1$, $\mathbf{e}_2$ are the unit Cartesian vectors.

Due to the heterogeneity of the medium, the physical properties G and ρ are represented by piecewise continuous functions of co-ordinates:

$$G(\mathbf{x}) = G^{(a)}, \quad \rho(\mathbf{x}) = \rho^{(a)}, \text{for} \quad \mathbf{x} \in \Omega^{(a)}, \quad \mathbf{x} = x_1\mathbf{e}_1 + x_2\mathbf{e}_2. \tag{2}$$

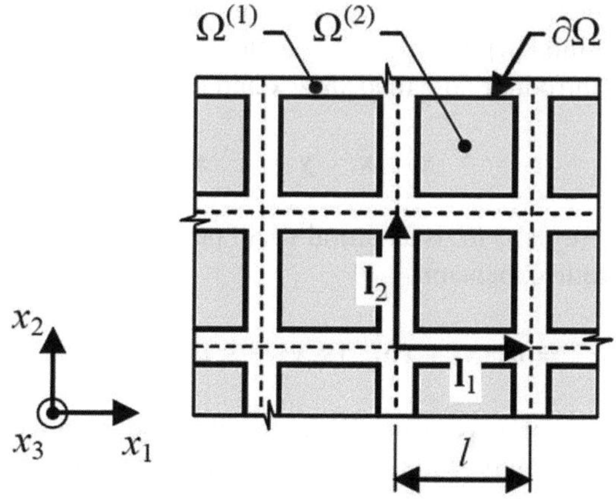

Fig. 1 Cancellous structure under consideration

Here and in the sequel the superscript (a) denotes different components of the structure $a=1,2$. In Eq. (2), $G^{(1)}$ is the real shear modulus of the elastic matrix and $G^{(2)}$ is the frequency-dependent imaginary shear modulus of the marrow. Following the linear theory of viscoelasticity, we can set $G^{(2)}=\mathrm{i}\omega\eta^{(2)}$, where ω is the frequency of a harmonic wave and $\eta^{(2)}$ is the viscosity of the marrow.

Equation (1) can be written in the equivalent form:

$$G^{(a)}\nabla_{xx}^2 u^{(a)} = \rho^{(a)}\frac{\partial^2 u^{(a)}}{\partial t^2}, \tag{3}$$

$$\left\{u^{(1)} = u^{(2)}\right\}\Big|_{\partial\Omega}, \quad \left\{G^{(1)}\frac{\partial u^{(1)}}{\partial \mathbf{n}} = G^{(2)}\frac{\partial u^{(2)}}{\partial \mathbf{n}}\right\}\Bigg|_{\partial\Omega}, \tag{4}$$

where $\nabla_{xx}^2 = \partial^2/\partial x_1^2 + \partial/\partial x_2^2$, $\partial/\partial\mathbf{n}$ is the normal derivative to the contour $\partial\Omega$. From the physical standpoint, Eqs. (4) means the perfect bonding conditions at the trabecula-marrow interface $\partial\Omega$.

3 Long-Wave Approach: Asymptotic Homogenization

3.1 Two-Scale Asymptotic Procedure

We start with the case when the wavelength L is essentially larger than the internal size l of the cancellous structure $l << L$. The original heterogeneous bone can be approximately substituted by a homogeneous one with a certain homogenized (*effective*) complex shear modulus G_0. Such an approach neglects local reflections and refractions of the waves on microlevel. The effect of dispersion is caused by the transmission of the mechanical energy of the acoustic wave to heat due to the viscosity of the marrow.

Let us study the input boundary value problem (3), (4) by the asymptotic homogenization method [4]. In order to separate macro- and microscale components of the solution we introduce so-called *slow* $\mathbf{x}$ and *fast* $\mathbf{y}$ co-ordinate variables

$$\mathbf{x} = \mathbf{x}, \quad \mathbf{y} = \varepsilon^{-1}\mathbf{x}, \tag{5}$$

where $\mathbf{y} = y_1\mathbf{e}_1 + y_2\mathbf{e}_2$, $\varepsilon = l/L$ is a natural small parameter, and search the displacement as an asymptotic expansion

$$u^{(a)} = u_0(\mathbf{x}) + \varepsilon u_1^{(a)}(\mathbf{x},\mathbf{y}) + \varepsilon^2 u_2^{(a)}(\mathbf{x},\mathbf{y}) + \ldots. \tag{6}$$

The first term u_0 of expansion (6) represents the homogenized part of the solution; it changes slowly within the whole sample of the bone and does not depend on the fast co-ordinates ($\partial u_0/\partial y_1 = \partial u_0/\partial y_2 = 0$). The next terms, $u_i^{(a)}$, $i = 1, 2, 3, \ldots$, provide corrections of the orders ε^i and describe local variations of the displacements on microlevel.

The differential operators read

$$\nabla_x = \nabla_x + \varepsilon^{-1}\nabla_y, \quad \nabla^2_{xx} = \nabla^2_{xx} + 2\varepsilon^{-1}\nabla^2_{xy} + \varepsilon^{-2}\nabla^2_{yy}, \tag{7}$$

where $\nabla_y = \mathbf{e}_1\partial/\partial y_1 + \mathbf{e}_2\partial/\partial y_2$, $\nabla^2_{xy} = \partial^2/(\partial x_1\partial y_1) + \partial^2/(\partial x_2\partial y_2)$, $\nabla^2_{yy} = \partial^2/\partial y_1^2 + \partial/\partial y_2^2$.

Splitting the input problems (3), (4) with respect to ε leads to a recurrent sequence of cell boundary value problems:

$$G^{(a)}\left(\nabla^2_{xx}u^{(a)}_{i-2} + 2\nabla^2_{xy}u^{(a)}_{i-1} + \nabla^2_{yy}u^{(a)}_i\right) = \rho^{(a)}\frac{\partial^2 u^{(a)}_{i-2}}{\partial t^2}, \tag{8}$$

$$\left\{u^{(1)}_i = u^{(2)}_i\right\}\Big|_{\partial\Omega},$$

$$\left\{G^{(1)}\left(\frac{\partial u^{(1)}_{i-1}}{\partial \mathbf{n}} + \frac{\partial u^{(1)}_i}{\partial \mathbf{m}}\right) = G^{(2)}\left(\frac{\partial u^{(2)}_{i-1}}{\partial \mathbf{n}} + \frac{\partial u^{(2)}_i}{\partial \mathbf{m}}\right)\right\}\Bigg|_{\partial\Omega}, \tag{9}$$

where $i = 1, 2, 3, \ldots$, $u^{(a)}_{-1} = 0$, $\partial/\partial\mathbf{m}$ is the normal derivative to the interface $\partial\Omega$ written in fast variables.

For a spatially periodic medium, the terms $u^{(a)}_i$ have to satisfy the conditions of periodicity

$$u^{(a)}_i(\mathbf{x}, \mathbf{y}) = u^{(a)}_i\left(\mathbf{x}, \mathbf{y} + \mathbf{L}_p\right), \tag{10}$$

and normalization

$$\left\langle u^{(a)}_i\right\rangle = 0, \tag{11}$$

where $\mathbf{L}_p = \varepsilon^{-1}\mathbf{l}_p$, $\mathbf{l}_p = p_1\mathbf{l}_1 + p_2\mathbf{l}_2$, $p_1, p_2 = 0, \pm 1, \pm 2, \ldots$, $\mathbf{l}_1$, $\mathbf{l}_2$ are the fundamental translation vectors of the cancellous structure (see Fig. 1),

$$\langle\cdot\rangle = \frac{1}{S_0}\left(\iint\limits_{\Omega^{(1)}_0}(\cdot)\,\mathrm{d}y_1\mathrm{d}y_2 + \iint\limits_{\Omega^{(2)}_0}(\cdot)\,\mathrm{d}y_1\mathrm{d}y_2\right) \tag{12}$$

is the homogenizing operator over the unit cell domain $\Omega_0 = \Omega^{(1)}_0 + \Omega^{(2)}_0$ (Fig. 2), and $S_0 = L^2$ is the area of the unit cell in the fast co-ordinates.

Conditions (10) and (11) can be approximately replaced by zero boundary conditions for same functions in the center and along the outer contour $\partial\Omega_0$ of the unit cell:

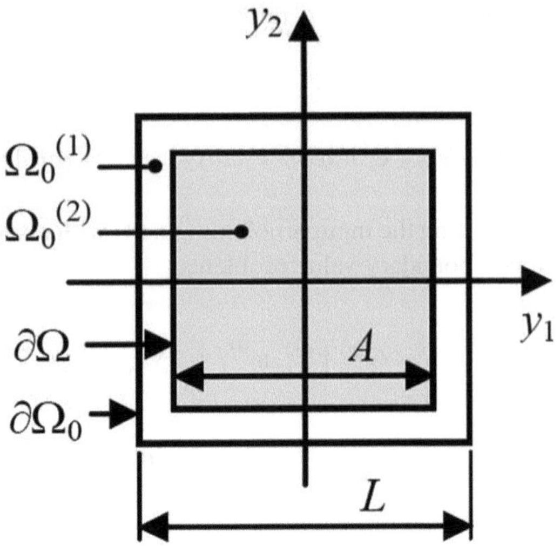

Fig. 2 Periodically repeated unit cell

$$\left\{u_i^{(2)} = 0\right\}\Big|_{x,y=0}, \quad \left\{u_i^{(1)} = 0\right\}\Big|_{\partial\Omega_0}. \tag{13}$$

For 1D problems, Eq. (13) appears entirely equivalent to Eqs. (10) and (11). For 2D problems, replacing the periodicity conditions by zero boundary conditions increases stiffness of the system and, thus, provides an upper bound for the effective properties. Analysis of numerical examples has shown that discrepancy between the final solutions in both cases is not essential [1, 3], whereas utilizing approximation (13) leads to a sufficient simplification of the cell problems.

Due to the periodicity of $u_i^{(a)}$ (10), Eqs. (8) and (9) can be considered within only one distinguished unit cell. Solution of the cell problems (8), (9), (13) at $i = 1$ determines the term $u_1^{(a)}$. In order to find the effective modulus G_0, the homogenizing operator (12) is applied to Eq. (8) at $i = 2$. The terms $u_2^{(a)}$ are eliminated by means of Green's theorem, which together with the boundary conditions (9) and the periodicity relation (10) implies

$$\left\langle G^{(a)}\left(\nabla_{xy}^2 u_i^{(a)} + \nabla_{yy}^2 u_{i+1}^{(a)}\right)\right\rangle = 0.$$

As a result, the homogenized wave equation of the order ε^0 is obtained:

$$\left\langle G^{(a)}\left(\nabla_{xx}^2 u_0 + \nabla_{xy}^2 u_1^{(a)}\right)\right\rangle = \left\langle \rho^{(a)}\right\rangle \frac{\partial^2 u_0}{\partial t^2}. \tag{14}$$

Substituting here expressions for $u_1^{(a)}$ evaluated below we shall come to a macroscopic wave equation

$$G_0 \nabla_{xx}^2 u_0 = \rho_0 \frac{\partial^2 u_0}{\partial t^2}, \tag{15}$$

where $\rho_0 = (1 - c^{(2)})\rho^{(1)} + c^{(2)}\rho^{(2)}$ is the effective mass density, $c^{(2)}$ is the volume fraction of the inclusions, $c^{(2)} = A/S_0$, A is the size of the inclusion (Fig. 2). The effective modulus G_0 can be derived after evaluation of the integrals in Eq. (14).

Below we find approximate solutions of the cell problem (8), (9), (13) and determine the effective shear modulus G_0 using a boundary shape perturbation and lubrication theory approaches.

3.2 The Case $c^{(2)} << 1$: Boundary Shape Perturbation

If the volume fraction $c^{(2)}$ of the marrow inclusions is relatively small, the square shapes of the domains $\Omega^{(1)}$, $\Omega^{(2)}$ can be approximately substituted by the equal circles of radii R_1, R_2 so that $c^{(2)} = R_2^2/R_1^2$ (Fig. 3). This simplification can be considered as the first approximation of the method of the boundary shape perturbation [1, 14].

Let us introduce in the unit cell the polar co-ordinates $r^2 = y_1^2 + y_2^2$, $\tan\theta = y_2/y_1$. Equations (8), (9), and (13) at $i = 1$ read

$$\frac{\partial^2 u_1^{(a)}}{\partial r^2} + \frac{1}{r}\frac{\partial u_1^{(a)}}{\partial r} + \frac{1}{r^2}\frac{\partial^2 u_1^{(a)}}{\partial \theta^2} = 0, \tag{16}$$

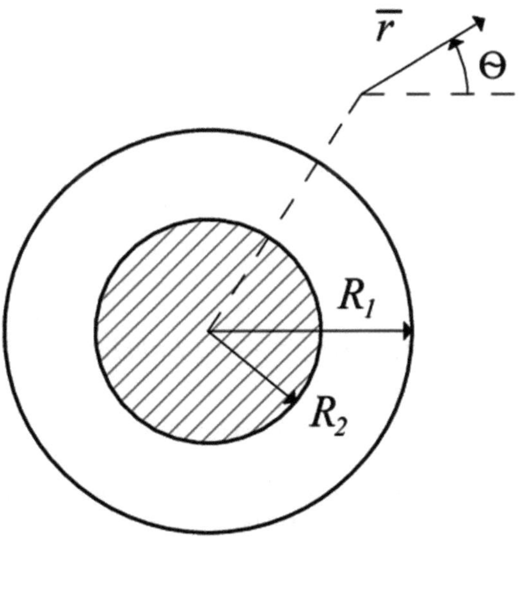

Fig. 3 Simplification of the unit cell in the case $c^{(2)} << 1$

$$\left\{ u_1^{(1)} = u_1^{(2)} \right\}\Big|_{r=R_2}, \\ \left\{ G^{(1)} \left(\frac{\partial u_0}{\partial \mathbf{n}} + \frac{\partial u_1^{(1)}}{\partial r} \right) = G^{(2)} \left(\frac{\partial u_0}{\partial \mathbf{n}} + \frac{\partial u_1^{(2)}}{\partial r} \right) \right\}\Big|_{r=R_2}, \tag{17}$$

$$\left\{ u_1^{(2)} = 0 \right\}\Big|_{r=0}, \quad \left\{ u_1^{(1)} = 0 \right\}\Big|_{r=R_1}, \tag{18}$$

where $\partial/\partial \mathbf{n} = \cos\theta\, \partial/\partial x_1 + \sin\theta\, \partial/\partial x_2$.

Solution of the simplified cell problem (16)–(18) is

$$u_1^{(a)} = \left(C_1^{(a)} r + C_2^{(a)} r^{-1} \right) \frac{\partial u_0}{\partial \mathbf{n}}, \tag{19}$$

$$C_1^{(1)} = \frac{\left(\lambda^{(2)} - 1\right) c^{(2)}}{\lambda^{(2)} + 1 - c^{(2)} \left(\lambda^{(2)} - 1\right)}, \quad C_2^{(1)} = -\frac{\left(\lambda^{(2)} - 1\right) R_2^2}{\lambda^{(2)} + 1 - c^{(2)} \left(\lambda^{(2)} - 1\right)},$$

$$C_1^{(2)} = -\frac{\left(\lambda^{(2)} - 1\right)\left(1 - c^{(2)}\right)}{\lambda^{(2)} + 1 - c^{(2)} \left(\lambda^{(2)} - 1\right)}, \quad C_2^{(2)} = 0,$$

where $\lambda^{(2)} = G^{(2)}/G^{(1)}$.

Substituting expressions (19) into the homogenized equation (14), we obtain the effective modulus G_0 in the closed analytical form:

$$\lambda_0 = \frac{\lambda^{(2)} + 1 + c^{(2)} \left(\lambda^{(2)} - 1\right)}{\lambda^{(2)} + 1 - c^{(2)} \left(\lambda^{(2)} - 1\right)}, \tag{20}$$

where $\lambda_0 = G_0/G^{(1)}$.

It should be noted that solution (20) is precisely the same as can be obtained by the composite cylinder assemblage model and by the generalized self-consistent scheme [8].

3.3 The Case $c^{(2)} \to 1$: Lubrication Theory

In the case of densely packed marrow inclusions, when the volume fraction $c^{(2)}$ is close to unit, $c^{(2)} \to 1$, an asymptotic solution of the cell problem can be obtained using as a natural small parameter the non-dimensional width $\delta = H/L$ of the trabecula (Fig. 4). Let us suppose $\delta << 1$. Being restricted by the $O(\delta^0)$ approximation,

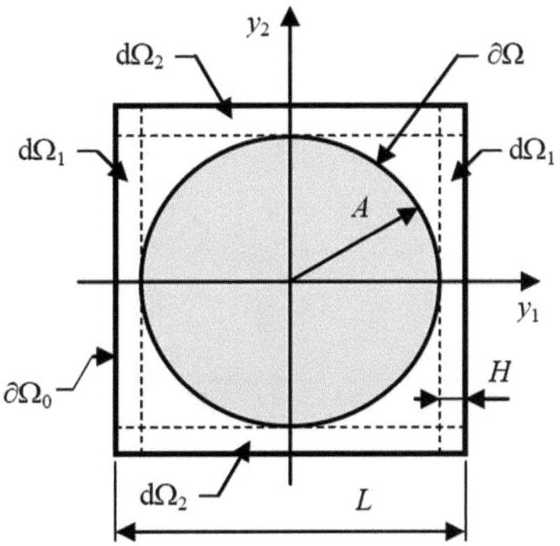

Fig. 4 Unit cell in the case $c^{(2)} \to 1$

for the matrix strips $d\Omega_1$, $d\Omega_2$, which separate neighbouring inclusions, one can show

$$\frac{\partial^2 u_1^{(1)}}{\partial y_1^2} >> \frac{\partial^2 u_1^{(1)}}{\partial y_2^2} \quad \text{for } \mathbf{y} \in d\Omega_1, \qquad \frac{\partial^2 u_1^{(1)}}{\partial y_1^2} << \frac{\partial^2 u_1^{(1)}}{\partial y_2^2} \quad \text{for } \mathbf{y} \in d\Omega_2. \tag{21}$$

The physical meaning of estimations (21) is that in the narrow strip $d\Omega_1$ the variation of local stresses in the direction y_1 is dominant and, hence, the term $\partial^2 u_1^{(1)}/\partial y_2^2$ can be neglected in comparison with $\partial^2 u_1^{(1)}/\partial y_1^2$. Vice versa, in the strip $d\Omega_2$, the dominant variation of the local stress field takes place in the direction y_2, so the term $\partial^2 u_1^{(1)}/\partial y_1^2$ can be neglected in comparison with $\partial^2 u_1^{(1)}/\partial y_2^2$. Such a simplification is similar to the basic idea of the well-known lubrication theory, which was used in the theory of composites for many years [8, 9].

Following estimations (21), in the $O(\delta^0)$ approximation, Eq. (8) reads

$$G^{(1)}\left(\frac{\partial^2 u_{i-2}^{(1)}}{\partial x_s^2} + 2\frac{\partial^2 u_{i-1}^{(1)}}{\partial x_s \partial y_s} + \frac{\partial^2 u_i^{(1)}}{\partial y_s^2}\right) = \rho^{(1)}\frac{\partial^2 u_{i-2}^{(1)}}{\partial t^2}, \tag{22}$$

$$G^{(2)}\left(\nabla_{xx}^2 u_{i-2}^{(2)} + 2\nabla_{xy}^2 u_{i-1}^{(2)} + \nabla_{yy}^2 u_i^{(2)}\right) = \rho^{(2)}\frac{\partial^2 u_{i-2}^{(2)}}{\partial t^2}$$

Solution of the simplified cell problems (9), (13), (22) at $i = 1$ is

$$u_1^{(2)} = -\frac{\left(1 - \sqrt{c^{(2)}}\right)\left(\lambda^{(2)} - 1\right)}{\lambda^{(2)} - \sqrt{c^{(2)}}\left(\lambda^{(2)} - 1\right)} y_s \frac{\partial u_0}{\partial x_s}, \tag{23}$$

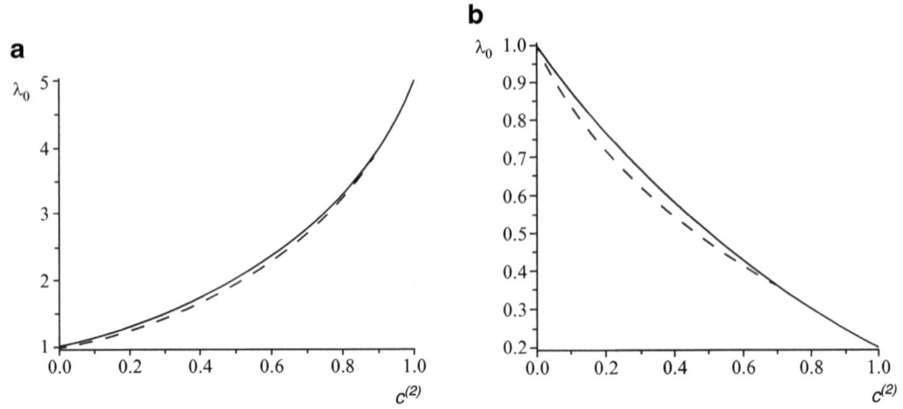

Fig. 5 Effective modulus in the elastic case. *Solids*—formula (20), *dashes*—formula (24). (**a**) $\lambda^{(2)} = 5$. (**b**) $\lambda^{(2)} = 0.2$

$$u_1^{(1)} = -\frac{\sqrt{c^{(2)}}\left(\lambda^{(2)}-1\right)}{\lambda^{(2)}-\sqrt{c^{(2)}}\left(\lambda^{(2)}-1\right)}\left(\frac{L}{2}-y_s\right)\frac{\partial u_0}{\partial x_s} \quad \text{at} \quad y_s > 0,$$

$$u_1^{(1)} = \frac{\sqrt{c^{(2)}}\left(\lambda^{(2)}-1\right)}{\lambda^{(2)}-\sqrt{c^{(2)}}\left(\lambda^{(2)}-1\right)}\left(\frac{L}{2}+y_s\right)\frac{\partial u_0}{\partial x_s} \quad \text{at} \quad y_s < 0,$$

where $s = 1, 2$.

For the effective shear modulus we obtain

$$\lambda_0 = \frac{\lambda^{(2)}-\sqrt{c^{(2)}}\left(1-\sqrt{c^{(2)}}\right)\left(\lambda^{(2)}-1\right)}{\lambda^{(2)}-\sqrt{c^{(2)}}\left(\lambda^{(2)}-1\right)}. \tag{24}$$

Numerical results, calculated by formulas (20) and (24), are very close (except the case $\lambda^{(2)} < 1$, $c^{(2)} \to 0$). This is illustrated at Fig. 5 for real values of $\lambda^{(2)}$, which correspond to elastic materials. Moreover, in the limit $c^{(2)} \to 1$, the approximate solutions (20) and (24) exhibit the same asymptotic behaviour and give identical expansions for λ_0 until the order $O[(1-c)^2]$:

$$\lambda_0 = \lambda^{(2)} - \frac{1}{2}\left(\lambda^{(2)}-1\right)\left(\lambda^{(2)}+1\right)\left(1-c^{(2)}\right) + O\left[\left(1-c^{(2)}\right)^2\right] \quad \text{at} \quad c^{(2)} \to 1.$$

This fact reveals that for the cancellous structure under consideration, expression (20), originally obtained for the case $c^{(2)} << 1$, provides a reasonable approximation in the whole region of the inclusions volume fraction $0 \le c^{(2)} \le 1$.

3.4 Propagation of Long Waves

Let us consider a harmonic wave

$$u_0 = U \exp(-\mathrm{i}\boldsymbol{\mu} \cdot \mathbf{x}) \exp(i\omega t), \tag{25}$$

where U is the amplitude, ω is the frequency, and $\boldsymbol{\mu} = \mu_1 \mathbf{e}_1 + \mu_2 \mathbf{e}_2$ is the wave vector and the direction of propagation is determined by the angle α, $\tan\alpha = \mu_2/\mu_1$.

Separating real $\boldsymbol{\mu}_R$ and imaginary $\boldsymbol{\mu}_I$ parts of the wave vector $\boldsymbol{\mu} = \boldsymbol{\mu}_R - \mathrm{i}\boldsymbol{\mu}_I$, expression (25) reads

$$u_0 = U \exp(-\boldsymbol{\mu}_I \cdot \mathbf{x}) \exp(-\mathrm{i}\boldsymbol{\mu}_R \cdot \mathbf{x}) \exp(\mathrm{i}\omega t). \tag{26}$$

Here $\mu_I = |\boldsymbol{\mu}_I|$ is the attenuation factor and $\mu_R = |\boldsymbol{\mu}_R| = 2\pi/L$ is the wave number.

For the viscoelastic composite medium, the effective complex modulus G_0, the attenuation coefficient μ_I and the phase velocity $v_p = \omega/\mu_R$ depend on the frequency of the travelling signal. Substituting expression (26) into the macroscopic wave equation (15), we obtain

$$(G_{0,R} + \mathrm{i}G_{0,I})(\mu_I + \mathrm{i}\mu_R)^2 = -\rho_0\omega^2, \tag{27}$$

where $G_{0,R}$, $G_{0,I}$ are, respectively, the real and the imaginary part of G_0, $G_0 = G_{0,R} + \mathrm{i}G_{0,I}$. Collecting in Eq. (27) the terms at 1 and i, after routine transformations we derive

$$\mu_I = \mu_R \tan(\varphi_0/2), \quad v_p^2 = \frac{v_0^2}{\cos(\varphi_0/2)^2}, \tag{28}$$

where $\tan(\varphi_0) = G_{0,I}/G_{0,R}$ is the effective loss tangent and $v_0 = \sqrt{|G_0|/\rho_0}$ is the effective velocity in the elastic case.

Adopting for G_0 the solution (20), we obtain

$$\begin{aligned} G_0 &= G^{(1)} \frac{G^{(1)}\left(1-c^{(2)}\right)+\mathrm{i}\omega\eta^{(2)}\left(1+c^{(2)}\right)}{G^{(1)}\left(1+c^{(2)}\right)+\mathrm{i}\omega\eta^{(2)}\left(1-c^{(2)}\right)}, \\ \tan(\varphi_0) &= \frac{4G^{(1)}c^{(2)}\omega\eta^{(2)}}{\left[1-\left(c^{(2)}\right)^2\right]\left[\left(G^{(1)}\right)^2+\left(\omega\eta^{(2)}\right)^2\right]}. \end{aligned} \tag{29}$$

In the numerical examples presented below we accept some rough estimations of the properties of the components following Bryant et al. [6], Guo [13], and Van Rietbergen and Huiskes [26]. The shear modulus of the trabeculae is $G^{(1)} = 3.85 \cdot 10^9$ Pa and the viscosity of the marrow is $\eta^{(2)} = 0.15$ Pa·s (at the room temperature of 20 °C) and $\eta^{(2)} = 0.05$ Pa·s (at the body temperature of 37 °C). The trabeculae volume fraction $c^{(1)} = 1 - c^{(2)}$ can vary from 0.05–0.1 for aged osteoporotic bones to 0.3–0.35 for young normal bones.

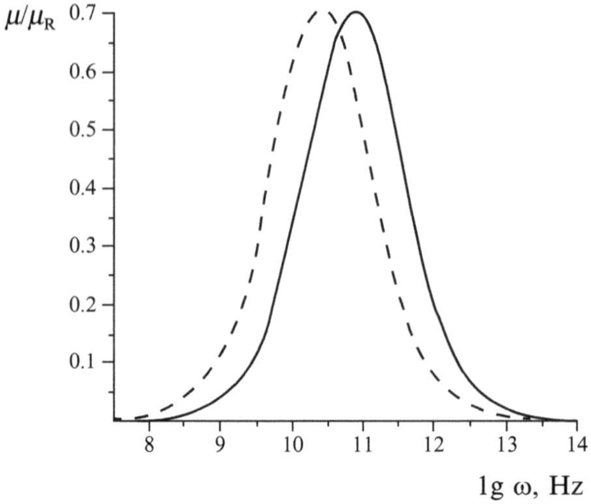

Fig. 6 Attenuation factor of a normal bone. *Solids*—$\eta^{(2)} = 0.05$ Pa · s, *dashes*—$\eta^{(2)} = 0.15$ Pa · s

Dependencies of the attenuation factor μ_I upon the frequency ω are displayed at Fig. 6 (normal bone, $c^{(1)} = 0.3$) and Fig. 7 (osteoporotic bone, $c^{(1)} = 0.1$). The dispersion effect vanishes (i) at $\omega \to 0$, when the deformation rate is small and the stiffness of the marrow is negligible, and (ii) at $\omega \to \infty$, when the deformation rate is high, so the marrow acts like a perfectly stiff medium. Decrease of the trabeculae volume fraction $c^{(1)}$ leads to the intensifying of the dispersion: the damping frequency region extends and the attenuation factor μ_I grows. Decrease in temperature (i.e. increase in the marrow viscosity $\eta^{(2)}$) leads to a reduction of the damping frequency. In any case, for physically meaningful values of the bone properties, the effect of viscoelastic damping can be observed starting from the frequencies of the order 100 MHz and higher.

It should be noted that in the long-wave limit ($l << L$) the cancellous structure under consideration is transversely orthotropic. The obtained solution for anti-plane shear waves is isotropic in the plane x_1x_2, so the parameters G_0, φ_0 do not depend on the direction of the wave propagation. The effect of anisotropy is predicted in the case of short waves (see Sect. 4).

4 Floquet-Bloch Approach: Plane-Wave Expansion Method

When the wavelength L is comparable to the internal size l of the cancellous structure, the effect of dispersion is caused by successive reflections and refractions of local waves at the trabecula-marrow interfaces. Decrease in the wavelength reveals a sequence of pass and stop frequency bands (so-called phononic bands)

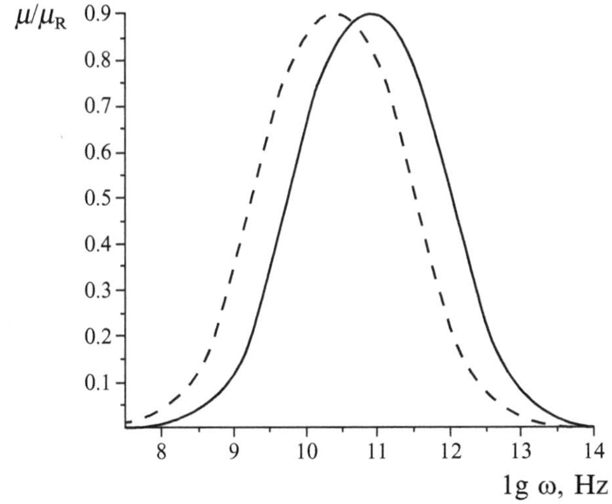

Fig. 7 Attenuation factor of an osteoporotic bone. *Solids*—$\eta^{(2)} = 0.05$ Pa·s, *dashes*—$\eta^{(2)} = 0.15$ Pa·s

[2, 20]. Thus, a heterogeneous bone can act as a discrete wave filter. If the frequency of the signal falls within a stop band, a stationary wave is excited and the neighbouring trabeculae vibrate in alternate directions. On macrolevel the amplitude of the global wave attenuates exponentially, so no propagation is possible.

In order to explore such a case, let us assume the threshold of the first stop band to be essentially lower than the viscoelastic damping frequencies. The marrow is not involved into the shear deformation, so we can set $G^{(2)} = 0$, $\rho^{(2)} = 0$.

Following the Floquet-Bloch theorem [5], a harmonic wave propagating through a periodic cancellous structure is represented in the form

$$w = F(\mathbf{x}) \exp(\mathrm{i}\boldsymbol{\mu} \cdot \mathbf{x}) \exp(i\omega t), \tag{30}$$

where $F(\mathbf{x})$ is a spatially periodic function, $F(\mathbf{x}) = F(\mathbf{x} + \mathbf{l}_p)$.

We use the plane-wave expansion method [20, 23] and express the function $F(\mathbf{x})$ and the material properties $G(\mathbf{x})$, $\rho(\mathbf{x})$ as infinite Fourier series:

$$\begin{aligned} F(\mathbf{x}) &= \sum_{k_1=-\infty}^{\infty} \sum_{k_2=-\infty}^{\infty} A_{k_1 k_2} \exp\left[\mathrm{i}\frac{2\pi}{l}(k_1 x_1 + k_2 x_2)\right], \\ G(\mathbf{x}) &= \sum_{k_1=-\infty}^{\infty} \sum_{k_2=-\infty}^{\infty} B_{k_1 k_2} \exp\left[\mathrm{i}\frac{2\pi}{l}(k_1 x_1 + k_2 x_2)\right], \\ \rho(\mathbf{x}) &= \sum_{k_1=-\infty}^{\infty} \sum_{k_2=-\infty}^{\infty} C_{k_1 k_2} \exp\left[\mathrm{i}\frac{2\pi}{l}(k_1 x_1 + k_2 x_2)\right], \end{aligned} \tag{31}$$

where

$$B_{k_1k_2} = \frac{1}{l^2}\iint\limits_{\Omega_0} G(\mathbf{x})\exp\left[-\mathrm{i}\frac{2\pi}{l}(k_1x_1 + k_2x_2)\right]\mathrm{d}x_1\mathrm{d}x_2,$$

$$C_{k_1k_2} = \frac{1}{l^2}\iint\limits_{\Omega_0} \rho(\mathbf{x})\exp\left[-\mathrm{i}\frac{2\pi}{l}(k_1x_1 + k_2x_2)\right]\mathrm{d}x_1\mathrm{d}x_2,$$

the operator $\iint\limits_{\Omega_0}(\cdot)\mathrm{d}x_1\mathrm{d}x_2$ denotes integration over a distinguished unit cell Ω_0.

Substituting Ansatz (30) and expansions (31) into the wave equation (1) and collecting the terms $\exp[\mathrm{i}2\pi l^{-1}(j_1x_1 + j_2x_2)]$, $j_1, j_2 = 0, \pm 1, \pm 2, \ldots$, we come to an infinite system of linear algebraic equations for the unknown coefficients $A_{k_1k_2}$:

$$\sum_{k_1=-\infty}^{\infty}\sum_{k_2=-\infty}^{\infty} A_{k_1k_2}\left\{B_{\substack{j_1-k_1,\\ j_2-k_2}}\left[\left(\tfrac{2\pi}{l}k_1 + \mu_1\right)\left(\tfrac{2\pi}{l}j_1 + \mu_1\right) + \right.\right.$$
$$\left.\left. + \left(\tfrac{2\pi}{l}k_2 + \mu_2\right)\left(\tfrac{2\pi}{l}j_2 + \mu_2\right)\right] - C_{\substack{j_1-k_1,\\ j_2-k_2}}\omega^2\right\} = 0. \tag{32}$$

System (32) has a nontrivial solution if and only if the determinant of the matrix of the coefficients is zero. Equating the determinant to zero, we derive a dispersion relation for ω and μ. It should be noted that the plane-wave expansion method does not use explicitly the bonding conditions (4), whereas they are "embedded" implicitly into Eq. (1) and expansions (31).

To illustrate the appearance of phononic band gaps, let us rewrite Ansatz (30) separating real $\boldsymbol{\mu}_R$ and imaginary $\boldsymbol{\mu}_I$ parts of the wave vector $\boldsymbol{\mu} = \boldsymbol{\mu}_R - \mathrm{i}\boldsymbol{\mu}_I$:

$$w = F(\mathbf{x})\exp(-\boldsymbol{\mu}_I\cdot\mathbf{x})\exp(\mathrm{i}\boldsymbol{\mu}_R\cdot\mathbf{x})\exp(i\omega t). \tag{33}$$

The imaginary part $\mu_I \equiv |\boldsymbol{\mu}_I|$ represents the attenuation factor. Frequency regions where $\mu_I \neq 0$ correspond to stop bands (signal (30) attenuates exponentially), while regions where $\mu_I = 0$ correspond to pass bands.

In numerical examples the dispersion relations are calculated approximately by the truncation of the infinite system (32) supposing $-j_{\max} \leq j_s \leq j_{\max}$. The number of the kept equations is $(2j_{\max} + 1)^2$. We expect that increase in $j_{\max}$ shall improve the accuracy of the solution. From the physical point of view, such a truncation means cutting off the higher frequencies.

Figure 8 displays dispersion curves for a normal bone with the following properties of the trabecular tissue: $G^{(1)} = 3.85\cdot 10^9$ Pa, $\rho^{(1)} = 1900$ kg/m^3, $c^{(1)} = 0.3$. Calculations are performed at $j_{\max} = 3$. The diagram consists of two (left and

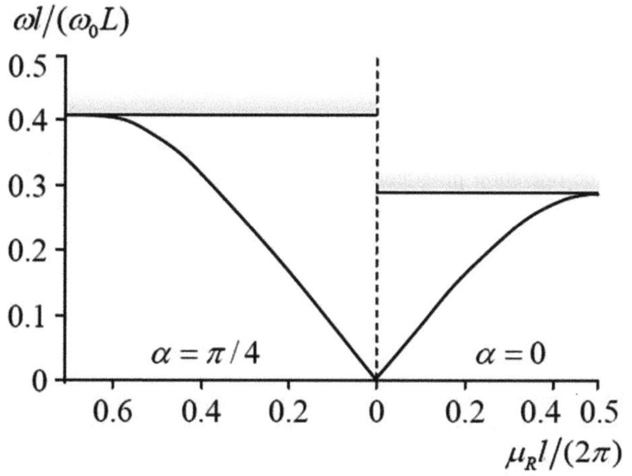

Fig. 8 Dispersion curves of a normal bone

right) parts separated by a vertical dash line. The right part displays a solution for the orthogonal direction ($\alpha = 0$) and the left part for the diagonal direction ($\alpha = \pi/4$) of the wave propagation. The results for the frequency ω are normalized to $\omega_0 = v_0\mu_R = 2\pi v_0/L$. We can observe that in the long-wave case ($\omega \to 0$, $l/L \to 0$) the solution is isotropic. However, with the increase in ω and decrease in L, the cancellous structure exhibits an anisotropic behaviour.

Shaded areas at Fig. 8 indicate the threshold of the first stop bands. Let us estimate the corresponding values ω_s of the frequency. We obtain $\omega_s l/(\omega_0 L) \approx 0.29$ at $\alpha = 0$ and $\omega_s l/(\omega_0 L) \approx 0.41$ at $\alpha = \pi/4$. The typical length of trabeculae is about $l \approx 10^{-3}$m. Taking into account $\omega_0 = 2\pi v_0/L$, $v_0 = \sqrt{G_0/\rho_0}$, we derive $\omega_s \approx 2.0$ MHz at $\alpha = 0$ and $\omega_s \approx 2.8$ MHz at $\alpha = \pi/4$.

5 Conclusions

For anti-plane shear waves the effect of Bloch dispersion, caused by the heterogeneity of cancellous bones, appears at essentially lower frequencies than the effect of viscoelastic damping, caused by the viscosity of the marrow. Bloch dispersion is expected to play the primary role in the processes of ultrasonic diagnostic, which usually deals with acoustic waves in the regions 1–10 MHz. The viscoelastic damping can be neglected until the frequency of about 100 MHz. Obtained results may be used for the development of new methods of non-invasive testing and diagnostic.

In the present paper, a perfectly regular arrangement of marrow inclusions is investigated. It is clear that the microstructure of real bones is not regular. At the same time, it has been shown in a number of studies [18, 22, 24] that regular structures exhibit the narrowest stop bands in comparing to disordered systems. Thus, the obtained solutions may be treated as theoretical bounds for the stop band thresholds that appear in randomly disordered bone tissues.

Acknowledgments This work is supported by the Alexander von Humboldt Foundation (Institutional academic co-operation programme, grant no. 3.4-Fokoop-UKR/1070297), the German Research Foundation (Deutsche Forschungsgemeinschaft, grant no. WE 736/30-1).

References

1. Andrianov, I.V., Danishevs'kyy, V.V., Guillet, A., Pareige, P.: Effective properties and micromechanical response of filamentary composite wires under longitudinal shear. Eur. J. Mech. A/Solids **24**, 195–206 (2005)
2. Andrianov, I.V., Bolshakov, V.I., Danishevs'kyy, V.V., Weichert, D.: Higher-order asymptotic homogenization and wave propagation in periodic composite materials. Proc. Roy. Soc. A Math. Phys. Eng. Sci. **464**, 1181–1201 (2008)
3. Andrianov, I.V., Bolshakov, V.I., Danishevs'kyy, V.V., Weichert, D.: Asymptotic study of imperfect interfaces in conduction through a granular composite material. Proc. Roy. Soc. A Math. Phys. Eng. Sci. **466**, 2707–2725 (2010)
4. Bakhvalov, N.S., Panasenko, G.P.: Homogenization: Averaging Processes in Periodic Media. Mathematical Problems in Mechanics of Composite Materials. Kluwer, Dordrecht (1989)
5. Brillouin, L.: Wave Propagation in Periodic Structures: Electric Filters and Crystal Lattices. Dover, New York (2003)
6. Bryant, J.D., David, T., Gaskell, P.H., King, S., Lond, G.: Rheology of bovine bone marrow. Proc. Inst. Mech. Eng. H J. Eng. Med. **203**(2), 71–75 (1989)
7. Cherkaev, E., Ou, M.-J.Y.: Dehomogenization: reconstruction of moments of the spectral measure of the composite. Inverse Probl. **24**, 065008 (2008)
8. Christensen, R.M.: Mechanics of Composite Materials. Dover, Mineola (2005)
9. Frankel, N.A., Acrivos, A.: On the viscosity of a concentrated suspension of solid spheres. Chem. Eng. Sci. **22**, 847–853 (1967)
10. Gałka, A., Telega, J.J., Tokarzewski, S.: A contribution to evaluation of effective moduli of trabecular bone with rod-like microstructure. J. Theor. Appl. Mech. **37**, 707–727 (1999)
11. Gałka, A., Telega, J.J., Tokarzewski, S.: Application of homogenization to evaluation of effective moduli of linear elastic trabecular bone with plate-like structure. Arch. Mech. **51**, 335–355 (1999)
12. Gibson, J.L., Ashby, M.P.: Cellular Solids: Structure and Properties. Pergamon, Oxford (1988)
13. Guo, X.E.: Mechanical properties of cortical bone and cancellous bone tissue. In: Cowin, S.C. (ed.) Bone Mechanics Handbook, pp. 10-1–10-23. CRC, Boca Raton (2001)
14. Guz, A.N., Nemish, Y.N.: Perturbation of boundary shape in continuum mechanics. Soviet Appl. Mech. **23**(9), 799–822 (1987)
15. Hollister, S.J., Fyhire, D.P., Jepsen, K.J., Goldstein, S.A.: Application of homogenization theory to the study of trabecular bone mechanics. J. Biomech. **24**, 825–839 (1991)
16. Hughes, E.R., Leighton, T.G., Petley, G.M., White, P.R., Chivers, R.C.: Estimation of critical and viscous frequencies for Biot theory in cancellous bone. Ultrasonics **41**, 365–368 (2003)
17. Jee, W.S.S.: Integrated bone tissue physiology: anatomy and physiology. In: Cowin, S.C. (ed.) Bone Mechanics Handbook, pp. 1-1–1-68. CRC, Boca Raton (2001)
18. Jensen, J.S.: Phononic band gaps and vibrations in one- and two-dimensional mass-spring structures. J. Sound Vib. **266**, 1053–1078 (2003)
19. Kasra, M., Grynpas, M.D.: Static and dynamic finite element analyses of an idealized structural model of vertebral trabecular bone. J. Biomech. Eng. **120**, 267–272 (1998)

20. Kushwaha, M.S., Halevi, P., Martinez, G., Dobrzynski, L., Djafari-Rouhani, B.: Theory of acoustic band structure of periodic elastic composites. Phys. Rev. B **49**, 2313–2322 (1994)
21. Odgaard, A.: Qualification of cancellous bone architecture. In: Cowin, S.C. (ed.) Bone Mechanics Handbook, pp. 14-1–14-19. CRC, Boca Raton (2001)
22. Schraad, M.W.: On the macroscopic properties of discrete media with nearly periodic microstructures. Int. J. Solids Struct. **38**, 7381–7407 (2001)
23. Sigalas, M.M., Economou, E.N.: Elastic and acoustic wave band structure. J. Sound Vib. **158**, 377–382 (1992)
24. Thorp, O., Ruzzene, M., Baz, A.: Attenuation and localization of wave propagation in rods with periodic shunted piezoelectric patches. Smart Mater. Struct. **10**, 979–989 (2001)
25. Tokarzewski, S., Telega, J.J., Gałka, A.: Prediction of torsional rigidities of bone filled with marrow: the application of multipoint Padé approximants. Acta Bioeng. Biomech. **2**, 560–566 (2000)
26. Van Rietbergen, B., Huiskes, R.: Elastic constants of cancellous bone. In: Cowin, S.C. (ed.) Bone Mechanics Handbook, pp. 15-1–15-24. CRC, Boca Raton (2001)
27. Williams, J.L., Lewis, J.L.: Properties and an anisotropic model of cancellous bone from the proximal tibial epiphysis. J. Biomech. Eng. **104**(1), 50–56 (1982)

Quasi-regular and Chaotic Dynamics of Postural Sway in Human

Natalya Kizilova, Elena Karpinska, and Michael Karpinsky

Abstract Parameters of different vertical two-leg stances are studied by the force platform measurements on a control group of young healthy individuals and two groups of patients with spine and joint pathologies. Body sway patterns are classified and analyzed and some new indexes are proposed for clinical diagnostics of the spine and joint pathologies. The dependencies between the body accelerations in the frontal and sagittal planes are proposed for the estimation of the resulting feedback force that provides stable vertical stance. The spectral and wavelet analyses are used for the detection of the quasi-regular and chaotic behavior in the dynamics of postural sway. It is shown that the amplitudes and power spectral densities of the center of the pressure coordinates are subjected to periodic oscillations, which are not in phase for different frequencies. The quasi-regular behavior is proper to the young healthy volunteers, while most of the patients with combined spine and joint impairments demonstrate the chaotic dynamics of the body sway.

1 Introduction

Body sway parameters are important for medical diagnostics of the musculoskeletal, neurological, visual, and balance system impairments. Steady stance of a human is provided by vestibular, visual, auditory, and tactile information relevant for the balance control. Degradation in the sensory integration is determined by age-related variations or specific diseases gradually leading to balance disorders, sudden fall, and trauma.

N. Kizilova (✉)
Department of Theoretical and Applied Mechanics, Kharkov National University, Svobody sq., 4, 61022 Kharkov, Ukraine

Interdisciplinary Centre for Mathematical and Computational Modeling, Warsaw University, ul. Krakowskie Przedmieście 26/28, 00-927 Warszawa, Polska
e-mail: n.kizilova@gmail.com

E. Karpinska • M. Karpinsky
Department of Biomechanics, M.I. Sytenko Institute of Spine and Joint Pathology, Pushkinska st., 80, 61024 Kharkov, Ukraine

J. Awrejcewicz (ed.), *Applied Non-Linear Dynamical Systems*, Springer Proceedings in Mathematics & Statistics 93, DOI 10.1007/978-3-319-08266-0_8

Computerized posturography is used for assessment, treatment, rehabilitation, and management of individuals with balance problems and implanted prosthesis. Computerized posturography is a simple and biomechanically tractable instrument for early diagnosis of the vestibular and musculoskeletal pathology. The technique assesses how well a given individual integrates muscular torque in joints and visual somatosensory and other information needed for the balance control. Computerized posturography is based on automatic measurements of the reaction forces $\left\{\vec{R}_i\right\}_{i=1}^{n}$ in n separate movable parts of the force platform or in films with continuously distributed pressure sensors. In this study the computer-assisted force platform "Statograph M006" that allows simultaneous measurement of four reaction forces (see Fig. 1a) and computes the coordinates (x_p, y_p) of the center of pressure (COP) was used. Since the body is composed of $\sim$200 bones and >700 muscles controlled by nervous, humoral, and local tissue regulation systems, the physiological cycles of different characteristic times are involved into the balance control and the trajectories $(x_p(t), y_p(t))$ are quite complex [6]. Physiological systems exhibit regular quasi-periodical or chaotic behavior. The former might be deterministic chaos or random behavior [9]. Quasi-periodicity is proper to physiological systems with complex spectra of oscillations at different scales, which is the case for human locomotor, cardiovascular, and nervous systems. Deterministic chaos is frequently observed in the physiological systems which are responding and adapting to external stimuli. For instance, for the musculoskeletal system, the visual, sensory, tactile, mechanical, and heat are the most important signals. Random behavior corresponds to uncontrolled and uncoordinated systems [9].

Some parameters of the posturographic curves and integral indexes like sway amplitudes in the frontal (0y) and sagittal (0x) planes and sway asymmetry are widely used in medicine for diagnostics of the musculoskeletal and nervous system disorders, balance control, and inner ear pathologies [8, 10]. Force platform is a useful tool for the stimulation of the coordination and balance functions in children, sportsmen, and elderly people during training or rehabilitation, because coordinated movement demands considerable practice at special conditions. Stabilography has gained widespread acceptance in the rehabilitation of thepatients with trauma, stroke, Parkinson's disease, and cerebral palsy for recovery and improvement of the locomotor function and speech after the stroke and for the development of the individual training regimes and optimal individual sport positions of biathlonists, weight lifters, figure skaters, and shooters [3–5]. Progressive decrease in the sway amplitude and body stabilization is observed during the force platform training of the sportsmen and the patients with balance impairments.

The posture stability may be maintained with respect to a moving preference point rather than a stationary one as it was suggested in [11, 12]. The sway amplitude is bigger and the sway pattern is more complicated at unstable one-leg stances and two-leg balance on the unstable or moving support. A method of decomposition of the sway into two time series, which are called rambling and trembling trajectories, can be used for checking the hypothesis [12]. The rambling component represents migration of the reference point, with respect to which the

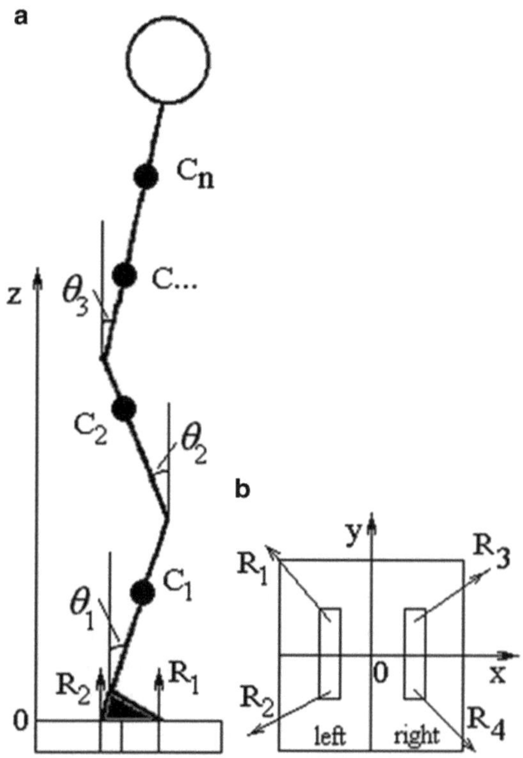

Fig. 1 An n-link model of a human upright stance on the force platform (**a**) and four measured reaction forces (**b**)

equilibrium is maintained. When the horizontal ground reaction force equals zero, the so-called instant equilibrium point is reached. Then the reference position at each time interval may be estimated by calculation of the coordinates (x_p, y_p) of COP [11]. The rambling trajectory may be obtained by spline approximation of the consecutive instant equilibrium points. The trembling component represents the oscillations of the COP relatively to the rambling trajectory. Analyses of the two sway components give new insight to understanding the balance control in normalcy and pathology. In the paper the results of the computer-assisted posturographic measurements on a control group of young healthy individuals and two groups of patients with spine and joint problems are analyzed and both chaotic and quasi-regular behaviors are detected and studied on the developed mathematical model.

2 Materials and Methods

A control group of 40 healthy subjects (weight 66.4 ± 19.7 kg, height 1.73 ± 0.14 m, and age 20–29 years) without marked neuromuscular disorders have been asked to keep 30-s quiet two-leg vertical stance (test I) on the force platform

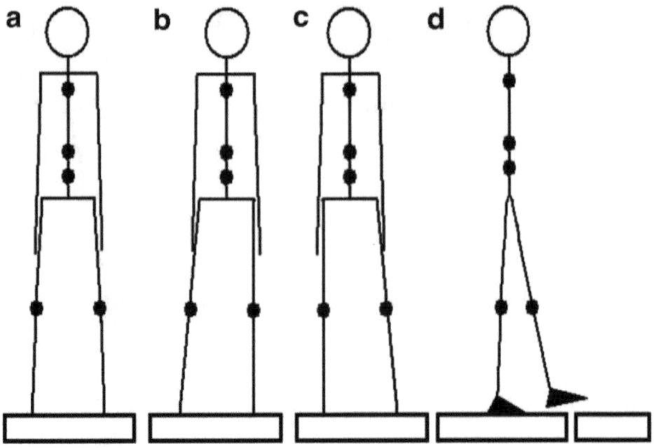

Fig. 2 Sketches of the stances studied experimentally on the force platform: symmetric two-leg stances (**a**); two-leg stances with body weight shifted onto the left (**b**) and right (**c**) foot; a step forward off from the force platform on the plate of the same thickness (**d**)

(Fig. 2a). The reaction forces $\left\{\vec{R}_i\right\}_{i=1}^{4}$ (Fig. 1b) have been measured for each foot and oscillations of COP $(x_p(t), y_p(t))$ have been automatically computed. The second set of tests (test II/III) was 40-s two-leg stance with body weight transferred onto the right/left leg (Fig. 2b, c correspondingly). The last set of tests was a step forward off the force platform on the plate of the same thickness (Fig. 2d). The healthy individuals were also asked to keep the balance standing on one (left/right, test IV) foot with 30-s rest between the tests.

A group of patients with spine and joint impairments such as osteochondrosis, arthritis, arthrosis, and trauma (weight 69.6 ± 27.5 kg, height 1.69 ± 0.11 m, and age 52–69 years) has been tested on the same series of experiments as the control group.

The calculated time series $(x_p(t), y_p(t))$ have been amplified and the low ($f < 0.01$ Hz) and high ($f > 10$ Hz) frequency components have been subtracted using the second-order Butterworth filter. Trend of the basic line has been eliminated by shifting the curves x(t) and y(t) relatively to the mean values $< x(t) >$ and $< y(t) >$. The first and the last 5-s portions of the data series have been deleted to eliminate the experimental errors [8].

3 Experimental Results and Analyses

The computed COP trajectories data have been classified in several types depending on the direction of the COP transfer. Different types of COP trajectories $(x_p(t), y_p(t))$ are presented in Fig. 3. The trajectories in the central box correspond to test I, while the trajectories in the left/right boxes correspond to the body mass shift onto the left/right leg (tests II/III). Very few individuals exhibited symmetric stance with $x_p \sim 0,\ y_p \sim 0$. In most cases a shift backward (Fig. 2b) has been observed, though

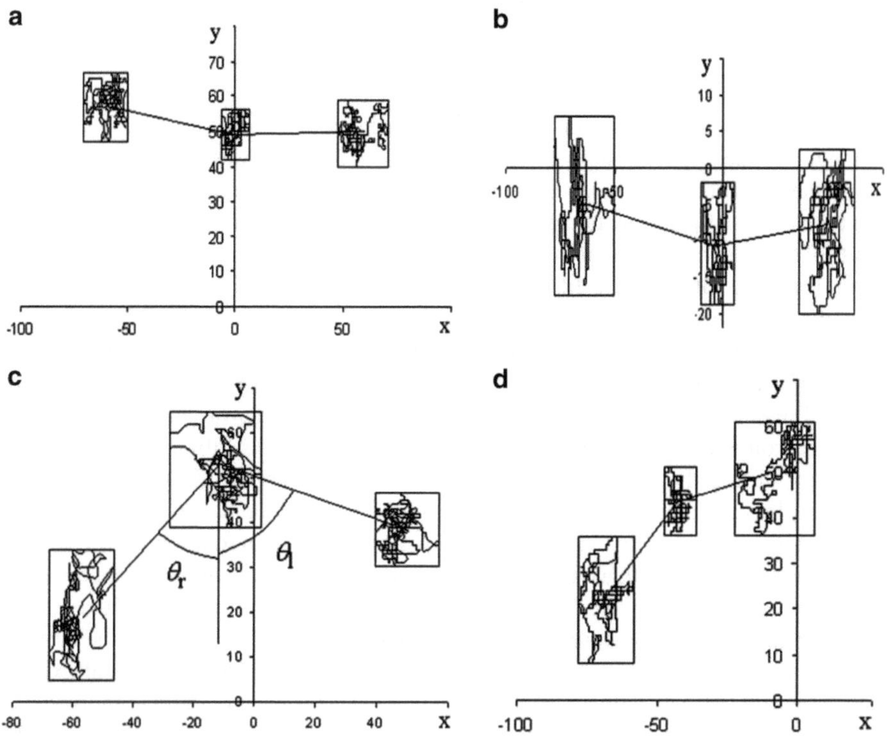

Fig. 3 Dynamic COP trajectories for the two-leg stances of healthy patients (**a**,**b**), a patient with coxartrosis (**c**) and osteochondrosis (**d**). The central box in each image allocates the normal stance, while the left and right boxes correspond to the stance with the body weight shift onto the left and right leg accordingly

some individuals demonstrated a shift forward (Fig. 3a) which is more proper to the military.

Young healthy individuals exhibited zero or small shift forward/backward while their body weight was transferred to one of the legs (Fig. 3a, b). It implies that body shift was not accompanied by body rotation or bending. The sway amplitudes were always lower during the first test; it means any two-leg stance with shifted body weight corresponds to a one-leg stance with additional support which is more unstable than the stance with full support of both feet, in spite of the total contact area remaining the same in test I and tests II and III (see Fig. 2a–c), though some exceptions have been observed (Fig. 3). In tests II and III the sway amplitudes could increase in comparison with test I in the sagittal only (A_x), frontal only (A_y), or in both (A_x and A_y) planes, either symmetrically or asymmetrically. Here A_x and A_y are the width and the height of the rectangular circumscribed around the sway trajectory (the boxes in Fig. 3). The corresponding asymmetry indexes

$$J^{II}_{x,y} = \frac{A^{II}_{x,y}}{A^{I}_{x,y}}, \qquad J^{III}_{x,y} = \frac{A^{III}_{x,y}}{A^{I}_{x,y}}$$

give valuable information on the postural balance of the individual during the change of the type of support. The ratio $J_{l/r} = J^{II}_{x,y} : J^{III}_{x,y}$ describes the left/right asymmetry of the body balance, which might be normal or pathological depending on the value of $J_{l/r}$. For instance, patients with spastic hemiparesis exhibit noticeable shift of COP toward the unaffected limb and keep this asymmetry while walking [7]. The directions of the shift differ in right- and left-sided hemispheric patients, and the changes correlate with the degree of the stance asymmetry $J_{l/r}$. More detailed diagnostic information can be obtained by gradual elimination of the visual (closed eyes [4]), auditory (soundless or high noise conditions [3]), and other kinds of control with consequent comparison of the trajectories and sway indexes.

The groups of patients with spine and joint pathology exhibited more complex behavior during the body weight transfer onto one of the legs (Fig. 3c, d). During the simple shift their bodies noticeably moved forward or backward in the same (Fig. 3c) or opposite (Fig. 3d) directions experiencing certain rotation and/or bend. It might be connected with asymmetric insufficiencies in the ligamentous apparatus or defense reaction of the body aimed at preventing the discomfort/pain in the injured joint or spine. As a measure of the disease the angles θ_r, θ_l between the straight lines connecting the centers of the corresponding rectangular and vertical axis 0y can be used [12]. Since the spine and joint pathologies are frequently combined, the differential diagnostics can be done using the indexes $J^{II,III}_{x,y}$, values, and signs of the angles θ_r and θ_l.

The trajectories $(x_p(t), y_p(t))$ are helpful for the estimation of the sway amplitudes and body shift in the frontal and sagittal planes, but more detailed information on the body balance can be obtained from the phase curves $\ddot{x}_p(\dot{x}_p)$, $\ddot{y}_p(\dot{y}_p)$ and acceleration curves $\ddot{y}_p(\ddot{x}_p)$. Since the acceleration being multiplied by the body mass gives the resulting force $\vec{R}_\Sigma$ produced by the musculoskeletal system to turn the body back to the equilibrium position, the curves $\ddot{y}_p(\ddot{x}_p)$ correspond to the feedback forces that provide a stable posture. Young healthy volunteers exhibited smooth circles around the center that is oscillating along the $\dot{x}_p$ axes (Fig. 4a). The moving centers and the cycles correspond to the gambling and trembling components of the body sway accordingly. The group of patients with spine and joint disorders mostly exhibited irregular dynamics when the cycles in the $\ddot{x}_p(\dot{x}_p)$ and $\ddot{y}_p(\dot{y}_p)$ planes become incomplete and non-smooth with rapid variations in the sway velocity and acceleration (Fig. 4b).

The curves $\ddot{y}_p(\ddot{x}_p)$ illustrate the respond $\vec{R}_\Sigma$ of the body to an arbitrary displacement $\vec{u}_P$ of COP. The simplest balance control function $\vec{R}_\Sigma = -k\vec{u}_P$ used in theoretical models has not been detected either in the control or in the two groups of patients. Instead of the restoring forces $\vec{R}_\Sigma(t) = \vec{R}_\Sigma\left(\vec{u}_P(t-\varsigma), \vec{v}_P(t-\varsigma)\right)$ where ς is the time delay, the COP acceleration, which differs from the force by the body mass multiplier, can be computed and analyzed. The young healthy volunteers

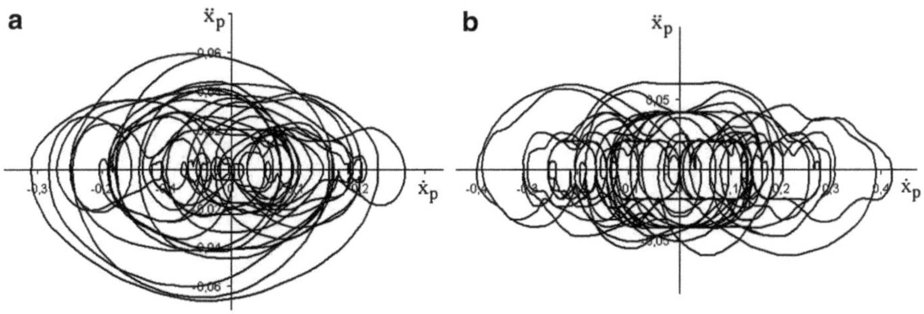

Fig. 4 Dependencies $\ddot{x}_p\left(\dot{x}_p\right)$ for a young healthy volunteer (**a**) and an elderly patient with coxarthrosis (**b**)

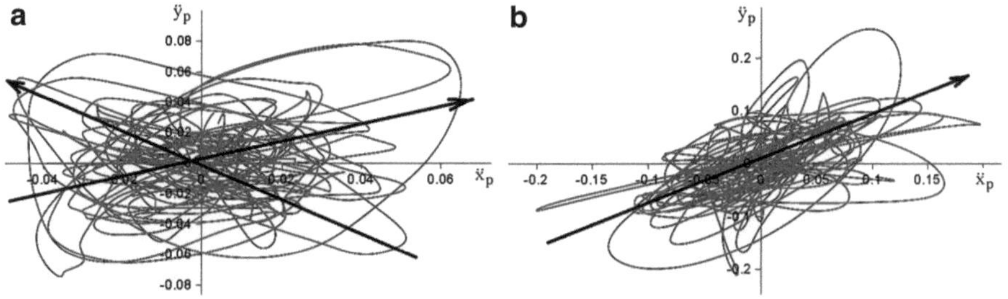

Fig. 5 Dependencies $\ddot{y}_p\left(\ddot{x}_p\right)$ for a patient with coxarthrosis of two (**a**) and one (**b**) hip joints

demonstrated uniform distribution of the accelerations in different directions, while the patients with joint problems exhibited the asymmetric $\ddot{y}_p\left(\ddot{x}_p\right)$ with one (Fig. 5b) or two (Fig. 5a) principal directions depending on the type of the disorder. Those directions depicted in Fig. 5a, b by the arrows have been computed as the most probable direction of the body deflection averaged over the time series. The balance control strategy can be based on the "scanning" of the area around the equilibrium point in different directions intentionally diverging the COP in the vicinity of the equilibrium point [2]. It means that the $\ddot{y}_p\left(\ddot{x}_p\right)$ asymmetry corresponds to the disorders when the balance control system avoids scanning some directions to avoid the pain/discomfort or exhibit the sway anisotropy.

Typical phase trajectories $\ddot{x}_p\left(\dot{x}_p\right)$ (the same for $\ddot{y}_p\left(\dot{y}_p\right)$) computed for a healthy young volunteer with two-leg stance with a body shift to one of the legs (Fig. 6a) and the one-leg stance on the same leg (Fig. 6b) differ in the values of velocities and accelerations. The stance on one leg is more unsteady than any two-leg stance and all the subjects experienced larger velocities and accelerations (forces). The maximal values of the velocities and accelerations in the sagittal and coronal planes may be easily estimated on the $\ddot{x}_p\left(\dot{x}_p\right)$ and $\ddot{y}_p\left(\dot{y}_p\right)$ curves, while it is hard to do on the COP trajectories (Fig. 3). The phase curves can be recommended for clinical diagnostics of the body balance problems.

The spectral and wavelet analyses have been carried out to detect the regular and chaotic dynamics of the body sway. The discrete short-time Fourier transform

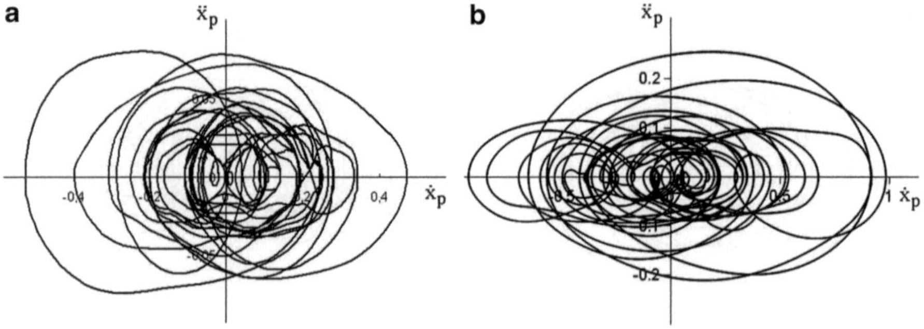

Fig. 6 Dependencies $\ddot{x}_p\left(\dot{x}_p\right)$ for a healthy subject standing on two legs with body mass transferred onto his left leg (**a**) and standing on his left leg (**b**)

(DSTFT) of the time series g_n

$$G(m, f) = \sum_{-\infty}^{+\infty} g_n W_{n-m} \exp(-i\omega n),$$

where m corresponds to the time shift and W[n − m] is the discrete windows function, allows simultaneous analysis of the amplitudes G or power spectral density (PSD) in the frequency and time domains. The Gaussian function was chosen as W[n − m]. Some results of DSTFT of the $x_p(t)$ and $y_p(t)$ time series are presented in Fig. 7.

The discrete wavelet transform (DWT)

$$\Psi(\tau, s) = \frac{1}{\sqrt{|s|}} \sum_{-\infty}^{+\infty} g_n \psi_k$$

where k = (n − m)/s, s is the scale, and ψ_k is the wavelet function.

DSTFT has a constant window size at any time and for any harmonics, and the problem of exactness of the simultaneous determination of frequency and time exists for it. DWT applied windows of different scaled sizes and thus allows study of appearance of the high and low harmonics in different times more precisely [1]. Some results of the DWT of the $x_p(t)$ and $y_p(t)$ time series are presented in Fig. 8.

It was found that 87 % of the healthy individuals exhibit quasi-periodic time oscillations of the PSD of different frequencies and the variations are slower in the frontal plane than in the sagittal plane. In the case presented in Fig. 7 two periods of the PSD oscillations in the frontal plane (Fig. 7a) and three periods in the sagittal plane (Fig. 7b) are visible at high frequencies f = 17.5 Hz. The same picture one can obtain by making slices of the function PSD(t, f) at different frequencies, though the oscillations at different frequencies are not in phase. Spectral analysis revealed

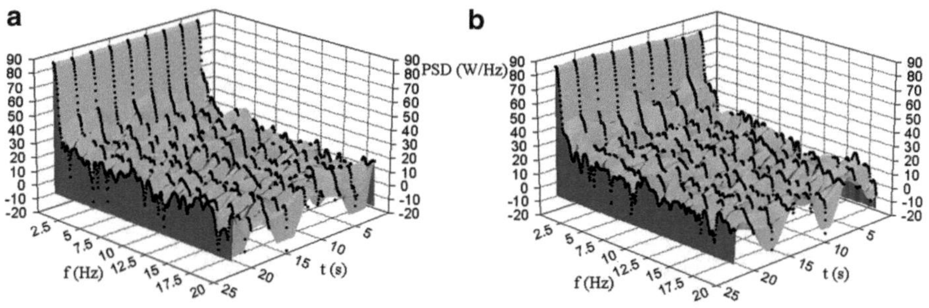

Fig. 7 Power spectral density of a healthy individual body sway in the frontal (**a**) and sagittal (**b**) planes

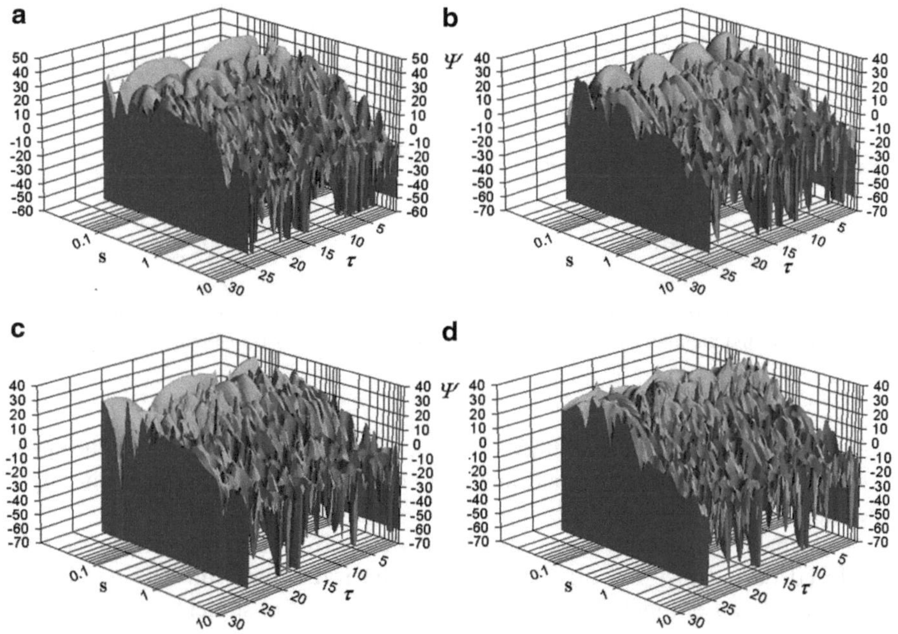

Fig. 8 Discrete wavelet transform $\Psi(\tau, s)$ of the postural sway of healthy (**a**, **b**) and sick (**c**, **d**) individuals in the frontal (**a**, **c**) and sagittal (**b**, **d**) planes

three main frequencies of the body sway [4] and DSTFT revealed the quasi-periodic behavior of the body sway, at least in the young healthy volunteers (see Fig. 7).

DWT confirms the slow variations of the amplitudes at upper scales. Two and four cycles at the basic scale (Fig. 8a, b correspondingly) are presented and the oscillations are slower in the coronary plane than in the sagittal plane. The patients with spine and joint problems demonstrate more complex chaotic dynamics with

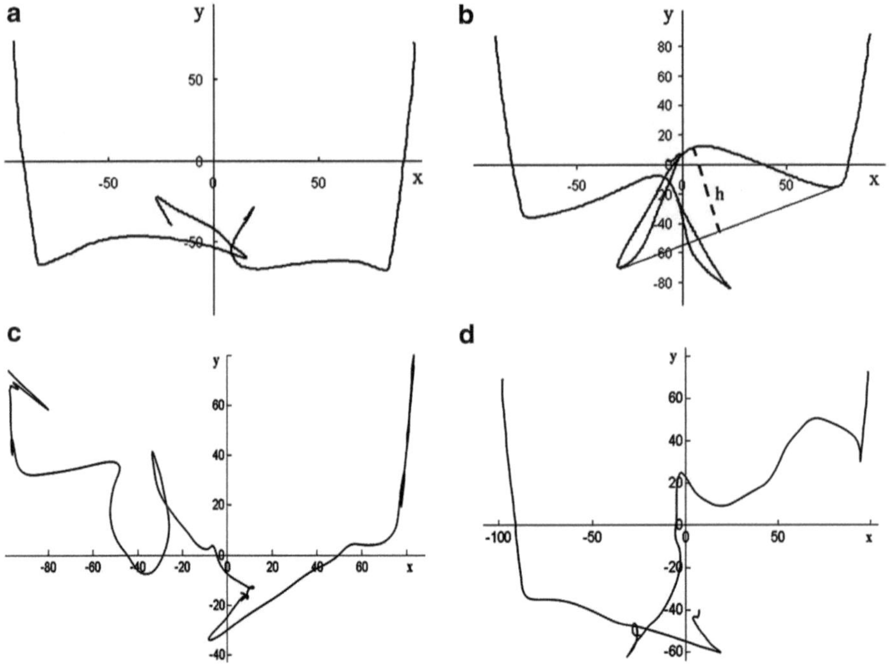

Fig. 9 COP trajectories of the step off forward from the force platform measured on the healthy individual (**a**) and patients with osteochondrosis (**b**) and coxarthrosis of the left (**c**) and right (**d**) knees

spike-type changes in the amplitudes (Fig. 8c, d) even at the basic scale. The spikes are not computational errors, and they did not disappear after iterative filtering of the signals.

The chaotic dynamics have also been detected in the trajectories of COP produced during the step off forward from the force platform. The test is very important because it includes a transition from the control strategy over the steady posture to the control over gait. In healthy individuals the trajectory starts with almost straight line connecting the COP positions for the two-leg stance and stance on the support foot followed by the body transfer forward onto another foot (Fig. 9a). The trajectories are reproducible during either short-term or long-term studies exhibiting steady personal body dynamics.

Patients with spine problems demonstrate different dynamics of the COP displacement during the step forward. First of all, they move their COP backward releasing the corresponding leg, then start to move along the straight direction to the final destination and on the almost half-way change the direction of motion toward the support foot transferring their body mass on it (Fig. 9b). The patterns of the COP trajectories are also patient-specific. The heights of the "harps" on the COP trajectories (h in Fig. 9b; the same sign on the other curve is not depicted for simplicity) and the ratio $h_r : h_l$ measured on the $y_p(x_p)$ curves for the steps forward with right and left legs can be recommended for diagnostics of the spine pathologies.

When patients with trauma or joint impairments who suffer of acute pain make a step with support on the affected leg, the COP demonstrates irregular trajectories with unpredictable acceleration and deceleration of the body in both frontal and sagittal planes (left foot in Fig. 9c and right foot in Fig. 9d). In this case the COP trajectories are irreproducible even being repeatedly measured during consequent steps in a short-term study. It implies that the control strategy is based on sudden changes in the locomotor programs depending on occasional combination of the space positions of the body segments resulting in either normal or painful reactions.

4 Conclusions

Human postural sway can exhibit quasi-regular and chaotic dynamics depending on the age, locomotor impairments, and balance system state. COP trajectories during the two-leg stances with different body weight shifts are characterized by variations of the body sway in the frontal and sagittal planes and/or variations in the sway asymmetry. The corresponding indexes together with the displacement of the COP forward or backward are helpful for the diagnostics of spine, joints, and combined pathology. Analysis of the phase curves is useful for understanding the quasi-regular and chaotic body sway dynamics and body control strategy. The quasi-regular behavior is more proper to the young healthy volunteers, while most of the patients with combined spine and joint impairments exhibited the chaotic dynamics with certain asymmetry of the body acceleration and the returning force produced by the muscle synergy. Some new indexes on the COP trajectories during the step forward off from the force platform are proposed for differential diagnostics spine and joint pathology.

References

1. Akansu, A.N., Smith, M.J.T.: Subband and Wavelet Transforms: Design and Applications. Kluwer Academic Publishers, Boston (1995)
2. Godi, M., Grasso, M., Guglielmetti, S., Nardone, A., Schieppati, M.P.: Changes of balance and gait in spastic hemiparesis are correlated with asymmetry of quiet stance. Gait Posture **28**, s6 (2008)
3. Griskevicius, J., Jarmaliene, E., Sesok, A., Daunoraviciene, K., Kizilova, N.: Evaluation of human postural balance in quiet standing by direct measurement of human body center of mass acceleration. J. Vibroeng. **11**, 556–561 (2009)
4. Karpinsky, M.Y., Kizilova, N.: Computerized posturography for data analysis and mathematical modelling of postural sway during different two-legged and one-legged human stance. J. Vibroeng. **9**, 118–124 (2007)
5. Kizilova, N., Karpinsky, M., Griskevicius, J., Daunoraviciene, K.: Posturographic study of the human body vibrations for clinical diagnostics of the spine and joint pathology. Mechanika **80**, 37–41 (2009)
6. Latash, M.L., Zatsiorsky, V.M. (eds.): Classics in Movement Sciences. Human Kinetics, Champaign (2001)
7. Mochizuki, L., Duarte, M., Amadio, A.C., Zatsiorsky, V.M., Latash, M.L.: Changes in postural sway and its fractions in conditions of postural instability different postural control mechanisms. J. Appl. Biomech. **22**, 51–60 (2006)

8. Norr, M.E., Forrez, G.: Posture testing (posturography) in the diagnosis of peripheral vestibular pathology. Arch. Otorhinolaryngol. **243**, 186–189 (1986)
9. Sharma, V.: Deterministic chaos and fractal complexity in the dynamics of cardiovascular behavior: perspectives on a new Frontier. Open Cardiovasc. Med. J. **3**, 110–123 (2009)
10. Sologubov, E.G., Yavorskii, A.B., Kobrin, V.I., Nemkova, S.A., Sinel'nikova, A.N.: Use of computer stabilography and computer-assisted biomechanical examination of gait for diagnosis of posture and movement disorders in patients with various forms of infantile cerebral paralysis. Biomed. Eng. **34**, 138–143 (2000)
11. Zatsiorsky, V.M., Duarte, M.: Instant equilibrium point and its migration in standing tasks: rambling and trembling components of the stabilogram. Motor Control **3**, 28–38 (1999)
12. Zatsiorsky, V.M., King, D.L.: An algorithm for determining gravity line location from posturographic recordings. J. Biomech. **31**, 161–164 (1998)

Modelling of Ropes with Consideration of Large Deformations and Friction by Means of the Rigid Finite Element Method

Iwona Adamiec–Wójcik, Jan Awrejcewicz, Lucyna Brzozowska, and Łukasz Drąg

Abstract This paper presents a model of a rope, which is an example of a slender system. The method used for modelling is the modified rigid finite element method, which enables us to consider not only bending flexibility but also longitudinal flexibility. A new approach presented in the paper is validated by comparison of our own results with an analytical solution for a catenary line. For this problem the influence of both the number of elements into which the link is discretised and various values of stiffness coefficient is analysed. Numerical simulations of the dynamics of a rope of an offshore crane lifting a load from a vessel are presented.

1 Introduction

Slender systems such as lines, ropes and pipes are very often used in offshore engineering applications. The analysis of these systems requires the use of methods enabling large deformations to be considered. Methods of discretising such systems have been developed over many years, yet the most popular is the finite element method [4, 10, 19]. Application of the finite segment method to modelling of planar systems is discussed in [18], while [9] presents the lumped mass method used for modelling spatial risers. A method containing features of both the lumped mass and finite segment methods is the rigid finite element method (RFEM) [15]. This enables slender links to be discretised by dividing them into rigid finite elements which assume the mass (inertial) characteristics of the link and massless, dimensionless spring-damping elements which reflect the flexible characteristics. The method is also used to analyse various systems in offshore engineering [16]. In the classical

I. Adamiec–Wójcik (✉) • L. Brzozowska • Ł. Drąg
University of Bielsko–Biała, ul. Willowa 2, 43-309 Bielsko–Biała, Poland
e-mail: i.adamiec@ath.bielsko.pl; lbrzozowska@ath.bielsko.pl; ldrag@ath.bielsko.pl

J. Awrejcewicz
Lodz University of Technology, ul. Stefanowskiego 1/15, 90-924 Lodz, Poland

Warsaw University of Techonology, ul. Narbutta 84, 02-524 Warsaw, Poland
e-mail: awrejcew@p.lodz.pl

J. Awrejcewicz (ed.), *Applied Non-Linear Dynamical Systems*, Springer Proceedings in Mathematics & Statistics 93, DOI 10.1007/978-3-319-08266-0_9

formulation of RFEM each rigid finite element (rfe) has six DOFs for a spatial model and three DOFs for a planar model. A modified formulation [1, 12, 17] limits considerations to the analysis of bending and torsional vibrations for the spatial systems and bending in one direction for the planar systems.

In modelling of cranes it is often assumed that the rope is nondeformable [5, 14] or that the rope is replaced by a substitute longitudinal stiffness [6–8]. In the analysis of planar vibrations of a rope fixed at one end the authors of [3] consider bending flexibility. However, wave phenomena in the rope can have an essential influence on the displacements of the load when considering ropes with large lengths. A complex model of the rope is necessary in order to enable us to consider such phenomena. This paper presents a planar model of the rope for the analysis of both bending and longitudinal vibrations.

2 Formulation of the Model

This section presents the formulation of the modified rigid finite element method (MRFEM) for planar slender systems in which both bending and longitudinal flexibilities are considered, without the need to connect rfes by means of constraint equations. Such an approach eliminates constraint reactions and reduces the size of the final system to $2(n + 2)$ equations.

Figure 1 shows a flexible link divided into $n + 1$ elements.

It is assumed that the motion is described by x_0, y_0, coordinates of point A_0; and Δ_i, φ_i, generalised coordinates of rfe $i(i = 0, 1, \ldots, n)$.

The number of elements of the vector of generalised coordinates is defined by the formula

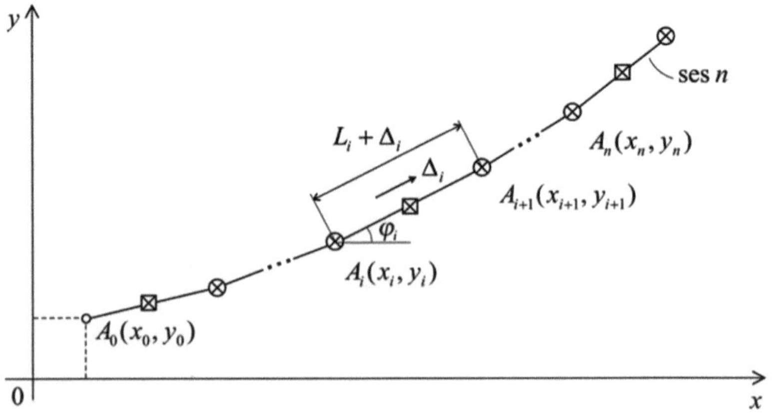

Fig. 1 Division of a flexible link into rfes and sdes

$$m = 2 + 2\,(n + 1) = 2n + 4. \tag{1}$$

Coordinates of point A_i can be calculated in the following way:

$$x_i = x_0 + \sum_{j=0}^{i-1} \left(l_j + \Delta_j\right) \cos\varphi_j, \quad y_i = y_0 + \sum_{j=0}^{i-1} \left(l_j + \Delta_j\right) \sin\varphi_j, \tag{2}$$

where l_j are lengths of rfes preceding rfe i and Δ_j, φ_j are defined in Fig. 1.

2.1 Kinetic Energy of a Link

The kinetic energy of rfe i can be expressed as follows:

$$E_i = \frac{1}{2} m_i^{(1)} \left[v_{c,i}^{(1)}\right]^2 + \frac{1}{2} J_{c,i}^{(1)} \dot{\varphi}_i^2 + \frac{1}{2} m_i^{(2)} \left[v_{c,i}^{(2)}\right]^2 + \frac{1}{2} J_{c,i}^{(2)}, \dot{\varphi}_i^2 \tag{3}$$

where $m_i^{(j)}$ is the mass of part (j) of rfe i, $v_{c,i}^{(j)}$ is the velocity of the centre of mass of part (j) of rfe i, and $J_{c,i}^{(j)}$ is the central inertial moment of part (j) of rfe i, $j = 1, 2$ (see Fig. 2).

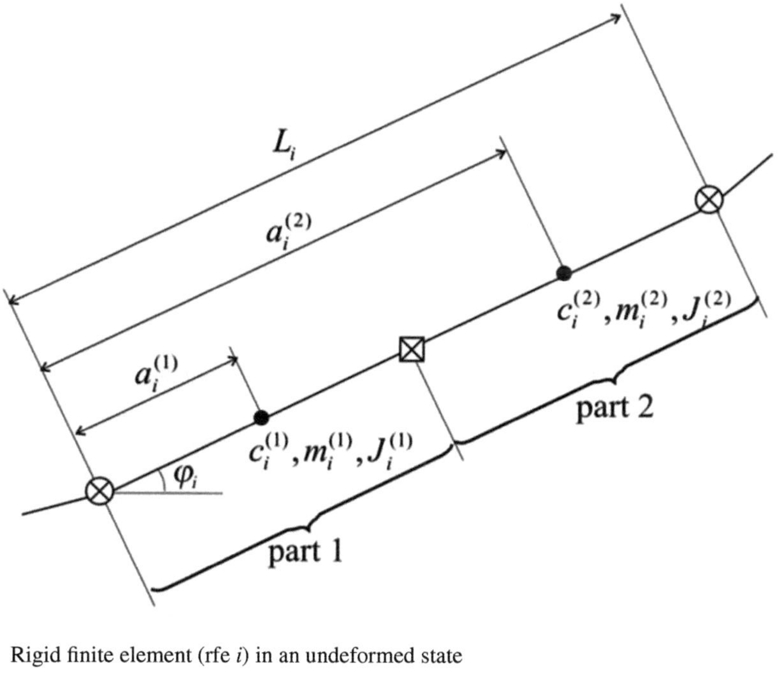

Fig. 2 Rigid finite element (rfe i) in an undeformed state

Coordinates of the centre of mass of rfe i can be defined by means of the formulae

$$x_{c,i}^{(1)} = x_i + a_i^{(1)} \cos\varphi_i, \quad y_{c,i}^{(1)} = y_i + a_i^{(1)} \sin\varphi_i \tag{4}$$

and

$$x_{c,i}^{(2)} = x_i + \left(a_i^{(2)} + \Delta_i\right) \cos\varphi_j, \quad y_{c,i}^{(2)} = y_i + \left(a_i^{(2)} + \Delta_i\right) \sin\varphi_i. \tag{5}$$

Differentiation of (4) and (5) yields

$$\begin{aligned}
\left[v_{c,i}^{(1)}\right]^2 &= \left[\dot{x}_{c,i}^{(1)}\right]^2 + \left[\dot{y}_{c,i}^{(1)}\right]^2 = \\
&= \dot{x}_i^2 + \dot{y}_i^2 + \left[a_i^{(1)}\right]^2 \dot{\varphi}_i^2 - 2a_i^{(1)}\dot{\varphi}_i\left[\dot{x}_i \mathrm{s}\varphi_i - \dot{y}_i \mathrm{c}\varphi_i\right], \\
\left[v_{c,i}^{(2)}\right]^2 &= \left[\dot{x}_{c,i}^{(2)}\right]^2 + \left[\dot{y}_{c,i}^{(2)}\right]^2 = \\
&= \dot{x}_i^2 + \dot{y}_i^2 + \dot{\Delta}_i^2 + \left(a_i^{(2)} + \Delta_i\right)^2 \dot{\varphi}_i^2 + 2\dot{\Delta}_i\left(\dot{x}_i \mathrm{c}\varphi_i + \dot{y}_i \mathrm{s}\varphi_i\right) - \\
&\quad -2\left(a_i^{(2)} + \Delta_i\right)\dot{\varphi}_i\left(\dot{x}_i \mathrm{s}\varphi_i - \dot{y}_i \mathrm{c}\varphi_i\right),
\end{aligned} \tag{6}$$

where $\mathrm{s}\varphi_i = \sin\varphi_i$, $\mathrm{c}\varphi_i = \cos\varphi_i$.

Substituting (6) into (3), the kinetic energy of rfe i can be obtained:

$$\begin{aligned}
E_i &= \tfrac{1}{2}m_i\dot{x}_i^2 + \tfrac{1}{2}m_i\dot{y}_i^2 + \tfrac{1}{2}m_i^{(2)}\dot{\Delta}_i^2 + \tfrac{1}{2}m_i^{(2)}\left(a_i^{(2)} + \Delta_i\right)^2\dot{\varphi}_i^2 + \tfrac{1}{2}J_i\dot{\varphi}_i^2 \\
&\quad + \left[-\dot{x}_i \mathrm{s}\varphi_\mathrm{i} + \dot{y}_i \mathrm{c}\varphi_\mathrm{i}\right]\dot{\varphi}_i\left[\alpha_i + m_i^{(2)}\Delta_i\right] + m_i^{(2)}\dot{\Delta}_i\left[\dot{x}_i \mathrm{c}\varphi_\mathrm{i} + \dot{y}_i \mathrm{s}\varphi_\mathrm{i}\right],
\end{aligned} \tag{7}$$

where $m_i = m_i^{(1)} + m_i^{(2)}$, $J_i = m_i^{(1)}[a_i^{(1)}]^2 + J_i^{(1)} + J_i^{(2)}$, $\alpha_i = m_i^{(1)}a_i^{(1)} + m_i^{(2)}a_i^{(2)}$.

Having differentiated (2), the following is calculated:

$$\begin{aligned}
\dot{x}_i &= \dot{x}_0 + \sum_{j=0}^{i-1}\dot{\Delta}_j \mathrm{c}\varphi_j - \sum_{j=0}^{i-1}\left(l_j + \Delta_j\right)\dot{\varphi}_j \mathrm{s}\varphi_j, \\
\dot{y}_i &= \dot{y}_0 + \sum_{j=0}^{i-1}\dot{\Delta}_j \mathrm{s}\varphi_j + \sum_{j=0}^{i-1}\left(l_j + \Delta_j\right)\dot{\varphi}_j \mathrm{c}\varphi_j
\end{aligned} \tag{8}$$

and

$$\begin{aligned} \ddot{x}_i &= \ddot{x}_0 + \sum_{j=0}^{i-1} \ddot{\Delta}_j c\varphi_j - \sum_{j=0}^{i-1} \left(l_j + \Delta_j\right) \ddot{\varphi}_j s\varphi_j - \\ &\quad - 2\sum_{j=0}^{i-1} \dot{\Delta}_j \dot{\varphi}_j s\varphi_j - \sum_{j=0}^{i-1} \left(l_j + \Delta_j\right) \dot{\varphi}_j^2 c\varphi_j, \\ \ddot{y}_i &= \ddot{y}_0 + \sum_{j=0}^{i-1} \ddot{\Delta}_j s\varphi_j + \sum_{j=0}^{i-1} \left(l_j + \Delta_j\right) \ddot{\varphi}_j c\varphi_j + \\ &\quad + 2\sum_{j=0}^{i-1} \dot{\Delta}_j \dot{\varphi}_j c\varphi_j - \sum_{j=0}^{i-1} \left(l_j + \Delta_j\right) \dot{\varphi}_j^2 s\varphi_j. \end{aligned} \tag{9}$$

In view of (8) the kinetic energy of rfe i depends on $x_0, y_0, \Delta_0, \varphi_0, \ldots, \Delta_i, \varphi_i$ and their derivatives.

The Lagrange operators for kinetic energy E_i of the rfe can be written in the following form:

$$\begin{aligned} \varepsilon_{x_0}(E_i) &= \frac{d}{dt}\frac{\partial E_i}{\partial \dot{x}_0} - \frac{\partial E_i}{\partial x_0} = m_i \ddot{x}_0 + \sum_{j=0}^{i} m_{ij} \ddot{\Delta}_j c\varphi_j - \\ &\quad - \sum_{j=0}^{i} n_{ij} \ddot{\varphi}_j s\varphi_j - 2\sum_{j=0}^{i} m_{ij} \dot{\Delta}_j \dot{\varphi}_j s\varphi_j - \sum_{j=0}^{i-1} n_{ij} \dot{\varphi}_j^2 c\varphi_j, \\ \varepsilon_{y_0}(E_i) &= \frac{d}{dt}\frac{\partial E_i}{\partial \dot{y}_0} - \frac{\partial E_i}{\partial y_0} = m_i \ddot{y}_0 + \sum_{j=0}^{i} m_{ij} \ddot{\Delta}_j s\varphi_j + \\ &\quad + \sum_{j=0}^{i} n_{ij} \ddot{\varphi}_j c\varphi_j + 2\sum_{j=0}^{i} m_{ij} \dot{\Delta}_j \dot{\varphi}_j c\varphi_j - \sum_{j=0}^{i} n_{ij} \dot{\varphi}_j^2 s\varphi_j \end{aligned} \tag{10}$$

for $j = 1, \ldots, i{-}1$:

$$\begin{aligned} \varepsilon_{\Delta_j}(E_i) &= \frac{d}{dt}\frac{\partial E_i}{\partial \dot{\Delta}_j} - \frac{\partial E_i}{\partial \Delta_j} = m_i \ddot{x}_0 c\varphi_j + m_i \ddot{y}_0 s\varphi_j + \\ &\quad + \sum_{k=0}^{i} m_{ik} \ddot{\Delta}_k \cos\left(\varphi_j - \varphi_k\right) + \sum_{k=0}^{i} n_{ik} \ddot{\varphi}_k \sin\left(\varphi_j - \varphi_k\right) + \\ &\quad + 2\sum_{k=0}^{i} m_{ik} \dot{\Delta}_k \dot{\varphi}_k \sin\left(\varphi_j - \varphi_k\right) - \sum_{k=0}^{i} n_{ik} \dot{\varphi}_k^2 \cos\left(\varphi_j - \varphi_k\right), \\ \varepsilon_{\varphi_j}(E_i) &= \frac{d}{dt}\frac{\partial E_i}{\partial \dot{\varphi}_j} - \frac{\partial E_i}{\partial \varphi_j} = \left(l_j + \Delta_j\right) \Bigg[- m_i \ddot{x}_0 s\varphi_j + m_i \ddot{y}_0 c\varphi_j - \\ &\quad - \sum_{k=0}^{i} m_{ik} \ddot{\Delta}_k \sin\left(\varphi_j - \varphi_k\right) + \sum_{k=0}^{i} n_{ik} \ddot{\varphi}_k \cos\left(\varphi_j - \varphi_k\right) + \\ &\quad + 2\sum_{k=0}^{i} m_{ik} \dot{\Delta}_k \dot{\varphi}_k \cos\left(\varphi_j - \varphi_k\right) + \sum_{k=0}^{i} n_{ik} \dot{\varphi}_k^2 \sin\left(\varphi_j - \varphi_k\right) \Bigg], \end{aligned}$$

$$m_{ik}=\begin{cases} m_i & \text{for } k<i \\ m_i^{(2)} & \text{for } k=i \end{cases}, \quad n_{ik}=\begin{cases} m_i\left(l_k+\Delta_k\right) & \text{for } k<i \\ \left.\alpha_i+m_i^{(2)}\Delta_k\right) & \text{for } k=i \end{cases},$$

$$\begin{aligned} \varepsilon_{\Delta_i}\left(E_i\right)=\frac{d}{dt}\frac{\partial E_i}{\partial\dot{\Delta}_i}-\frac{\partial E_i}{\partial\Delta_i}=m_i^{(2)}\Bigg[&\ddot{x}_0 c\varphi_i+\ddot{y}_0 s\varphi_i+ \\ &+\sum_{k=0}^{i}\ddot{\Delta}_k\cos\left(\varphi_i-\varphi_k\right)+\sum_{k=0}^{i}\left(l_k+\Delta_k\right)\ddot{\varphi}_k\sin\left(\varphi_i-\varphi_k\right)+ \\ &+2\sum_{k=0}^{i}\dot{\Delta}_k\dot{\varphi}_k\sin\left(\varphi_i-\varphi_k\right)-\sum_{k=0}^{i}\left(l_{ik}+\Delta_i\right)\dot{\varphi}_k^2\cos\left(\varphi_i-\varphi_k\right)\Bigg], \end{aligned} \tag{11}$$

$$\begin{aligned} \varepsilon_{\varphi_i}\left(E_i\right)=\frac{d}{dt}\frac{\partial E_i}{\partial\dot{\varphi}_i}-\frac{\partial E_i}{\partial\varphi_i}=&-\left(\alpha_i+m_i^{(1)}\Delta_i\right)\left(\ddot{x}_0 s\varphi_i-\ddot{y}_0 c\varphi_i\right)- \\ &-\sum_{k=0}^{i}\alpha_{ik}\ddot{\Delta}_k\sin\left(\varphi_i-\varphi_k\right)+\sum_{k=0}^{i}\beta_{ik}\ddot{\varphi}_k\cos\left(\varphi_i-\varphi_k\right)+ \\ &+2\sum_{k=0}^{i}\alpha_{ik}\dot{\Delta}_k\dot{\varphi}_k\cos\left(\varphi_i-\varphi_k\right)+\sum_{k=0}^{i}\beta_{ik}\dot{\varphi}_k^2\sin\left(\varphi_i-\varphi_k\right), \end{aligned}$$

$$l_{ik}=\begin{cases} l_k+\Delta_k & \text{for } k<i \\ a_i^{(2)} & \text{for } k=i \end{cases}, \quad \alpha_{ik}=\begin{cases} \alpha_i+m_i\Delta_i & \text{for } k<i \\ m_i^2\left(a_i^{(2)}+\Delta_j\right) & \text{for } k=i \end{cases},$$

$$\beta_{ik}=\begin{cases} \left(\alpha_i+m_i^{(2)}\Delta_i\right)\left(l_k+\Delta_k\right) & \text{for } k<i \\ m_i^{(2)}\left(a_k^{(2)}+\Delta_i\right)^2+J_i & \text{for } k=i \end{cases}.$$

Formulae (11) can be recast to the following compact form for $j = 1, \ldots, i–1, i$:

$$\begin{aligned} \varepsilon_{\Delta_j}\left(E_i\right)=&\,m_{ij}\ddot{x}_0 c\varphi_j+m_{ij}\ddot{y}_0 s\varphi_j+\sum_{k=0}^{i}m_{ijk}\ddot{\Delta}_k\cos\left(\varphi_j-\varphi_k\right)+ \\ &+\sum_{k=0}^{i}n_{ijk}\ddot{\varphi}_k\sin\left(\varphi_j-\varphi_k\right)+2\sum_{k=0}^{i}m_{ijk}\dot{\Delta}_k\dot{\varphi}_k\sin\left(\varphi_j-\varphi_k\right)- \\ &-\sum_{k=0}^{i}n_{ijk}\dot{\varphi}_k^2\cos\left(\varphi_j-\varphi_k\right), \\ \varepsilon_{\varphi_j}\left(E_i\right)=&-n_{ij}\ddot{x}_0 s\varphi_j+n_{ij}\ddot{y}_0 c\varphi_j-\sum_{k=0}^{i}\alpha_{ijk}\ddot{\Delta}_k\sin\left(\varphi_j-\varphi_k\right)+ \\ &+\sum_{k=0}^{i}\beta_{ijk}\ddot{\varphi}_k\cos\left(\varphi_j-\varphi_k\right)+2\sum_{k=0}^{i}\alpha_{jik}\dot{\Delta}_k\dot{\varphi}_k\cos\left(\varphi_j-\varphi_k\right)+ \\ &+\sum_{k=0}^{i}\beta_{ijk}\dot{\varphi}_k^2\sin\left(\varphi_j-\varphi_k\right), \end{aligned} \tag{12}$$

$$m_{ijk} = \begin{cases} m_{ik} & \text{for } j < i \\ m_i^{(2)} & \text{for } j = i \end{cases}, \quad n_{ijk} = \begin{cases} n_{ik} & \text{for } j < i \\ m_i^{(2)}\left(l_{ik} + \Delta_i\right) & \text{for } j = i \end{cases},$$

$$\alpha_{ijk} = \begin{cases} \left(l_j + \Delta_j\right) m_{ik} & \text{for } j < i \\ \alpha_{ik} & \text{for } j = i \end{cases}, \quad \beta_{ijk} = \begin{cases} \left(l_j + \Delta_j\right) n_{ik} & \text{for } j < i \\ \beta_{ik} & \text{for } j = i \end{cases}.$$

Since the kinetic energy of the link is a sum of the kinetic energies of rfes $0, \ldots, n$, i.e.:

$$E = \sum_{i=0}^{n} E_i, \tag{13}$$

and bearing in mind that $E_i = E_i(x_0, y_0, \Delta_0, \varphi_0, \ldots, \Delta_i, \varphi_i)$, one gets

$$\varepsilon_{x_0}(E) = \sum_{i=0}^{n} \varepsilon_{x_0}(E_i), \qquad \varepsilon_{y_0}(E) = \sum_{i=0}^{n} \varepsilon_{y_0}(E_i) \tag{14}$$

and for $j = 0, 1, \ldots, n$:

$$\varepsilon_{\Delta_j}(E) = \sum_{i=j}^{n} \varepsilon_{\Delta_j}(E_i), \qquad \varepsilon_{\varphi_j}(E) = \sum_{i=j}^{n} \varepsilon_{\varphi_j}(E_i). \tag{15}$$

Taking into account

$$\begin{aligned} &\sum_{i=0}^{n}\sum_{k=0}^{i} A_{ik} a_k = \sum_{k=0}^{n} a_k \sum_{i=k}^{n} A_{ik}, \\ &\sum_{i=j}^{n}\sum_{k=0}^{i} A_{ijk} a_k b_{jk} = \sum_{k=0}^{n} a_k b_{jk} \sum_{i=\max\{j,k\}}^{n} A_{ijk}, \end{aligned} \tag{16}$$

the following relations are obtained:

$$\begin{aligned} \varepsilon_{x_0}(E) &= m\ddot{x}_0 + \sum_{k=0}^{n} \ddot{\Delta}_k A_k c\varphi_k - \sum_{k=0}^{n} \ddot{\varphi}_k B_k s\varphi_k - \\ &\quad - 2\sum_{k=0}^{n} \dot{\Delta}_k \dot{\varphi}_k A_k s\varphi_k - \sum_{k=0}^{n} \dot{\varphi}_k^2 B_k c\varphi_k, \\ \varepsilon_{y_0}(E) &= m\ddot{y}_0 + \sum_{k=0}^{n} \ddot{\Delta}_k A_k s\varphi_k + \sum_{k=0}^{n} \ddot{\varphi}_k B_k c\varphi_k + \\ &\quad + 2\sum_{k=0}^{n} \dot{\Delta}_k \dot{\varphi}_k A_k c\varphi_k - \sum_{k=0}^{n} \dot{\varphi}_k^2 B_k s\varphi_k, \end{aligned} \tag{17}$$

where $A_k = \sum_{i=k}^{n} m_{ik}$, $B_k = \sum_{i=k}^{n} n_{ik}$, $m = \sum_{i=0}^{n} m_i$ and for $j = 0, 1, \ldots, n$:

$$\begin{aligned}
\varepsilon_{\Delta_j}(E) &= \ddot{x}_0 A_j c\varphi_j + \ddot{y}_0 A_j s\varphi_j + \sum_{k=0}^{n} \ddot{\Delta}_k a_{jk}^{\Delta} \cos(\varphi_j - \varphi_k) + \\
&+ \sum_{k=0}^{n} \ddot{\varphi}_k b_{jk}^{\Delta} \sin(\varphi_j - \varphi_k) + 2\sum_{k=0}^{n} \dot{\Delta}_k \dot{\varphi}_k a_{jk}^{\Delta} \sin(\varphi_j - \varphi_k) - \\
&- \sum_{k=0}^{n} \dot{\varphi}_k^2 b_{jk}^{\Delta} \cos(\varphi_j - \varphi_k), \\
\varepsilon_{\varphi_j}(E) &= -\ddot{x}_0 B_j s\varphi_j + \ddot{y}_0 B_j c\varphi_j - \sum_{k=0}^{n} \ddot{\Delta}_k a_{jk}^{\varphi} \sin(\varphi_j - \varphi_k) + \\
&+ \sum_{k=0}^{n} \ddot{\varphi}_k b_{jk}^{\varphi} \cos(\varphi_j - \varphi_k) + 2\sum_{k=0}^{n} \dot{\Delta}_k \dot{\varphi}_k a_{jk}^{\varphi} \cos(\varphi_j - \varphi_k) - \\
&- \sum_{k=0}^{n} \dot{\varphi}_k^2 b_{jk}^{\varphi} \sin(\varphi_j - \varphi_k),
\end{aligned} \tag{18}$$

where $a_{jk}^{\Delta} = \sum_{i=\max\{j,k\}}^{n} m_{ijk}$, $b_{jk}^{\Delta} = \sum_{i=\max\{j,k\}}^{n} n_{ijk}$, $a_{jk}^{\varphi} = \sum_{i=\max\{j,k\}}^{n} \alpha_{ijk}$, $b_{jk}^{\varphi} = \sum_{i=\max\{j,k\}}^{n} \beta_{ijk}$.

Let us introduce the following notations:

$$\mathbf{q} = \begin{bmatrix} \mathbf{r} \\ \mathbf{q}_0 \\ \mathbf{q}_1 \\ \\ \mathbf{q}_i \\ \\ \mathbf{q}_n \end{bmatrix}, \quad \mathbf{r} = \begin{bmatrix} x_0 \\ y_0 \end{bmatrix}, \quad \mathbf{q}_j = \begin{bmatrix} \Delta_j \\ \varphi_j \end{bmatrix}, \tag{19}$$

where $\mathbf{q}$ is a vector of generalised coordinates of the link, and

$$\boldsymbol{\varepsilon}_{\mathbf{q}}(E) = \begin{bmatrix} \boldsymbol{\varepsilon}_{\mathbf{r}}(E) \\ \boldsymbol{\varepsilon}_0(E) \\ \boldsymbol{\varepsilon}_1(E) \\ \vdots \\ \boldsymbol{\varepsilon}_i(E) \\ \vdots \\ \boldsymbol{\varepsilon}_n(E) \end{bmatrix}, \quad \boldsymbol{\varepsilon}_{\mathbf{r}}(E) = \begin{bmatrix} \varepsilon_{x_0}(E) \\ \varepsilon_{y_0}(E) \end{bmatrix}, \quad \boldsymbol{\varepsilon}_j(E) = \begin{bmatrix} \varepsilon_{\Delta_j}(E) \\ \varepsilon_{\varphi_j}(E) \end{bmatrix}. \tag{20}$$

Finally, the following second-order ODEs are obtained

$$\boldsymbol{\varepsilon}_{\mathbf{q}}(E) = \mathbf{M}\ddot{\mathbf{q}} + \mathbf{h}, \tag{21}$$

where

$$\mathbf{M} = \begin{bmatrix} \mathbf{M}_{\mathbf{rr}} & \mathbf{M}_{\mathbf{r}0} & \dots & \mathbf{M}_{\mathbf{r}j} & \dots & \mathbf{M}_{\mathbf{r}n} \\ \mathbf{M}_{0\mathbf{r}} & \mathbf{M}_{00} & \dots & \mathbf{M}_{0j} & \dots & \mathbf{M}_{0n} \\ \vdots & \vdots & & & & \vdots \\ \mathbf{M}_{j\mathbf{r}} & \mathbf{M}_{j0} & \dots & \mathbf{M}_{jj} & \dots & \mathbf{M}_{jn} \\ \vdots & \vdots & & \vdots & & \vdots \\ \mathbf{M}_{n\mathbf{r}} & \mathbf{M}_{n0} & \dots & \mathbf{M}_{nj} & \dots & \mathbf{M}_{nn} \end{bmatrix}, \quad \mathbf{h} = \begin{bmatrix} \mathbf{h}_{\mathbf{r}} \\ \mathbf{h}_0 \\ \vdots \\ \mathbf{h}_j \\ \vdots \\ \mathbf{h}_n \end{bmatrix},$$

$$\mathbf{M}_{\mathbf{rr}} = \begin{bmatrix} m & 0 \\ 0 & m \end{bmatrix}, \quad \mathbf{M}_{\mathbf{r}k} = \mathbf{M}_{k\mathbf{r}}^T = \begin{bmatrix} A_k \mathrm{c}\varphi_k & -B_k \mathrm{s}\varphi_k \\ A_k \mathrm{s}\varphi_k & B_k \mathrm{c}\varphi_k \end{bmatrix}$$

$$\mathbf{M}_{jk} = \begin{bmatrix} a_{jk}^{\Delta} \cos(\varphi_j - \varphi_k) & b_{jk}^{\Delta} \sin(\varphi_j - \varphi_k) \\ -a_{jk}^{\varphi} \sin(\varphi_j - \varphi_k) & b_{jk}^{\phi} \cos(\varphi_j - \varphi_k) \end{bmatrix},$$

$$\mathbf{h}_{\mathbf{r}} = \begin{bmatrix} -2\sum_{k=0}^{n} \dot{\Delta}_k \dot{\varphi}_k A_k \mathrm{s}\varphi_k - \sum_{k=0}^{n} \dot{\varphi}_k^2 B_k \mathrm{c}\varphi_k \\ 2\sum_{k=0}^{n} \dot{\Delta}_k \dot{\varphi}_k A_k \mathrm{c}\varphi_k - \sum_{k=0}^{n} \dot{\varphi}_k^2 B_k \mathrm{s}\varphi_k \end{bmatrix},$$

$$\mathbf{h}_j = \begin{bmatrix} 2\sum_{k=0}^{n} \dot{\Delta}_k \dot{\varphi}_k a_{jk}^{\Delta} \sin(\varphi_j - \varphi_k) - \sum_{k=0}^{n} \dot{\varphi}_k^2 b_{jk}^{\Delta} \cos(\varphi_j - \varphi_k) \\ 2\sum_{k=0}^{n} \dot{\Delta}_k \dot{\varphi}_k a_{jk}^{\varphi} \cos(\varphi_j - \varphi_k) - \sum_{k=0}^{n} \dot{\varphi}_k^2 b_{jk}^{\varphi} \sin(\varphi_j - \varphi_k) \end{bmatrix}.$$

Elements of matrix **M** depend on the generalised coordinates **q** of the link, while the elements of vector **h** depend on **q** and $\dot{\mathbf{q}}$. It should be noted that if the longitudinal flexibility is left out of consideration ($\Delta_j \equiv 0$, for $j = 0, 1, \ldots, n$), then the formulae for elements of matrix **M** and vector **h** (having erased respective rows and columns) coincide with those presented in [17].

2.2 Potential Energy of the Link

The potential energy of gravity forces of rfe i can be presented in the form

$$V_i^g = m_i^{(1)} g y_{c,i}^{(1)} + m_i^{(2)} g y_{c,i}^{(2)}, \tag{22}$$

where g stands for the gravity acceleration.

Bearing in mind (4), (5) and (2), one finds that

$$\begin{aligned} \frac{\partial V_i^g}{\partial \mathbf{r}} &= \mathbf{G}_{i\mathbf{r}} = \begin{bmatrix} 0 \\ m_i g \end{bmatrix}, \\ \frac{\partial V_i^g}{\partial \mathbf{q}_j} &= \mathbf{G}_{ij} = \begin{cases} \begin{bmatrix} m_i g s\varphi_j \\ m_i g \left(l_j + \Delta_j\right) c\varphi_j \end{bmatrix} & \text{for } j < i \\ \begin{bmatrix} m_i^{(2)} g s\varphi_i \\ g \left[m_i^{(1)} a_i^{(1)} + m_i^{(2)} \left(a_i^{(2)} + \Delta_i \right) c\varphi_i \right] \end{bmatrix} & \text{for } j = i. \end{cases} \end{aligned} \tag{23}$$

Since the potential energy of the link is a sum of the energies of all rfes

$$V^g = \sum_{i=0}^{n} V_i^g, \tag{24}$$

then using the following relations

$$\frac{\partial V^g}{\partial \mathbf{r}} = \sum_{i=0}^{n} \frac{\partial V_i^g}{\partial \mathbf{r}}, \qquad \frac{\partial V^g}{\partial \mathbf{q}_j} = \sum_{i=j}^{n} \frac{\partial V_i^g}{\partial \mathbf{q}_j}, \tag{25}$$

one obtains

$$\frac{\partial V^g}{\partial \mathbf{q}} = \mathbf{G} = \begin{bmatrix} \mathbf{G}_\mathbf{r} \\ \mathbf{G}_0 \\ \vdots \\ \mathbf{G}_j \\ \vdots \\ \mathbf{G}_n \end{bmatrix}, \tag{26}$$

where $\mathbf{G_r} = \begin{bmatrix} 0 \\ mg \end{bmatrix}$, $\mathbf{G}_j = \sum_{i=j}^{n} \mathbf{G}_{ij}$ for $j = 0, 1, \ldots, n$.

The energy of spring deformation of the link in the case of linear characteristics of sdes can be written in the following form:

$$V^s = \frac{1}{2}\sum_{i=1}^{n} c_i^b (\varphi_i - \varphi_{i-1})^2 + \frac{1}{2}\sum_{i=0}^{n} c_i^l \Delta_i^2. \tag{27}$$

Thus, the following is obtained:

$$\frac{\partial V^s}{\partial \mathbf{r}} = 0, \qquad \frac{\partial V^s}{\partial \mathbf{q}_j} = \mathbf{S}_j, \tag{28}$$

where $\mathbf{S}_0 = \begin{bmatrix} c_0^l \Delta_0 \\ -c_0^b (\varphi_1 - \varphi_0) \end{bmatrix}$, $\mathbf{S}_j = \begin{bmatrix} c_j^l \Delta_j \\ c_j^b (\varphi_j - \varphi_{j-1}) - c_{j+1}^b (\varphi_{j+1} - \varphi_j) \end{bmatrix}$,

for $j = 0, 1, \ldots, n-1$, $\mathbf{S}_n = \begin{bmatrix} c_n^l \Delta_n \\ c_n^b (\varphi_n - \varphi_{n-1}) \end{bmatrix}$.

The above relations can be written in the equivalent compact form as follows:

$$\frac{\partial V^s}{\partial \mathbf{q}} = \mathbf{S} = \begin{bmatrix} 0 \\ \mathbf{S}_0 \\ \vdots \\ \mathbf{S}_j \\ \vdots \\ \mathbf{S}_n \end{bmatrix}. \tag{29}$$

The motion of the link discretised in the way described above can be caused by either external forces (force input) or by definition of the motion of point A_0 (kinematic input). In the next section, both approaches are presented.

2.3 Synthesis of the Equations

In the case of a force input the equations of motion follow:

$$\mathbf{M}\ddot{\mathbf{q}} = \mathbf{f}, \tag{30}$$

where $\mathbf{M}$ is defined in (21) and

$$\mathbf{f} = -\mathbf{h} - \frac{\partial V^g}{\partial \mathbf{q}} - \frac{\partial V^s}{\partial \mathbf{q}} + \mathbf{Q}.$$

Here $\mathbf{Q}$ denotes the vector of generalised forces resulting from the external forces.

On the other hand, in the case of kinematic input, the following relations hold:

$$\begin{cases} x_0 = x_0(t), & \dot{x}_0 = \dot{x}_0(t), & \ddot{x}_0 = \ddot{x}_0(t), \\ y_0 = y_0(t), & \dot{y}_0 = \dot{y}_0(t), & \ddot{y}_0 = \ddot{y}_0(t), \end{cases} \tag{31}$$

and thus the components of the vector

$$\mathbf{r} = \mathbf{r}(t) \tag{32}$$

and its derivatives with respect to time are known. Then equations (30) can be cast to the following form:

$$\begin{aligned} \mathbf{M}_{\mathbf{rr}}\ddot{\mathbf{r}} + \sum_{k=0}^{n} \mathbf{M}_{\mathbf{r}k}\ddot{\mathbf{q}}_k &= \mathbf{f}_\mathbf{r} = -\mathbf{h}_\mathbf{r} - \tfrac{\partial V^g}{\partial \mathbf{r}} + \mathbf{R}_0, \\ \mathbf{M}_{j\mathbf{r}}\ddot{\mathbf{r}} + \sum_{k=0}^{n} \mathbf{M}_{jk}\ddot{\mathbf{q}}_k &= \mathbf{f}_j = -\mathbf{h}_j - \tfrac{\partial V^g}{\partial \ddot{\mathbf{q}}_j} - \tfrac{\partial V^s}{\partial \ddot{\mathbf{q}}_j}, \quad j = 0, 1, \ldots, n, \end{aligned} \tag{33}$$

where $\mathbf{R}_0 = \begin{bmatrix} \mathbf{R}_0^x \\ \mathbf{R}_0^y \end{bmatrix}$ are the components of an unknown force acting at point A_0 ensuring the realisation of (31) and (32).

Since vector $\ddot{\mathbf{r}}$ is known in the case of kinematic input, the second part of (33) can be written in the form

$$\sum_{k=0}^{n} \mathbf{M}_{jk}\ddot{\mathbf{q}}_k = \mathbf{f}_k - \mathbf{M}_{j\mathbf{r}}\ddot{\mathbf{r}} = \mathbf{P}_k \tag{34}$$

or equivalently

$$\overline{\mathbf{M}}\ddot{\overline{\mathbf{q}}} = \mathbf{P}, \tag{35}$$

where $\overline{\mathbf{M}} = \begin{bmatrix} \mathbf{M}_{00} \ldots \mathbf{M}_{0j} \ldots \mathbf{M}_{0n} \\ \vdots \quad\quad \vdots \quad\quad \vdots \\ \mathbf{M}_{j0} \ldots \mathbf{M}_{jj} \ldots \mathbf{M}_{jn} \\ \vdots \quad\quad \vdots \quad\quad \vdots \\ \mathbf{M}_{n0} \ldots \mathbf{M}_{nj} \ldots \mathbf{M}_{nn} \end{bmatrix}, \quad \overline{\mathbf{q}} = \begin{bmatrix} \mathbf{q}_0 \\ \vdots \\ \mathbf{q}_j \\ \vdots \\ \mathbf{q}_n \end{bmatrix}, \quad \mathbf{P} = \begin{bmatrix} \mathbf{P}_0 \\ \vdots \\ \mathbf{P}_j \\ \vdots \\ \mathbf{P}_n \end{bmatrix}.$

Having calculated $\ddot{\mathbf{q}}$ from (35), the components of reaction vector $\mathbf{R}_0$ can be calculated from the first part of (33):

$$\mathbf{R}_0 = \mathbf{M}_{\mathbf{rr}}\ddot{\mathbf{r}} + \sum_{k=0}^{n} \mathbf{M}_{\mathbf{r}k}\ddot{\mathbf{q}}_k + \mathbf{h}_{\mathbf{r}} + \frac{\partial V^g}{\partial \mathbf{r}}. \tag{36}$$

Thus, in the case of kinematic input, the calculation of the acceleration vector requires the solution of the system of $2(n+1)$ linear algebraic equations. In the case of force input the number of unknowns in system (30) is larger by 2 and is equal to $2(n+1)+2$.

3 Validation of the Model

If in equations (33) we omit components connected with accelerations and velocities, we obtain a system of nonlinear algebraic equations enabling the static problem to be solved. In order to verify the algorithms and computer programs elaborated, calculations of the shape of the line fixed and loaded as in Fig. 3 are carried out.

Coordinates of point P from the analytical solution of the problem, when bending flexibility is omitted, are calculated according to the formulae [9]:

$$\begin{aligned} x(s) &= \frac{F_x^E s}{EA} + \frac{F_x^E}{w}\left[\operatorname{arcsinh}\frac{F_y^E\text{-wL+ws}}{F_x^E} - \operatorname{arcsinh}\frac{F_y^E\text{-wL}}{F_x^E}\right], \\ y(s) &= \frac{s}{EA}\left(F_y^E - wL + \tfrac{1}{2}ws\right) + \\ &+ \frac{F_x^E}{w}\left\{\left[1+\left(\frac{F_y^E-wL+ws}{F_x^E}\right)^2\right]^{\frac{1}{2}} - \left[1+\left(\frac{F_y^E-wL}{F_x^E}\right)^2\right]^{\frac{1}{2}}\right\}, \end{aligned} \tag{37}$$

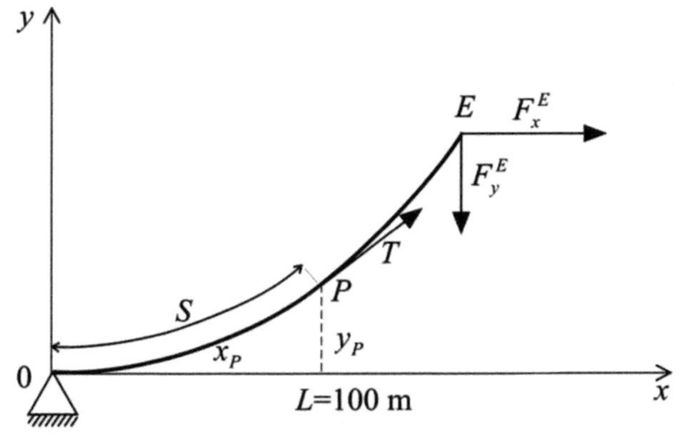

Fig. 3 Catenary line

and force T takes the form

$$T(s) = \left[\left(F_x^E\right)^2 + \left(F_y^E - wL + ws\right)^2\right]^{\frac{1}{2}}, \tag{38}$$

where F_x^E, and F_y^E are vertical and horizontal components of the force acting on the line at point E, E is the Young modulus of the line, $w = \rho \cdot A \cdot g$ is the weight of the length unit of the line, L is the length of the line and A is the area of the cross section.

In order to calculate the shape of the catenary line shown in Fig. 3, we use discretisation by means of the MRFEM presented above. Forces F_x^E and F_y^E are introduced into the equations of motion (30) as the generalised forces. The coordinates of point E are as follows:

$$x_E = x_0 + \sum_{i=0}^{n} \left(l_i + \Delta_i\right) c\varphi_i, \qquad y_E = y_0 + \sum_{i=0}^{n} \left(l_i + \Delta_i\right) s\varphi_i. \tag{39}$$

Therefore, we get

$$Q_{x_0}\left(F_x^E, F_y^E\right) = F_x^E, \qquad Q_{y_0}\left(F_x^E, F_y^E\right) = F_y^E \tag{40}$$

and

$$\begin{aligned} Q_{\Delta_i}\left(F_x^E, F_y^E\right) &= F_x^E c\varphi_i + F_y^E s\varphi_i, \\ Q_{\varphi_i}\left(F_x^E, F_y^E\right) &= -F_x^E \left(l_i + \Delta_i\right) s\varphi_i + F_y^E \left(l_i + \Delta_i\right) c\varphi_i. \end{aligned} \tag{41}$$

In order to define the accuracy of the method, calculations have been carried out for a varying number n of elements into which the line with the parameters $L = 100$ m, $A = 0.25\pi D^2$, $D = 0.035$ m, $E = 10^{11}$ N/m^2, $\rho = 2{,}000$ kg/m^3, $F_x^E = 1{,}000$ N, $F_y^E = 2{,}000$ N has been divided.

Table 1 presents the comparison of results obtained according to formulae (37) and the MRFEM when it has been assumed that

$$C_i^b = 0, \quad \text{for } i = 1, \ldots, n, \tag{42}$$

which means that bending flexibility is omitted. Quantity $e\,\%$ is defined by the formula

$$e\% = \left|\frac{w_a - w_m}{w_a}\right| \cdot 100\%, \tag{43}$$

where w_a is calculated analytically and w_m is calculated using MRFEM.

Table 1 Comparison of displacements and axial force of the catenary line

x_P

		s = 25 m		s = 50 m		s = 75 m		s = 100 m	
n		Value	e %	Value	e %	Value	e %	Value	e %
Analytical solution (37), (38)	x_P	23.478	–	42.828	–	58.175	–	70.540	–
	y_P	8.046	–	23.743	–	43.436	–	65.149	–
	T	1,158.2	–	1,454.5	–	1,826.2	–	2,236.1	–
10	x_P	23.493	0.366	42.760	0.158	58.181	0.101	70.541	0.014
	y_P	8.070	3.323	23.869	0.530	43.462	0.265	65.216	0.116
	T	1,158.2	0	1,454.5	0	1,826.2	0	2,236.1	0
20	x_P	23.460	0.075	42.812	0.038	58.163	0.022	70.541	0.003
	y_P	8.090	0.559	23.751	0.136	43.462	0.059	65.166	0.029
	T	1,158.2	0	1,454.5	0	1,826.2	0	2,236.1	0
40	x_P	23.473	0.018	42.826	0.009	58.172	0.005	70.540	0.001
	y_P	8.049	0.138	23.751	0.034	43.443	0.015	65.154	0.007
	T	1,158.2	0	1,454.5	0	1,826.2	0	2,236.1	0
80	x_P	23.477	0.005	42.827	0.002	58.175	0.001	70.540	0
	y_P	8.049	0.035	26.745	0.008	43.438	0.004	65.150	0.002
	T	1,158.2	0	1,454.5	0	1,826.2	0	2,236.1	0
160	x_P	23.478	0.001	42.828	0.001	58.175	0	70.540	0
	y_P	8.047	0.009	23.743	0.002	43.437	0.001	65.149	0
	T	1,158.2	0	1,454.5	0	1,826.2	0	2,236.1	0

The analysis of results from Table 1 allows to conclude that for the MRFEM an exactness of 0.1 % is obtained even for the division of the line into $n = 40$ elements. It is characteristic, however, that the smaller the parameter s, the larger the error.

Figure 4 presents the shape of the catenary according to the analytical solution with the points representing values obtained by the MRFEM for $n = 20$. Very good compatibility of the results can be observed.

Numerical simulations have also been carried out in the case when the stiffness coefficient of bending is defined according to the following formula:

$$C_i^b = \frac{EI}{d}, \quad \text{for } i = 1, 2, \ldots, n; \tag{44}$$

where I is inertial moment of the cross-sectional area of the rope and $d = L/n$.

The calculations have been carried out for $n = 400$, which means that the line has been divided into segments of 0.25 m in the primary division.

Table 2 presents the results obtained when formula (42) has been used so that the bending flexibility of the line has been omitted (line W), with the consideration of the bending stiffness using formulae (44) (line P). For the second case it has been additionally assumed that $\varphi_0 \approx 0$ by introduction of the generalised force:

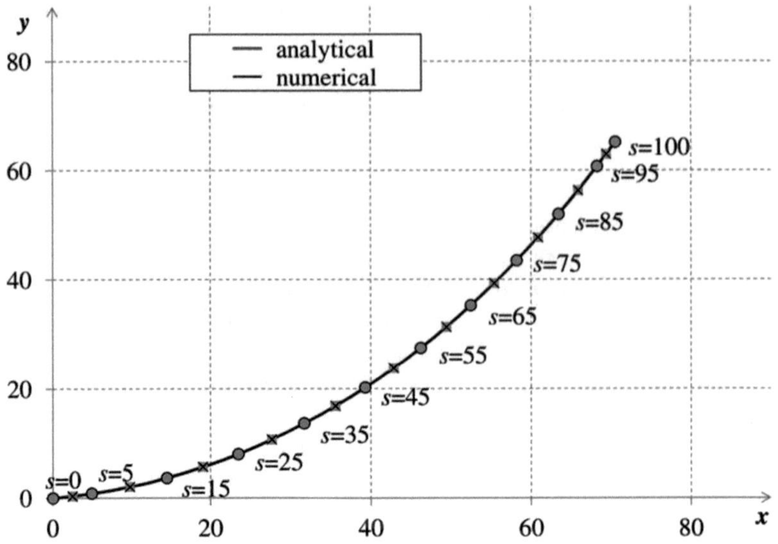

Fig. 4 Catenary line - comparison of results

Table 2 Influence of bending stiffness

	x_P			y_P			T		
s [m]	$C_i^b = 0$ line W	$C_i^b = \frac{EI}{d}$ line P	e %	$C_i^b = 0$ line W	$C_i^b = \frac{EI}{d}$ line P	e %	$C_i^b = 0$ line W	$C_i^b = \frac{EI}{d}$ line P	e %
0.25	0.248	0.250	0.63	0.028	0.005	87.2	1,006.8	1,002.8	0.40
0.50	0.497	0.500	0.64	0.058	0.013	77.0	1,007.4	1,004.0	0.33
1.00	0.993	0.999	0.65	0.121	0.039	67.8	1,008.6	1,005.2	0.23
2.00	1.983	1.996	0.64	0.260	0.121	53.3	1,011.2	1,010.1	0.11
4.00	3.955	3.976	0.57	0.593	0.388	34.6	1,017.5	1,017.2	0.08
8.00	7.856	7.889	0.42	1.473	1.219	17.3	1,034.1	1,034.1	0
16.00	15.429	15.468	0.26	4.035	3.759	6.8	1,082.4	1,082.4	0
32.00	29.349	29.346	0.18	11.858	11.561	2.5	1,230.1	1,230.1	0
64.00	51.851	51.922	0.14	34.439	34.121	0.9	1,656.4	1,656.4	0
100.00	70.540	70.632	0.13	65.149	64.820	0.5	2,236.1	2,236.1	0

$$Q_{\varphi_0} = -C_i^b \varphi_0. \tag{45}$$

Percentage error $e\%$ has been calculated as

$$e\% = \left| \frac{w_P - w_W}{w_P} \right| \cdot 100\%, \tag{46}$$

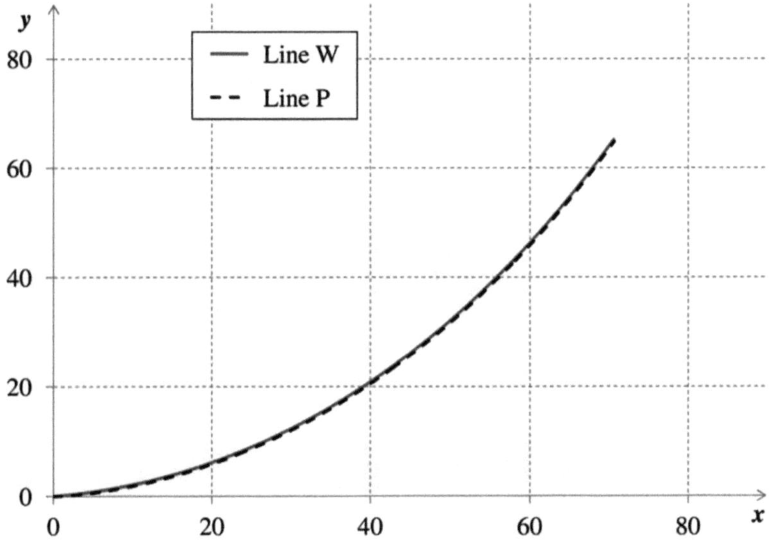

Fig. 5 The shape of slender line W ($C_i^b = 0$) and flexible line P ($C_i^b = \frac{EI}{d}$)

where $w_P \in \{x_P, y_P, T\}$ is calculated for the rope with the assumption $C_i^b = \frac{EI}{d}$ and $w_W \in \{x_P, y_P, T\}$ is calculated for the rope with the assumption $C_i^b = 0$.

Results presented in Table 2 show that models W and P give considerably different results for coordinate y_P when coordinate s is small, while for $s > 1/2\,L$, those differences decrease to 1 %. Part of the graph for $s \leq 16$ m showing the shapes of lines W and P is presented in Fig. 5.

From the results presented above, it can be concluded that the consideration of bending flexibility for long ropes is not necessary. It may be of importance in the case of detailed analysis.

4 Numerical Example

One of the potential applications of the models presented is the analysis of failures, which may occur during the work of cranes mounted on drilling platforms (Fig. 6).

The motion of the vessel due to the sea waves or floating away despite disconnecting the load from the crane can cause slippage of the load from the deck. As a result this can cause large loading of the rope system considerably larger than nominal loading. A simplified model of such a system is presented in Fig. 7. The system consists of:

(i) A rope with length L_r mounted to a fixed point A_0, where the origin of the inertial coordinate system A_0xy is placed

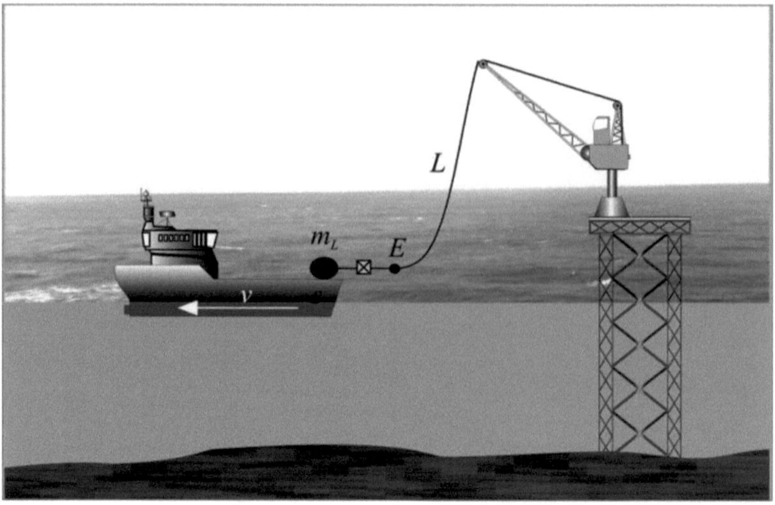

Fig. 6 Crane lifting a load from a supply vessel

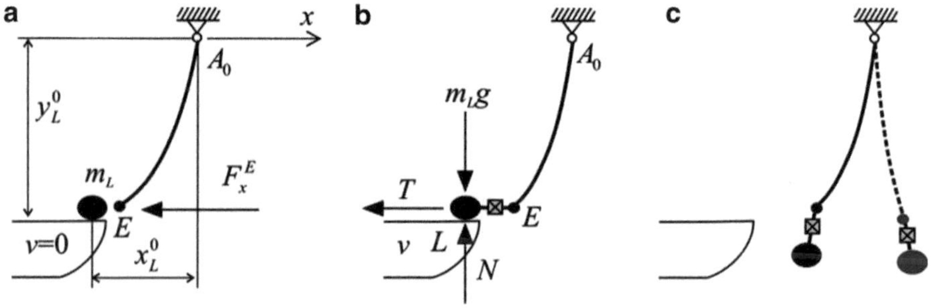

Fig. 7 Phases of motion of the load: (**a**) static problem; (**b**) movement of the load with the vessel; (**c**) free vibrations of the rope with the load

(ii) A sling with length L_n connecting end E of the rope with load L, treated as a massless spring-damping element (sde)
(iii) A load with mass m_L treated as a lumped mass

For numerical analysis three essential phases are distinguished.

Phase 1 In the situation presented in Fig. 7a the initially vertical rope is displaced under the influence of horizontal force F_x^E, so that coordinate x_E of the end of the rope achieves given value x_L^0. In this phase mass m_L is motionless, and the load is not connected with end E of the rope. As a result, the value y_L^0 is obtained, assuming that

$$y_L^0 = |y_E|_{x_E=x_L^0}. \tag{47}$$

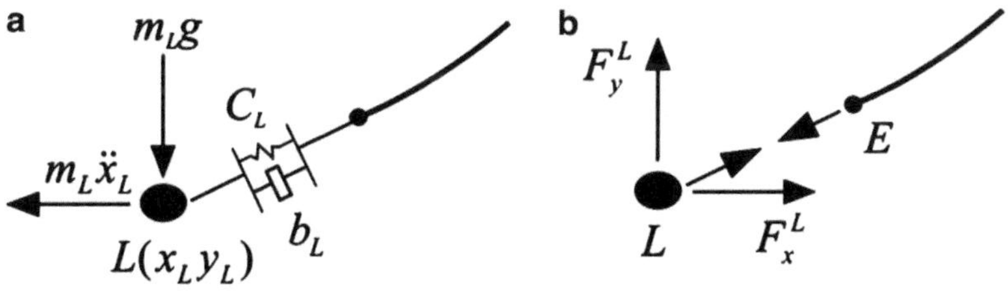

Fig. 8 Force transferred by the sling: (**a**) inertial force; (**b**) force F_y^L

The task is solved as a static problem by the solution of a system of nonlinear algebraic equations.

Phase 2 A sling connects the load with end E of the rope. Mass m_L moves together with the vessel (Fig. 7b). This phase lasts until one of the following conditions is satisfied:

$$T \geq \mu N \text{ (case P1)}, \quad N \leq 0 \text{ (case P2)}. \tag{48}$$

In the first case (P1) the load and the rope are subjected to free vibrations as a result of the friction slip. In the second case (P2) the phase of free vibrations starts, when the vertical contact of the load to the vessel is lost. Forces N and T can be calculated from the relations (Fig. 8):

$$N = m_L g - F_y^L, \qquad T = m_L \dot{v}_L - F_x^L, \tag{49}$$

where $\dot{v}_L = \frac{dv}{dt}$ is the acceleration of the vessel and F_x^L and F_y^L are components of the force in the sling.

Phase 3 At this stage (Fig. 7c) the load and rope connected by the sling are subjected to free vibrations caused by the initial tension of the system due to the forces of gravity.

Force F^L and its components F_x^L, and F_y^L can be calculated in the following way:

$$F^L = c_L \Delta_{EL} + b_L \dot{\Delta}_{EL}, \quad F_x^L = F^L \frac{x_E - x_L}{|EL|}, \quad F_y^L = F^L \frac{y_E - y_L}{|EL|}, \tag{50}$$

where $\Delta_{EL} = |EL| - L_n$, $|EL| = \left[(x_E - x_L)^2 + (y_E - y_L)^2\right]^{\frac{1}{2}}$; x_E, y_E are defined in (39); x_L, y_L are coordinates of the load; c_L, b_L are stiffness and damping coefficients of the sling; $\dot{\Delta}_{EL} = \frac{d|EL|}{dt}$.

During the second phase, coordinates x_L, y_L and $\dot{x}_L$, $\dot{y}_L$ can be calculated from the motion of the vessel, while during the third phase these are the generalised coordinates of the load. Hence, the equations of motion take the form

Table 3 Numerical parameters of the system

Notation	Description	Value	Unit
L	Length of the rope	80	m
D	Diameter of the rope	0.04	m
E	The Young modulus	$1 \cdot 10^{11}$	N/m^2
ρ	Density	6,500	kg/m^3
m_L	Mass of the load	5,000	kg
x_L^0	Horizontal displacement	$2D$	m
v	Velocity of the vessel $t \le 20 t > 20$	$-1/4$ t-5	m/s m/s
c_L	Stiffness coefficient of the sling	10^6	N/m
b_L	Damping coefficient of the sling	$2 \cdot 10^2$	N·s/m

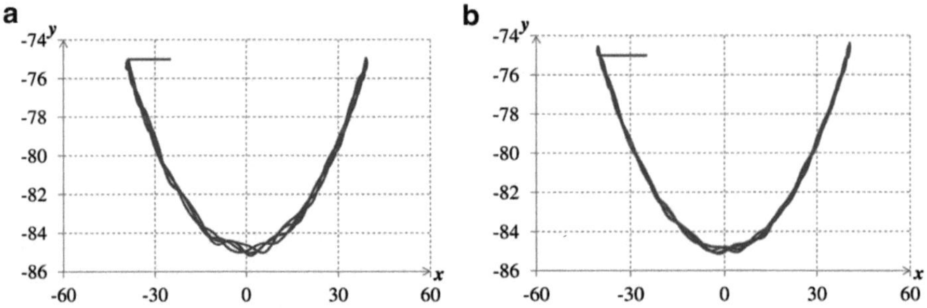

Fig. 9 Trajectory of the load: (**a**) P1, (**b**) P2

$$m_L \ddot{x}_L = F_x^L, \qquad m_L \ddot{y}_L = F_y^L. \tag{51}$$

The equations of motion presented earlier have to be modified in order to take into account force F^L. Numerical parameters of the system analysed are reported in Table 3.

Calculations in both cases, P1 and P2, have been carried out for the division of the rope into $n = 40$ elements. The equations of motion have been integrated by the Runge–Kutta method of the fourth order with the constant integration step $2 \cdot 10^{-4}$ s. Trajectories of mass m_L in both cases are shown in Fig. 9.

The degree of rope loading during the motion of the load is well represented by the dynamic coefficient as presented in [8]:

$$\eta(b) = \frac{F^L}{[m_L + w(L - s)]\, g}, \tag{52}$$

where F^L is the force in the rope, w is the unit mass and s is the coordinate of the position of the point.

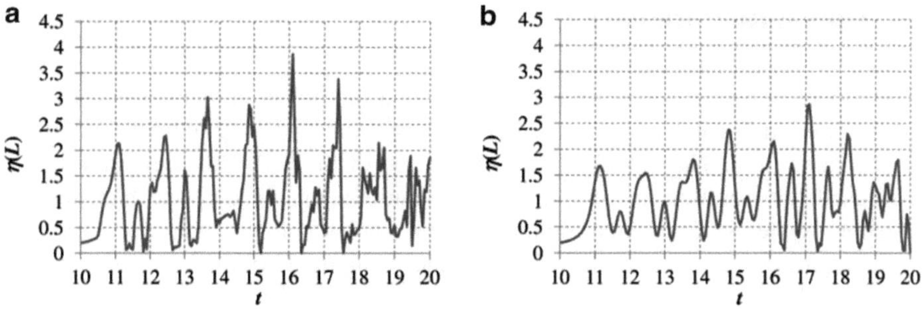

Fig. 10 Time history of function $\eta(L)$: (**a**) case P1, (**b**) case P2

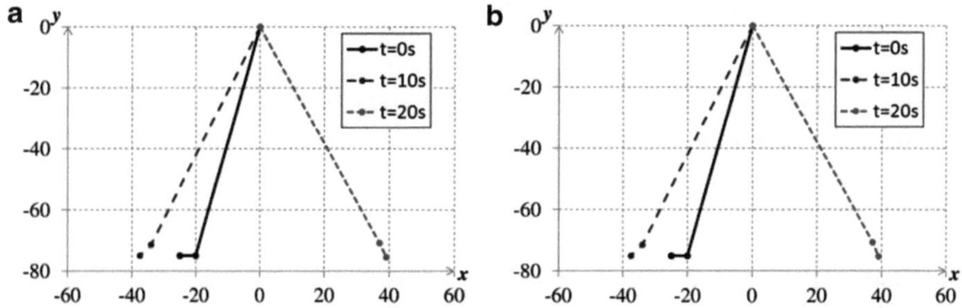

Fig. 11 Position of the system at different time instants: (**a**) P1, (**b**) P2

This coefficient relates the actual value of the force in the rope to the nominal static load. Table 4 gives maximal values of coefficient η for $s = 0$ (for point A_0) and for $s = L$ (point E). For both cases simulation time is $T = 50$ s.

Table 4 Maximal values of dynamic coefficient η

Case	$\eta(0)$	$\eta(L)$
P1	3.860	3.715
P2	2.861	2.813

The values of coefficient η are larger for case P1. The reason is that when the friction force exceeds μN, the vertical component of force F^L is still small, and thus, when the load loses contact to the deck of the vessel, a significant decrease in mass m_L occurs. In case P2 the force F_y^E is balanced by the gravity force of the load. Figure 10 presents fragments of time histories of function η for both cases.

Figure 11 shows the system at the end of phases 1 and 2, and at time instant $t = 20$ s in phase 3.

5 Concluding Remarks

The model of discretisation of slender links is a modification of models presented up to now and enables us to consider both bending and longitudinal flexibilities at the same time. The model is a generalisation of the MRFEM presented in [17] and used for vibration analysis of cranes, manipulators, linkage mechanisms and offshore systems. Computer implementation enabled us to carry out validation by comparing our own results with an exact analytical solution for the catenary line. Good compatibility of results has been achieved. The model is then used for vibration analysis in the case of sudden loss of contact of the load lifted from the vessel deck by a crane or A–frame.

A disadvantage of the method is a full mass matrix, which makes it impossible to use quick algorithms for solutions of linear algebraic equations with sparse matrices. The formulation presented in [2] is without this disadvantage.

It is important to note that in the modified formulation of the RFEM it is easy to take into account physical nonlinearities, which is essential in the case of pipe-laying process [13]. In such a case the procedure described in detail in [11, 12, 16] can be used.

Acknowledgments This work has been supported by National Centre of Science under the grant MAESTRO 2, No. 2012/04/A/ST8/00738 for the years 2012–2015.

References

1. Adamiec–Wójcik, I.: Modelling Dynamics of Multibody Systems, Use of Homogenous Transformations and Joint Coordinates. Lambert Academic Publishing, Köln (2009)
2. Adamiec–Wójcik, I., Wittbrodt, E., Wojciech, S.: Rigid finite element method in modelling of bending and longitudinal vibrations of ropes. Int. J. Appl. Mech. Eng. **12**(3), 665–676 (2012)
3. Fritzkowski, P., Kamiński, H.: A discrete model of a rope with bending stiffness or viscous damping. Acta Mech. Sinica **27**(1), 108–113 (2011)
4. Kane, T.R., Levinson, D.A.: Dynamics: Theory and Applications. McGraw Hill, New York (1985)
5. Kłosiński, J.: Swing–free stop control of the slewing motion of a mobile crane. Control Eng. Pract. **13**(4), 451–460 (2005)
6. Maczyński, A.: Positioning and Stabilisation of Load of Jib Cranes. Publishing Office of the University of Bielsko–Biała, Bielsko–Biała (2005) (in Polish)
7. Maczyński, A., Wojciech, S.: Dynamics of a mobile crane and optimisation of the slewing motion of its upper structure. Nonlinear Dyn. **32**(3), 259–290 (2003)
8. Maczyński, A., Wojciech, S.: Stabilization of load's position in offshore cranes. J. Offshore Mech. Arct. Eng. **134**, 1–10 (2012)
9. Raman-Nair, W., Baddour, R.: Three-dimensional dynamics of a flexible marine riser undergoing large elastic deformations. Multibody Syst. Dyn. **10**, 393–423 (2003)
10. Shabana, A.A.: Dynamics of Multibody Systems. Cambridge University Press, Cambridge (1998)
11. Szczotka, M.: Pipe laying simulation with an active reel drive. Ocean Eng. **37**, 539–548 (2010)

12. Szczotka, M.: The Rigid Finite Element Method in Modelling of Nonlinear Offshore Systems. Gdańsk University of Technology Publishing Office, Gdańsk (2011) (in Polish)
13. Szczotka, M., Wojciech, S., Maczyński, A.: Mathematical model of a pipelay spread. Arch. Mech. Eng. **54**(1), 27–46 (2007)
14. Tomski, L., Szmidla, J., Podgórska–Brzdękiewicz, I., Sochacki, W., Kukla, S., Posiadała, B., Przybylski, J., Uzny, S.: Vibrations and Stability of Slender Systems. WNT, Warsaw (2004) (in Polish)
15. Wittbrodt, E., Adamiec–Wójcik, I., Wojciech, S.: Dynamics of Flexible Multibody Systems: Rigid Finite Element Method. Springer, Berlin (2006)
16. Wittbrodt, E., Szczotka, M., Maczyński, A., Wojciech, S.: Rigid Finite Element Method in Analysis of Dynamics of Offshore Structures. Springer, Berlin (2013)
17. Wojciech, S.: Dynamics of Planar Linkage Mechanisms with Considerations of Both Flexible Links and Friction as Well as Clearance in Joints. Publishing Office of the Technical University of Łódź, Łódź (1984) (in Polish)
18. Xu, X., Wang, S.: A flexible-segment-model-based dynamics calculation method for free hanging marine risers in re-entry. China Ocean Eng. **26**, 139–152 (2012)
19. Zienkiewicz, O.C., Taylor, R.L.: The finite element method. In: Solid Mechanics, vol. 2. Butterworth–Heinemann, Oxford (2000)

Dynamical Response of a Van der Pol–Duffing System with an External Harmonic Excitation and Fractional Derivative

Arkadiusz Syta and Grzegorz Litak

Abstract We examine the Van der Pol–Duffing system with external forcing and a memory possessing a fractional damping term. The system exhibits broad spectrum of nonlinear behavior including transitions from the periodic to nonperiodic motion. Replacing a first-order derivative damping term by a fractional damping one, we include to the system memory effect which increases the dimension of the dynamical system. As a consequence of such assumptions, the quantitative nonlinear analysis meets some limitations in this case. Instead of the well-known Lyapunov exponent treatment, we advocate to use the 0–1 test that combines both statistical and frequency properties of the attractor but does not depend on the dimension of the state space. The results have been confirmed by quantitative nonlinear analysis: bifurcation diagrams, phase portraits, Poincare sections, and the maximal Lyapunov exponent estimated in the limited two-dimensional phase space.

1 Introduction

Recently, dynamical systems with fractional damping have brought the attention of scientists from different research fields with the expectation to describe complex system behavior and/or complex material dynamical responses [2, 6, 13–15, 18–20]. This approach differs from the linear and nonlinear damping given by a function of velocity [3, 18]. Introducing fractional damping to a dynamical system is made by replacing an integer order derivative with a fractional operator. There are many definitions of fractional operator but the most common are Grünwald–Letnikov and Riemman–Liouville. Both of them are derived from general fractional q order operator; more precisely, the first one represents the q order derivative, while

A. Syta (✉)
Lublin University of Technology, Institute of Technological Systems of Information, Nadbystrzycka 36, PL-20-618 Lublin, Poland
e-mail: a.syta@pollub.pl

G. Litak
Department of Applied Mechanics, Lublin University of Technology, Nadbystrzycka 36, PL-20-618 Lublin, Poland
e-mail: g.litak@pollub.pl

J. Awrejcewicz (ed.), *Applied Non-Linear Dynamical Systems*, Springer Proceedings in Mathematics & Statistics 93, DOI 10.1007/978-3-319-08266-0_10

the second represents the q fold integral. It is worth to note that the Grünwald–Letnikov definition requires the function to be $m+1$ continuously differentiable while the Riemman–Liouville definition, only to be integrable which may play the role in scope of application. However, for function defined by Grünwald–Letnikov class both definitions are equivalent. Below, we show briefly how to construct the fractional derivative in the meaning of Grünwald–Letnikov definition which is fairly straightforward from a numerical point of view. Let us consider the Leibniz definition of the first and second derivative:

$$f'(t) = \lim_{h\to 0} \frac{f(t) - f(t-h)}{h} \tag{1}$$

$$f''(t) = \lim_{h\to 0} \frac{f'(t) - f'(t-h)}{h}$$

$$= \lim_{h\to 0} \frac{f(t) - 2f(t-h) + f(t-2h)}{h^2}$$

Now, basing on Eq. (1) one can write a general form of the nth order derivative:

$$f^n(t) = \lim_{h\to 0} \frac{1}{h^n} \sum_{j=0}^{n} (-1)^j \binom{n}{j} f(t-jh) \;\; n \in \mathbb{N}. \tag{2}$$

By generalizing the integer order in the derivative in Eq. (2) one can obtain the Grünwald–Letnikov definition of fractional operator ${}_aD_t^q f(t)$ [17]:

$$\frac{d^q f}{dt^q} = \lim_{h\to 0} \frac{1}{h^q} \sum_{j=0}^{[\frac{t-a}{h}]} (-1)^j \binom{q}{j} f(t-jh) ={}_a D_t^q f(t), \tag{3}$$

where $q > 0$, and binomial coefficients can be extended to complex numbers by use of the Euler Gamma function:

$$\binom{q}{j} = \frac{q!}{j!(q-j)!} = \frac{\Gamma(q+1)}{\Gamma(j+1)\Gamma(q-j+1)}; \tag{4}$$

a pair of square brackets [.] appearing in the upper limit of the sum denotes the integer part, while a is the length of the memory. Generally, dynamical systems with memory effects pose long transients, but in this case it is possible to shorten the length of the memory (a) in Eq. (3) using the short memory principle [16, 17]. Such a length, introduced to the system, reduces the computational effort of simulations while generally preserving the system dynamics. Thus, Eq. (3) becomes

$${}_L D_t^q f(t) = \lim_{h\to 0} \frac{1}{h^q} \sum_{j=0}^{[N(t)]} (-1)^j \binom{q}{j} f(t-jh), \tag{5}$$

where $N(t) = \min(\frac{\mathrm{t-L}}{\mathrm{h}}, \frac{\mathrm{L}}{\mathrm{h}})$.

2 The Model

We start from the standard Van der Pol–Duffing equation decomposed into the set of two differential equations of first degree:

$$\begin{aligned}\frac{dx}{dt} &= y(t),\\ \frac{dy}{dt} &= \alpha x(t) - \beta x(t)^3 + \epsilon(1 - x(t)^2)y(t) + A\cos(\omega t),\end{aligned} \tag{6}$$

where $\alpha = 0.5, \beta = 0.5, \epsilon = 0.1, A = 0.15$, and the frequency of external excitation, ω, is a control parameter. This model shows many interesting nonlinear features, e.g., periodic type solution with period doubling sequence, and transitions to quasiperiodic or chaotic solutions.

Let us compare the system response for various values of the excitation frequency $\omega = \{0.45, 0.515, 0.58\}$ with same initial conditions $(x(0), y(0)) = (0.0, 0.0)$. We calculated the solutions using Runge–Kutta of fourth order with fixed time step taking 200 periods (each with 200) after the transient with length of the memory $a = 5{,}000$.

Analyzing Fig. 1 one can see different types of motion: from quasiperiodic (a) with Poincare points located on circle like curve (closed curve would be obtained for longer computational time) to nonperiodic (b) with clouded Poincare points and periodic one (c) with single isolated Poincare point. Now, to study the influence of fractional damping to the model (6) we replaced first-order derivative damping with fractional damping:

$$\begin{aligned}{}_L D_t^q x(t) &= y(t)\\ {}_L D_t^1 y(t) &= \alpha x(t) - \beta x(t)^3 + \epsilon(1 - x(t)^2)y(t) + A\cos(\omega t).\end{aligned} \tag{7}$$

The set of equations can be written in the discretized form by the following fractional order Newton–Leipnik algorithm [16]:

$$x(t_k) = y(t_{k-1})h^q - \sum_{j=1}^{N-1} c_j^{(q)} x(t_{k-j}) \tag{8}$$

$$\begin{aligned}y(t_k) = {} & [\alpha x(t_k) - \beta x(t_k)^3 + \epsilon(1 - x(t_k)^2)y(t_{k-1})\\ & + A\cos(\omega t_{k-1})]h + y(t_{k-1}),\end{aligned} \tag{9}$$

where h is an integration step and coefficients $c_j^{(q)}$ satisfy the following recursive relations:

$$c_0^{(q)} = 1, \qquad c_j^{(q)} = \left(1 - \frac{1+q}{j}\right) c_{j-1}^{(q)}. \tag{10}$$

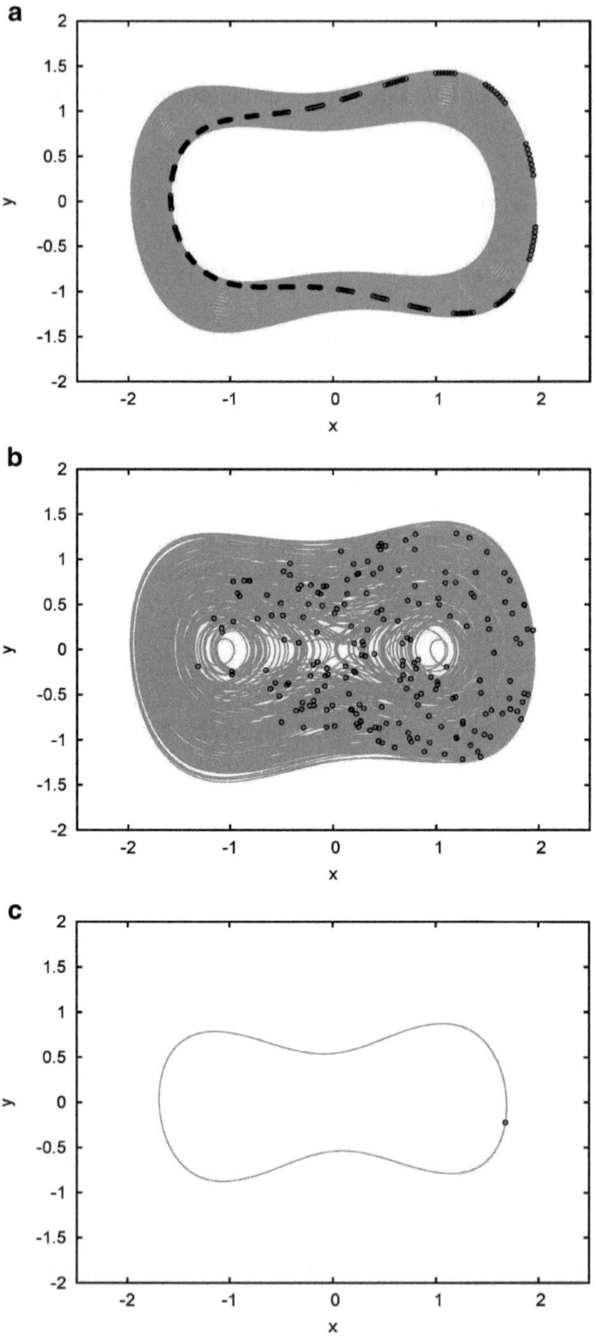

Fig. 1 Phase portraits and Poincare sections (as introduced in [1]) of system (6) solutions for (**a**) $\omega = 0.45$, (**b**) $\omega = 0.52$, (**c**) $\omega = 0.58$

Figure 2 presents bifurcation diagrams obtained for various fractional order damping $q = \{0.9, 0.95, 1.0, 1.05, 1.1\}$ and initial conditions $(x(0), y(0)) = (0.0, 0.0)$.

Comparing the bifurcation diagrams one can see a change of system response even in a short interval of ω: for $\omega \in (0.35 : 0.7)$ (Fig. 2a, b) there dominate periodic solutions with narrow nonperiodic regimes. When q increases (Fig. 2c) it is changing to quasiperiodic ($\omega \in (0.35, 0.48)$) and chaotic motion ($\omega \in (0.48, 0.55)$). Finally, for $q = \{1.05, 1.1\}$ (Fig. 2d, e) it becomes quasiperiodic or chaotic.

To quantify obtained results we use the 0–1 test [4, 7, 8, 10–12] with single time series and the maximal Lyapunov exponent (MLE) with projection into two dimension [9, 21]. The 0–1 test for chaos detection uses spectral and statistical properties of the single time series, while MLE uses geometrical properties of phase space. Both methods can distinguish between periodic and chaotic motion, but because of the memory effect introduced by fractional damping the phase space of the system has infinite dimension ($a = 5{,}000$) and in computing MLE only projection to finite and low dimension can be used. As a consequence we don't possess information about system dynamics in higher dimension. In this case the 0–1 test might be considered and we will show a brief description of it in a few steps. The first one is conversion of the coordinates from (x, y) to the new set (p, q) defined as follows:

$$p(n) = \sum_{j=1}^{n} \tilde{x}_j \cos(jc), \quad q(n) = \sum_{j=1}^{n} \tilde{x}_j \sin(jc), \tag{11}$$

where $\tilde{x} = [\tilde{x}_1, \tilde{x}_2, \tilde{x}_3, \ldots]$ is a discrete time series sampled from the originally simulated x using one forth of excitation period $T/4$ ($T = 2\pi/\omega$) which corresponds to the nodal autocorrelation function of excitation harmonic term $A\cos(\omega t)$. Equation (11) corresponds to the Fourier transform (in the limit of larger n) for chosen frequency c where $c \in (0, \pi)$. The next step is calculation of the mean square displacement (MSD) in the plane spanned by new coordinates $[p, q]$:

$$\mathrm{MSD}(c, j) = \frac{1}{n-j} \sum_{i=1}^{n-j} \{[p(i+j) - p(i)]^2 + [q(i+j) - q(i)]^2\} \tag{12}$$

where $0 \ll j \ll n$ (in practice $n/100 \leq j \leq n/10$). It is bounded for regular dynamics or unbounded for chaotic dynamics [4, 7, 8, 10–12]. The final quantity K is obtained as a asymptotic growth rate of MSD (here given by the correlation method):

$$K(c) = \frac{\mathrm{Cov}[j, \mathrm{MSD}(c, j)]}{\sqrt{\mathrm{Cov}[j, j] \cdot \mathrm{Cov}[\mathrm{MSD}(c, j), \mathrm{MSD}(c, j)]}}, \tag{13}$$

where j is based on series of natural numbers, $j = n/100, n/100+1 \ldots, n/10$, and $\mathrm{Cov}[x,y]$ denotes corresponding covariance of two series where the same arguments

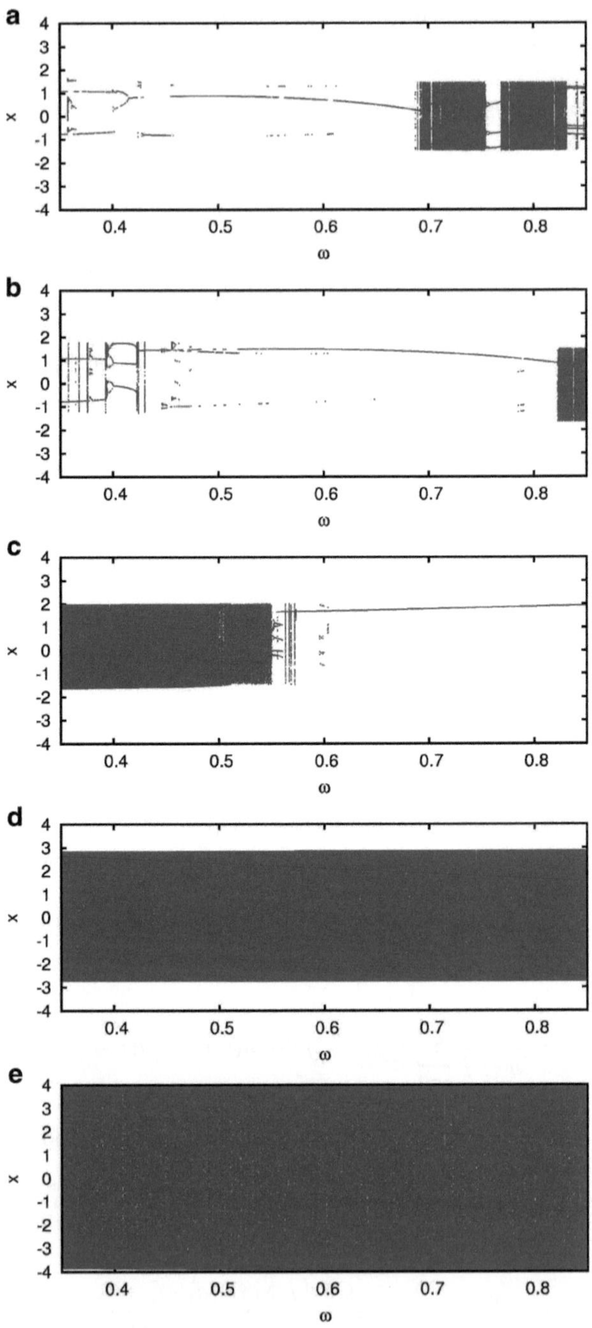

Fig. 2 Bifurcation diagrams with the sampling $\Delta\omega = 0.00025$ for different order of the damping: (**a**) $q = 0.9$, (**b**) $q = 0.95$, (**c**) $q = 1.0$, (**d**) $q = 1.05$, (**e**) $q = 1.1$. Note that Fig. 2d, e shows quasiperiodic system response in the whole investigated frequency interval

$x = y$ mean variance while for different ($x = j$ and $y = \mathrm{MSD}(c, j)$) it can be expressed in terms of the expectation value E[.]:

$$\mathrm{Cov}[j, \mathrm{MSD}(c, j)] = \mathrm{E}[[j - \mathrm{E}[j]] \cdot [\mathrm{MSD}(c, j) - \mathrm{E}[\mathrm{MSD}(c, j)]]]. \tag{14}$$

The $K \approx 0$ indicates periodic motion while $K \approx 1$ indicates nonperiodic behavior. The value of the quantificator might vary with respect of c, so we choose 100 values of c equally spaced from the interval $(0.1, \pi - 0.1)$ and computed K as the median; see Fig. 3.

Comparing the K values (Fig. 2) to the bifurcation diagrams (Fig. 3) one can see that this indicator stays close to 0 for periodic or quasiperiodic motion while it oscillates near 1 for chaotic motion. Interestingly, for fractional order of the damping $q = \{1.05, 1.1\}$ in considered interval of excitation frequency ($\omega \in (0.35, 0.85)$), the motion is quasiperiodic.

We have also calculated MLE which is commonly used to distinguish between different types of response of dynamical systems, but in the case of our system this indicator has no direct meaning because the history effect provided by fractional damping makes the dimension of the system undetermined. Having that in mind, one cannot use the well-known Wolf [21] method (with linearized system) nor Kantz [9] algorithm (with phase space reconstructed from time series). Instead of that we measured the distance $d(i)$ between reference and test orbits projected in two dimension (x, y) and disturbed by some small initial perturbation $d_0(i)$, $i = 1, \ldots, N$, where i denotes the subsequent interval which is smaller compared to the excitation period. In that case, the value of MLE can be approximated by

$$\mathrm{MLE} = \frac{1}{\Delta t} \sum_{i=1}^{N} \ln\left(d/d_0\right). \tag{15}$$

The corresponding results are presented in Fig. 4.

By analyzing MLE values, one notices fairly good agreement with K values obtained for $q = \{0.9, 0.95, 1.0, 1.05\}$, but for $q = 1.1$, MLE is grater then 0.2 for all $\omega \in (0.35, 0.85)$ which indicates chaotic motion and is contrary to the K values. The reason might be connected with lack of information about the distance between trajectories in additional dimensions introduced by the memory effect of the fractional damping term. For better clarity, Fig. 5 shows (x, y) coordinates of the phase space for fractional damping order $q = 1.05$.

3 Conclusions

Our results show that in the case of Van der Pol–Duffing potential with a fractional damping, various bifurcations can occur. However, the most common solutions are quasiperiodic with short intervals of chaotic solutions. These effects agree with

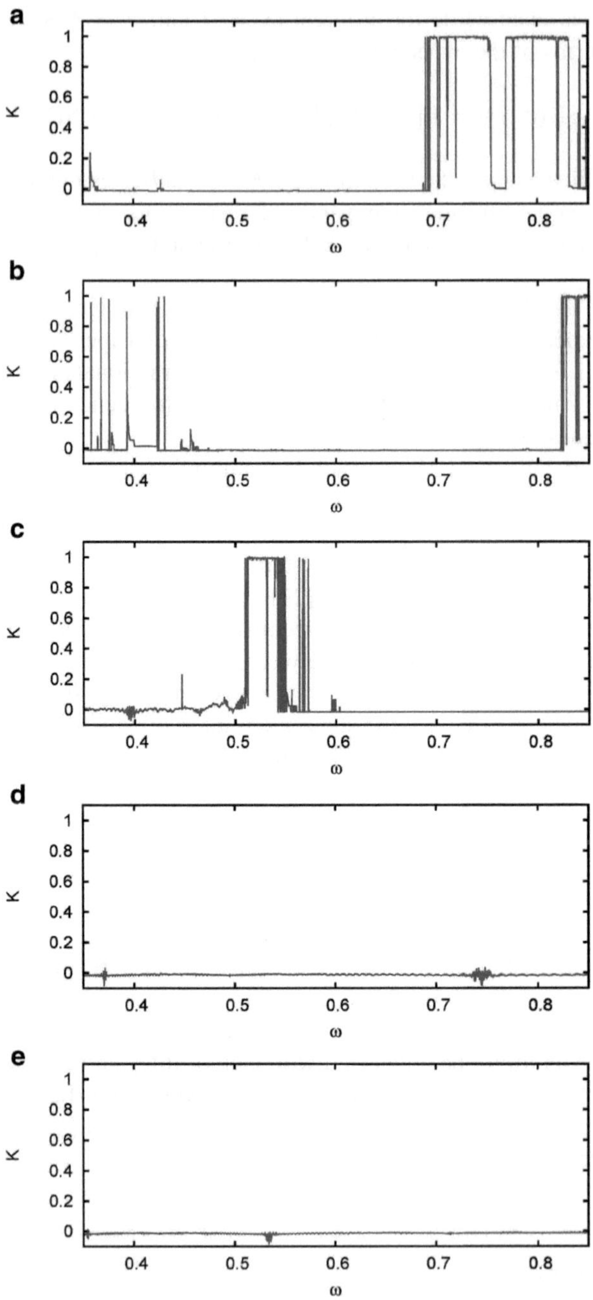

Fig. 3 K versus ω with the sampling $\Delta\omega = 0.00025$ for (**a**) $q = 0.9$, (**b**) $q = 0.95$, (**c**) $q = 1.0$, (**d**) $q = 1.05$, (**e**) $q = 1.1$. $K \approx 0$ indicates regular while $K \approx 1$ chaotic behavior

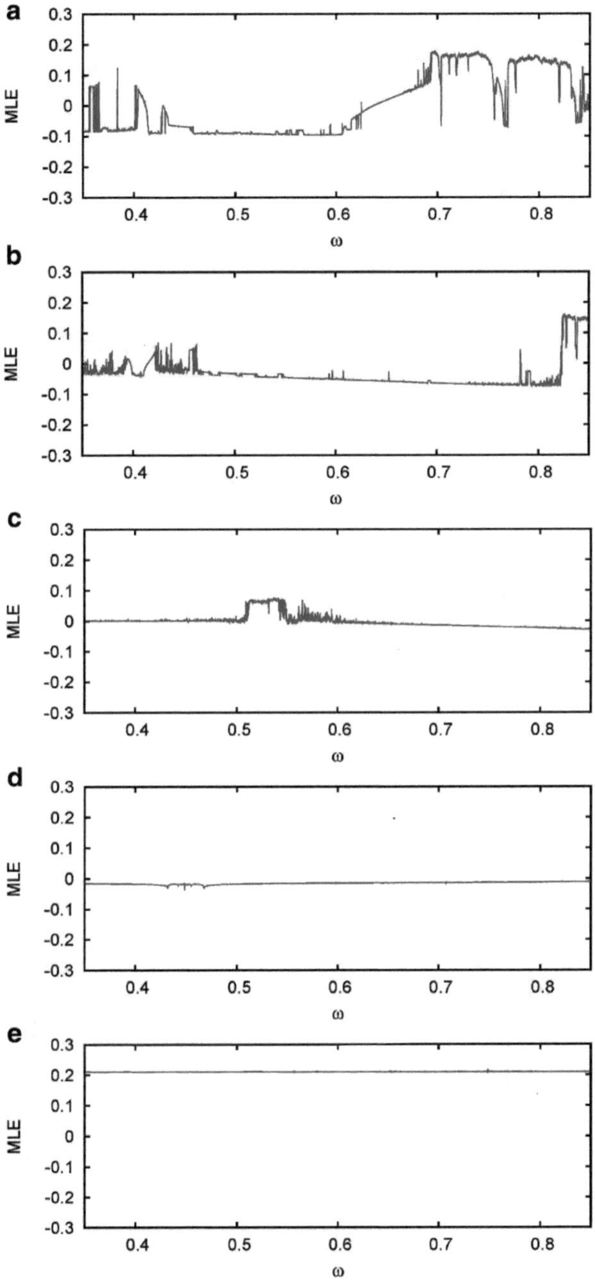

Fig. 4 MLE versus ω with the sampling $\Delta\omega = 0.00025$ for (**a**) $q = 0.9$, (**b**) $q = 0.95$, (**c**) $q = 1.0$, (**d**) $q = 1.05$, (**e**) $q = 1.1$. Note that MLE > 0 implies chaos. In many cases, it is not consistent with bifurcation diagrams (Fig. 2) and K values (Fig. 3)

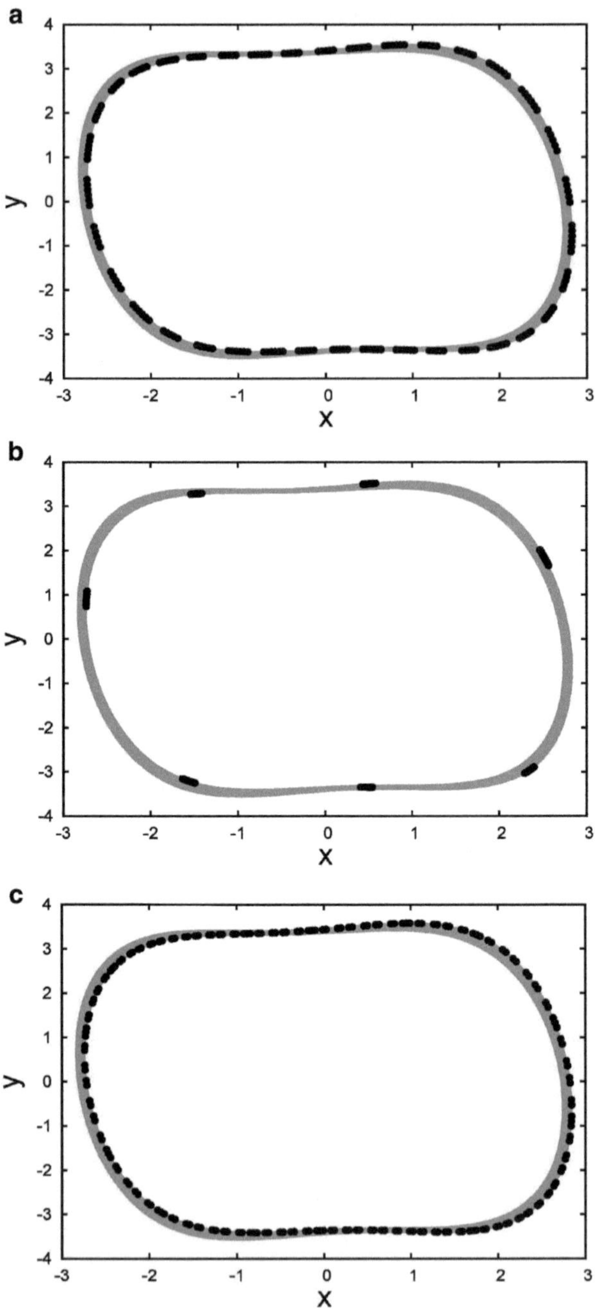

Fig. 5 Phase portraits and Poincare sections of Eq. (7) solutions for (**a**) $\omega = 0.45$, (**b**) $\omega = 0.52$, (**c**) $\omega = 0.58$ and $q = 1.05$

expectations in fairly large dimension of the dynamical system with delays [5]. It is not surprising that the simply defined MLE detects chaos wrongly in particular cases (see Fig. 4e). This must be an effect of higher dimension. Note that for the same quasiperiodic solution, the 0–1 test gives the correct state identification (see Fig. 3e). The results have been also confirmed by phase portraits and Poincare series. On the other hand, in the system with memory the role of transient solutions increases. In the specific situations transient chaotic motion can be also stabilized. However, to tell more about such a tendency, more systematic studies should be done to explain the evolution of the corresponding basins of attractions.

Acknowledgements This paper was supported by the Polish National Science Center under the grant agreement No. 2012/05/B/ST8/00080.

References

1. Awrejcewicz, J., Olejnik, P.: Stick-slip dynamics of a two-degree-of-freedom system. Int. J. Bifurcat. Chaos **13**, 843–861 (2003)
2. Barbosa, R.S., Machado, J.A.T., Vinagre, B.M., Calderon, A.J.: Analysis of the Van der Pol oscillator containing derivatives of fractional order. J. Vib. Control. **13**, 1291 (2007)
3. Borowiec, M., Litak, G., Syta, A.: Vibration of the Duffing oscillator: effect of fractional damping. Shock Vib. **14**, 29 (2007)
4. Falconer, I., Gottwald, G.A., Melbourne, I., Wormnes, K.: Application of the 0–1 test for chaos to experimental data. SIAM J. Appl. Dyn. Syst. **6**, 395 (2007)
5. Farmer, J.D.: Chaotic attactors of an infinite-dimensional dynamical system. Physica D **4**, 366 (1982)
6. Gao, X., Yu, J.: Chaos in the fractional order periodically forced complex Duffing's systems. Chaos Solitons Fractals **24**, 1097 (2005)
7. Gottwald, G.A., Melbourne, I.: A new test for chaos in deterministic systems. Proc. Roy. Soc. A **460**, 603 (2004)
8. Gottwald, G.A., Melbourne, I.: Testing for chaos in deterministic systems with noise. Physica D **212**, 100 (2005)
9. Kantz, H.: A robust method to estimate the maximal Lyapunov exponent of a time series. Phys. Lett. A **185**, 77 (1994)
10. Litak, G., Syta, A., Wiercigroch, M.: Identification of chaos in a cutting process by the 0–1 test. Chaos Solitons Fractals **40**, 2095 (2009)
11. Litak, G., Syta, A., Budhraja, M., Saha, L.M.: Detection of the chaotic behaviour of a bouncing ball by the 0–1 test. Chaos Solitons Fractals **42**, 1511 (2009)
12. Litak, G., Schubert, S., Radons, G.: Nonlinear dynamics of a regenerative cutting process. Nonlinear Dyn. **69**, 1255 (2012)
13. Machado, J.A.T., Silva, M.F., Barbosa, R.S., Jesus, I.S., Reis, C.M., Marcos, M.G., Galhano, A.F.: Some applications of fractional calculus in engineering. Math. Prob. Eng. **2010**, 639801 (2010)
14. Mikens, R.E., Oyedeji, K.O., Rucker, S.A.: Analysis of the simple harmonic oscillator with fractional damping. J. Sound Vib. **268**, 842 (2003)
15. Padovan, J., Sawicki, T.: Nonlinear vibration of fractionally damped systems. Nonlinear Dyn. **16**, 321–336 (1998)
16. Petras, I.: Fractional-Order Nonlinear Systems: Modeling, Analysis and Simulation. Springer, New York (2010)
17. Podlubny, I.: Fractional Differential Equations. Academic, San Diego (1999)

18. Rossikhin, Y.A., Shitikova, M.V.: Application of fractional calculus for dynamic problems of solid mechanics: novel trends and recent results. Appl. Mech. Rev. **63**, 010801 (2010)
19. Seredynska, M., Hanyga, A.: Nonlinear differential equations with fractional damping with application to the 1dof and 2dof pendulum. Acta Mech. **1760**, 169 (2005)
20. Sheu, L.J., Chen, H.K, Tam, L.M.: Chaotic dynamics of the fractionally damped duffing equation. Chaos Solitons Fractals **32**, 1459 (2007)
21. Wolf, A., Swift, J.B., Swinney, H.L.: Determining Lyapunov exponents from a time series. Physica D **16**, 285 (1985)

Vortex Structure Around the Cylinder at a Flow of Viscous Fluid

Rustyam G. Akhmetov and Ruslan R. Kutluev

Abstract The problem of a stationary viscous incompressible fluid flow around the cylinder has been analyzed by means of the asymptotic methods. The fluid flow equations in the variables "stream function-a vortex" are considered. One component of a vortex vector has remained in case of two-dimensional motion. Having applied the matching method the vortex asymptotics has been investigated in an interface near the cylinder. The equation of an interior boundary layer for stream function has been made with the help of using the method of matched asymptotic expansions and a matching condition with the solution for vortex. The received equation is investigated by means of numerical methods for great values of Reynolds number.

1 Introduction

In case of two-dimensional motion equation of the fluid flow is set down in the variables "stream function-a vortex.' The equation of incompressibility is carried out as the result of introduction of stream function ψ. So for a stationary problem in the dimensionless variables we have the equations (see, e.g., [14], Chap. XI, p. 471, Eq. (135), (136))

$$\frac{1}{Re}\Delta\Omega = u\frac{\partial\Omega}{\partial x} + v\frac{\partial\Omega}{\partial y}, \tag{1}$$

$$\Delta\psi = -\Omega, \tag{2}$$

where Δ is the Laplace operator, Re is the Reynolds number, and Ω is the scalar function:

R.G. Akhmetov (✉) • R.R. Kutluev
Bashkortostan State Pedagogical University, Faculty of Physics and Mathematics, ul. Oktya'brskoy Revolutsii 3a, 450000 Ufa, Russia
e-mail: akrust@mail.ru; rus4652@yandex.ru

J. Awrejcewicz (ed.), *Applied Non-Linear Dynamical Systems*, Springer Proceedings in Mathematics & Statistics 93, DOI 10.1007/978-3-319-08266-0_11

$$\Omega = \Omega_z = \frac{\partial v}{\partial x} - \frac{\partial u}{\partial y} = \frac{\partial}{\partial x}\left(-\frac{\partial \psi}{\partial x}\right) - \frac{\partial}{\partial y}\left(\frac{\partial \psi}{\partial y}\right) = -\Delta\psi,\ u = \frac{\partial \psi}{\partial y},\ v = -\frac{\partial \psi}{\partial x}.$$

One of the ways of equations researching (1), (2) is the transition to polar coordinates $r,\ \varphi$. For stream function we specify conditions

$$\psi = \frac{\partial \psi}{\partial r} = 0 \text{ for } r = 1 \tag{3}$$

and on infinity we specify conditions

$$\frac{1}{r}\frac{\partial \psi}{\partial \varphi} \to \cos\varphi,\ \frac{\partial \psi}{\partial r} \to \sin\varphi \text{ for } r \to \infty. \tag{4}$$

For a vortex on infinity

$$\Omega \to 0 \text{ for } r \to \infty \tag{5}$$

is obtained. Tom or Woods approximate conditions are usually set at numerical calculations (see [14], Chap. XI, p. 472, formulas (137), (138)). Problems analogous to (1)–(5) and a broader class of problems were considered by many authors (e.g., see [1, 5, 6, 10–12] and references).

2 Boundary Layer Near the Cylinder

In view of this work the method of the research consists in the following. In the capacity of the first approximation of stream function outside the wake behind the cylinder (in area $r \cdot \sin^2\varphi > const$) we will consider

$$\psi_1(r,\varphi) = \left(r - \frac{1}{r} - \frac{3\ln r}{r} + \frac{\ln r}{r^2}\right)\sin\varphi,\ \text{where} \tag{6}$$

$$\Delta\psi_1 = O(r^{-3}\sin\varphi),\ \text{as } r \to \infty \text{ and } \psi_1(1,\varphi) = (\psi_1)'_r(1,\varphi) = 0.$$

Thus, $Re = U_\infty d/\nu$, where d is the diameter of the cylinder and ν is the viscosity coefficient.

Then, we are to consider that $Re \in [10, A]$. Further, calculations show that $A = O(10^4)$. We investigate a vortex asymptotics Ω applying a method of matched asymptotic expansions [15] near the cylinder. Equations (1) and (2) are considered in [3], when stream function is $\psi(r,\varphi) = \left(r - r^{-1}\right)\sin\varphi$ (in area $r \cdot \sin^2\varphi > const$).

The small parameter $\epsilon^3 = 1/Re$ is introduced for convenience, and (1) is rewritten as

$$\epsilon^3\left(\frac{\partial^2\zeta}{\partial r^2}+\frac{1}{r}\frac{\partial\zeta}{\partial r}+\frac{1}{r^2}\frac{\partial^2\zeta}{\partial\varphi^2}\right)-\frac{1}{r}\frac{\partial\psi}{\partial\varphi}\frac{\partial\zeta}{\partial r}+\frac{1}{r}\frac{\partial\psi}{\partial r}\frac{\partial\zeta}{\partial\varphi}=0, \tag{7}$$

$$\zeta\to 0,\ r\to\infty, \tag{8}$$

where $\zeta(r,\varphi)=\Omega(x,y)$. The function ψ_1 (see (6)) at $r=1$ has the asymptotic expansion in a series

$$\psi_1=\left((r-1)^2+O\left((r-1)^3\right)\right)\sin\varphi. \tag{9}$$

The asymptotics ζ is sought in the form of a series

$$\zeta=\zeta_0(\rho,\varphi)+\epsilon\zeta_1(\rho,\varphi)+\cdots,\ \text{where } \rho=(r-1)/\epsilon.$$

Then from (7) in variables ρ,φ we receive the equation for the main member

$$\frac{\partial^2\zeta_0}{\partial\rho^2}-\rho^2\cos\varphi\frac{\partial\zeta_0}{\partial\rho}+2\rho\sin\varphi\frac{\partial\zeta_0}{\partial\varphi}=0. \tag{10}$$

In (10) we should substitute variables $x=\rho\sqrt{\sin\varphi}$, $\tau=\frac{1}{2}\int_\varphi^\pi\sqrt{\sin y}dy$ (see [8], Chap. 3, p. 110, 111). Then the equation

$$\frac{\partial^2\zeta_0}{\partial x^2}-x\frac{\partial\zeta_0}{\partial\tau}=0 \tag{11}$$

is obtained.

The solution of (11) satisfying condition (8) has the form

$$\zeta_0=c\cdot\Gamma^{-1}\left(\frac{1}{3}\right)\Gamma\left(\frac{1}{3},\frac{x^3}{9\tau}\right),\ \text{where} \tag{12}$$

$$\Gamma(\alpha,x)=\int_x^\infty e^{-y}y^{\alpha-1}dy$$

and the constant c is determined from the boundary conditions for the ζ_0 where $r=1$. To find the approximate condition at the border, where $r\to 1$ for ζ_0, we should use (2) in the vicinity of the boundary. Then, taking into account formulas (9), (12), we have

$$\frac{\partial^2\psi_1(1,\varphi)}{\partial r^2}=2\sin\varphi\approx-\zeta_0(0,\tau)=-c\ \text{at}\ r\to 1.$$

It is followed the formula is:

$$\zeta_0(x,\tau) \approx -2\sin\varphi \cdot \Gamma^{-1}\left(\frac{1}{3}\right)\Gamma\left(\frac{1}{3},\frac{x^3}{9\tau}\right) \text{ at } \tau \to \tau_0 \ (\varphi \to 0), \text{ where} \tag{13}$$

$$\tau_0 = \frac{1}{2}\int_0^{\pi} \sqrt{\sin y}dy \approx 1,19814.$$

3 Interior Boundary Layer

Behind the cylinder there is an additional boundary layer. The asymptotics of the solution is regarded in terms of the variables ψ, $s = r - 1$. And for the main member we receive the equation

$$\frac{\partial \zeta}{\partial s} = 0. \tag{14}$$

From (14) we receive $\zeta = \zeta(\psi)$. Then from (2), we have

$$\Delta\psi = -\zeta(\psi). \tag{15}$$

Matching condition with solution (13) for $\zeta(\psi)$ is represented as

$$\zeta(\psi) - \zeta_0\left(\frac{\psi_1^{1/2}(r,\varphi)}{\epsilon(9\tau_0)^{1/3}}\right) = O\left(\epsilon^2\right) \text{ for } \varphi \in O(\epsilon). \tag{16}$$

Thus it is natural to obtain the explicit function $\zeta(\psi)$

$$\zeta(\psi) = -2\epsilon \cdot \Gamma^{-1}\left(\frac{1}{3}\right)\Gamma\left(\frac{1}{3},\frac{\psi^{3/2}}{\epsilon^3 9\tau_0}\right) \text{ for } \varphi \in O(\epsilon). \tag{17}$$

Equation (15) and formula (17) in the interior boundary layer for the flow function are written out in work [2].

The equation similar to (15) is usually written out during the study of vortex motions of ideal fluid $\Delta\psi = f(\psi)$ (see, e.g., [13], Chap. 7, Sect. 165). Besides $f(\psi)$ is an arbitrary function. Our consideration is different as the (15) is written out in an internal around the cylinder by a streamlined flow of a viscous incompressible fluid boundary layer, using the method of matched asymptotic expansions [15], system (1), (2), and matching condition (16) with solution (13).

4 Numerical Solution

For the solution of (15) where the right part looks like (17), taking into account (3), (4), it is necessary to set additional boundary conditions $\partial\psi/\partial\varphi = 0$, for $\varphi = 0$, and matching condition with solution (13) at $r - 1 = O(Re^{-1/3})$, $\varphi = O(Re^{-1/3})$.

Let us consider a special case, when $\psi = \psi(r)$. In this case from (15) we receive the equation

$$\psi''(r) + \frac{1}{r}\psi'(r) = -\zeta(\psi(r)). \tag{18}$$

Taking into account expression (16) and setting initial conditions in a point r_0

$$\psi(r_0) = \psi_{01},\ \psi'(r_0) = \psi_{02}. \tag{19}$$

We receive the solution by numerical methods (see Figs. 1, 2, 3, 4, and 5).

The existence of a solution to problem (18) and (19) at $\psi'(r) = 0$ follows from Poincare–Bendixson theorem (see [9], Chap. I, p. I.16). Phase trajectories of the solution to problem (18) and (19) (see Fig. 1) are natural, taking into account existence in the left part of the component $\frac{1}{r}\psi'(r)$ (see, e.g., [4], Chap. 3, Sect. 10). This is a special point of focus type.

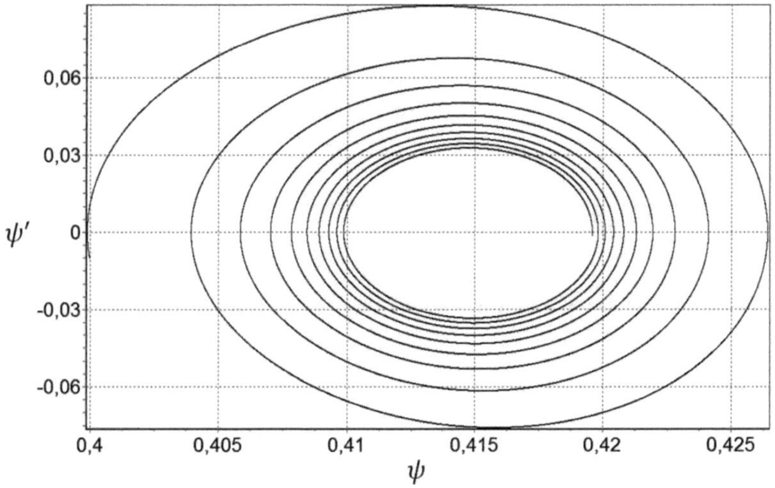

Fig. 1 Phase trajectories for $r_0 = 1, 1$, $\psi_{01} = 0, 4$, $\psi_{02} = -0, 01$, $Re = 300$

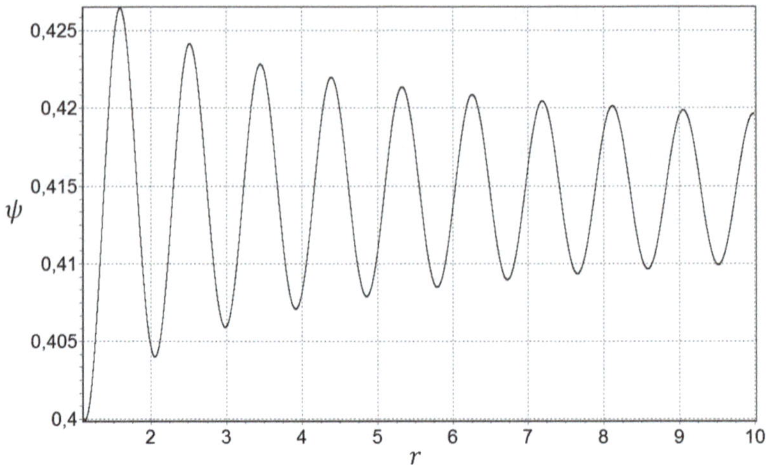

Fig. 2 Dependence of the solution on r at $Re = 300$

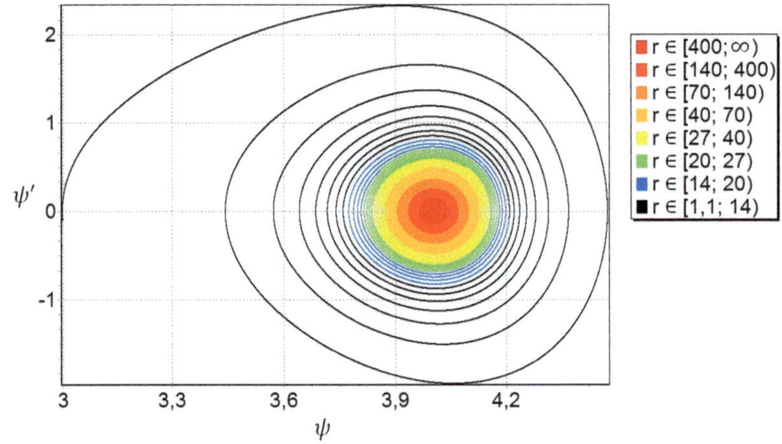

Fig. 3 Phase trajectories for $r_0 = 1, 1$, $\psi_{01} = 3$, $\psi_{02} = -0, 1$, $Re = 10$

Remark

Equations (1) and (2) are written in dimensionless variables (see [14], Chap. XI, p. 472, formulas (137), (138)). It follows that function (13) is also dimensionless for $\psi(r) = O(\epsilon^2)$.

In the work [5], the stationary problem flow of the circular cylinder for Reynolds numbers in an interval from 1/8 to 100 is investigated. The results of numerical calculations are presented. Formula (13) gives a quite satisfactory approximation compared with the results of numerical calculation [5] outside the wake of a cylinder $Re \in [10; 100]$ (see, e.g., [5], p. 195, Fig. 8 and $Re = 100$).

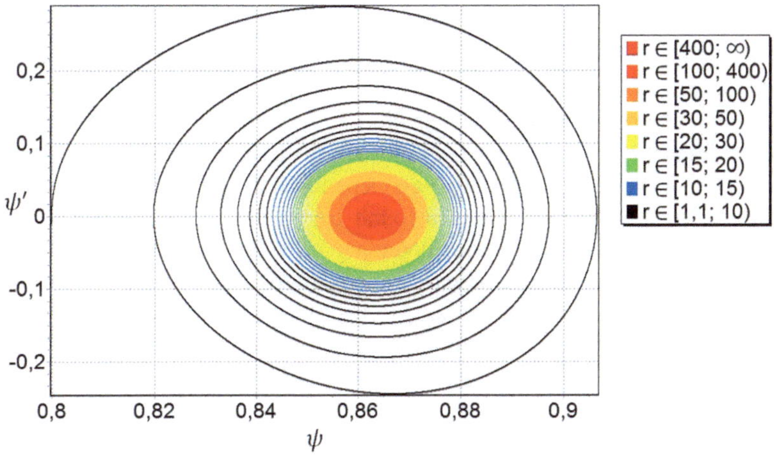

Fig. 4 Phase trajectories for $r_0 = 1,1$, $\psi_{01} = 0,8$, $\psi_{02} = -0,01$, $Re = 100$

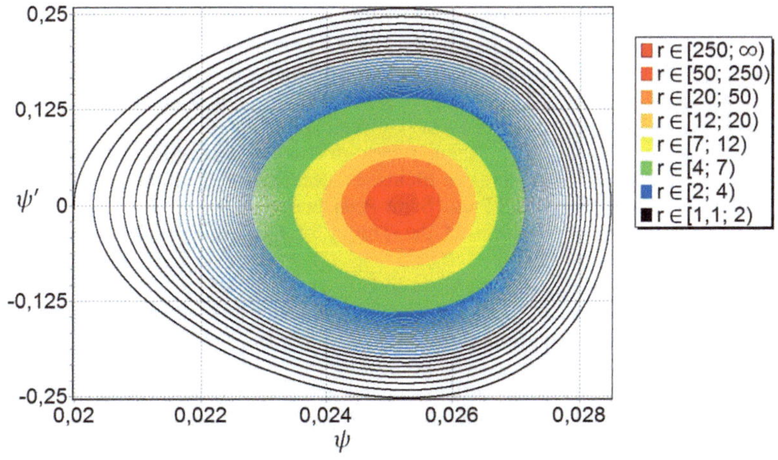

Fig. 5 Phase trajectories for $r_0 = 1,1$, $\psi_{01} = 0,02$, $\psi_{02} = -0,005$, $Re = 2 \cdot 10^4$

In this work [6, 7] a two-dimensional problem near the cylinder for Reynolds numbers in an interval from $2 \cdot 10^4$ to 10^6 was considered. The phenomenon of the pressure crisis in the boundary layer and the separation of the boundary layer were analyzed. The formation of the boundary layer (and the separation of the boundary layer) starts near the midship section of the cylinder (see [7], Fig. 2a). It is noted that because of the rise of Reynolds number, the boundary layer becomes thinner in case first vortex ([7], Fig. 2). This article demonstrates that the increase of Re leads to a reduction of the size of the boundary layer (see Figs. 3, 4, and 5, component ψ).

Conclusion

Outside the wake of a cylinder the flow function is known and it corresponds to the flow around the cylinder by the viscous fluid; the equation for the vortex component near the cylinder is investigated. It is illustrated that there is an interior boundary layer where the vortex component part depends only on the flow function. The structure of the vortex component is found by means of the method of matched asymptotic expansions. The interior boundary layer equation for the flow function is investigated by means of numerical methods at great values of Reynolds number. The properties of the flow function are studied in this boundary layer. The solution in this boundary layer has a special point of focus type.

References

1. Afendikov, A.L., Babenko, K.I.: Mathematical modelling of turbulence in flows of a viscous incompressible fluid (Russia). Math. Model. Moscow **1**(8), 45–74 (1989)
2. Akhmetov, R.G.: About a matching method in the problem of viscous fluid flow around a cylinder (Russia). In: Theory of Functions, Its Applications and Related Issues. Proceedings of International Conference, 22–28 Aug 2013. Proceedings of the Mathematical Center of N.I. Lobachevsky, vol. 46, pp. 106–108. Kazan University, Kazan (2013)
3. Akhmetov, R.G., Kutluev, R.R.: About vortex structure at the cylinder flow viscous fluid (Russia). In: Fluxes and Structures in Fluids. Proceedings of International Conference, pp. 10–13, 25–28 June 2013. MAKS Press, Saint-Petersburg (2013)
4. Arnold, V.I.: Additional Chapters of the Theory of the Ordinary Differential Equations (Russia). Nauka, Moscow (1978)
5. Babenko K.I., Vvedenskaya N.D., Orlova M.G.: Calculation of the steady flow of a viscous fluid past a circular cylinder. Zhurnal Vychislitel'noi Matematiki and Matematicheskoi Fiziki **15**(1), 183–196 (1975) [Translate from Russian to English in USSR Comput. Math. Math. Phys. **15**(1), 176–190 (1975)]
6. Dynnikova, G.Ya.: Calculation of flow around a circular cylinder on the basis of two-dimensional Navier-Stokes equations at large Reynolds numbers with high resolution in a boundary layer. Doklady Akad. Nauk **422**(6), 755–757 (2008) [Translate from Russian to English in Doklady Phys. **53**(10), 544–547 (2008)]
7. Dynnikova, G.Ya.: Fast technique for solving the N-body problem in flow simulation by vortex methods. Zhurnal Vychislitel'noi Matematiki and Matematicheskoi Fiziki **49**(8), 1458–1465 (2009) [Translate from Russian to English in Comput. Math. Math. Phys. **49**(8), 1389–1396 (2009)]
8. Gupalo, Yu.P., Polyanin, A.D., Ryazantzev, Yu.S.: Mass and Heat Exchange of Reactive Particles with the Flow (Russia). Nauka, Moscow (1985)
9. Hairer, E., Norsett, S.P., Wanner, G.: Solving Ordinary Differential Equations, Nonstiff Problems. Springer, Berlin (1987)
10. Krasnikov, Yu.G., Solovyev, V.R.: Finding of approximate analytical solutions of the equations of Navier-Stokes for a stationary flow of the cylinder incompressible liquid (Russia). News Acad. Sci. Mech. Liquid Gas **4**, 22–33 (1999)
11. Ladyzhenskaya, O.A.: Mathematical Problems of the Dynamics of a Viscous Incompressible Fluid (Russia). Nauka, Moscow (1970)

12. Ladyzhenskaya, O.A.: Sixth Problem of the Millennium: Navier–Stokes Equations, Existence and Smoothness, vol. 58(2), pp. 45–78. Uspekhi Matematicheskikh Nauk, Moscow (2003) [Translate from Russian to English in Russian Math. Surv. **58**(2), 251–286 (2003)]
13. Lamb, G.: Hydrodynamics. Gostekhizdat, Moskow (1947) [English original text in Cambridge University Press, New York (1932)]
14. Loytsyansky, L.G.: Mechanics of liquids and gases (Russia). Bustard, Moskow (2003)
15. Van Dyke, M.: Perturbation methods in fluid mechanics. World, Moskow (1967) [English original text in Academic Press, New York (1964)]

Asymptotic Analysis and Limiting Phase Trajectories in the Dynamics of Spring Pendulum

Jan Awrejcewicz, Roman Starosta, and Grażyna Sypniewska-Kamińska

Abstract Spring pendulum is a widely discussed two degree-of-freedom (DOF) mechanical systems in numerous references. In this paper the asymptotic approach and limiting phase trajectories (LPT) have been applied to analyze the two DOF mathematical model of a spring pendulum. The LPT and multiple timescale (MTS) methods are effective tools of the investigation of non-linear systems. Some interesting and important aspects of dynamics of the system are discussed. The main attention is focused on the non-steady-state vibrations when the energy is intensively exchanged. Then with increasing values of the selected parameters, a sudden change in the character of vibrations is observed. These phenomena are very well described by the LPT. The method allows to determine the critical values of the parameters responsible for the mentioned transitions. Our analytical studies are verified by numerical calculations.

1 Introduction

The steady-state vibrations are, in general, mainly observed in engineering practice. However, in some cases of sharp resonance, the transient stage of the oscillatory process and its relaxation can last a long time. Energy exchange and non-stationary processes appear in many dynamical systems and they are of great interest of many researchers. This problem has been widely discussed in [2, 6]. It is usually studied numerically due to occurred essential mathematical difficulties [2]. However, in recent years, one may observe a great interest in successful application of modern asymptotic methods to engineering-oriented problems [1, 4]. In particular, a novel

J. Awrejcewicz
Department of Automatics and Biomechanics, Technical Universityof of Łódź,
ul. Stefanowskiego, 90-924 Łódź, Poland
e-mail: awrejcew@p.lodz.pl

R. Starosta • G. Sypniewska-Kamińska (✉)
Poznań University of Technology, Institute of Applied Mechanics,
ul. Piotrowo 3, 60-965 Poznań, Poland
e-mail: roman.starosta@put.poznan.pl; grazyna.sypniewska-kaminska@put.poznan.pl

J. Awrejcewicz (ed.), *Applied Non-Linear Dynamical Systems*, Springer Proceedings in Mathematics & Statistics 93, DOI 10.1007/978-3-319-08266-0_12

idea for an effective study of non-linear dynamical systems is linked with a concept of the so-called limiting phase trajectories (LPT) (see [3]).

The analysis of the non-linear spring pendulum is carried out in the paper. The unsteady-state oscillations near resonance are discussed. The pendulum-type mechanical systems with non-linear and parametric interactions exhibit a rich behaviour, and hence their understanding and prediction are important both from a point of view of the theory and application. Pendulums are relatively simple systems; neverthelessthey can be used to simulate the dynamics of a wide variety of engineering devices and machine parts. The coupling of the equations of motion causes possibility of autoparametric excitation and is connected to the energy exchange between modes of vibrations [8]. The energy transfer is well known in dynamics of multi-degree-of-freedom systems and is widely discussed by many authors [5, 7]. A key role either for theoretical- or application-oriented analysis is played by prediction and determination of thresholds (critical set of parameters), where transitions of system dynamics take place from a periodic quasi-linear to strongly non-linear behaviour. It can be observed in the neighbourhood of a resonance. Such critical value of non-linear parameter of the spring pendulum is determined in the paper.

2 Formulation of the Problem

Let us consider the planar motion of a mass attached to the massless non-linear spring. The examined system is shown in Fig. 1.

The Lagrangian of the system is given by

$$\mathsf{L} = mg\cos\phi\,(L_0 + Z) - \frac{1}{2}k_1 Z^2 - \frac{1}{4}k_2 Z^4 + \frac{1}{2}m\left(\dot{Z}^2 + (L_0 + Z)^2\dot{\phi}^2\right) \tag{1}$$

where m is the mass of the pendulum, L_0 is the length of the nonstretched spring, k_1 and k_2 are the stiffness coefficients, g is the Earth's acceleration and $Z(t)$ and $\phi(t)$ are generalized coordinates (see Fig. 1). The magnitudes of the forces $\mathbf{F}_1$ and $\mathbf{F}_2$ acting on the mass along and transversally to the pendulum are $F_1(t) = F_1\cos(\Omega_1\, t)$ and $F_2(t) = F_2\cos(\Omega_2 t)$. Forces of linear viscous damping are considered to be present in both longitudinal and swing motions of the pendulum (C_1 and C_2 are viscous coefficients).

The equations of motion have been obtained using Lagrange equations of the second type. Their non-dimensional form follows

$$\ddot{z}+c_1\dot{z}+z+\alpha\, z^3 + 3\alpha\, z_r^2 z + 3\alpha z_r z^2 + w^2\,(1-\cos\varphi) - (z+1)\,\dot{\varphi}^2 = f_1\cos(p_1\tau) \tag{2}$$

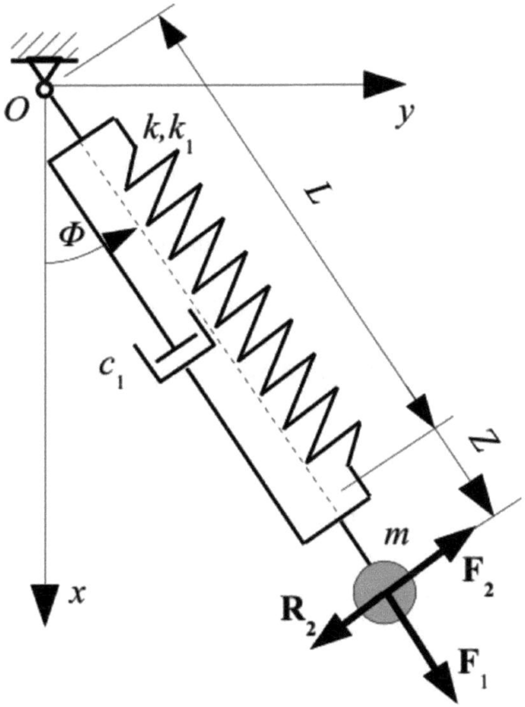

Fig. 1 The spring pendulum

$$(z+1)\left((z+1)\ddot{\varphi}+w^2\sin\varphi+\dot{\varphi}\left(c_2+2\dot{z}\right)\right)=(z+1)\,f_2\cos\left(p_2\tau\right) \tag{3}$$

where $z = Z/L$, $L = L_0 + Z_r$, $c_1 = C_1/m\omega_1$, $c_2 = C_2/L^2 m\omega_1$, $w = \omega_2/\omega_1$, $\omega_2 = \sqrt{g/L}$, $\omega_1 = \sqrt{k_1/m}$, $p_1 = \Omega_1/\omega_1$, $p_2 = \Omega_2/\omega_1$, $f_1 = F_1/Lm\omega_1^2$, $f_2 = F_2/Lm\omega_1^2$ and dimensionless time $\tau = t\omega_1$. Now z and ϕ are functions of τ, whereas z_r denotes the elongation of the spring at the static equilibrium position and fulfils the equation

$$\alpha\, z_r^3 + z_r = w^2 \tag{4}$$

The second equation of motion (3) gives, among others, a trivial solution $z = -1$ which has no physical meaning and should be rejected.

Vibrations of the system are investigated in the neighbourhood of the equilibrium position; hence, the trigonometric functions can be substituted by their power series approximations

$$\sin\varphi \approx \varphi - \varphi^3/6, \quad \cos\varphi \approx 1 - \varphi^2/2 \tag{5}$$

which limit the angle about to π/6 with precision of four significant digits.

The above remarks lead to a new form of the equations of motion

$$\ddot{z} + c_1\dot{z} + z + \alpha\, z^3 + 3\alpha\, z_r^2 z + 3\alpha\, z_r z^2 + \frac{1}{2}w^2\varphi^2 - (z+1)\,\dot{\varphi}^2 = f_1\cos(p_1\tau) \tag{6}$$

$$(z+1)\,\ddot{\varphi} + w^2\left(\varphi - \frac{\varphi^3}{6}\right) + \dot{\varphi}\,(c_2 + 2\dot{z}) = f_2\cos(p_2\tau) \tag{7}$$

Let us assume the homogeneous initial conditions

$$z(0) = 0,\;\; \dot{z}(0) = 0,\;\; \varphi(0) = 0,\;\; \dot{\varphi}(0) = 0 \tag{8}$$

The above initial problem described by the coupled and non-linear equations is investigated.

2.1 Complex Representation of the Problem

Let us introduce the phase space coordinates $\dot{z}(\tau) = v(\tau)$ and $\dot{\varphi}(\tau) = \beta(\tau)$ into (6)–(8) and then rewrite the problem in the form

$$\dot{v} + c_1 v + z + \alpha\, z^3 + 3\alpha\, z_r^2 z + 3\alpha z_r z^2 + \frac{1}{2}w^2\varphi^2 - (z+1)\,\beta^2 = f_1\cos(p_1\tau)\,, \tag{9}$$

$$(z+1)\,\dot{\beta} + w^2\left(\varphi - \frac{\varphi^3}{6}\right) + \beta\,(c_2 + 2\varpi) = f_2\cos(p_2\tau)\,, \tag{10}$$

$$z(0) = 0,\;\; v(0) = 0,\;\; \varphi(0) = 0,\;\; \beta(0) = 0. \tag{11}$$

Then, the approach proposed in the paper [3] is applied. Introduction of the complex-valued functions

$$\Psi_z = v + i\,z,\quad \Psi_\varphi = \beta + i\,w\,\varphi,\;\; \overline{\Psi}_z = v - i\,z,\quad \overline{\Psi}_\varphi = \beta - i\,w\,\varphi \tag{12}$$

converts the problems (9)–(11) to the complex form

$$\begin{aligned}
&\tfrac{1}{2}\left(\dot{\Psi}_z + \dot{\overline{\Psi}}_z\right) + \tfrac{c_1}{2}\left(\Psi_z + \overline{\Psi}_z\right) - \tfrac{1}{2}i\left(\Psi_z - \overline{\Psi}_z\right) + \tfrac{1}{8}i\;\alpha\left(\Psi_z - \overline{\Psi}_z\right)^3 \\
&- \tfrac{3}{2}\alpha\, z_r^2\left(\Psi_z - \overline{\Psi}_z\right) - \tfrac{3}{4}\alpha\, z_r\left(\Psi_z - \overline{\Psi}_z\right)^2 - \tfrac{1}{8}\left(\Psi_\varphi - \overline{\Psi}_\varphi\right)^2 \\
&+ \tfrac{1}{4}\left(\Psi_\varphi + \overline{\Psi}_\varphi\right)^2\left(\tfrac{1}{2}i\left(\Psi_z - \overline{\Psi}_z\right) - 1\right) = f_1\cos(p_1\tau)\,,
\end{aligned} \tag{13}$$

$$\frac{1}{2}\left(1-\frac{1}{2}i\left(\Psi_z-\overline{\Psi}_z\right)\right)\left(\dot{\Psi}_\varphi+\dot{\overline{\Psi}}_\varphi\right)+\frac{1}{2}\left(\Psi_\varphi+\overline{\Psi}_\varphi\right)\left(c_2+\left(\Psi_z+\overline{\Psi}_z\right)\right)$$
$$+w^2\left(-\frac{i\left(\Psi_\varphi-\overline{\Psi}_\varphi\right)^3}{48w^3}-\frac{i\left(\Psi_\varphi-\overline{\Psi}_\varphi\right)}{2w}\right)=f_2\cos\left(p_2\tau\right), \tag{14}$$

$$\Psi_z(0)=0,\ \ \Psi_\varphi(0)=0, \overline{\Psi}_z(0)=0,\ \ \overline{\Psi}_\varphi(0)=0 \tag{15}$$

The complex conjugate equations similar to (13) and (14) are also derived. They and all the consequent formulas are not written for greater clarity.

Afterwards the exponential form of the functions $\Psi_z(\tau)=\psi_z(\tau)e^{i\tau}$ and $\Psi_\varphi(\tau)=w\psi_\varphi(\tau)e^{iw\tau}$ is postulated which leads to the new form of the governing equations

$$\dot{\psi}_z+\frac{1}{8}i\ e^{-4i\tau}\alpha\left(e^{2i\tau}\psi_z-\overline{\psi}_z\right)^3+\frac{1}{2}c_1\left(\psi_z+\overline{\psi}_z e^{-2i\tau}\right)-\frac{3}{2}i\alpha\ z_r^2\left(\psi_z-\overline{\psi}_z e^{2i\tau}\right)$$
$$-\frac{3}{4}\alpha\ z_r e^{-3i\tau}\left(\overline{\psi}_z-e^{2i\tau}\psi_z\right)^2-\frac{3}{8}w^2e^{-i\tau(1+2w)}\left(\overline{\psi}_\varphi^2+\psi_\varphi^2e^{4i\tau w}\right)$$
$$+\frac{1}{4}ie^{2i\tau}w^2\psi_\varphi\overline{\psi}_\varphi\left(e^{2i\tau}\psi_z-\overline{\psi}_z+e^{i\tau}\right)$$
$$+\frac{1}{8}ie^{-2i\tau(1+w)}w^2\left(e^{2i\tau}\psi_z-\overline{\psi}_z\right)\left(e^{4iw\tau}\psi_\varphi^2+\overline{\psi}_\varphi^2\right)=f_1e^{-i\tau}\cos\left(p_1\tau\right), \tag{16}$$

$$w\dot{\psi}_\varphi+\frac{1}{2}ie^{-i\tau}w\dot{\psi}_\varphi\left(\overline{\psi}_z-e^{2i\tau}\psi_z\right)-\frac{1}{2}c_2w\left(\psi_\varphi-e^{-2iw\tau}\overline{\psi}_\varphi\right)+\frac{1}{4}e^{i\tau}w\left(w+2\right)\psi_\varphi\psi_z$$
$$-\frac{1}{4}e^{i\tau(1-2w)}w\left(w-2\right)\overline{\psi}_\varphi\psi_z-\frac{1}{4}e^{-i\tau}w\left(w-2\right)\psi_\varphi\overline{\psi}_z+\frac{1}{4}e^{-i\tau(1+2w)}w\left(w+2\right)\overline{\psi}_\varphi\overline{\psi}_z$$
$$-\frac{1}{48}iw^2e^{-4i\tau w}\left(\psi_\varphi e^{2i\tau w}-\overline{\psi}_\varphi\right)^3=f_2e^{-i\tau w}\cos\left(p_2\tau\right), \tag{17}$$

with the initial conditions

$$\psi_z(0)=0,\ \ \psi_\varphi(0)=0, \overline{\psi}_z(0)=0,\ \ \overline{\psi}_\varphi(0)=0 \tag{18}$$

3 Asymptotic Solution

The problems (16)–(18) can be efficiently solved by the asymptotic multiple scale method. The assumptions of smallness of the parameters are proposed in the form

$$c_1=\tilde{c}_1\varepsilon^2,\ \ c_2=\tilde{c}_2\varepsilon^2,\ \ z_r=\tilde{z}_r\varepsilon,\ \ f_1=\tilde{f}_1\varepsilon^3\ \ f_2=\tilde{f}_2\varepsilon^3, \tag{19}$$

where ε is the so-called small parameter.

Adopting three timescales in the analysis the solutions are searched in the following form of series with respect to the small parameter:

$$\psi_z(\tau;\varepsilon) = \sum_{k=1}^{k=3} \varepsilon^k \xi_{zk}(\tau_0, \tau_1, \tau_2) + O\left(\varepsilon^4\right),$$
$$\psi_\varphi(\tau;\varepsilon) = \sum_{k=1}^{k=3} \varepsilon^k \xi_{\varphi k}(\tau_0, \tau_1, \tau_2) + O\left(\varepsilon^4\right), \tag{20}$$

and the differential operator has the form

$$\frac{d}{d\tau} = \frac{\partial}{\partial \tau_0} + \varepsilon \frac{\partial}{\partial \tau_1} + \varepsilon^2 \frac{\partial}{\partial \tau_2} + \ldots. \tag{21}$$

3.1 Motion Near Resonance

Let us focus attention on the case of main resonance $p_2 \approx w$ and $p_1 \approx 1$. In order to deal this case the following substitutions have been done

$$p_1 = 1 + \sigma_1 \quad \text{and} \quad p_2 = w + \sigma_2, \tag{22}$$

where $\sigma_1 = \tilde{\sigma}_1 \varepsilon^2$ and $\sigma_2 = \tilde{\sigma}_2 \varepsilon^2$ are detuning parameters.

Introducing now (19), (20) and (22) into (16) and (17) and replacing the ordinary derivatives by the differential operator (21) we obtain two equations in which the small parameter ε appears. These equations should be satisfied for any value of the small parameter, so after sorting them with respect to the powers of ε we get

(i) the equations of order ε^1

$$\frac{\partial \xi_{z1}}{\partial \tau_0} = 0, \tag{23}$$

$$\frac{\partial \xi_{\varphi 1}}{\partial \tau_0} = 0, \tag{24}$$

(ii) the equations of order ε^2

$$\frac{\partial \xi_{z2}}{\partial \tau_0} + \frac{\partial \xi_{z1}}{\partial \tau_1} - \frac{1}{4} w^2 e^{-i\tau_0} \xi_{\varphi 1} \overline{\xi}_{\varphi 1} - \frac{3}{8} w^2 e^{-i\tau_0(1+2w)} \left(e^{4i\tau_0 w} \xi_{\varphi 1}^2 - \overline{\xi}_{\varphi 1}^2 \right) = 0, \tag{25}$$

$$\begin{aligned} & w\tfrac{\partial \xi_{\varphi 2}}{\partial \tau_0} - \tfrac{1}{2} i e^{i\tau_0} w \tfrac{\partial \xi_{\varphi 1}}{\partial \tau_0} \xi_{z1} + \tfrac{1}{2} i e^{-i\tau_0} w \tfrac{\partial \xi_{\varphi 1}}{\partial \tau_0} \overline{\xi}_{z1} + w \tfrac{\partial \xi_{\varphi 1}}{\partial \tau_1} + \tfrac{1}{4} e^{i\tau_0} w \xi_{\varphi 1} \xi_{z1} (w+2) \\ & - \tfrac{1}{4} e^{i\tau_0(1-2w)} w^2 \overline{\xi}_{\varphi 1} \xi_{z1} - \tfrac{1}{4} e^{-i\tau_0} w \xi_{\varphi 1} \overline{\xi}_{z1} (w-2) + \tfrac{1}{4} e^{-i\tau_0(1+2w)} w^2 \overline{\xi}_{\varphi 1} \overline{\xi}_{z1} \\ & + \tfrac{1}{2} e^{i\tau_0(1-2w)} w \overline{\xi}_{\varphi 1} \xi_{z1} + \tfrac{1}{2} e^{-i\tau_0(1+2w)} w \overline{\xi}_{\varphi 1} \overline{\xi}_{z1} = 0, \end{aligned} \tag{26}$$

(iii) the equations of order ε^3

$$\frac{\partial\xi_{z3}}{\partial\tau_0}+\frac{\partial\xi_{z2}}{\partial\tau_1}+\frac{\partial\xi_{z1}}{\partial\tau_2}-\frac{3}{2}i\tilde{\alpha}\,z_r^2\left(\xi_{z1}-\overline{\xi}_{z1}e^{-2i\tau_0}\right)-\frac{3}{4}\tilde{\alpha}\,z_re^{-3i\tau_0}\left(\overline{\xi}_{z1}-\xi_{z1}e^{2i\tau_0}\right)^2$$
$$+\frac{1}{8}i\tilde{\alpha}e^{-4i\tau_0}\,z_r^2\left(\xi_{z1}e^{2i\tau_0}-\overline{\xi}_{z1}\right)^3+\frac{\tilde{c}_1}{2}\left(\overline{\xi}_{z1}e^{-2i\tau_0}+\xi_{z1}\right)-\frac{1}{2}\tilde{f}_1e^{i\tilde{\sigma}_1\tau_2}$$
$$-\frac{1}{4}w^2e^{-i\tau_0}\left(\xi_{\varphi2}\overline{\xi}_{\varphi1}+\overline{\xi}_{\varphi2}\xi_{\varphi1}\right)-\frac{1}{4}w^2|\xi_{\varphi1}|^2\left(\xi_{z1}-\overline{\xi}_{z1}e^{-2i\tau_0}\right)$$
$$-\frac{3}{4}w^2\left(\xi_{\varphi1}\xi_{\varphi2}e^{i\tau_0(2w-1)}+\overline{\xi}_{\varphi2}\xi_{\varphi1}e^{-i\tau_0(2w+1)}\right)-\frac{1}{2}\tilde{f}_1e^{-i\tilde{\sigma}_1\tau_0-i\tilde{\sigma}_1\tau_2}$$
$$+\frac{1}{8}iw^2\left(\xi_{\varphi1}^2\xi_{z1}e^{2i\tau_0w}+\overline{\xi}_{\varphi1}^2\xi_{z1}e^{+2i\tau_0w}+\xi_{\varphi1}^2\overline{\xi}_{z1}e^{2i\tau_0(w-1)}+\overline{\xi}_{\varphi1}^2\overline{\xi}_{z1}e^{2i\tau_0(w+1)}\right)=0, \tag{27}$$

$$\frac{\partial\xi_{\varphi3}}{\partial\tau_0}-\frac{1}{2}iwe^{-i\tau_0}\left[\left(\xi_{z1}e^{2i\tau_0}-\overline{\xi}_{z1}\right)\left(\frac{\partial\xi_{\varphi2}}{\partial\tau_0}-\frac{\partial\xi_{\varphi1}}{\partial\tau_1}\right)-\left(\xi_{z2}e^{2i\tau_0}-\overline{\xi}_{z2}\right)\frac{\partial\xi_{\varphi1}}{\partial\tau_0}\right]$$
$$+w\frac{\partial\xi_{\varphi2}}{\partial\tau_1}+w\frac{\partial\xi_{\varphi1}}{\partial\tau_2}+\frac{\tilde{c}_2}{2}w\left(\xi_{\varphi1}+\overline{\xi}_{\varphi1}e^{-2iw\tau_0}\right)-\frac{1}{2}\tilde{f}_2e^{i\tilde{\sigma}_2\tau_2}-\frac{1}{2}\tilde{f}_2e^{-i(2w\tau_0+\tilde{\sigma}_2\tau_2)}$$
$$+\frac{1}{4}e^{i\tau_0}w(2+w)\left(\xi_{z2}\xi_{\varphi1}+\xi_{z1}\xi_{\varphi2}\right)-\frac{1}{4}e^{-i\tau_0}w(w-2)\left(\overline{\xi}_{z2}\xi_{\varphi1}+\overline{\xi}_{z1}\xi_{\varphi2}\right)$$
$$-\frac{1}{48}iw^2e^{-4iw\tau_0}\left(\xi_{\varphi1}e^{2iw\tau_0}-\overline{\xi}_{\varphi1}\right)^3-\frac{1}{4}w(w-2)\,e^{i\tau_0(1-2w)}\xi_{z2}\overline{\xi}_{\varphi1}$$
$$+\frac{1}{4}w(w+2)\,e^{-i\tau_0(1+2w)}\overline{\xi}_{z2}\overline{\xi}_{\varphi1}+\frac{1}{2}we^{i\tau_0(1-2w)}\xi_{z1}\overline{\xi}_{\varphi2}\left(1-\frac{w}{2}\right)$$
$$+\frac{1}{2}we^{-i\tau_0(1+2w)}\overline{\xi}_{z1}\overline{\xi}_{\varphi2}\left(1+\frac{w}{2}\right)=0. \tag{28}$$

The requirement of zeroing of secular terms in (23)–(26) causes the functions $\xi_{z1}(\tau_2), \overline{\xi}_{z1}(\tau_2), \xi_{\varphi1}(\tau_2), \overline{\xi}_{\varphi1}(\tau_2)$ to depend only on the slowest timescale τ_2.
Solutions of the second-order equations (25) and (26)

$$\xi_{z2}=G_1(\tau_1,\tau_2)+\frac{3iw^2}{8(1-2w)}\left(e^{i\tau_0(2w-1)}\xi_{\varphi1}^2+e^{-i\tau_0(2w+1)}\overline{\xi}_{\varphi1}^2\right)+\frac{1}{4}iw^2e^{-i\tau_0}\xi_{\varphi1}\overline{\xi}_{\varphi1}, \tag{29}$$

$$\xi_{\varphi2}=G_2(\tau_1,\tau_2)+\frac{1}{4}ie^{i\tau_0}(2+w)\,\xi_{z1}\xi_{\varphi1}+\frac{1}{4}ie^{-i\tau_0}(w-2)\,\overline{\xi}_{z1}\xi_{\varphi1}$$
$$+\frac{i(w-2)\,e^{i\tau_0(1-2w)}\xi_{z1}\overline{\xi}_{\varphi1}}{8w-4}-\frac{i(w+2)\,e^{-i\tau_0(1+2w)}\overline{\xi}_{z1}\overline{\xi}_{\varphi1}}{8w+4}, \tag{30}$$

are then introduced into equations of the third order (27) and (28). According to the initial conditions (18), $G_1=0$ and $G_2=0$.

Assuming that the system vibrates far from the internal resonance $2w-1=0$, the requirement that the solutions should be limited in time leads to the equations

$$\frac{\partial\xi_{z1}}{\partial\tau_2}+\frac{\tilde{c}_1}{2}\xi_{z1}-\frac{3}{2}i\,\tilde{\alpha}\,\tilde{z}_r^2\xi_{z1}p-\frac{3}{8}i\,\tilde{\alpha}\,|\xi_{z1}|\,\xi_{z1}+\frac{3i\,w^2\left(w^2-1\right)}{4-16w^2}\left|\xi_{\varphi1}\right|\,\xi_{z1}=\frac{1}{2}\tilde{f}_1e^{i\tau_2\tilde{\sigma}_1}, \tag{31}$$

$$w\frac{\partial\xi_{\varphi 1}}{\partial\tau_2}+\frac{\tilde{c}_2}{2}w\xi_{\varphi 1}-\frac{1}{4}i\,w^2\left|\xi_{z1}\right|\,\xi_{\varphi 1}+\frac{i\,w^2\left(w^2+2\right)}{16w^2-4}\left|\xi_{z1}\right|\,\xi_{\varphi 1}$$
$$+\frac{i\,w^2\left(8w^4-7w^2-1\right)}{64w^2-16}\left|\xi_{\varphi 1}\right|\,\xi_{\varphi 1}=\frac{1}{2}\tilde{f}_2e^{i\tau_2\tilde{\sigma}_2}. \tag{32}$$

Now real representation of the functions ξ_{z1} and $\xi_{\varphi 1}$ of the following form

$$\xi_{z1}\left(\tau_2\right)=\tilde{a}_1\left(\tau_2\right)e^{i\delta_1(\tau_2)},\quad \xi_{\varphi 1}\left(\tau_2\right)=\tilde{a}_2\left(\tau_2\right)e^{i\delta_2(\tau_2)},\quad \tilde{a}_i=a_i\varepsilon\quad \text{for}\quad i=1,2 \tag{33}$$

is introduced to the above secular terms (31) and (32).

Then we go back to the original denotations according to (19) and take advantage of the definition (21). Comparison of the real and imaginary parts of both sides of (31) and (32) leads to four modulation equations with respect to amplitudes a_1, a_2 and modified phases θ_1, θ_2:

$$\frac{da_1}{d\tau}=-\frac{1}{2}c_1a_1+\frac{1}{2}f_1\cos\theta_1, \tag{34}$$

$$a_1\frac{d\theta_1}{d\tau}=-\frac{3}{2}z_r^2\alpha\,a_1+\sigma_1a_1-\frac{3}{8}\alpha\,a_1^3+\frac{3w^2\left(w^2-1\right)}{4-16w^2}a_1a_2^2-\frac{1}{2}f_1\sin\theta_1, \tag{35}$$

$$\frac{da_2}{d\tau}=-\frac{1}{2}c_2a_2+\frac{1}{2w}f_2\cos\theta_2, \tag{36}$$

$$a_2\frac{d\theta_2}{d\tau}=\sigma_2a_2+\frac{3w\left(w^2-1\right)}{4-16w^2}a_2a_1^2+\frac{w\left(8w^4-7w^2-1\right)}{64w^2-16}a_2^3-\frac{1}{2w}f_2\sin\theta_2, \tag{37}$$

where modified phases θ_1, θ_2 are defined as follows:

$$\delta_1\left(\tau_2\right)=\tau_2\sigma_1-\theta_1\left(\tau_2\right),\quad \delta_2\left(\tau_2\right)=\tau_2\sigma_2-\theta_2\left(\tau_2\right). \tag{38}$$

The above definitions cause the systems (34)–(37) to become an autonomous one. It describes the dynamics of the non-linear spring pendulum near simultaneously occurring main resonances.

4 Examples

The LPT concept allows to describe the intensive energy exchange between the degrees of freedom and external sources. The system examined in the paper is especially sensitive to changes of the value of the parameter α responsible for non-linear characteristics of the spring. One can observe a critical value $\alpha=\alpha_{LPT}$ for which the character of the vibrations dramatically changes. The value of α_{LPT}

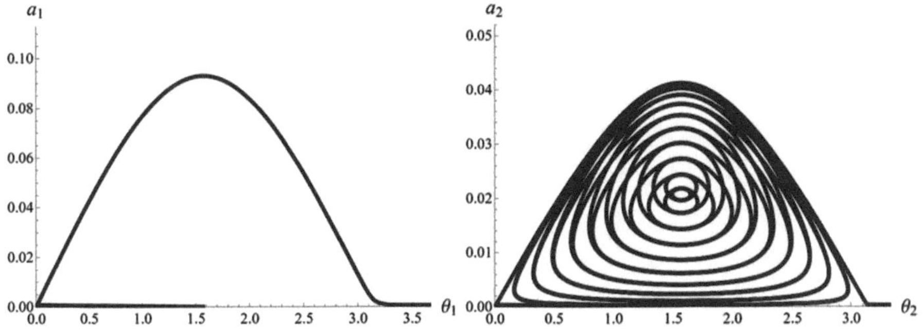

Fig. 2 Limiting phase trajectories for longitudinal and swing vibrations; $\alpha = 0.3 < \alpha_{LPT}$

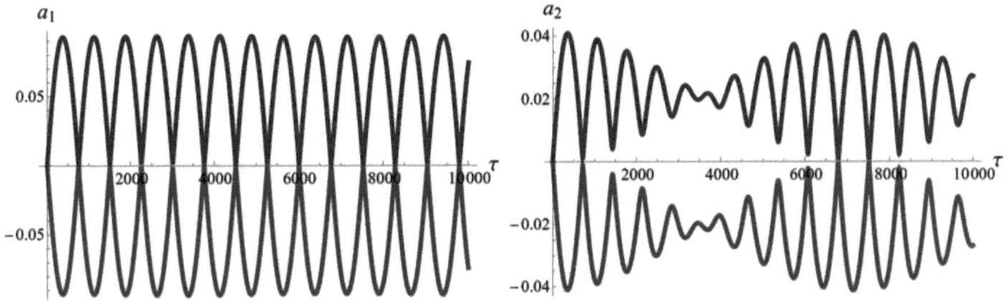

Fig. 3 Amplitudes modulations; $\alpha = 0.3 < \alpha_{LPT}$

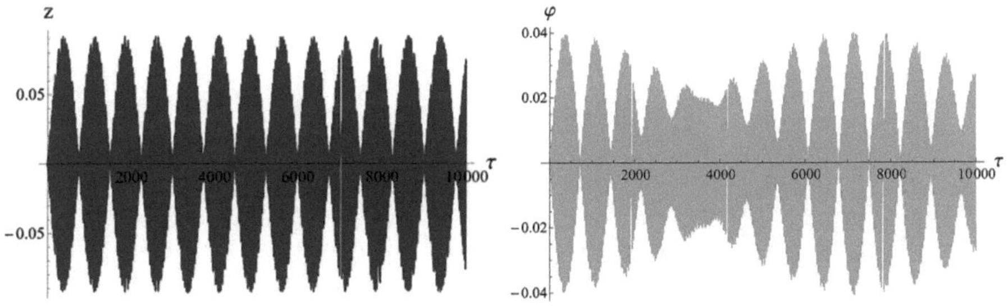

Fig. 4 Time histories obtained numerically; $\alpha = 0.3 < \alpha_{LPT}$

depends on all the parameters of the system. In the case of certain one degree-of-freedom (DOF) systems α_{LPT} can be obtained analytically [2, 3]. When the number of DOF is higher than one and couplings appear in the equations of modulation, α_{LPT} can be obtained approximately.

The results of calculations for the chosen values of parameters $\sigma_1 = 0.01$, $\sigma_2 = 0.01, f_1 = 0.0008, f_2 = 0.00008$, $c_1 = 0$, $c_2 = 0$, $w = 0.21$ are presented below. For these parameters $\alpha_{LPT} \approx 0.654$. Figures 2, 3 and 4 show some graphs concerning the case when $\alpha = 0.3 < \alpha_{LPT}$.

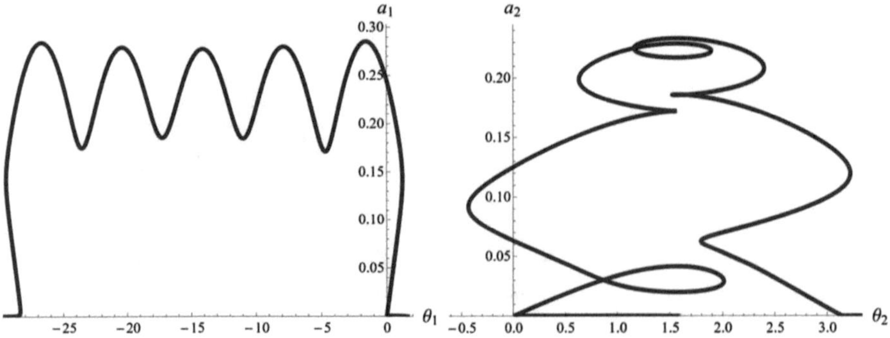

Fig. 5 Limiting phase trajectories for spring and swing vibrations; $\alpha = 0.7 > \alpha_{LPT}$

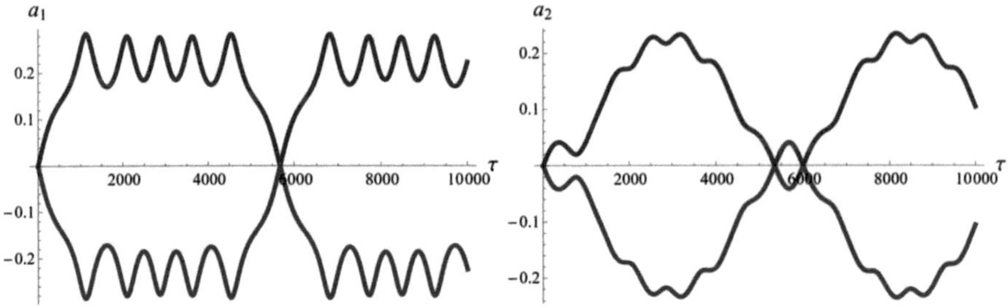

Fig. 6 Amplitudes modulations; $\alpha = 0.7 > \alpha_{LPT}$

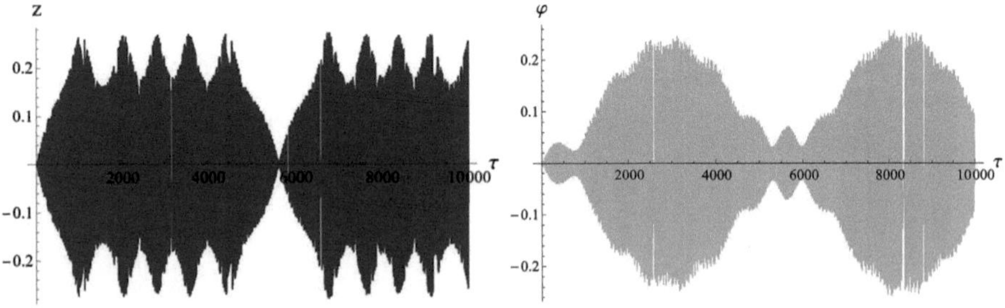

Fig. 7 Time histories obtained numerically; $\alpha = 0.7 > \alpha_{LPT}$

In Figs. 2, 3 and 4 the intensive energy exchange between the system and its surrounding can be observed. The amplitudes are relatively small and the vibrations are quasi-linear. The time histories presented in Fig. 4 have been received by numerical solution to the problem (6)–(8). The amplitude modulations and time histories (compare Figs. 3 and 4) are highly compatible.

For $\alpha > \alpha_{LPT}$ some non-linear effects occur. In Figs. 5, 6 and 7 amplitudes are much larger than in the case when $\alpha < \alpha_{LPT}$ and the vibrations become strongly non-linear.

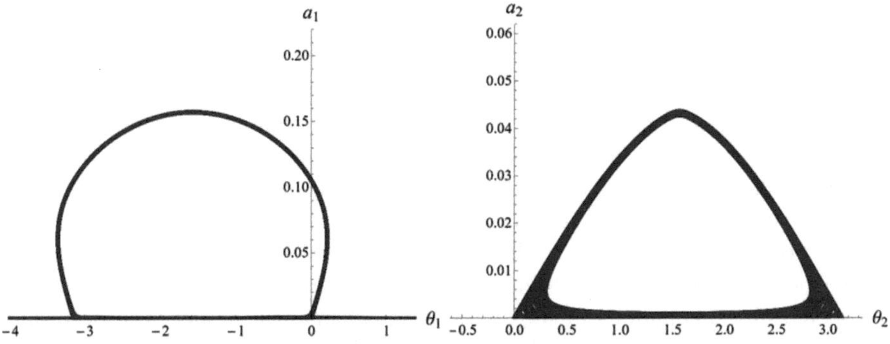

Fig. 8 Limiting phase trajectories for spring and swing vibrations; $\alpha = 2.0 >> \alpha_{LPT}$

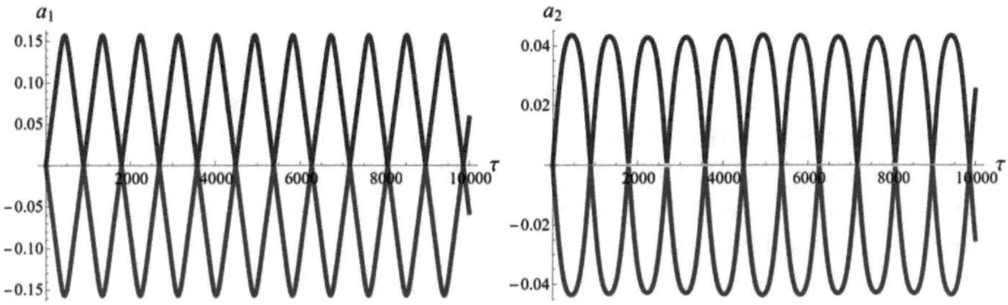

Fig. 9 Amplitudes modulations; $\alpha = 2.0 >> \alpha_{LPT}$

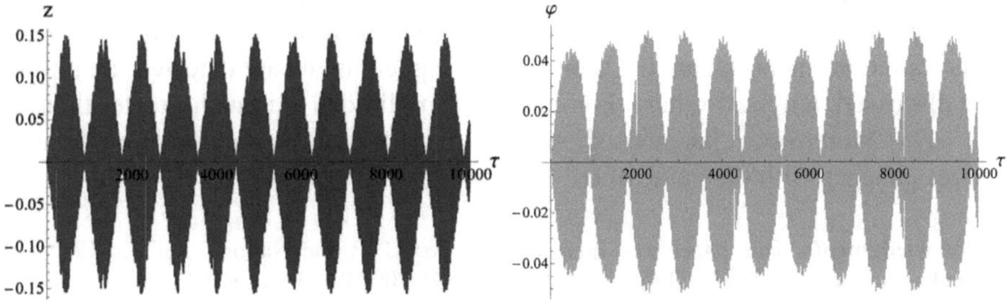

Fig. 10 Amplitudes modulations; $\alpha = 2.0 >> \alpha_{LPT}$

In the case $\alpha >> \alpha_{LPT}$ modulations of amplitudes of both coordinates become again more regular, as is shown in Figs. 9 and 10. However, their shape indicates the non-linear effects. The longitudinal vibration tends to sawtooth form with the increase of non-linearity parameter. A synchronization between the amplitude modulations of both general coordinates in the slow timescale is observable (Fig. 9). The oscillations of the system tend to steady state what is seen not only in the time history but also in the phase – amplitude plane (Fig. 8).

5 Conclusions

Analytical study of the non-linear spring pendulum in planar motion has been carried out. After transformation of the governing equations of motion to the complex representation, the asymptotic analysis with the help of multiple timescale (MTS) method has been applied. That approach leads to obtain a set of differential equations of simpler form than the original ones. It is worth to note that, thanks to the application of the MTS variant with three timescales, the non-linear terms as well as the most important coupled terms between the generalized coordinates have been preserved in the equations of the simplified mathematical model. The performed investigations have been focused on the nonsteady vibrations of the forced system near the simultaneously occurring external resonances. The modulation equations concerning this case have been derived from the equations of first order as well as from the requirement of vanishing the secular terms of the equations of higher order. The solutions obtained analytically from the equations of modulation of amplitudes and phases have been verified by comparing them with the solutions which are received numerically from the original equations of motion. Their high accuracy has been confirmed in all performed numerical simulations.

The main advantage of the asymptotic solutions consists in achieving qualitative information about the dynamics of the considered system.

Analysis of the curves which represent the dynamical behaviour of the system in the plane phase-amplitude gives evidence of very interesting features of dynamics of the system. The shape of these curves depends strongly on the values of parameter α, which is connected with the spring non-linearity. The most intensive energy transfer between the system and its surroundings is governed by the so-called LPT. Important non-linear dynamical transition-type phenomena are detected, monitored and discussed, amongst others. For $\alpha > \alpha_{LPT}$ amplitudes are much greater than for $\alpha < \alpha_{LPT}$. Moreover, it has been shown that the shape of the amplitude modulation curves changes with the value of α.

Acknowledgments This paper was financially supported by the National Science Centre of Poland under the grant MAESTRO 2, No. 2012/04/A/ST8/00738, for years 2013–2016.

References

1. Awrejcewicz, J., Krysko, V.A.: Introduction to Asymptotic Methods. Chapman and Hall, Boca Raton (2006)
2. Manevitch, E.L., Manevitch, L.I.: Limiting phase trajectories (LPT) in 1 dof asymmetric system with damping and 1:1 resonance. In: DSTA 10th Conference, Łódź, pp. 559–568, 2009
3. Manevitch, L.I., Musienko, A.I.: Limiting phase trajectories and energy exchange between anharmonic oscillator and external force. Nonlinear Dyn. **58**, 633–642 (2009)
4. Sado, D.: Energy Transfer in the Nonlinear Coupled Systems of Two-Degree of Freedom (in Polish). OWPW, Warszawa (1997)

5. Sado, D., Gajos, K.: Analysis of vibrations of three-degree-of-freedom dynamical system with double pendulum. J. Theor. Appl. Mech. **46**(1), 141–156 (2008)
6. Shivamoggi, B.K.: Perturbation Methods for Differential Equations. Birkhauser, Boston (2002)
7. Starosta, R., Sypniewska-Kamińska, G., Awrejcewicz, J.: Resonances in kinematically driven nonlinear spring pendulum. In: DSTA 11th Conference, pp. 103–108, 2011
8. Starosvetsky, Y., Gendelman, O.V.: Dynamics of a strongly nonlinear vibration absorber coupled to a harmonically excited two-degree-of-freedom system. J. Sound Vib. **312**, 234–256 (2008)

Vibration Surveillance System with Variable Stiffness Holder for Milling Flexible Details

Krzysztof J. Kaliński, Marek Chodnicki, Michał R. Mazur, and Marek A. Galewski

Abstract Efficient milling of the flexible details (i.e. rotor blades, thin-walled elements) using slender ball-end tools is a difficult task due to possibility of vibration occurrence. Because of the existence of certain conditions (small depths of cutting, regeneration phenomena), cutting process may lose stability and self-excited chatter vibration may appear. Frequency of the chatter vibration is close to dominant natural frequency of the workpiece or the tool. One of the methods of chatter vibration avoidance is matching the spindle speed to the optimal phase shift between subsequent cutting edge passes (i.e. the Liao–Young condition). In previous works the authors successfully implemented the idea of optimal speeds map where optimal speed was calculated for every point of the machined surface based on the dominant natural frequencies for local areas. During milling, spindle speed was set according to the map. However, changing spindle speed during tool pass may reduce surface quality in speed change point and is difficult to perform it in some milling centres. The article presents the idea of a new workpiece holder with adjustable stiffness. Milling process will be performed with constant spindle and feed speed. In order to avoid vibration, stiffness of the specially designed workpiece holder will be modified off-line. Stiffness changes modify natural frequencies of the workpiece and thus, it is possible to modify dynamic properties of the workpiece in such a way that arbitrary chosen, constant spindle speed will be optimal, due to the Liao–Young condition performance. This will need calculation of the optimal stiffness map (referred to different spindle speeds), which will be performed before milling based on the workpiece's modal identification results and Finite Element Model simulations.

1 Introduction

Milling flexible details (i.e. rotor blades, thin-walled elements) using slender ball-end tools is a frequent task for contemporary machining centres. It is also usually demanded that the milling treatment should be "final" which means that no

K.J. Kaliński (✉) • M. Chodnicki • M.R. Mazur • M.A. Galewski
Gdansk University of Technology, G. Narutowicza 11/12, 80-233 Gdansk, Poland
e-mail: kkalinsk@o2.pl; marchodn@pg.gda.pl; mazur.m.r@gmail.com; margalew@pg.gda.pl

J. Awrejcewicz (ed.), *Applied Non-Linear Dynamical Systems*, Springer Proceedings in Mathematics & Statistics 93, DOI 10.1007/978-3-319-08266-0_13

additional finishing operations are required for the surface quality being sufficiently high. On the other hand, milling flexible details is also a challenging problem because such objects can be prone to vibration especially when the workpiece is difficult to support or cannot be supported during milling processes. In such a case, tool–workpiece relative vibration plays a principal role because, in certain conditions, it may lead to a loss of stability and cause a generation of self-excited chatter vibration [8]. One of the methods of the chatter reduction depends on the selection of the spindle speed matched to the optimal phase shift between subsequent passes of the tool cutting edges [7]. This optimal spindle speed can be calculated using the generalised Liao–Young condition [1]:

$$\frac{zn_\alpha}{60} = \frac{f_\alpha}{0.25 + m}, \qquad m = 0, 1, \;\; 2, \;\; \ldots, \tag{1}$$

where:

f_α is the natural frequency no. α of the workpiece (Hz),
n_α is the requested spindle speed (rev/min),
z is the number of cutting edges of the milling tool.

This means that the only unknown in Eq. 1 is the dominant natural frequency f_α. It may be identified, for example, by the modal test performance. As the result, a set of optimal spindle speeds is obtained. From this set, one speed is selected for performing milling operation. The problem is that the speed, which is optimal from the point of view of Liao–Young condition, may not be optimal from the point of view of the other requirements, i.e. cutting process efficiency. As an example, for $f_\alpha = 83.3$ Hz and $z = 2$, the set of optimal spindle speeds is $n_\alpha \in <10{,}000, 2{,}000, 1{,}111, 769, \ldots>$ (rev/min). One can see that there is a huge gap between the first and the second speed. For example, in case when milling centre can achieve speeds only up to 8,000 rev/min, the highest optimal (in terms of vibration reduction) speed, which can be selected, is 2,000 rev/min. This means that in this situation it is not possible to make the most of milling centre potential.

Additionally, in case of complex-shaped workpieces, there may be different dominant natural frequencies for different workpiece zones. This problem can be solved by creating a map of optimal spindle speeds. In order to create such a map, a Finite Element Model of a workpiece is created and tuned according to the experimental modal tests. Then, the model is used for calculation of dominant natural frequencies for the selected workpiece points. For these points optimal spindle speeds are calculated. The whole procedure is described in details in [5, 6] and successfully implemented for milling complex-shaped flexible workpieces. However, changing spindle speed during tool pass may reduce surface quality in speed change point and is difficult to perform in some milling centres. This happens especially when milling centre controller does not allow speed changes without stopping the tool or workpiece feed. The problem can be avoided by milling a whole tool pass with only one spindle speed. However, this also means that in some zones this speed will not be optimal.

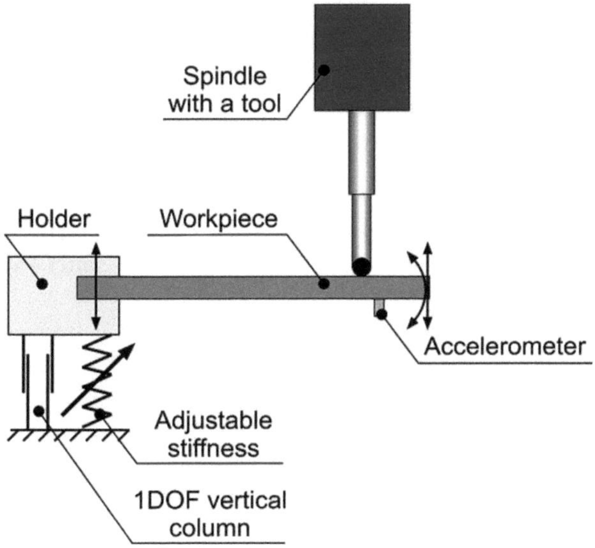

Fig. 1 Workpiece holder with adjustable stiffness

2 Variable Stiffness Workpiece Holder

According to Eq. (1), optimal spindle speed depends on dominant natural frequency of the workpiece. In turn, this frequency depends on the workpiece's dynamic properties, which may be considered as time-invariant. However, the workpiece is mounted in a holder. Usually the holder is considered as of a stiff support. In the presented case, a holder is one degree of freedom (DOF) structure with adjustable stiffness (Fig. 1). Thanks to this, it is possible to modify dynamic properties of the whole system (consisting of the holder and the workpiece) and to adjust its natural frequency. This foresees a possibility of tuning the natural frequency in such a way that any arbitrary spindle speed will be optimal, in accordance with Eq. (1). For example, choosing $n_\alpha = 10{,}000$ rev/min implies that natural frequency should be one of set $f_\alpha \in <83, 416, 750, \ldots>$ (Hz). This means that the stiffness of the holder must be suitably adjusted.

3 Holder, Workpiece and Cutting Process Models

Dynamics of cutting process can be described using a proportional model [2, 4]. In this model cutting force components depend proportionally on cutting layer thickness and also depend on variable in time depth of cutting. Resultant cutting force lies in orthogonal plane. Depending on the direction of action three force components can be separated, i.e. the one acting along nominal cutting velocity

(Eq. 2), thrust force acting along change in cutting layer thickness (Eq. 3) and the force acting along tool axis which is equal to 0 (Eq. 4):

$$F_{yl1}(t) = \begin{cases} k_{dl} a_p(t) \left[h_{Dl}(t) - \Delta h_l(t) + \Delta h_l (t - \tau_l)\right], & h_{Dl}(t) - \Delta h_l(t) + \Delta h_l (t - \tau_l) > 0, \\ 0 & , h_{Dl}(t) - \Delta h_l(t) + \Delta h_l (t - \tau_l) \leq 0, \end{cases} \tag{2}$$

$$F_{yl2}(t) = \begin{cases} \mu_l k_{dl} a_p(t) \left[h_{Dl}(t) - \Delta h_l(t) + \Delta h_l (t - \tau_l)\right], & h_{Dl}(t) - \Delta h_l(t) + \Delta h_l (t - \tau_l) > 0, \\ 0 & , h_{Dl}(t) - \Delta h_l(t) + \Delta h_l (t - \tau_l) \leq 0, \end{cases} \tag{3}$$

$$F_{yl3}(t) = 0 \tag{4}$$

where:

k_{dl} is the average dynamic specific cutting pressure,
$a_p(t)$ is the depth of cutting,
$h_{Dl}(t)$ is the desired cutting layer thickness,
$\triangle h_l(.)$ is the dynamic change in cutting layer thickness,
μ_l is the cutting force ratio,
τ_l is the time-delay.

In order to idealise dynamics of non-stationary milling process of a flexibly supported rectangular plate, a hybrid approach is adopted [2, 3]. As a result of the milling process modelling, the hybrid system is obtained which assembles three subsystems (Fig. 2):

- *Modal subsystem*, i.e. a stationary model of one-side-supported flexible plate, which displaces itself with feed speed v_f
- *Structural subsystem*, i.e. non-stationary discrete model of ball-end mill and cutting process
- *Connective subsystem*, i.e. conventional contact point S between the tool and the workpiece

Matrix dynamic equation of non-stationary model of the milling process in hybrid co-ordinates has a form [2] (Eq. 5)

$$\begin{bmatrix} \mathbf{M} & \mathbf{0} \\ \mathbf{0} & \mathbf{I} \end{bmatrix} \ddot{\boldsymbol{\xi}} + \begin{bmatrix} \mathbf{L} & \mathbf{0} \\ \mathbf{0} & 2\mathbf{Z}\boldsymbol{\Omega} \end{bmatrix} \dot{\boldsymbol{\xi}} + \begin{bmatrix} \mathbf{K} + \sum_{l=1}^{i_l} \mathbf{T}_l^T(t)\mathbf{D}_{Pl}(t)\mathbf{T}_l(t) & -\sum_{l=1}^{i_l} \mathbf{T}_l^T(t)\mathbf{D}_{Pl}(t)\mathbf{W}_{ml}(t) \\ -\sum_{l=1}^{i_l} \mathbf{W}_{ml}^T(t)\mathbf{D}_{Pl}(t)\mathbf{T}_l(t) & \boldsymbol{\Omega}^2 + \sum_{l=1}^{i_l} \mathbf{W}_{ml}^T(t)\mathbf{D}_{Pl}(t)\mathbf{W}_{ml}(t) \end{bmatrix} \boldsymbol{\xi} =$$
$$= \begin{bmatrix} \sum_{l=1}^{i_l} \mathbf{T}_l^T(t)\mathbf{F}_l^0(t) \\ -\sum_{l=1}^{i_l} \mathbf{W}_{ml}^T(t)\mathbf{F}_l^0(t) \end{bmatrix} + \begin{bmatrix} \sum_{l=1}^{i_l} \mathbf{T}_l^T(t)\mathbf{D}_{Ol}(t)\boldsymbol{\Delta}\mathbf{w}(t-\tau_l) \\ -\sum_{l=1}^{i_l} \mathbf{W}_{ml}^T(t)\mathbf{D}_{Ol}(t)\boldsymbol{\Delta}\mathbf{w}(t-\tau_l) \end{bmatrix}, \tag{5}$$

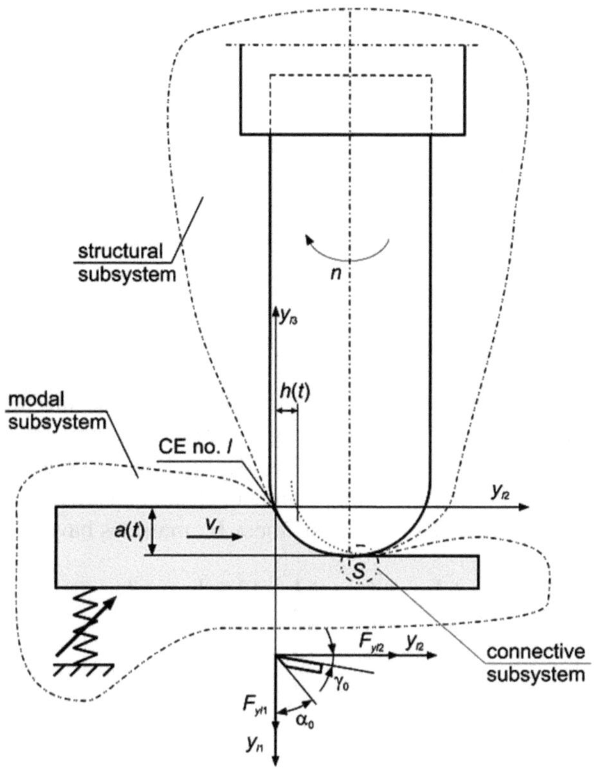

Fig. 2 Hybrid system of the milling process

where:

$\boldsymbol{\xi} = \begin{Bmatrix} \mathbf{q} \\ \mathbf{a} \end{Bmatrix}$ is the vector of hybrid co-ordinates of the hybrid system,

M, L, K are the matrices of inertia, damping and stiffness of the decoupled structural subsystem,

a is the vector of modal co-ordinates of the modal subsystem,

q is the vector of generalised co-ordinates of the structural subsystem,

$\mathbf{T}_l$(t) is the matrix of transformation of displacements from co-ordinate system of the cutting tool, to co-ordinate system y_{l1}, y_{l2}, y_{l3} of CE no. l,

$\mathbf{W}_{ml}(t)$ is the matrix of constraints between displacements in modal co-ordinates **a** of the modal subsystem and displacements in co-ordinate system y_{l1}, y_{l2}, y_{l3} of CE no. l,

$\mathbf{D}_{Pl}$(t), $\mathbf{D}_{Ol}$(t) are the matrices of proportional and delayed feedback of CE no. l,

Z is the matrix of dimensionless damping coefficients of the modal subsystem,

Ω is the matrix of angular natural frequencies of the modal subsystem,

$\mathbf{F}_l^0(t)$ is the vector of the desired forces of CE no. l,

$\Delta\mathbf{w}_l(t-\tau_l)$ is the vector of deflections of CE no. l for time-instant $t-\tau_l$,

Fig. 3 Scheme of the simplified workpiece model, the considered DOF are denoted with *grey arrows*

i_l is the number of "active" coupling elements.

In order to identify modal model of the machined plate, the mass normalised matrix of normal modes $\boldsymbol{\Psi}$ corresponding to angular natural frequencies of the modal subsystem must be determined. This can be done using a simplified model based on the Euler–Bernoulli bar finite element [2] and considering only three DOF (Fig. 3). The appropriate mass $\mathbf{M}_e$ and stiffness $\mathbf{K}_e$ matrices have forms

$$\mathbf{M}_e = \rho_e l_e \begin{bmatrix} \frac{13A_e}{35} + \frac{6I_e}{5l_e^2} & \frac{9A_e}{70} - \frac{6I_e}{5l_e^2} & \frac{13A_e l_e}{420} - \frac{I_e}{10l_e} \\ & \frac{13A_e}{35} + \frac{6I_e}{5l_e^2} & \frac{11A_e l_e}{210} + \frac{I_e}{10l_e} \\ sym. & & \frac{A_e l_e^2}{105} + \frac{2I_e}{15} \end{bmatrix} + \begin{bmatrix} m_l & 0 & 0 \\ 0 & 0 & 0 \\ 0 & 0 & 0 \end{bmatrix}, \tag{6}$$

$$\mathbf{K}_e = \frac{E}{l_e} \begin{bmatrix} \frac{12I_e}{l_e^2} & -\frac{12I_e}{l_e^2} & -\frac{6I_e}{l_e} \\ & \frac{12I_e}{l_e^2} & \frac{6I_e}{l_e} \\ sym. & & 4I_e \end{bmatrix} + \begin{bmatrix} k_s & 0 & 0 \\ 0 & 0 & 0 \\ 0 & 0 & 0 \end{bmatrix} \tag{7}$$

where:

ρ_e is the density of the workpiece material,
E is the Young modulus of the workpiece material,
l_e is the length of the plate workpiece,
A_e is the cross-section area of the plate workpiece,
I_e is the cross-section area moment of inertia of the plate workpiece,
k_s is the adjustable stiffness of the holder's support,
m_l is the lumped mass of the holder's moving part.

4 Simulations

At first, simulations are performed in order to confirm that variable stiffness of the holder modifies normal modes of the workpiece. For the selected settings of the holder stiffness adjusting mechanism, i.e. position of the micrometre screw (mm), the corresponding stiffness is calculated and then the related first natural frequencies and corresponding normal modes of the system described by matrices (Eq. 6) and (Eq. 7) are determined. Additionally, using the generalised Liao–Young condition (Eq. 1), relevant optimal spindle speeds are calculated. Computation is performed for the following data: workpiece material—aluminium alloy EN AW-6101A, $\rho_e = 2.7 \times 10^3$ kg/m^3, $E = 7 \times 10^{10}$ N/m^2, $l_e = 0.160$ m, $A_e = 2.5 \times 10^{-4}$ m^2, $I_e = 5.2083 \times 10^{-10}$ m^4, $m_l = 2$ kg. The results are shown in Table 1.

It is worthy of note that not all of the holder stiffness adjustments assure successful result of the workpiece vibration surveillance. The optimal spindle speed is really obtained, when the mass normalised normal mode of the workpiece:

- Has a similar form to one of the stiff support cases, i.e. relevant components of the two compared normal modes have the same signs
- Does not concern the rigid body movement, i.e. first and second component of the normal mode significantly differ with each other

Table 1 Holder stiffness influence on first normal mode of the workpiece and corresponding optimal spindle speed

Micrometre screw position (mm)	Holder stiffness k_s (N/m)	Natural frequency f_α (Hz)	Normal mode (mass normalised) Ψ_1			Optimal spindle speed n (rev/min)	Result of surveillance
Stiff support		161.4	0.0000	6.1439	−52.8957	–	–
20	2.3100×10^6	151.0	−0.4002	−4.8077	37.7958	18,120	Negative
21	1.9955×10^6	145.4	−0.5070	−3.9404	29.3819	17,450	Negative
22	1.7355×10^6	138.5	0.5787	3.1180	−21.6776	16,623	Successful
23	1.5189×10^6	131.2	0.6189	2.5081	−16.0883	15,743	Successful
24	1.3368×10^6	124.0	0.6412	2.0884	−12.2956	14,880	Successful
25	1.1827×10^6	117.2	0.6544	1.7975	−9.6923	14,064	Successful
26	1.0514×10^6	110.9	0.6627	1.5896	−7.8452	13,305	Successful
27	9.3888×10^5	105.0	0.6682	1.4361	−6.4888	12,603	Successful
28	8.4184×10^5	99.6	0.6722	1.3194	−5.4617	11,956	Successful
29	7.5772×10^5	94.7	0.6750	1.2284	−4.6637	11,358	Successful
30	6.8444×10^5	90.1	0.6772	1.1559	−4.0299	10,812	Successful
31	6.2032×10^5	85.8	0.6789	1.0970	−3.5173	10,298	Successful
32	5.6396×10^5	81.9	0.6802	1.0486	−3.0962	9,826	Negative
35	4.3102×10^5	72.0	0.6829	0.9451	−2.1995	8,604	Negative
40	2.8875×10^5	59.0	0.6853	0.8488	−1.3685	7,056	Negative

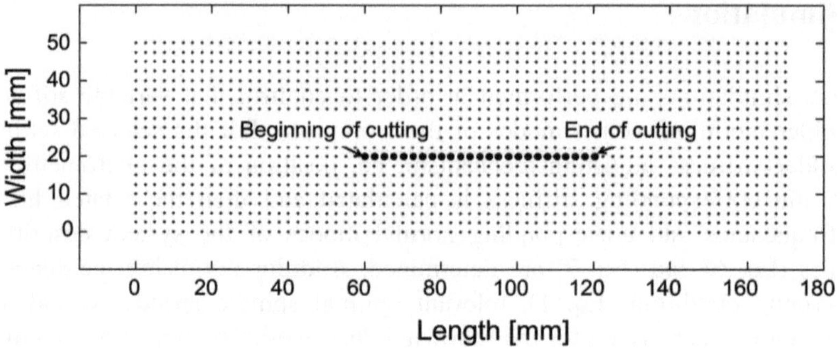

Fig. 4 The selected path of the tool along the workpiece length

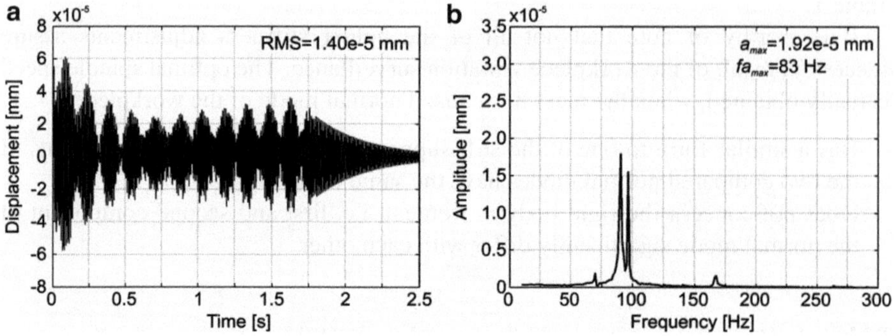

Fig. 5 (**a**) Vibration displacements, (**b**) corresponding amplitude spectrum for $n = 10{,}000$ rev/min and $k_s = 5.845 \times 10^5$ N/m, optimal case, $f_\alpha = 83$ Hz

The purpose of the next step of simulations is to show that for a chosen stiffness value the calculated spindle speed is really optimal and, on the other hand, for a given spindle speed, this stiffness is optimal as well. Also for this couple, i.e. optimal stiffness—optimal spindle speed, the workpiece vibration level should be the lowest. For every simulation the root mean square (RMS) of the free (unsupported) workpiece end displacement is observed as well as the maximum value of vibration spectrum. The process of milling 60 mm long groove on the workpiece surface (Fig. 4) is simulated using the hybrid model of the milling process (Eq. 5). Some selected cases of simulation (Figs. 5, 6 and 7) as well as the summary of the results for various cases are presented (Fig. 8).

The results of simulation (Fig. 8) confirm expectations of the holder stiffness influence on first normal mode of the workpiece and corresponding optimal spindle speed for a wide range of spindle speeds, i.e. between 10,000 and 14,064 rev/min. The only one stiffness setting is optimal and gives both the lowest value of RMS of displacement as well as the lowest maximum amplitude in spectrum. In such cases the generalised Liao–Young condition is really fulfilled (Eq. 1). Additionally, for some nonoptimal stiffness-spindle speed pairs, vibrations may be very high. For

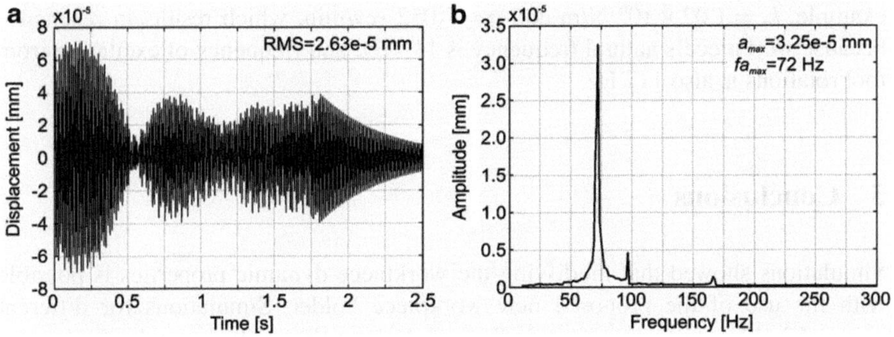

Fig. 6 (**a**) Vibration displacements, (**b**) corresponding amplitude spectrum for $n = 10{,}000$ rev/min and $k_s = 6.844 \times 10^5$ N/m, nonoptimal case, higher stiffness, $f_\alpha = 72$ Hz

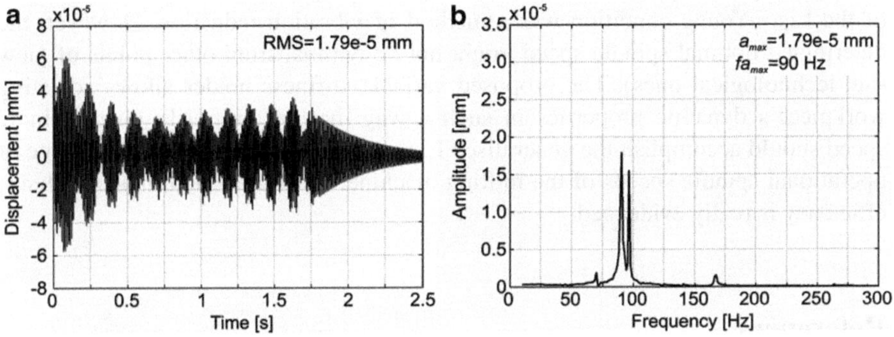

Fig. 7 (**a**) Vibration displacements, (**b**) corresponding amplitude spectrum for $n = 10{,}000$ rev/min and $k_s = 4.310 \times 10^5$ N/m, nonoptimal case, lower stiffness, $f_\alpha = 90$ Hz

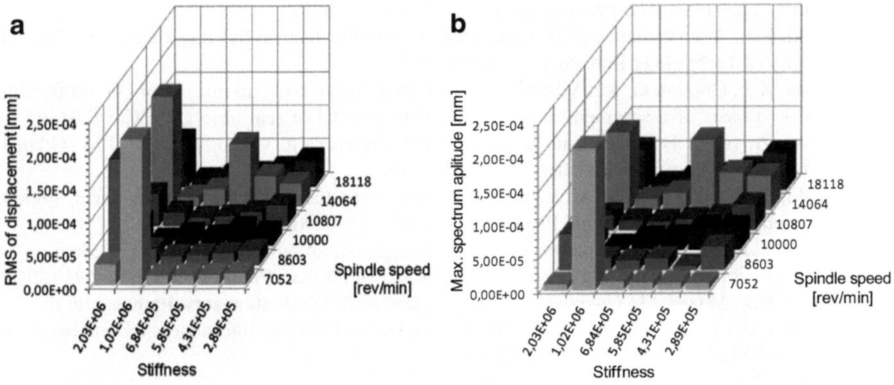

Fig. 8 (**a**) Summary of vibration displacements, (**b**) maximum value of corresponding amplitude spectrum, for chosen simulated pairs of n and k_s settings

example, $k_s = 1.02 \times 10^6$ N/m and $n = 7{,}052$ rev/min, which results in resonance because workpiece's natural frequency is 117 Hz, and frequency of excitation from tool rotations is also 117 Hz.

5 Conclusions

Simulations showed that modifying the workpiece dynamic properties is possible with the use of the proposed new workpiece holder. Simulations for different pairs of holder stiffness and spindle speed show that only in case of a proper optimal combination of these two parameters, vibrations are the lowest. These results mean that the proposed variable stiffness holder has a great potential. In a standard perfectly stiff holder, choosing the proper spindle speed on a basis of the Liao–Young condition was a method of vibration reduction. However, the determined optimal spindle speed might not be optimal from other points of view (i.e. technological ones). The proposed variable stiffness holder allows adjusting workpiece's dynamic properties in such a way that an arbitrarily given spindle speed should accomplish the generalised Liao–Young condition, in a wide range of operational spindle speeds of the milling machine. Thus the vibration surveillance efficiency is really evidenced.

References

1. Kalinski, K.J.: Modelling and investigation of vibration surveillance during ball end milling of curved flexible details. In: Uhl, T. (ed.) Chosen Problems of the Modal Analysis of Mechanical Structures, pp. 91–99. The Publication of AGH, Cracow (2006)
2. Kaliński, K.J.: A Surveillance of Dynamic Processes in Mechanical Systems (in Polish). Gdańsk University of Technology Publishers, Gdańsk (2012)
3. Kaliński, K.J., Chodnicki, M.: Vibration surveillance during ball end milling of curved flexible details on a basis of assessment of assurance of the model of real object. In: Cieśliński, J.T., Barylski, A. (eds.) Developments in mechanical engineering, vol. 3, pp. 117–124. Gdańsk University of Technology Publishers, Gdańsk (2009)
4. Kaliński, K.J., Galewski, M.A.: Chatter vibration surveillance by the optimal-linear spindle speed control. Mech. Syst Signal Process. **25**(1), 383–399 (2011)
5. Kalinski, K.J., Mazur, M., Galewski, M.A.: The optimal spindle speed map for reduction of chatter vibration during milling of bow thruster blade. Solid State Phenom. **198**, 686–691 (2013)
6. Kalinski, K.J., Mazur, M., Galewski, M.: High speed milling vibration surveillance with the use of the map of optimal spindle speeds. In: Proceedings of the 8th International Conference on High Speed Machining, Metz, France, 2010
7. Liao, Y.S., Young, Y.C.: A new on-line spindle speed regulation strategy for chatter control. Int. J. Mach. Tools Manuf. **36**, 651–660 (1996)
8. Tomkow, J.: Vibrostability of Machine Tools (in Polish). The Scientific and Technical Publications, Warsaw (1997)

Diversities in the Inverse Dynamics Problem for Underactuated Mechanical Systems Subject to Servo-constraints

Wojciech Blajer

Abstract Underactuated mechanical systems are featured with less control inputs than degrees of freedom. Their performance goal may then be realization of specified in time outputs whose number coincides the number of inputs. A solution to the inverse simulation problem (servo-constraint problem), that is, determination of an input control strategy that forces the underactuated system to complete the partly specified motion, is a challenging task. Since systems may be "underactuated" in several ways and the servo-constraint realization may range from orthogonal to tangential, diverse formulations and analysis methods of the servo-constraint problem arise. The diversity is discussed with reference to some simple case studies. The governing equations are handled in two ways. A direct formulation in configuration coordinates is first motivated and is then compared to a setting in which the actuated coordinates are replaced with the outputs. The governing equations arise either as ODEs (ordinary differential equations) or DAEs (differential-algebraic equations). Some computational issues related to the ODE and DAE formulations are discussed, and simulation results for the sample case studies are reported.

1 Introduction

Underactuated mechanical systems are featured with less control inputs than degrees of freedom, represented, e.g., by cranes [1, 16], aircrafts [8, 25], manipulators with passive and/or elastic joints [12, 22], and flexible robots [3]; see also [13, 24] for other representatives. A possible performance goal of such systems is realization of specified in time outputs whose number is equal to the number of inputs. The motion specifications are treated as servo-constraints on the system [6, 11, 15], with the control forces referred to as reactions of the constraints. An inverse dynamics problem (servo-constraint problem), that is, determination of an input control strategy that forces an underactuated system to complete the

W. Blajer (✉)
Faculty of Mechanical Engineering, University of Technology and Humanities in Radom, ul. Krasickiego 54, 26-600 Radom, Poland
e-mail: w.blajer@uthrad.pl

J. Awrejcewicz (ed.), *Applied Non-Linear Dynamical Systems*, Springer Proceedings in Mathematics & Statistics 93, DOI 10.1007/978-3-319-08266-0_14

partly specified motion, may in general be a challenging task. As systems may be "underactuated" in several ways [13, 24] and the servo-constraint realization may range from orthogonal to tangential [5], diverse formulations and analysis methods of the servo-constraint problem arise. The solution may also be purely algebraic or may include some internal (unspecified) dynamics [21, 23].

The diversities in the servo-constraint problem for underactuated systems are discussed in this paper with reference to the way the servo-constraints are realized and are illustrated with some simple case studies. A standard formulation in configuration coordinates is first motivated and then compared to a setting in which the actuated coordinates are replaced with the outputs. Depending on the way of servo-constraint realization, the arising governing equations are formulated either as ODEs (ordinary differential equations) or DAEs (differential-algebraic equations). Solutions of the servo-constraint problem for *differentially flat systems* [14, 18] (with no internal dynamics) and *non-flat systems* (with internal dynamics) are also discussed. The computational issues related to the ODE and DAE formulations are finally discussed, and some simulation results for the sample case studies are reported.

2 Formulation of the Problem

We consider an f-degree-of-freedom underactuated mechanical system, described by f configuration coordinates $\mathbf{q} = [q_1 \ \cdots \ q_f]^T$ and actuated by m inputs $\mathbf{u} = [u_1 \ \cdots \ u_m]^T$, $m < f$. The generic dynamic equations of the system are

$$\mathbf{M}(\mathbf{q})\ \ddot{\mathbf{q}} + \mathbf{k}(\mathbf{q}, \dot{\mathbf{q}}) = \mathbf{g}(\mathbf{q}, \dot{\mathbf{q}}) + \mathbf{B}(\mathbf{q})\ \mathbf{u} \tag{1}$$

where $\mathbf{M}$ is the $f \times f$ generalized mass matrix, $\mathbf{k}$ is the f-vector of Coriolis, gyroscopic, and centrifugal dynamic terms, $\mathbf{g}$ is the f-vector of applied forces, and $\mathbf{B}$ is the $f \times m$ maximum rank matrix of distribution of the m control inputs $\mathbf{u}$ on the f directions of $\mathbf{q}$.

A solution to the inverse dynamics problem for underactuated systems is possible only if the motion is specified by as many outputs as inputs [5, 6, 19, 21–23]. The m outputs $\mathbf{y} = [y_1 \ \cdots \ y_m]^T$ can be expressed in terms of $\mathbf{q}$ as $\mathbf{y} = \boldsymbol{\Phi}(\mathbf{q})$, where $\boldsymbol{\Phi} = [\Phi_1 \ \cdots \ \Phi_m]^T$ are appropriately differentiable functions. Given specified in time outputs, $\mathbf{y} = \tilde{\mathbf{y}}(t)$, m servo-constraints [6, 11, 15] on the underactuated system arise, whose equations, at the position, velocity, and acceleration levels, are

$$\boldsymbol{\varphi}(\mathbf{q}, t) = \boldsymbol{\Phi}(\mathbf{q}) - \tilde{\mathbf{y}}(t) = \mathbf{0} \tag{2}$$

$$\boldsymbol{\gamma}(\dot{\mathbf{q}}, \mathbf{q}, t) = \mathbf{H}(\mathbf{q})\ \dot{\mathbf{q}} - \dot{\tilde{\mathbf{y}}}(t) = \mathbf{0} \tag{3}$$

$$\boldsymbol{\eta}\left(\ddot{\mathbf{q}},\dot{\mathbf{q}},\mathbf{q},t\right)=\mathbf{H}\left(\mathbf{q}\right)\ \ddot{\mathbf{q}}+\mathbf{h}\left(\mathbf{q},\dot{\mathbf{q}}\right)-\ddot{\mathbf{y}}(t)=\mathbf{0}\tag{4}$$

where the $m \times f$ Jacobian matrix $\mathbf{H} = \partial\mathbf{\Phi}/\partial\mathbf{q}$ is of maximum rank, and $\mathbf{h} = \dot{\mathbf{H}}\ \dot{\mathbf{q}}$ is the m-vector of accelerations induced by the output relationship $\mathbf{y} = \mathbf{\Phi}(\mathbf{q})$.

As noticed in [5, 6], the servo-constraints introduced in Eq. (5) are mathematically equivalent to m passive (holonomic and rheonomic) constraints on the system, $\varphi(\mathbf{q}, t) = \mathbf{0}$. By analogy to the constrained system dynamics, the generalized actuating force $\mathbf{g}_u = \mathbf{B}\ \mathbf{u}$ can be viewed as a generalized reaction force of the servo-constraints. As opposed to the reactions of (ideal) passive constraints, in the classical sense of smooth hard surfaces, joints, and supports, whose reactions are orthogonal to constraint manifold [4], the control reaction $\mathbf{g}_u$ may be arbitrary oriented with respect to the servo-constraint manifold [5, 6], and in the extreme, it may be tangent to the manifold. The diversity in realization of servo-constraints can be deduced from the input–output relationship, obtained from Eq. (4) after using $\ddot{\mathbf{q}}$ from Eq. (1), i.e.,

$$\ddot{\mathbf{y}}(t)=\mathbf{H}\ \mathbf{M}^{-1}\left(\mathbf{g}-\mathbf{k}\right)+\mathbf{H}\ \mathbf{M}^{-1}\mathbf{B}\ \mathbf{u}+\mathbf{h}\tag{5}$$

An important consequence of arbitrary orientation of $\mathbf{g}_u$ with respect to the servo-constraint manifold is that some (or even all) of the system outputs $\mathbf{y}$ might not be directly regulated by the inputs $\mathbf{u}$. A measure of the control singularity is possible deficiency in rank of the $m \times m$ matrix $\mathbf{Y} = \mathbf{H}\,\mathbf{M}^{-1}\mathbf{B}$, which expresses the projection of the control reactions onto the directions of the constraint gradients (represented as rows of $\mathbf{H}$). It can also be viewed, in the system configuration space $\mathcal{Q}$, as the intersection, the orthogonal subspace $\mathcal{H}$ (spanned by the constraint gradients), and the actuated subspace $\mathcal{B}$ (spanned by the vectors represented as columns of $\mathbf{B}$), $\mathcal{Y} = \mathcal{H} \cap \mathcal{B}$ [4]. According to the rank of $\mathbf{Y}$ (dimension of $\mathcal{Y}$) three types of servo-constraint realization can be distinguished [5, 6]:

$$\operatorname{rank}\left(\mathbf{Y}\right)=\operatorname{rank}\left(\mathbf{H}\ \mathbf{M}^{-1}\mathbf{B}\right)=p\begin{cases}p=m & -\textit{orthogonal}\\ 0<p<m & -\textit{orthogonal-tangential}\\ p=0 & -\textit{tangential}\end{cases}\tag{6}$$

The *orthogonal realization* ($p = m$) denotes that all of the system outputs $\mathbf{y}$ can directly be regulated by the inputs $\mathbf{u}$. This is usually a nonideal orthogonal (skew-orthogonal) realization in which the control reactions, being explicitly represented in $\mathcal{H}$, project also in the tangent subspace $\mathcal{G}$ with respect to the servo-constraint manifold, $\mathcal{H} \cup \mathcal{G} = \mathcal{Q}$ and $\mathcal{H} \cap \mathcal{G} = 0$; see Fig. 1a for an illustration. The mixed *orthogonal–tangential realization* ($0 < p < m$) and pure *tangential realization* ($p = 0$) show that at most p (none) outputs can be actuated directly (in the orthogonal way), whereas the realization of the remaining $m - p$ (all) outputs is performed without direct influence of the actuating forces. The geometry of the tangential realization of servo-constraints is illustrated in Fig. 1b.

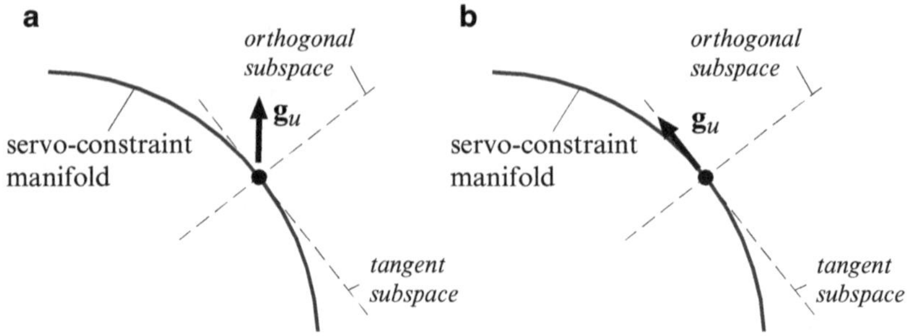

Fig. 1 Geometrical illustration of (**a**) skew-orthogonal realization of servo-constraints, (**b**) tangential realization of servo-constraints

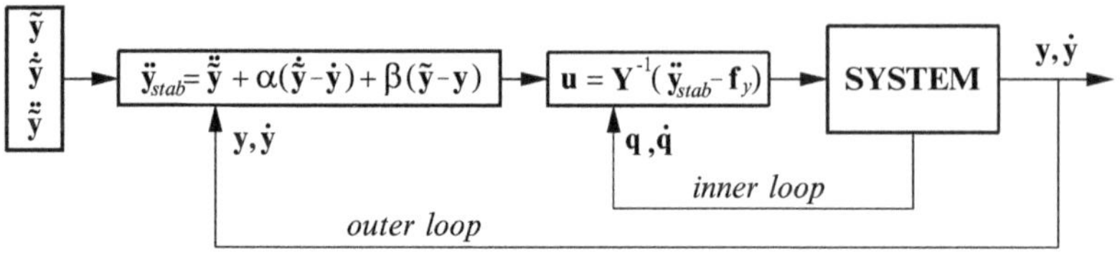

Fig. 2 Structure of control for orthogonal realization of servo-constraints

3 Orthogonal Realization of Servo-Constraints

For the case of (skew-) orthogonal realization of servo-constraints, conditioned upon rank($\mathbf{H}\,\mathbf{M}^{-1}\mathbf{B}$) $= m =$ max, the relationship of Eq. (5) directly leads to the output–input inverse dynamics relationship as follows

$$\mathbf{u}\,(t, \mathbf{q}, \dot{\mathbf{q}}) = \mathbf{Y}^{-1}\left(\ddot{\tilde{\mathbf{y}}} - \mathbf{f}_y\right) \tag{7}$$

where $\mathbf{Y}(\mathbf{q}) = \mathbf{H}\,\mathbf{M}^{-1}\mathbf{B}$ and $\mathbf{f}_y\,(\mathbf{q}, \dot{\mathbf{q}}) = \mathbf{H}\,\mathbf{M}^{-1}\,(\mathbf{g} - \mathbf{k}) + \mathbf{h}$. The above feedforward controller, required for reproduction of $\tilde{\mathbf{y}}(t)$, can then be enhanced by a feedback loop after replacing $\ddot{\tilde{\mathbf{y}}}$ with $\ddot{\mathbf{y}}_{stab}$ in the form (PD controller)

$$\ddot{\mathbf{y}}_{stab} = \ddot{\tilde{\mathbf{y}}} + \boldsymbol{\alpha}\,\left(\dot{\tilde{\mathbf{y}}} - \dot{\mathbf{y}}\right) + \boldsymbol{\beta}\,\left(\tilde{\mathbf{y}} - \mathbf{y}\right) \tag{8}$$

where $\boldsymbol{\alpha}$ and $\boldsymbol{\beta}$ are the diagonal matrices of gain values, and $\mathbf{y}$ and $\dot{\mathbf{y}}$ are the actual outputs and their time derivatives. The structure of control with feedback is seen in Fig. 2, which is the feedforward controller structure after setting $\boldsymbol{\alpha} = \mathbf{0}$ and $\boldsymbol{\beta} = \mathbf{0}$. The governing equations are then $2f$ ODEs in $\mathbf{q}$ and $\mathbf{v} = \dot{\mathbf{q}}$

$$\begin{aligned} \dot{\mathbf{q}} &= \mathbf{v} \\ \dot{\mathbf{v}} &= \mathbf{M}^{-1}(\mathbf{g}-\mathbf{k}) + \mathbf{M}^{-1}\mathbf{B}\,\mathbf{Y}^{-1}\left(\ddot{\mathbf{y}}_{stab} - \mathbf{f}_y\right) \end{aligned} \tag{9}$$

The controller $\mathbf{u} = \mathbf{Y}^{-1}\left[\ddot{\tilde{\mathbf{y}}} + \boldsymbol{\alpha}\left(\dot{\tilde{\mathbf{y}}} - \dot{\mathbf{y}}\right) + \boldsymbol{\beta}\,(\tilde{\mathbf{y}} - \mathbf{y}) - \mathbf{f}_y\right]$ is equivalent in structure to the computed torque control scheme [17] for fully actuated systems executing a fully specified motion. Its application to underactuated systems in partly specified motion is not straightforward, however. Firstly, it is limited to the orthogonal realization of servo-constraints. Then, assumed $\det(\mathbf{Y}) \neq 0$, in addition to the output–input inverse dynamic model represented (in the orthogonal subspace $\mathcal{H}$) by the algebraic formula of Eq. (7), an internal dynamics (in the tangent subspace $\mathcal{G}$) exists since the system motion is only partly specified by the outputs. For the prevailing skew-orthogonal realization of servo-constraints, the evolution in time of the internal dynamics and the inverse dynamics control are mutually dependent. Of critical importance for the control strategy is therefore that the internal dynamics remains bounded, which may require a proper design of the mechanical system [5] and/or careful definition of the imposed motion specifications.

Case study 1: Internal dynamics stability problem. Let us consider a rotational arm (Fig. 3a) consisting of two links, $\mathbf{q} = [\theta_1\ \theta_2]^T$, connected at the passive joint A with a spring–damper combination and supported at the active joint O with the actuating torque, $\mathbf{u} = [\tau\ ]$. The lengths, masses, and central mass moment of inertia of the links are l_i, m_i and J_{Ci}, $i = 1,\ 2$, and the locations of the mass centers are s_1 (OC_1) and s_2 (OC_2). The spring and damper constants are c and d, respectively, and the vanishing torque of the torsion spring is achieved for $\theta_1 = \theta_2$. The arm moves in the horizontal plane perpendicular to the direction of gravity. The dynamic equations of the system, related to Eq. (1), are defined by

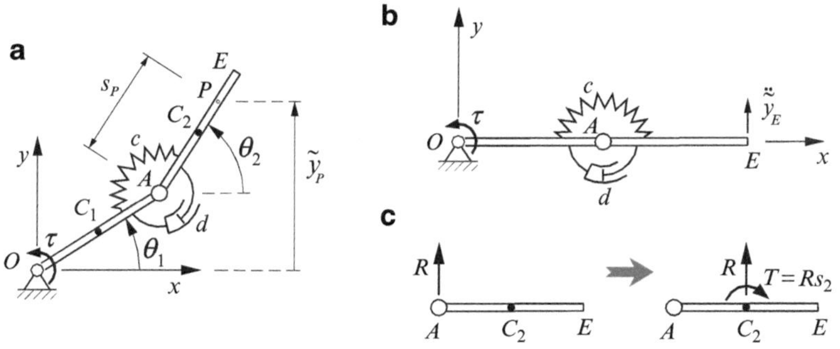

Fig. 3 Illustration of (**a**) the rotational arm with one active and one passive joints, (**b**) the arm initial configuration, and (**c**) the action of torque τ on link 2 at the initial rest position

$$\mathbf{M} = \begin{bmatrix} J_{C1} + m_1 s_1^2 + m_2 l_1^2 & m_2 s_2 l_1 \cos(\theta_2 - \theta_1) \\ m_2 s_2 l_1 \cos(\theta_2 - \theta_1) & J_{C2} + m_2 s_2^2 \end{bmatrix}; \quad \mathbf{k} = \begin{bmatrix} -m_2 s_2 l_1 \, \dot{\theta}_2^2 \sin(\theta_2 - \theta_1) \\ m_2 s_2 l_1 \, \dot{\theta}_1^2 \sin(\theta_2 - \theta_1) \end{bmatrix};$$
$$\mathbf{g} = \begin{bmatrix} c\ (\theta_2 - \theta_1) + d\ \left(\dot{\theta}_2 - \dot{\theta}_1\right) \\ -c\ (\theta_2 - \theta_1) - d\ \left(\dot{\theta}_2 - \dot{\theta}_1\right) \end{bmatrix}; \quad \mathbf{B} = \begin{bmatrix} 1 \\ 0 \end{bmatrix} \tag{10}$$

The output is assumed as the vertical position of a point P on link 2, defined by s_P (distance AP), $\mathbf{y} = [y_P]$. The servo-constraint equation, defined in Eq. (2), and the Jacobian matrix $\mathbf{H} = \partial\mathbf{\Phi}/\partial\mathbf{q}$ are then

$$\boldsymbol{\varphi}(\mathbf{q}, t) = l_1 \sin\theta_1 + s_P \sin\theta_2 - \tilde{y}_P(t) = 0; \qquad \mathbf{H} = \begin{bmatrix} l_1 \cos\theta_1 & s_P \cos\theta_2 \end{bmatrix} \tag{11}$$

The matrix $\mathbf{Y} = \mathbf{H}\,\mathbf{M}^{-1}\mathbf{B}$, used in Eqs. (5)–(7) and (9), is

$$\mathbf{Y} = [Y] = \left[\frac{\left(J_{C2} + m_2 s_2^2\right)\ l_1 \cos\theta_1 - m_2 s_2 l_1 s_P \cos(\theta_2 - \theta_1)\cos\theta_2}{\det(\mathbf{M})} \right] \tag{12}$$

where, by assumption, $\det(\mathbf{M}) > 0$. As seen, depending on the inertial and geometrical properties of the system, its current configuration (coordinates θ_1 and θ_2), and the placement of point P (distance s_P), the numerator of the above expression, and as such Y, can be either negative or positive or may occasionally diminish to zero. While the latter case makes orthogonal realization of the servo-constraint impossible, let us concentrate on the cases $Y < 0$ and $Y > 0$. Specifically, assume for a while that both the links are homogeneous and identical, i.e., $s_i = l_i/2$, $J_{Ci} = m_i l_i^2/12$ $(i = 1, 2)$, and $l_1 = l_2 = l$, $m_1 = m_2 = m$, and that the point P coincides with the end point E of link 2, $P \to E$. The value of Y is then

$$Y = \frac{m l^3}{\det(\mathbf{M})} \left(\frac{1}{3}\cos\theta_1 - \frac{1}{2}\cos(\theta_2 - \theta_1)\cos\theta_2 \right) \tag{13}$$

from which one can deduce that $Y < 0$ for all the configurations where θ_1 and θ_2 are not too far from each other and especially for $\theta_1 \approx \theta_2$. Looking at the output–input inverse dynamics relationship of Eq. (7), for the case at hand, a negative value of Y denotes that an assumed positive (upward) $\ddot{\tilde{y}}_E$ might require a negative (clockwise) direction of τ, which is against intuition and, actually, leads to instability of the internal dynamics (inapplicability of the inverse dynamics control).

To be more specific about the physical reasons for the aforementioned instability, let us consider that the arm is in rest position as seen in Fig. 3b, $\theta_1 = \theta_2 = 0$ and $\dot{\theta}_1 = \dot{\theta}_2 = 0$, and the task is to move the end point E in the upward direction, which yields $\ddot{\tilde{y}}_E(t_0) > 0$ at the beginning of the maneuver. Since $\mathbf{k}$, $\mathbf{g}$, and $\mathbf{h}$ are equal to zero at the initial state and are negligible just after the maneuver begins, which results in $\mathbf{f}_y = \mathbf{0}$ and $\mathbf{f}_y \approx \mathbf{0}$, respectively, the inverse control formula of Eq. (7), for the case at hand, simplifies to $\tau = \ddot{\tilde{y}}_E / Y$. Intuitively one can expect a positive sense of τ required to initiate the upward motion of the end point E. This is not the case, however. More strictly, the effect of a positive τ on link 2 is an upward reaction R in joint A (Fig. 3c), which moved to the link mass center C_2 isassociated with a couple of clockwise (negative) torque $T = R/s_2$. For the geometrical and inertial settings stated above, the acceleration of point E is then

$$\ddot{y}_E = \frac{R}{m_2} - \frac{R\,s_2}{J_{C2}}(l_2 - s_2) = \frac{R}{m} - \frac{R\,l}{2}\frac{12}{m\,l^2}\frac{l}{2} = -\frac{2R}{m} \tag{14}$$

The action of a positive τ has thus the "opposite" effect from what is expected—a negative $\ddot{y}_E$ is produced. The feature is related to the negative value of Y in the output–input inverse dynamics model $\tau = \ddot{\tilde{y}}_E/Y$, and a positive $\ddot{\tilde{y}}_E$ will result in the negative (clockwise) sense of τ. This "reverse" control is improper and, when continued, leads to the collapse of the servo-constraint problem execution. A proper solution to the problem requires therefore $Y > 0$.

The desired effect $Y > 0$ can be achieved in two ways. Firstly, as seen from Eq. (12), one can redesign the system to modify appropriately its geometrical and inertial characteristics; see also [21–23] where a similar approach was used. In particular, for the case at hand, one may play with the values of m_2, J_{C2} and s_2, and $Y > 0$ can most effectively be achieved by decreasing s_2 and increasing J_{C2}. A redesign of an existing system may not always be possible, however. The other possibility is therefore a reformulation of the imposed task $\tilde{\mathbf{y}}(t)$ and, in particular, a reduction of s_P value; see Eq. (12). The reference point should thus not be the end point E ($s_P = l_2$) but an inner point P on link 2, and a reasonable choice is $P \to C_2$ ($s_P = l_2/2$). Using this setting, a rest-to-rest maneuver was modeled as

$$\tilde{y}_P(t) = y_{P0} + \left[126\left(\frac{t}{\tau}\right)^5 - 420\left(\frac{t}{\tau}\right)^6 + 540\left(\frac{t}{\tau}\right)^7 - 315\left(\frac{t}{\tau}\right)^8 + 70\left(\frac{t}{\tau}\right)^9\right]\left(y_{Pf} - y_{P0}\right) \tag{15}$$

where $y_{P0} = 0$ m and $y_{Pf} = 0.75$ m are the initial ($t_0 = 0$ s) and final ($t_f = 3$ s) y coordinates of P, and $\tau = t_f - t_0 = 3$ s. The system geometrical and inertial data used in calculations were $l_1 = l_2 = 1$ m, $m_1 = m_2 = 2$ kg, $s_i = l_i/2$, and $J_{Ci} = m_i l_i^2/12$, and the spring and damper coefficients were $c = 5$ Nm/rad and $d = 0.1$ Nms/rad. Selected simulation results obtained using the scheme illustrated in Fig. 2 (without the outer loop), with the integration time step $\Delta t = 0.001$ s, are reported in Fig. 4. As seen from the graphs, the internal dynamics is bounded, and the system asymptotically approaches the final rest position after the maneuver is finished.

4 DAE Formulation of the Governing Equations

The output–input dynamics model of Eq. (7) applies to the orthogonal realization of servo-constraints, conditioned upon $\det(\mathbf{Y}) \neq 0$. A general formulation of the servo-constraint problem should relate to all the realization types, including the mixed orthogonal–tangent and pure tangent ones, for which $\det(\mathbf{Y}) = 0$. This aim

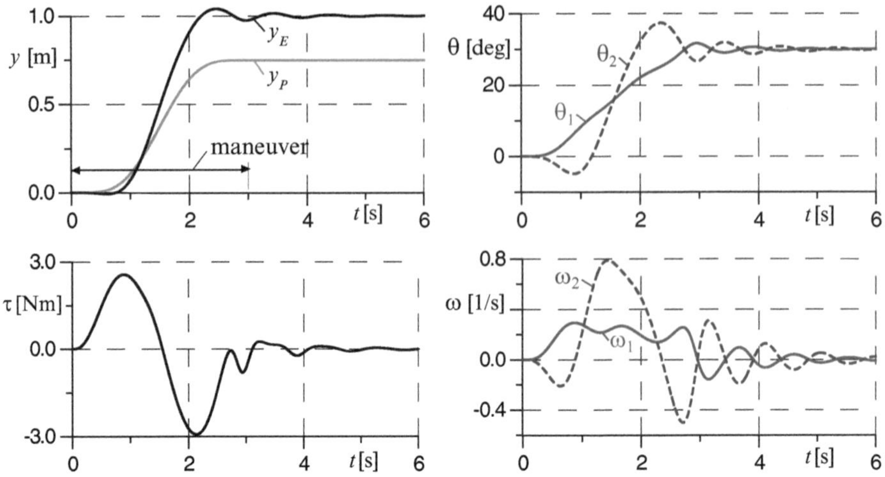

Fig. 4 Simulation results of the rotational arm executing a rest-to-rest y-motion of point P, with $s_P = l_2/2$ ($P \rightarrow C_2$), and the links modeled as identical homogeneous bars

can be achieved by formulating the governing equations as DAEs. In [5, 6] such a formulation was proposed in the configuration coordinates $\mathbf{q}$, resulted in $2f+m$ DAEs in the same number of variables: $\mathbf{q}$, $\mathbf{v} = \dot{\mathbf{q}}$, and $\mathbf{u}$. The initial DAEs, composed of f kinematic relationships $\dot{\mathbf{q}} = \mathbf{v}$, f dynamic equations (1), and m servo-constraint equation (2), were then transformed to a numerically more tractable form with index [2, 10] reduced by two. The reduction is consequent to the projection of the dynamic equations into the orthogonal $\mathcal{H}$ and tangent $\mathcal{G}$ subspaces and application of the servo-constraint equation (4) at the acceleration level. With the use of an $f \times (f-m)$ matrix $\mathbf{G}$ which is an orthogonal complement to $\mathbf{H}$, i.e., $\mathbf{H}\,\mathbf{G} = \mathbf{0} \iff \mathbf{G}^T\mathbf{H}^T = \mathbf{0}$, and whose columns represent the spanning vectors of $\mathcal{G}$, the reduced index DAEs are (see [5, 6, 23] for details)

$$\begin{aligned} \dot{\mathbf{q}} &= \mathbf{v} \\ \mathbf{G}^T\mathbf{M}\,\dot{\mathbf{v}} &= \mathbf{G}^T(\mathbf{g}-\mathbf{k}) + \mathbf{G}^T\mathbf{B}\,\mathbf{u} \\ \mathbf{0} &= \mathbf{H}\,\mathbf{M}^{-1}(\mathbf{g}-\mathbf{k}) + \mathbf{H}\,\mathbf{M}^{-1}\mathbf{B}\,\mathbf{u} + \mathbf{h} - \ddot{\tilde{\mathbf{y}}}(t) \\ \mathbf{0} &= \mathbf{\Phi}(\mathbf{q}) - \tilde{\mathbf{y}}(t) \end{aligned} \iff \begin{aligned} \dot{\mathbf{q}} &= \mathbf{v} \\ \mathbf{A}(\mathbf{q})\,\dot{\mathbf{v}} &= \mathbf{a}(\mathbf{q},\mathbf{v},\mathbf{u}) \\ \mathbf{0} &= \boldsymbol{\eta}(\mathbf{q},\mathbf{v},\mathbf{u},t) \\ \mathbf{0} &= \boldsymbol{\varphi}(\mathbf{q},t) \end{aligned} \tag{16}$$

where $\mathbf{A}$ and $\mathbf{a}$ are of dimensions $(f-m)\times f$ and $(f-m)\times 1$, respectively.

Another DAE formulation of the governing equations can be obtained by applying a coordinate transformation, exploited in [21, 22] and used hereafter in a little modified way. The configuration coordinates are first partitioned into m actuated and $f-m$ unactuated coordinates, written symbolically as $\mathbf{q} = [\mathbf{q}_a^T\ \mathbf{q}_u^T]^T$, and the partition is conditioned upon the $m \times m$ matrix $\mathbf{B}_a$, gathering the rows of $\mathbf{B}$ related to $\mathbf{q}_a$, and is of maximal rank, $\det(\mathbf{B}_a) \neq 0$. The new set of coordinates is then $\mathbf{q}' = [\mathbf{y}^T\ \mathbf{q}_u^T]^T$, where, compared to $\mathbf{q} = [\mathbf{q}_a^T\ \mathbf{q}_u^T]^T$, the outputs $\mathbf{y}$ are introduced instead of $\mathbf{q}_a$. At the acceleration level, the transformation formula is

$$\ddot{\mathbf{q}}' = \begin{bmatrix} \ddot{\mathbf{y}} \\ \ddot{\mathbf{q}}_u \end{bmatrix} = \begin{bmatrix} \mathbf{H}\,\ddot{\mathbf{q}} \\ \ddot{\mathbf{q}}_u \end{bmatrix} + \begin{bmatrix} \mathbf{h} \\ \mathbf{0} \end{bmatrix} = \begin{bmatrix} \mathbf{H}_a & \mathbf{H}_u \\ \mathbf{0} & \mathbf{I} \end{bmatrix} \begin{bmatrix} \ddot{\mathbf{q}}_a \\ \ddot{\mathbf{q}}_u \end{bmatrix} + \begin{bmatrix} \mathbf{h} \\ \mathbf{0} \end{bmatrix} = \mathbf{H}'(\mathbf{q})\;\ddot{\mathbf{q}} + \mathbf{h}'(\mathbf{q}, \dot{\mathbf{q}}) \tag{17}$$

where $\mathbf{H}_a$ and $\mathbf{H}_u$ are of dimensions $m \times m$ and $m \times (f - m)$, $\mathbf{H} = [\mathbf{H}_a \,\vdots\, \mathbf{H}_u]$, and $\mathbf{H}'$ is the $f \times f$ matrix of transformation, whose inverse is conditioned upon $\det(\mathbf{H}_a) \neq 0$. By applying $\ddot{\mathbf{q}} = \mathbf{M}^{-1}(\mathbf{g} - \mathbf{k} + \mathbf{B}\,\mathbf{u})$ obtained from Eq. (1), the underactuated system dynamics in the new coordinates can then be formulated in the following resolved form:

$$\begin{aligned} \ddot{\mathbf{y}} &= \mathbf{H}\,\mathbf{M}^{-1}(\mathbf{g} - \mathbf{k}) + \mathbf{h} + \mathbf{H}\,\mathbf{M}^{-1}\mathbf{B}\,\mathbf{u} \\ \ddot{\mathbf{q}}_u &= \begin{bmatrix} \mathbf{0} & \vdots & \mathbf{I} \end{bmatrix} \mathbf{M}^{-1}(\mathbf{g} - \mathbf{k}) + \begin{bmatrix} \mathbf{0} & \vdots & \mathbf{I} \end{bmatrix} \mathbf{M}^{-1}\mathbf{B}\,\mathbf{u} \end{aligned} \quad \Longleftrightarrow \quad \begin{aligned} \ddot{\mathbf{y}} &= \mathbf{f}_y(\mathbf{q}, \dot{\mathbf{q}}) + \mathbf{Y}(\mathbf{q})\;\mathbf{u} \\ \ddot{\mathbf{q}}_u &= \mathbf{f}_q(\mathbf{q}, \dot{\mathbf{q}}) + \mathbf{Q}(\mathbf{q})\;\mathbf{u} \end{aligned} \tag{18}$$

The arising governing equations are then $(f - m) + (f - m) + m + m + m = 2f + m$ DAEs in $\mathbf{q}$, $\mathbf{v}$, and $\mathbf{u}$ (see [9, 23] for more details):

$$\begin{aligned} \dot{\mathbf{q}}_u &= \mathbf{v}_u \\ \dot{\mathbf{v}}_u &= \mathbf{f}_q(\mathbf{q}, \dot{\mathbf{q}}) + \mathbf{Q}(\mathbf{q})\;\mathbf{u} \\ \mathbf{0} &= \mathbf{f}_y(\mathbf{q}, \dot{\mathbf{q}}) + \mathbf{Y}(\mathbf{q})\;\mathbf{u} - \ddot{\tilde{\mathbf{y}}}(t) \\ \mathbf{0} &= \mathbf{H}(\mathbf{q})\;\dot{\mathbf{q}} - \dot{\tilde{\mathbf{y}}}(t) \\ \mathbf{0} &= \boldsymbol{\Phi}(\mathbf{q}) - \tilde{\mathbf{y}}(t) \end{aligned} \quad \Longleftrightarrow \quad \begin{aligned} \dot{\mathbf{q}}_u &= \mathbf{v}_u \\ \dot{\mathbf{v}}_u &= \mathbf{b}(\mathbf{q}, \dot{\mathbf{q}}, \mathbf{u}) \\ \mathbf{0} &= \boldsymbol{\eta}(\mathbf{q}, \dot{\mathbf{q}}, \mathbf{u}, t) \\ \mathbf{0} &= \boldsymbol{\gamma}(\mathbf{q}, \dot{\mathbf{q}}, t) \\ \mathbf{0} &= \boldsymbol{\varphi}(\mathbf{q}, t) \end{aligned} \tag{19}$$

The two DAE formulations, introduced in Eqs. (16) and (19), are equivalent and lead to same solutions of the servo-constraint problem. They have the same dimension $2f + m$ and are expressed in $\mathbf{q}$, $\mathbf{v}$, and $\mathbf{u}$ and are applicable irrespective of the servo-constraint realization type: orthogonal ($p = m$), orthogonal–tangent ($0 < p < m$), or tangent ($p = 0$); see Eq. (6). They have also the same index (equal to one for the orthogonal realization of servo-constraints) and exceeding one for the non-orthogonal realizations [5, 6, 9, 21, 23]. Finally, the third and fourth algebraic equations of DAEs (16) are identical to the third and fifth ones in DAEs (19). A difference is that the f kinematic relationships $\dot{\mathbf{q}} = \mathbf{v}$ in DAEs (16) are replaced in DAEs (19) with $f - m$ kinematic relationships $\dot{\mathbf{q}}_u = \mathbf{v}_u$ and m servo-constraint equations $\boldsymbol{\gamma}(\mathbf{q}, \dot{\mathbf{q}}, t) = \mathbf{0}$ at the velocity level. The $f - m$ differential equations $\mathbf{A}(\mathbf{q})\;\dot{\mathbf{v}} = \mathbf{a}(\mathbf{q}, \mathbf{v}, \mathbf{u})$ describing the system dynamics in $\mathcal{G}$ subspace are also replaced with $f - m$ differential equations $\dot{\mathbf{v}}_u = \mathbf{b}(\mathbf{q}, \dot{\mathbf{q}}, \mathbf{u})$ that govern the system dynamics in the unactuated coordinate directions.

A simple numerical code was proposed in [5] to solve DAEs (16), based on Euler backward differentiation scheme in which the variable time derivatives are approximated with their backward differences with respect to the integration time step Δt. The same can be applied to DAEs (19), resulting in

$$\begin{aligned} (\mathbf{q}_u)_{n+1} - (\mathbf{q}_u)_n - \Delta t \ \ (\mathbf{v}_u)_{n+1} &= \mathbf{0} \\ (\mathbf{v}_u)_{n+1} - (\mathbf{v}_u)_n - \Delta t \ \ \mathbf{b}(\mathbf{q}_{n+1}, \mathbf{v}_{n+1}, \mathbf{u}_{n+1}) &= \mathbf{0} \\ \boldsymbol{\eta}(\mathbf{q}_{n+1}, \mathbf{v}_{n+1}, \mathbf{u}_{n+1}, t_{n+1}) &= \mathbf{0} \\ \boldsymbol{\gamma}(\mathbf{q}_{n+1}, \mathbf{v}_{n+1}, t_{n+1}) &= \mathbf{0} \\ \boldsymbol{\varphi}(\mathbf{q}_{n+1}, t_{n+1}) &= \mathbf{0} \end{aligned} \tag{20}$$

[see [5, 6] for a similar numerical code related to DAEs (16)]. Given $\mathbf{q}_n$ and $\mathbf{v}_n$ at time t_n ($\mathbf{u}_n$ are not required), the values $\mathbf{q}_{n+1}$, $\mathbf{v}_{n+1}$ and $\mathbf{u}_{n+1}$ at time t_{n+1} can then be obtained as a solution of the above nonlinear equations, and in this way the solution is advanced from t_n to t_{n+1}. The rough computational scheme is of acceptable accuracy for appropriately small values of Δt. The accuracy can be improved by applying more precise higher-order backward difference approximation methods or other specialized DAE solvers [2]. The controller structure based on the solution to DAEs (16)/(19) is seen in Fig. 5.

Case study 2: From orthogonal to tangential realization of servo-constraints. Let us consider another two-degree-of-freedom system (Fig. 6a) that consists of a carriage of mass m_1, moving horizontally, and a slider of mass m_2, moving along a stick mounted on the carriage and inclined horizontally at angle α. The actuating force F is applied to the carriage, $\mathbf{u} = [F]$, and the specified output is a desired horizontal position of the slider with respect to the ground, $\tilde{\mathbf{y}}(t) = [\tilde{x}_2(t)]$. The spring and damper coefficients are, respectively, c and d, and b is the value of s_2 in the rest state of the system. Friction in the system is neglected. Choosing $\mathbf{q} = [x_1 \ s_2]^T$, where x_1 is the carriage position with respect to the inertial frame and s_2 is the slider position with respect to the carriage, the components of the dynamic equations of the system, corresponding to Eq. (1), are

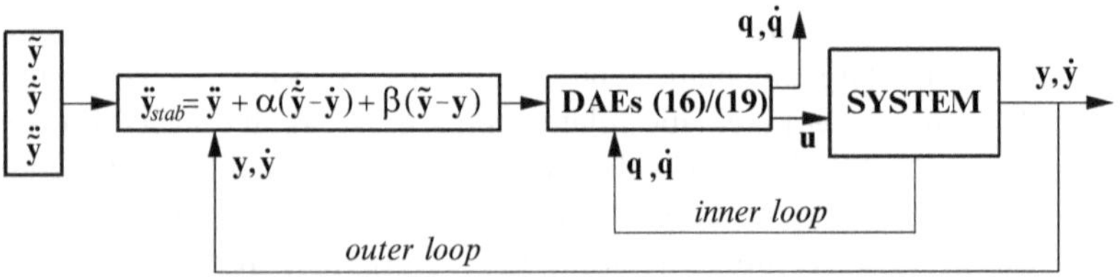

Fig. 5 Structure of control based on DAE formulation

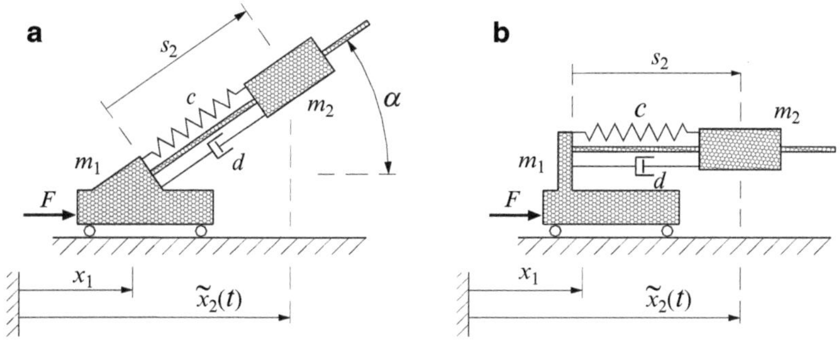

Fig. 6 The spring–mass system mounted on a movable carriage, the cases of (**a**) skew-orthogonal and (**b**) tangential realization of the servo-constraint imposed

$$\mathbf{M} = \begin{bmatrix} m_1 + m_2 & m_2 \cos\alpha \\ m_2 \cos\alpha & m_2 \end{bmatrix}; \mathbf{k} = \mathbf{0}; \mathbf{g} = \begin{bmatrix} 0 \\ -c\ (s_2 - b) - d\ \dot{s}_2 \end{bmatrix}; \mathbf{B} = \begin{bmatrix} 1 \\ 0 \end{bmatrix} \tag{21}$$

The servo-constraint equations, introduced in Eqs. (2)–(4), are then defined by

$$\varphi\,(\mathbf{q}, t) = x_1 + s_2 \cos\alpha - \tilde{x}_2(t) = 0; \qquad \mathbf{H} = \ [1 \quad \cos\alpha]\,; \qquad \mathbf{h} = \ [0] \tag{22}$$

and the matrices $\mathbf{Y} = \mathbf{H}\,\mathbf{M}^{-1}\mathbf{B}$ and $\mathbf{Q} = \ [\mathbf{0} \ \vdots\ \mathbf{I}\,]\,\mathbf{M}^{-1}\mathbf{B}$, introduced in Eq. (18), are

$$\mathbf{Y} = \ [Y] \ = \left[\frac{m_2 \sin^2\alpha}{m_1 m_2 + m_2^2 \sin^2\alpha}\right]; \quad \mathbf{Q} = \ [Q] \ = \ [-m_2 \cos\alpha] \tag{23}$$

Assuming $\alpha \in <0, \pi/2>$, it is evident from Eq. (23) that for $\alpha = \pi/2$ one has $Y = 1/(m_1 + m_2)$, and an ideal orthogonal realization of the servo-constraint is observed in which the output-inverse dynamics (motion of the two masses along x_1) and the internal dynamics (relative vertical motion of the slider) are decoupled, $Q = 0$. For $0 < \alpha < \pi/2$ a skew-orthogonal realization arises, $Y > 0$ and $Q \neq 0$, in which the inverse dynamics control and the internal dynamics are mutually related. Finally, for $\alpha = 0$ (Fig. 6b), there is no direct input–output relationship, $Y = 0$, and the tangential realization of the servo-constraint occurs.

Test simulations were carried out for $m_1 = 1$ kg, $m_2 = 2$ kg, $c = 5$ N/m, and $b = 1$ m, and different values of the inclination angle α and damping coefficient d were used. The specified output $\tilde{x}_2(t)$ was a rest-to-rest maneuver according to Eq. (15) with $x_{20} = 0.5$ m and $x_{2f} = 2.5$ m instead of y_{P0} and y_{Pf}, respectively, and the maneuver duration was $\tau = 6$ s. The integration time step for the DAE formulation (19) was $\Delta t = 0.001$ s.

The results shown in Fig. 7 are for $\alpha = 40°$ (skew-orthogonal realization of the servo-constraint). As seen, realization of the specified slider motion induces

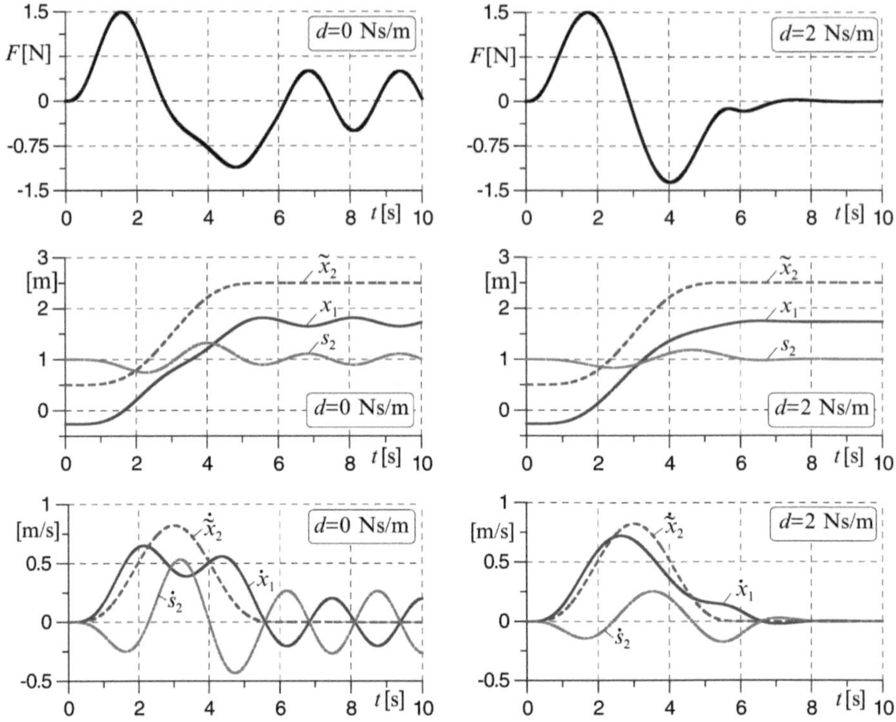

Fig. 7 Simulation results for $\alpha = 40°$ (orthogonal realization) and for $d = 0$ Ns/m (*left column*) and $d = 2$ Ns/m (*right column*)

substantial internal motion. When not damped, for $t > \tau$, after the maneuver is finished and the slider horizontal position is kept constant, the interaction from the internal motion (the carriage moves to and fro, together with the slider raising up and down) must be compensated by appropriate variations in $F(t)$. Nonetheless, the internal motion is bounded (stable). For $d = 2$ Ns/m the residual motion is damped and the required $F(t)$ goes to zero.

In Fig. 8, the results obtained for $\alpha = 0°$ (tangential realization of the servo-constraint) are shown. As seen, the case with damping (right columns) is similar to those obtained for $\alpha = 40°$—in the present case, for $t > \tau$, the residual motion is slightly quicker damped to the final state. The case without damping is now qualitatively different from that obtained for $\alpha = 40°$ in Fig. 7. The motion is completely finished at $t = \tau$. This is because the problem is differentially flat, and no internal dynamics is left, which is explained shortly below.

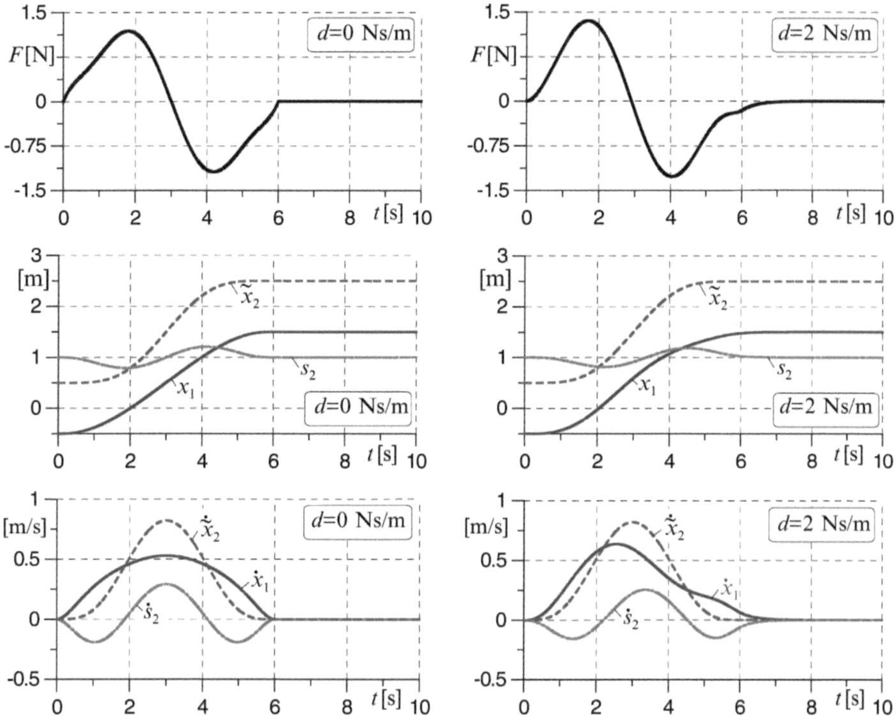

Fig. 8 Simulation results for $\alpha = 0°$ (tangential realization) and for $d = 0$ Ns/m (*left column*) and $d = 2$ Ns/m (*right column*)

5 Differentially Flat and Non-flat Servo-constraint Problems

The servo-constraint problems can be differentially flat [14, 18, 20] or non-flat. In case of flatness of the inverse dynamics problems described in Eqs. (1) and (2), all m control inputs $\mathbf{u}$ and $2f$ state that $\mathbf{q}$ and $\mathbf{v} = \dot{\mathbf{q}}$ can be algebraically expressed in terms of the desired m outputs $\tilde{\mathbf{y}}(t)$ and their time derivatives up to a certain order r, which is by one smaller than the index of the initial DAEs. The inverse dynamics solution is then explicit in time and unique, denoted

$$\begin{aligned}\tilde{\mathbf{q}}(t) &= \mathbf{q}\left(\tilde{\mathbf{y}}, \dot{\tilde{\mathbf{y}}}, \ldots, \tilde{\mathbf{y}}^{r-2}\right)\\ \tilde{\mathbf{v}}(t) &= \mathbf{v}\left(\tilde{\mathbf{y}}, \dot{\tilde{\mathbf{y}}}, \ldots, \tilde{\mathbf{y}}^{r-1}\right)\\ \tilde{\mathbf{u}}(t) &= \mathbf{u}\left(\tilde{\mathbf{y}}, \dot{\tilde{\mathbf{y}}}, \ldots, \tilde{\mathbf{y}}^{r}\right),\end{aligned} \tag{24}$$

and there is no internal dynamics in the system [5, 7, 9, 20–23]. The flatness-based analytical solution of Eq. (24) is often featured by substantial complexity, which may render its obtainment difficult/impractical for more complex problems of technical relevance. By contrast, the non-flat servo-constraint problems involve some (unspecified) internal dynamics, which needs to be bounded.

For fully actuated systems in fully prescribed motion, $m = f$, the inverse dynamics problem is always flat with the order $r = 2$. For underactuated systems in partly specified motion, $m < f$, in the case of orthogonal realization of the servo-constraints imposed, the servo-constraint problem is always non-flat (is a dynamic problem), and of critical importance is to assure stability of the remained internal dynamics; see Sect. 3. The servo-constraint problems for the cases of mixed orthogonal–tangential and pure tangential realizations of servo-constraints can then be either flat or non-flat. Examples of the differentially flat inverse dynamics problems are cranes executing a load prescribed motion [5–7] and aircrafts in prescribed trajectory flight [8], both characterized by mixed orthogonal–tangential realization of servo-constraints, and the trajectory tracking problem for flexible joint manipulators [3, 12, 24] (pure tangential realization of servo-constraints). A simple illustration of the differentially flat servo-constraint problem is given below.

Case study 3: Differential flatness. Let us reconsider the servo-constraint problem for the variant $\alpha = 0$ (Fig. 6b) of the cart–mass system. As already stated, the case results in the tangential realization of the servo-constraint defined in Eq. (22), $Y = 0$, and the output $\tilde{\mathbf{y}}(t) = [\tilde{x}_2(t)]$ tracking is achieved through the spring and damper forces (the actuating force F is not directly involved). One can then become convinced that the problem is differentially flat for $d = 0$ (no damping) and is non-flat for $d > 0$ (with damping). The conclusion can be deduced from the simulation results shown in Fig. 8. For $d = 0$ (left column graphs), both the states and actuation of the system, determined from the inverse dynamics, achieve the final steady state values exactly at the instant the maneuver is finished (at $t = 6$ s) and must thus be consequent only upon the specified output $\tilde{x}_2(t)$ and its time derivatives (no internal dynamics remains). For $d > 0$ there is some internal dynamics, exemplified in the fact that the system achieves the final steady state values at $t > 6$ s (after the specified maneuver is finished). For the flat case, the inverse dynamics solution as in Eq. (24) is the following:

$$\begin{aligned} \tilde{x}_1(t) &= \tilde{x}_2 + \frac{m_2}{c}\ddot{\tilde{x}}_2 - b; & \tilde{s}_2(t) &= b - \frac{m_2}{c}\ddot{\tilde{x}}_2; \\ \dot{\tilde{x}}_1(t) &= \dot{\tilde{x}}_2 + \frac{m_2}{c}\dddot{\tilde{x}}_2; & \dot{\tilde{s}}_2(t) &= -\frac{m_2}{c}\dddot{\tilde{x}}_2; \\ \tilde{F}(t) &= \frac{m_1 m_2}{k}\ddddot{\tilde{x}}_2 + (m_1 + m_2)\ \ddot{\tilde{x}}_2 \end{aligned} \tag{25}$$

with the order parameter $r = 4$. There is no such a solution for the non-flat case.

6 Summary and Conclusions

The diversity in the inverse simulation problem for underactuated mechanical system in partly specified motion, referred here to as servo-constraint problem, concerns numerous issues. Of fundamental relevance is the way the servo-constraints on the system are realized. By contrast to a system subject to (ideal) passive constraints, whose reactions are orthogonal to the constraint manifold, the orientation of actuating forces of the system (control reactions) with respect to the servo-constraint manifold may vary from orthogonal to tangential. Only for the orthogonal realization, conditioned upon the existence of the output–input relationship of Eq. (7), $\det(\mathbf{Y}) \neq 0$, the governing equations can be formulated as ODEs (9). For the mixed orthogonal–tangent and pure tangent realizations of servo-constraints, the governing equations must be formulated as DAEs. Two variants of the DAE formulations were motivated in this paper: DAEs (16) that use the system configuration coordinates $\mathbf{q}$ and DAEs (19) that are based on the coordinate transformation $\mathbf{q} = [\mathbf{q}_a^T \ \mathbf{q}_u^T]^T \rightarrow \mathbf{q}' = [\mathbf{y}^T \ \mathbf{q}_u^T]^T$. An advantage of using the DAE formulations is that they are valid for all possible types of realization of servo-constraints.

The servo-constraint problem for the case of orthogonal realization of servo-constraints is always non-flat. More strictly, in addition to the output–input inverse dynamics model of Eq. (7), a residual internal dynamics in the system remains. The internal dynamics is usually influenced by the inverse dynamics control, and, vice versa, the enforced internal dynamics has an impact on the output–input relationship. Stability of the internal dynamics is therefore of critical importance for applicability of the servo-constraint problem solution, which can be achieved either by an optimized design of the underactuated system or appropriate selection of the output variables. The servo-constraint problems for the cases of mixed orthogonal–tangential and pure tangential realizations of servo-constraints can then be either non-flat or flat. In case of flatness, the servo-constraint problem becomes pure algebraic, in which all the state and control variables can explicitly be determined in terms of the specified in time outputs and its time derivatives.

The feedforward control of underactuated systems in partly specified motion can effectively be designed following the schemes seen in Fig. 2 (for orthogonal realization of servo-constraints) or in Fig. 5 (for all types of realization of servo-constraints), where $\boldsymbol{\alpha} = \mathbf{0}$ and $\boldsymbol{\beta} = \mathbf{0}$. The design of a feedback controller, to provide stable tracking of the specified outputs in the presence of external disturbances and modeling uncertainties, is in general more challenging. For the case of orthogonal realization of servo-constraints, provided that the assisted internal dynamics is bounded, the feedback controller can be designed by applying the stabilized output form defined in Eq. (8), which is as a direct generalization of the control laws used for fully actuated systems. Similar feedback controllers have also been applied to the differentially flat problems with tangent and orthogonal–tangent realizations of servo-constraints [5–7]. Robustness of the controllers has however been verified only through numerical tests and for "smooth" external disturbances during the simulated tracking of the reference trajectories. Due to the high index of the governing DAEs (16) used there, equal to three, "higher-order" controllers, discussed in [12], might be more appropriate. These issues are not addressed in this contribution.

References

1. Abdel-Rahman, E.M., Nayfeh, A.H., Masoud, Z.N.: Dynamics and control of cranes: a review. J. Vib. Contr. **9**, 863–908 (2003)
2. Ascher, U.M., Petzold, L.R.: Computer Methods for Ordinary Differential Equations and Differential-Algebraic Equations. SIAM, Philadelphia (1998)
3. Benosman, M., Le Vey, G.: Control of flexible manipulators: a survey. Robotica **22**, 533–545 (2004)
4. Blajer, W.: A geometrical interpretation and uniform matrix formulation of multibody system dynamics. Z. Angew. Math. Mech. **81**, 247–259 (2001)
5. Blajer, W., Kołodziejczyk, K.: A geometric approach to solving problems of control constraints: theory and a DAE framework. Multibody Syst. Dyn. **11**, 343–364 (2004)
6. Blajer, W., Kołodziejczyk, K.: Control of underactuated mechanical systems with servo-constraints. Nonlinear Dyn. **50**, 781–791 (2007)
7. Blajer, W., Kołodziejczyk, K.: Improved DAE formulation for inverse dynamics simulation of cranes. Multibody Syst. Dyn. **25**, 131–143 (2011)
8. Blajer, W., Graffstein, J., Krawczyk, M.: Modeling of aircraft prescribed trajectory flight as an inverse simulation problem. In: Awrejcewicz, J. (ed.) Modeling, Simulation and Control of Nonlinear Engineering Dynamical Systems, pp. 153–162. Springer, Netherlands (2009)
9. Blajer, W., Seifried, R., Kołodziejczyk, K.: Diversity of servo-constraint problems for underactuated mechanical systems: a case study illustration. Solid State Phenom. **198**, 473–482 (2013)
10. Campbell, S.L., Gear, C.W.: The index of general nonlinear DAEs. Numer. Math. **72**, 173–196 (1995)
11. Chen, Y.-H.: Equations of motion of mechanical systems under servoconstraints: the Maggi approach. Mechatronics **18**, 208–217 (2008)
12. De Luca, A.: Trajectory control of flexible manipulators. In: Siciliano, B., Valavanis, K.P. (eds.) Control Problems in Robotics and Automation, pp. 83–104. Springer, London (1998)
13. Fantoni, I., Lozano, R.: Non-linear Control for Underactuated Mechanical Systems. Springer, London (2002)
14. Fliess, M., Lévine, J., Martin, P., Rouchon, P.: Flatness and defect of nonlinear systems: introductory theory and examples. Int. J. Control **61**, 1327–1361 (1995)
15. Kirgetov, V.I.: The motion of controlled mechanical systems with prescribed constraints (servoconstraints). J. Appl. Math. Mech. USS **31**, 433–466 (1967)
16. Lee, H.-H.: Modeling and control of a three-dimensional overhead crane. J Dyn. Syst. Meas. Control T ASME **120**, 471–476 (1998)
17. Paul, R.P.: Robot Manipulators: Mathematics, Programming, and Control. MIT, Cambridge, MA (1981)
18. Rouchon, P.: Flatness based control of oscillators. Z. Angew. Math. Mech. **85**, 411–421 (2005)
19. Sahinkaya, M.N.: Inverse dynamic analysis of multiphysics systems. Proc. Inst. Mech. Eng. I J. Syst. Control Eng. **218**, 13–26 (2004)
20. Sara-Ramirez, H., Agrawal, S.K.: Differentially Flat Systems. Marcel Dekker, New York (2004)
21. Seifried, R.: Integrated mechanical and control design of underactuated multibody systems. Nonlinear Dyn. **67**, 1539–1557 (2012)

22. Seifried, R.: Two approaches for feedforward control and optimal design of underactuated multibody systems. Multibody Syst. Dyn. **27**, 75–93 (2012)
23. Seifried, R., Blajer, W.: Analysis of servo-constraint problems for underactuated multibody systems. Mech. Sci. **4**, 113–129 (2013)
24. Spong, M.W.: Underactuated mechanical systems. In: Siciliano, B., Valavanis, K.P. (eds.) Control Problems in Robotics and Automation, pp. 135–150. Springer, London (1998)
25. Thompson, D., Bradley, R.: Inverse simulation as a tool for flight dynamics research: principles and applications. Progr. Aero. Sci. **42**, 174–210 (2006)

Rare Phenomena and Chaos in Switching Power Converters

Dmitry Pikulin

Abstract This paper demonstrates the general applicability of the method of complete bifurcation groups to provide the global analysis of nonlinear dynamics of simple hybrid electronic systems. Our research is devoted to the investigation of complex nonlinear phenomena, including rare attractors and chaotic modes of operation in one of the most widely used electronic switching power converters (SPC)—boost one. Firstly, the discrete-time model, describing the operation of the mentioned power converter, is defined. Then the obtained model is used, providing numerical investigation of the dynamics of SPC on the basis of the method of complete bifurcation groups: the bifurcation map is constructed in two-parameter plane, the complete one-parameter bifurcation diagrams for several sections of the bifurcation map are obtained, and the most significant features of nonlinear dynamics of boost converter are studied. The results obtained in this paper prove that the design of reliable switching power converters is possible effectively applying the method of complete bifurcation groups, which allows the prediction and avoidance of occurrence of undesirable regimes (such as chaotic modes of operation, variety of rare attractors, complex protuberances, regions of unstable periodic infinitium) in the operation of these devices.

1 Introduction

Nowadays the existence of complex nonlinear phenomena, such as subharmonic oscillations and chaotic attractors, exhibited during the operation of widely used switching power converters (SPC), is commonly accepted and proved fact [2, 7, 8]. Sudden appearance of undesirable nonlinear operating regimes could damage not only the SPC itself, but also lead to the collapse of the whole system it supports [8]. Realizing the importance of the problem a variety of different methodologies has been proposed, providing the tools for the prediction of the first bifurcation patterns and subsequent evolution to chaotic oscillations [1, 8]. One of the most important

D. Pikulin (✉)
Faculty of Electronics and Telecommunication, Riga Technical University, Riga, Latvia
e-mail: dmitrijs.pikulins@rtu.lv

J. Awrejcewicz (ed.), *Applied Non-Linear Dynamical Systems*, Springer Proceedings in Mathematics & Statistics 93, DOI 10.1007/978-3-319-08266-0_15

plots used for the analysis of the dynamics of nonlinear systems with variable parameters is the bifurcation diagram [3]. The majority of methodologies utilize the so-called "brute-force" bifurcation diagrams, depicting only stable operating regimes and discarding the information about existing unstable modes of operation as unnecessary. However it has been shown that the disregard of the branches of bifurcation diagram corresponding to the unstable orbits may lead to the loss of very important information about the global dynamics of the system [5, 6, 9, 10].

Taking into consideration the mentioned disadvantages of commonly used techniques, this research utilizes the relatively new approach—method of complete bifurcation groups (MCBG)—developed in the Institute of Mechanics of Riga Technical University [6, 9, 10], for the investigation of the global dynamics of one of the most widely used electronic devices, boost SPC. During the analysis of the constructed complete bifurcation diagrams, the significance of the topology of unstable branches for the exploration of complex smooth and non-smooth bifurcation patterns is explored.

2 The Model of the Boost Switching Power Converter

An elementary boost power converter consists of a basic RLC circuit, diode, and switching element (see Fig. 1). The aim of this circuit is to preserve preferred output voltage v_c higher than that provided by the input source E. This can be achieved by appropriately turning on and off the switch S, so that the circuit is repeatedly forced by the external voltage source [7].

The position of the switching element is defined by parameters of the implemented voltage feedback loop that compares the output voltage to the reference value, amplifies the difference (see gain k in the Fig. 1), and generates the corresponding pulse-width modulated signal.

Although SPC, being inherently hybrid systems, are modeled using switched state-space models, discrete-time models are also found to be useful in analytical

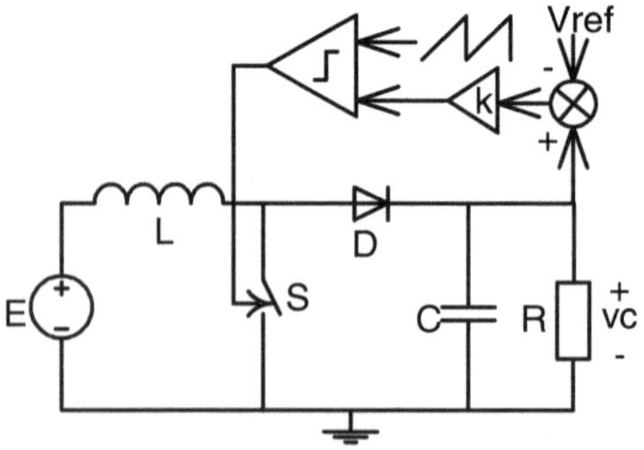

Parameter	Value
Switching period, T	333 (μs)
Inductance, L	208 (μH)
Load resistance, R	12.5 (Ω)
Capacity, C	222 (μF)
Input voltage, E	11-19 (V)
Reference voltage, V_{ref}	25(V)
Feedback gain, k	0.1-0.5

Fig. 1 Simplified model of the boost SPC and its general parameters [7]

and especially numerical investigation of nonlinear dynamics [4]. The most widely used discrete-time model for switching converters is called iterative map, where the state variables are periodically sampled at time instants t = nT (T is the switching period of SPC). The map F is then the relationship that links the previous and the current state variable samples: x(n + 1) = F(x(n)). This kind of mappings is used as an effective tool for locating bifurcation points and constructing bifurcation diagrams.

Tse in his book [7] proves that the operation of the boost converter, operating in discontinuous current mode (DCM) under voltage-mode control (shown in the Fig. 1), could be described by simple one-dimensional map in closed form for the capacitor voltage:

$$v_C\,(n+1) = \alpha v_C(n) + \beta E^2\big(H\,\big(D - k\,\big(v_C(n) - V_{ref}\big)\big)\big)^2 / \left(v_C(n) - E\right) \quad (1)$$

where

$$D = \sqrt{V_C\,(1-\alpha)\,(V_C - E)\,/\beta E^2}; \alpha = 1 - T/RC + T^2/2R^2C^2; \beta = T^2/2LC, \quad (2)$$

D, steady-state duty cycle; V_C, steady-state output voltage; $H(.)$, limits the range of duty cycle between 0 and 1; V_{ref}, reference voltage; k, small signal gain; and T, switching period. Other notations and parameter values of the boost SPC under test are depicted in Fig. 1.

The discrete model defined in (1) and (2) will be used during the complete bifurcation analysis of the boost converter, as it has been shown in [7]—even this simple one-dimensional map is capable of predicting a variety of nonlinear phenomena in the operation of SPC with high level of accuracy.

3 The Construction of Bifurcation Map

At the beginning of the investigation the primary and secondary bifurcation parameters (input voltage E and small signal feedback gain k), defining the qualitative changes in the dynamics of SPC, as well as dimensions and location of regions with different types of periodic (or chaotic) regimes in the parameter plane, are selected. Then the bifurcation map in the plane $\{(k, E){:}0.05 < k < 0.4;\ 11 < E < 19\}$ is obtained, providing the information about the division of the parameter plane into regions with different periodic and chaotic operating regimes. Figure 2 depicts the bifurcation map obtained for the parameters of the system defined in Fig. 1. Each hatching style in the diagram represents different periodic (or chaotic) operating regimes, specified in the legend (up to P10).

The obtained bifurcation map allows concluding that in the certain range of parameter space the boost converter operates in the stable P1 (the only desirable

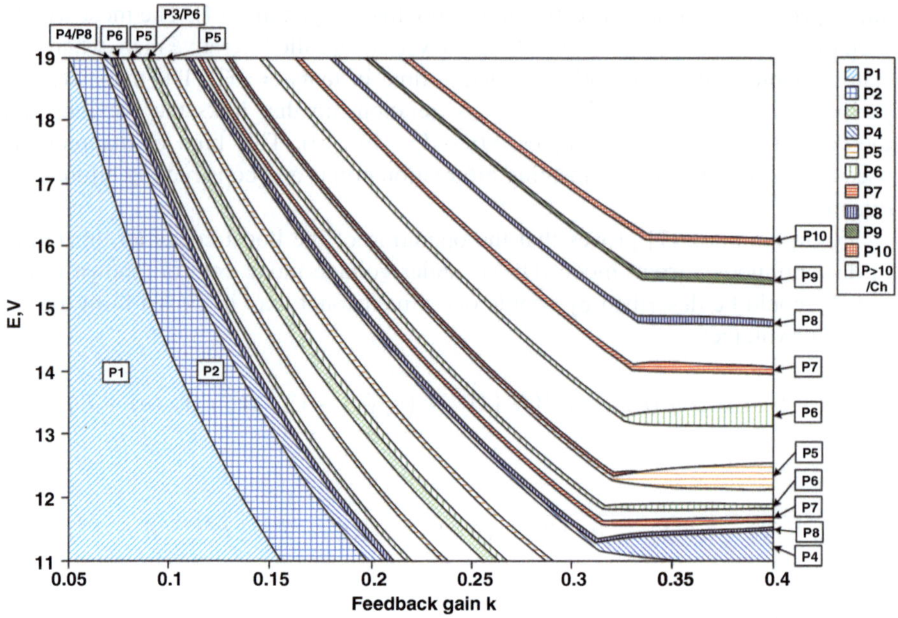

Fig. 2 The bifurcation map of the boost switching power converter in the k−E parameter plane

operating mode for SPC) that undergoes period-doubling bifurcation cascade with subsequent transition to chaotic mode of operation. The chaotic region is not homogenous, as there exists a great variety of periodic windows that are formed by certain types of rare attractors (RA, periodic or chaotic regimes, being stable in narrow range of parameter values [9]). The existence of RA is proved and the appropriate identification is provided, during the construction of complete one-parameter bifurcation diagrams.

The closer look on the topology of different periodic regions of the bifurcation map allows ascertaining the fact that for the small values of the feedback gain k, only smooth bifurcations and transitions to chaos could be observed. However, for $E < 14$ V and $k > 0.3$ some kind of non-smooth phenomena and sudden changes in the bifurcation structures could be expected.

The next section represents the complete bifurcation diagrams, constructed as the cross sections of the bifurcation map in Fig. 2, selecting the small signal feedback gain k as the primary bifurcation parameter. These diagrams provide the in-depth information about the structure of bifurcation paths, types of RA, and chaotization scenarios.

4 Periodic Skeleton and Complete Bifurcation Diagrams

Prior to the construction of the first bifurcation diagram, the topology and approximate structure of the diagram could be estimated, obtaining the periodic skeleton [9] for $E = 16$ V and $k = 0.4$ (other parameter values are defined in Fig. 1). The data obtained in the process of construction of periodic skeleton are summarized in Table 1.

Each regime in the periodic skeleton is described by the coordinates of its fixed point and multiplier, defining the stability of the regime. As it could be seen, all found periodic regimes (up to P6) are unstable, so in this case the converter operates either in stable periodic regime nP with $n > 6$ or in chaotic mode of operation. As it is shown in Table 1, several periodic regimes form different bifurcation groups (BG, stable and unstable branches of bifurcation diagram, having common bifurcation points [9])—one of the cornerstones of the MCBG. For example, P1–P2–P4 regimes define the origin of period-doubling cascade of 1 T BG.

The information obtained in the periodic skeleton is used in the process of numerical continuation (path-following) for the construction of complete bifurcation diagrams, depicting the most significant nonlinear phenomena, observed in boost converter and showing main paths to chaos.

Table 1 Periodic skeleton (up to P6) of the boost converter under study for $E = 16$ V and $k = 0.4$

Bifurcation group (BG)	Period	Fixed point position ($v_{C,n}$*)	Multiplier λ	Stability
1 T	P1	25	−7.2880	Unstable
	P2	27.8377	−10.1033	Unstable
	P4	24.6659	−515.7154	Unstable
3 T	P3	27.5612	−11.8149	Unstable
		25.0486	70.3031	Unstable
	P6	24.6573	−792.7106	Unstable
4 T	P4	27.3159	−12.9183	Unstable
		24.3970	76.7561	Unstable
$5\ T_1$	P5	24.66123	−714.3610	Unstable
		24.4269	118.0974	Unstable
$5\ T_2$	P5	25.1289	77.8386	Unstable
		27.0880	−13.6282	Unstable
$6\ T_1$	P6	24.6978	5233.2455	Unstable
		25.0391	−820.4291	Unstable
$6\ T_2$	P6	24.1754	−581.3176	Unstable
		24.7462	126.6094	Unstable
$6\ T_3$	P6	23.9490	75.5339	Unstable
		26.8706	−14.0650	Unstable

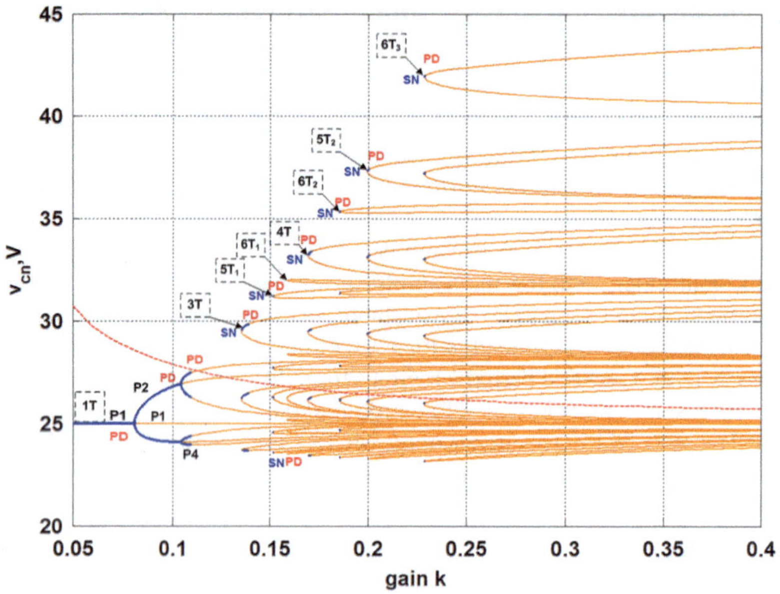

Fig. 3 The complete bifurcation diagram of the boost converter for E = 16 (V)

The obtained bifurcation diagrams (shown in Figs. 3 and 4) allow ascertaining that for small values of parameter k, the system operates in a stable P1 regime that loses its stability as a result of classical period-doubling, leading to the subsequent formation of chaotic region of 1 T BG (designated as UPI-1). Thus the increase of small signal feedback gain leads to the smooth transition to chaotic mode of operation. However this chaotic regime is not robust, as the great variety of tip type rare attractors, corresponding to 3–6 T_3 bifurcation groups, could be observed (see, e.g., small stable regions in Fig. 3). The topology of these RA is similar to the structure of 1 T BG shown in Fig. 4, including the complete period-doubling route to chaos. The overall dynamics of the system, including the route to chaos, in this case is completely defined by smooth period-doubling and saddle-node bifurcations, caused by the transition of the corresponding multipliers outside unit circle.

It should be noted that in all subsequent complete bifurcation diagrams dark lines represent stable periodic regimes and light lines, the unstable ones and all bifurcation points are accordingly marked.

5 Non-smooth Bifurcations and Submerged Isles

As it has been already mentioned for small values of input voltage E, the non-smooth bifurcations, manifesting as sudden jumps of characteristic multipliers, determine the dynamics of the converter. Thus, changing the value of secondary bifurcation

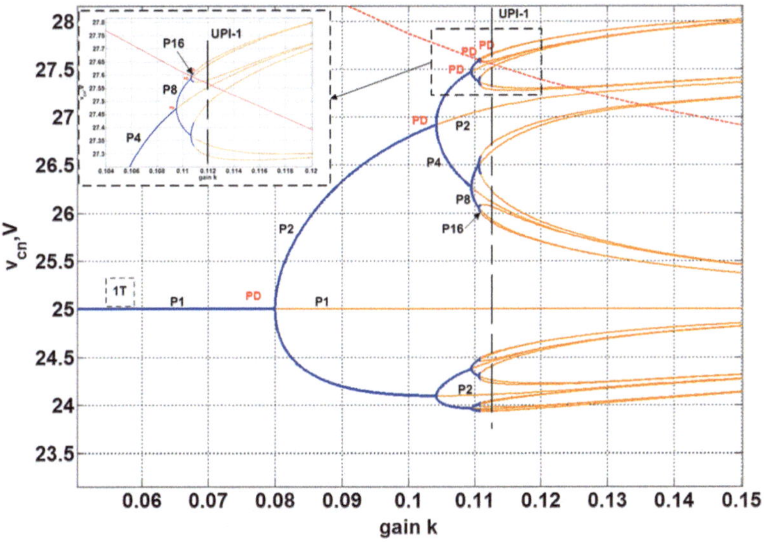

Fig. 4 The detailed structure of 1 T bifurcation group

parameter E allows the construction of bifurcation diagrams with more complex topological bifurcation groups, including uncommon protuberances, submerged isles, and non-smooth bifurcations. Figure 5 shows an example of detailed structure of complex protuberances of 4 T bifurcation group and submerged isles of 12 T bifurcation group for $E = 11.4$ V.

The topology of protuberances and the submerged islands (see [9]) in this case is defined by complex interaction of smooth saddle-node and period-doubling bifurcations (with classical transition to chaotic mode of operation) from the left side, as well as non-smooth bifurcations from the right, leading to abrupt changes in the branches of bifurcation diagrams and sudden loss of stability of definite groups of periodic regimes. Thus the chaotization of the system as well as the resumption of stable P8 operation is governed by simultaneous loss of stability through smooth period-doubling cascade and rapid changes in the dynamics, defined by the appearance of non-smooth bifurcations. It should be noted that in spite of rather uncommon structure, the submerged isles include small regions of stable periodic regimes—tip type rare attractors—coexisting with stable P8 regime (see Fig. 5).

6 Conclusions

In the present paper it has been demonstrated that the method of complete bifurcation groups could be successfully applied to the investigation of complex nonlinear dynamics of widely used electronic systems—SPC.

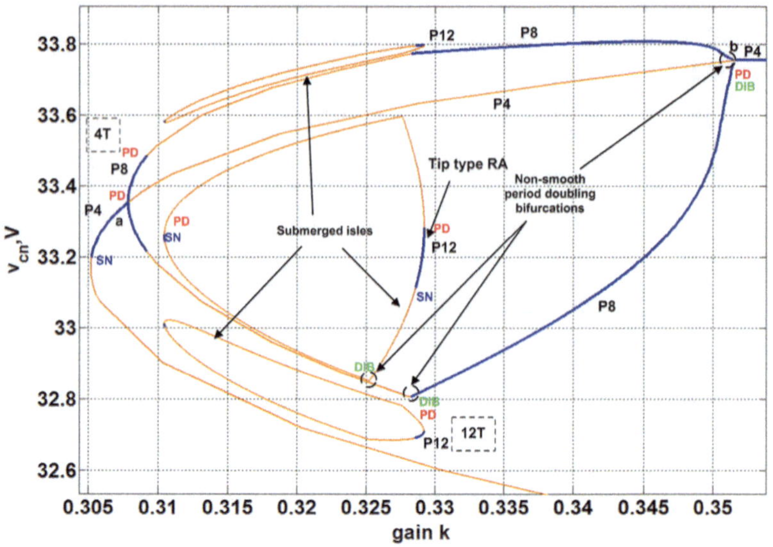

Fig. 5 The detailed structure of 4 and 12 T bifurcation groups for E = 11.4 V

It has been shown that the use of MCBG allows carrying out global bifurcation analysis of mentioned piecewise-smooth hybrid systems, exploring new nonlinear effects, periodic and chaotic regimes, rare attractors, protuberances, and submerged isles. The results include the construction of bifurcation map, obtaining periodic skeleton, and the subsequent construction of bifurcation diagrams depicting stable and unstable periodic regimes, as well as regions of unstable periodic infinitium.

The obtained results establish that in SPC within the definite parameter range rather specific types of non-smooth bifurcations define the global topology of bifurcation patterns, leading to unexpected and abrupt changes in the system dynamics.

References

1. Awrejcewicz, J.: Bifurcations and Chaos in Coupled Oscillators. World Scientific, Singapore (1991)
2. Banerjee, S., Verghese, G.C. (eds.): Nonlinear Phenomena in Power Electronic: Attractors, Bifurcations, Chaos and Nonlinear Control. Wiley, Hoboken (2001)
3. Di Bernardo, M., Budd, C.J., Champeneys, A.R., Kowalczyk, P.: Piecewise-Smooth Dynamical Systems: Theory and Application. Springer, Berlin (2008)
4. Hammil, D.C., Deane, J.H.B., Jeferries, D.J.: Modeling of chaotic dc/dc converters by iterative nonlinear mappings. IEEE Trans. Circ. Syst. Pt I **35**, 25–36 (1992)
5. Pikulins, D.: Effects of non-smooth phenomena on the dynamics of DC-DC converters. Sci. J. RTU Ser. 4 **29**, 119–122 (2011)

6. Schukin, I., Zakrzhevsky, M.: Application of software spring and method of complete bifurcation groups for the bifurcation analysis of nonlinear dynamical systems. J. Vibroeng. **10**, 510–518 (2008)
7. Tse, C.K.: Complex Behavior of Switching Power Converters. CRC, Boca Raton (2003)
8. Vilamitjana, E.R., Alarcon, E., El Aroudi, A.: Chaos in Switching Converters for Power Management. Springer, New York (2012)
9. Zakrzhevsky, M.: New concepts of nonlinear dynamics: complete bifurcation groups, protuberances, unstable periodic infinitiums and rare attractors. J. Vibroeng. **10**, 421–441 (2008)
10. Zakrzhevsky M.: Bifurcation theory of nonlinear dynamics and chaos. In: Periodic Skeletons and Rare Attractors. Proceedings of the 2nd international symposium RA'11, pp. 26–30. RTU, Riga (2011)

Constrained Motion of Mechanical Systems and Tracking Control of Nonlinear Systems

Firdaus E. Udwadia and Harshavardhan Mylapilli

Abstract This paper aims to expose the interrelations and connections between constrained motion of mechanical systems and tracking control of nonlinear mechanical systems. The interrelations between the imposition of constraints on a mechanical system and the trajectory requirements for tracking control are exposed through the use of a simple example. It is shown that given a set of constraints, d'Alembert's principle corresponds to the problem of finding the optimal tracking control of a mechanical system for a specific control cost function that Nature seems to choose. Furthermore, the general equations for constrained motion of mechanical systems that do not obey d'Alembert's principle yield, through this duality, the entire set of continuous controllers that permit exact tracking of the trajectory requirements. The way Nature seems to handle the tracking control problem of highly nonlinear systems suggests ways in which we can develop new control methods that do not make any approximations and/or linearizations related to the nonlinear system dynamics or its controllers. More general control costs are used and Nature's approach is thereby extended to general control problems. A simple, unified methodology for modeling and control of mechanical systems emerges. Examples drawn from diverse areas of control are provided dealing with synchronization of multiple nonlinear gyroscopes, design of optimal Lyapunov stable controllers for nonautonomous nonlinear systems, and energy control of nonhomogeneous Toda chains.

F.E. Udwadia (✉)
Departments of Aerospace and Mechanical Engineering, Civil and Environmental Engineering, Mathematics, Systems Architecture Engineering, and Information and Operations Management, University of Southern California, Los Angeles, CA 90089, USA
e-mail: fudwadia@usc.edu

H. Mylapilli
Department of Aerospace and Mechanical Engineering, University of Southern California, Los Angeles, CA 90089, USA
e-mail: mylapill@usc.edu

J. Awrejcewicz (ed.), *Applied Non-Linear Dynamical Systems*, Springer Proceedings in Mathematics & Statistics 93, DOI 10.1007/978-3-319-08266-0_16

1 Introduction

Sir Isaac Newton described the field of mechanics in his preface to the Principia in the following words [12]:

> In this sense rational mechanics will be the science of motions resulting from any forces whatsoever, and the forces required to produce any motions, accurately proposed and demonstrated.

Today, while the first part of Newton's definition of mechanics has become our usual understanding of this field, the second part is usually relegated primarily to the field of control theory. Indeed, the problem that Newton famously solved was a control problem: the determination of the forces required to be acting on the planets so that their motions obey the observed motions described by Kepler's first two laws. [The third law that relates the periods of the planets to their mean radial distances, which took Kepler a long time to ferret out, is not essential to obtain the inverse square law of gravitation, though it is consistent with it.] To make certain that he considered the control of mechanical systems within the purview of what he thought of as mechanics, Newton went on in his preface to say that:

> . . . I offer this work as the mathematical principles of philosophy, for the whole burden of philosophy seems to consist in this–from the phenomena of motions to investigate the forces of nature, and then from these forces to demonstrate the other phenomena [of motions] and to this end the first and second books are directed.

In the above, the bracketed words have been added to the original quotation. To illustrate the viewpoint of Newton, let us consider an elementary example, the problem of finding the equations of motion of a spherical pendulum like the one shown in Fig. 1.

The problem of finding the equation of motion of this simple system, which consists of a particle of mass m constrained to move so that it is always at a fixed distance, L, from its fixed point of support, O, in a nonuniform gravitational field, can alternatively be looked at from the dual (the word "dual" in this paper is used as in ordinary parlance and not in the restricted sense used in optimization theory) standpoint of tracking control in the following way.

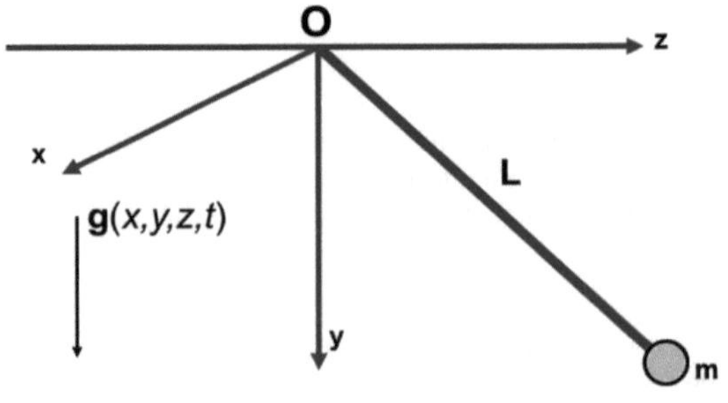

Fig. 1 A spherical pendulum

Consider a particle of mass m moving in a nonuniform gravitational field; it is now required to determine the control force that needs to be applied to this particle so that it is constrained to lie, at each instant of time t, on the sphere S^2 defined by the relation

$$\varphi(x, y, z, t) := x^2(t) + y^2(t) + z^2(t) - L^2 = 0. \tag{1}$$

We will show that this control problem can be handily approached using the theory of constrained motion of mechanical systems. Let us denote the 3 by 1 vector (the 3-vector) $q := [x\, y\, z]^T$. Clearly, the equation of motion of the particle as it freely moves in the nonuniform gravitational field in which the acceleration due to gravity at any point is $g(x, y, z, t)$ (see Fig. 1) is simply given by

$$M\,\ddot{q}(t) := \begin{bmatrix} m & 0 & 0 \\ 0 & m & 0 \\ 0 & 0 & m \end{bmatrix} \begin{bmatrix} \ddot{x} \\ \ddot{y} \\ \ddot{z} \end{bmatrix} = \begin{bmatrix} 0 \\ mg(x, y, z, t) \\ 0 \end{bmatrix} := Q\ , \tag{2}$$

so that the acceleration of the particle at any time t can be written as the 3-vector $a(q, t) = [0\; g\; 0]^T$. [From here on, we shall drop the arguments of the various quantities, unless needed for clarity.] We shall refer to Eq. (2) as the *unconstrained* (or uncontrolled) *equation of motion* for the mechanical system. A control theorist may prefer to call the equation a description of the "plant" whose trajectories need to be controlled so that they satisfy the control requirement stated in (1). We observe that from a physical viewpoint, the initial conditions, $q(0)$ and $\dot{q}(0)$, can have any arbitrary numbers for their respective components, depending, of course, on the location and the initial velocity of the particle.

However, this particle's motion described by Eq. (2) will not, in general, satisfy the constraint—or alternatively put, our trajectory requirement—namely, that its motion lies on the surface of the sphere described by Eq. (1). In order to achieve this, an additional force will need to be applied to the particle so that its acceleration is altered from $a(q, t)$, and its equation of motion now becomes

$$M\,\ddot{q} = Q + Q^C. \tag{3}$$

This additional force, Q^C, which is a 3-vector, that needs to be applied to the constrained system can be viewed as the force of constraint that ensures that Eq. (1) is satisfied. It can also, from a dual perspective, be seen as the control force that must be applied to the system described by Eq. (2), so that it satisfies the trajectory requirement (1) that is imposed on it. The addition of a control force Q^C on the right-hand side of Eq. (2) is not the only alteration needed in our update from the unconstrained system (2) to the constrained system that now satisfies Eq. (1).

The initial conditions $q(0)$, and $\dot{q}(0)$ whose components could be chosen arbitrarily in the case of system (2) can no longer be chosen arbitrarily. Instead the components of $q(t)$ at each instant of time (and hence also at the initial time)

Table 1 Analogous concepts in analytical dynamics and control theory

Analytical dynamics	Control theory
Unconstrained system	Uncontrolled system or plant
Constrained system	Controlled system
Constraints	Trajectory requirements
Constraint force	Control force or control

must satisfy relation (1); also, the components of $\dot{q}(t)$ must satisfy at each instant of time (and hence also at the initial time) the following relation:

$$x(t)\dot{x}(t) + y(t)\dot{y}(t) + z(t)\dot{z}(t) = 0, \tag{4}$$

which is obtained by differentiating Eq. (1) with respect to time. One may want to further differentiate Eq. (4) to obtain the relation

$$x(t)\ddot{x}(t) + y(t)\ddot{y}(t) + z(t)\ddot{z}(t) = -\dot{x}^2(t) - \dot{y}^2(t) - \dot{z}^2(t), \tag{5}$$

which can be written in matrix-vector form as

$$A\,\ddot{q} = b, \tag{6}$$

where $A := [x\ y\ z]^T$, and $b = -\dot{x}^2(t) - \dot{y}^2(t) - \dot{z}^2(t)$. We note that for a given set of initial conditions that satisfy Eqs. (1) and (4) at $t = 0$, Eq. (6) is equivalent to Eq. (1).

This simple example thus illustrates the connections between the problem of constrained motion and the problem of tracking control. Specifically, we find the following analogous concepts given in Table 1 above.

As we go along, we will extend and refine this preliminary table. In what follows, we will move back and forth between these dual concepts allowing ourselves to be aided in our understanding of constrained motion to expose new insights into trajectory control and vice versa.

2 General Constrained Mechanical Systems and the Trajectory Control Problem

Our spherical pendulum problem was an illustrative "toy problem" created simply to provide some insights into the connections that we are trying to establish. The problem could, of course, have been made considerably more challenging by requiring that the point of support, O, moves over a surface, say $\phi(q, \dot{q}, t) = 0$, and/or requiring that the pendulum's length varies in a prescribed manner so that $L(t) = f(q, \dot{q}, t)$. Though we will continue to use our toy problem to gain further insights, we can now frame the general problem of constrained motion in analytical dynamics as follows:

1. Consider an unconstrained (uncontrolled) nonlinear nonautonomous mechanical system described by the equation

$$M(q,t)\ \ddot{q} = Q(q, \dot{q}, t)\,, q(0) = q_0 \text{ and } \dot{q}(0) = \dot{q}_0, \tag{7}$$

   where M is a positive definite n by n matrix and q is an n-vector.
2. We require this system to satisfy the m consistent constraints (trajectory requirements) given by the relations

$$\phi_i(q,t) = 0,\ \ i = 1, 2, \ldots, h, \text{and,} \tag{8}$$

$$\psi_i(q, \dot{q}, t) = 0,\ \ i = h+1, h+2, \ldots, m. \tag{9}$$

3. We need to find the constraint (control) force, Q^C, so that the constrained (controlled) system described by

$$M(q,t)\ \ddot{q} = Q(q, \dot{q}, t) + Q^C(q, \dot{q}, t)\,, \qquad q(0) = q_0, \text{and}\ \ \dot{q}(0) = \dot{q}_0, \tag{10}$$

   exactly satisfies the trajectory requirements (8) and (9).

The uncontrolled system (or "plant") described by Eq. (7) that we shall be dealing with is nonlinear and nonautonomous. We shall initially assume that q_0 and $\dot{q}_0$ satisfy the trajectory requirements (8) and (9) at time $t = 0$. Later on, we will relax this condition and allow for the possibility that the initial conditions q_0 and $\dot{q}_0$ may not lie on the constraint manifold described by (8) and (9). We define the acceleration of the uncontrolled (unconstrained) system by

$$a(q, \dot{q}, t) = M^{-1}(q,t)\, Q(q, \dot{q}, t)\,. \tag{11}$$

Also, assuming sufficient smoothness, we can differentiate the h equations in the set (8) twice with respect to time (as we just did in our toy problem, see (5)), and the $(m - h)$ equations in the set (9) once with respect to time, to obtain the relation

$$A(q, \dot{q}, t)\ \ddot{q} = b(q, \dot{q}, t)\,, \tag{12}$$

where A is an m by n matrix of rank k. Each row of the matrix A corresponds to one of the trajectory requirements in the sets (8) or (9). We permit the trajectory requirements to be functionally dependent, though always consistent.

3 The Control Force Q^C

Having now laid out some of the underlying concepts relevant to the connections between the problem of constrained motion and the problem of tracking control, let us concentrate in this section on how one might determine the control force Q^C.

Before we embark on this, it might be worthwhile going back to our toy problem and investigating if such a force Q^C indeed exists, so that the trajectory requirement (1) is always satisfied, and if so, whether it can be uniquely found. That such a force Q^C exists is obvious, because we know the equation of motion of a pendulum, and so we know that a right-hand side for Eq. (3) exists so that the constraint (1) is *exactly* satisfied for all time, given that the initial conditions satisfy this constraint. So there most likely exists a control that is Lipschitz continuous, as we require in mechanics so that the solution of (3) is unique and it concurs with practical observations of the motions of a pendulum. Our next question is then, can Q^C be uniquely found?

Unfortunately, not! For the spherical pendulum, at each instant of time, we have the following six unknowns: the three components of the 3-vector $\ddot{q}$ and the three components of the 3-vector Q^C (see Eq. (3)). At each instant of time, starting with a given state $(q, \dot{q})$ of the system, we have the three equations given by the set (3) and an additional equation of constraint (1) (or alternately Eq. (6))—a total of 4 equations. The number of unknowns exceeds the number of equations by two, and hence, at each instant of time, the problem of finding the unknowns (accelerations and control forces) in the system is underdetermined! To get them uniquely we would need to have two more independent equations. Moving to our dual vision of the problem as one of trajectory control, there must then be an infinity of control forces (controllers) Q^C that can *exactly* track the trajectory expressed by Eq. (1)!

However, the equation of motion of a spherical pendulum, which satisfies the constraint (trajectory requirement), *is* unique—hence Q^C is unique—and its motion pretty well agrees with what is in fact physically observed. So clearly, Nature must then be picking the constraint force (control force) Q^C in such a manner so as to satisfy some additional criterion—one which somehow yields the (additional) two missing equations in our toy problem and yields a unique answer for the control force! The next subsection discusses this.

3.1 D'Alembert's Principle, Gauss's Principle, and the Cost Function

Flipping back to our understanding of constrained motion, we may then ask, how does Nature pick the constraint force Q^C so that the motion of our spherical pendulum matches our physical observations? This is a problem that was first attacked by d'Alembert and later on, more generally, by Lagrange [10]. Lagrange came up with the precise statement of what is today called d'Alembert's principle or prescription. D'Alembert's prescription is as follows.

The constraint force Q^C is such that for all vectors $v(t) \neq 0$ that satisfy the relation $A\,v = 0$, Nature seems to require that $v^T Q^C = 0$.

The nonzero vectors v that satisfy the relation $Av = 0$ are called virtual displacements, and the quantity $W^C = v^T Q^C$ is referred to as the total work done by the forces of constraint under virtual displacements. And this prescription, somewhat miraculously—for any general mechanical system—generates the correct number of additional equations so that the constraint force Q^C in Eq. (10) can be uniquely found at each instant of time!

To see how this works for our spherical pendulum, observe that the rank of our matrix A in Eq. (6) is 1, and so the null space of this 1 by 3 matrix is 2. Thus at each time there are two linearly independent 3-vectors v_1 and v_2 that satisfy the relation $Av = 0$ which we can find. D'Alembert's prescription then requires that ${v_1}^T Q^C = 0$ and ${v_2}^T Q^C = 0$. These two additional equations used with the four equations (the three equations in set (3) and Eq. (6)) that we had previously yield the six equations needed for finding the six unknowns—$\ddot{q}$ and Q^C—at each instant of time.

What is more astonishing—and indeed a tribute to the genius of Lagrange—is that d'Alembert's prescription yields the constraint force Q^C which when used in Eq. (10) yields the motion, $q(t)$, of the mechanical system that is fairly close, in numerous situations, to what is *actually observed* in the physical world, hence its enormous value in modeling physical systems.

To summarize, the equations of motion of the unconstrained system and the constraints cannot in general determine the constraint force Q^C uniquely. D'Alembert's principle generates additional equations (exactly the right number) to give us a unique Q^C at each instant of time, which causes the constrained system to move in a manner that is in concert with physical observations. Now it turns out that this prescription of d'Alembert regarding the constraint force Q^C is exactly the same as the following condition on the constraint (control) force Q^C from the dual viewpoint [24]. This condition, called Gauss's principle [5], is the following:

From all those control (constraint) forces Q^C that can exactly satisfy the trajectory requirements (8) and (9), Nature chooses that control force Q^C that minimizes the control cost

$$J(t) = \left[Q^C(q,\dot{q},t)\right]^T M^{-1}(q,t)\; Q^C(q,\dot{q},t) = \left\|Q^C\right\|^2_{M^{-1}} \tag{13}$$

at *each instant of time*. As seen from (13), $J(t)$ is simply the square of the weighted L_2 norm of the control force, Q^C. So we see that d'Alembert's prescription in mechanics—a prescription that causes mathematical models of constrained mechanical system to suitably predict the physically observed motions of these systems—has a dual that says that Nature appears to be constantly solving an optimal control problem, minimizing the cost function $J(t)$ given in (13). But unlike most control engineers today, who would prefer to minimize $\int_0^T J(t)dt$, where T is some final time over which the control is executed, Nature seems to do this minimization *at each instant of time*. Also, the so-called weighting matrix that she uses in the cost function is M^{-1}. This is indeed clever! For example, imagine a multi-body system, with several masses, that is described by Eq. (7). Say we want to control this system so that it satisfies some given trajectory requirements given by relations (8) and (9). Realizing that the larger masses require larger forces to be exerted on them to cause them to move, Nature attempts to satisfy these requirements (constraints) on this multi-body system, by being in favor of applying forces to the smaller masses—hence, the weighting by the matrix M^{-1}. It is only recently that control engineers in robotics have realized, mainly by trial and error, that this weighting matrix works well in most robotic applications and leads to robust control designs [15].

Having underscored the duality between the problem of constrained motion and the problem of trajectory tracking, we find then that insights from mechanics do shed some light on control theory. But while so far we have only been talking about the properties of the constraint force Q^C, we have not yet answered the question: what is it? Can one find it explicitly, in closed form, for a general unconstrained mechanical system?

3.2 *Closed-Form Solution to the Optimal Tracking Control Problem for Nonlinear Nonautonomous Mechanical Systems Using the Theory of Constrained Motion*

The problem of finding the constraint force Q^C that Nature uses has a long and varied history. The problem was first formulated by Lagrange [10] and has been attempted by numerous scientists and engineers [1, 4–6, 8, 16]. A simple expression for the explicit form of the control force was obtained in 1992, and it is given by [23]

$$Q^C = -M^{1/2}\left(AM^{-1/2}\right)^+ (Aa - b)\,. \tag{14}$$

Here, X^+ denotes the Moore-Penrose inverse of the matrix X [11, 14]. The equation of motion of the constrained system, which may be thought of as the fundamental equation of mechanics, can thus be explicitly written *in extensio*, using relation (10), as

$$\begin{aligned} M(q,t)\,\ddot{q} = Q(q,\dot{q},t) - M^{1/2}(q,t)\left[A(q,\dot{q},t)\,M^{-1/2}(q,t)\right]^+ \\ \left[A(q,\dot{q},t)\,a(q,\dot{q},t) - b(q,\dot{q},t)\right]. \end{aligned} \tag{15}$$

What now might be gleaned from a controls point of view from relation (15)? First, we observe that $a(q,\dot{q},t)$ (see Eq. (11)) is the acceleration of the uncontrolled (unconstrained) system. However, to track the given trajectory described by the set of Eqs. (8) and (9), the acceleration of the system needs to satisfy the trajectory requirement (12). Hence the extent to which the acceleration, a, of the uncontrolled system does not satisfy this trajectory requirement is simply

$$e(q,\dot{q},t) := \left[A(q,\dot{q},t)\,a(q,\dot{q},t) - b(q,\dot{q},t)\right]. \tag{16}$$

This is in fact the error in the satisfaction of the trajectory constraint at time t by the acceleration (at that time) of the uncontrolled system. And so Eq. (15) says that this error signal is fed back to the system (7), just the way a modern-day control engineer might want to do negative feedback control! Second, we observe that Nature seems to choose a control gain matrix whose elements are, in general, highly nonlinear functions of $q, \dot{q}$, and t. It is given explicitly by

$$K(q,\dot{q},t) := M^{1/2}(q,t)\left[A(q,\dot{q},t)\,M^{-1/2}(q,t)\right]^{+}. \tag{17}$$

Thus the control methodology used by Nature, so that the uncontrolled system (7) exactly tracks the trajectory requirements stated in sets (8) and (9), can be encapsulated by the relation

$$M(q,t)\;\ddot{q} = Q(q,\dot{q},t) - K(q,\dot{q},t)\,e(q,\dot{q},t), \tag{18}$$

where K is the gain matrix and e is the error signal. Lastly, we point out that Nature appears to use an error signal for its feedback control law that is related to accelerations and not to displacements, nor to velocities, or to integrals of the displacement, as is commonly done in control theory; recall PID control! She appears to be doing *acceleration feedback*. The tracking controller represented by Eq. (18) is not only optimal in that it minimizes the control cost $J(t)$ given in Eq. (13), but it yields exact tracking; for the set of equations (8) and (9) are the integrals of motion of the nonlinear system described by Eq. (15) (or (18)). The minimal control cost at each instant of time is explicitly given by

$$J(t) = \left[Q^{C}\right]^{T} M^{-1} Q^{C} = \left\|\left(AM^{-1/2}\right)^{+}(Aa-b)\right\|^{2}. \tag{19}$$

As mentioned before, the closed-form expression for the control force Q^C that Nature uses when satisfying trajectory requirements is given by relation (14). She gets this unique control force by minimizing the control cost $J(t)$ given in (13), which is simply the square of the weighted L_2 norm of control force, Q^C. Nature picks the weighting matrix to be the positive definite matrix $M^{-1}(q,t)$. This choice produces control forces (forces of constraint) that are in conformity with the physically observed motions of constrained systems.

What if the control engineer wants to use a different weighting matrix in the cost function? Namely, suppose he/she wants to minimize at each instant of time the control cost

$$J(t) = \left[Q^{C}(q,\dot{q},t)\right]^{T} N(q,t)\;Q^{C}(q,\dot{q},t) = \left\|Q^{C}\right\|_{N}^{2}, \tag{20}$$

where $N(q,t)$ is a positive definite matrix. Using our dual perspective, this may also be thought of as a generalization of Gauss's principle (in mechanics), wherein we use a weighting matrix in our control cost minimization that may be different from M^{-1}. It turns out that the unique control that minimizes this control cost is given (instead of Eq. (14)) by [21]

$$Q^C = -N(q,t)^{-1/2} A_N^+ (Aa - b) = -N^{-1} M^{-1} A^T \left[A (MNM)^{-1} A^T \right]^+ (Aa - b), \tag{21}$$

where $A_N = A(q,\dot{q},t)\, M(q,t)^{-1} N(q,t)^{-1/2}$. The last equality above follows from the identity $X^+ = X^T(XX^T)^+$. When $N = M^{-1}$, we get back Eq. (14), as expected.

There is one last remaining point that is worth mentioning. Recall that we had assumed that the initial conditions of the controlled system satisfy the trajectory requirements (8) and (9). What if the initial conditions do not lie on the so-called manifold described by the trajectory requirements? If one is close to the trajectory manifold, then instead of thinking of the trajectory requirements (8) and (9) as $\phi_i(q,t) = 0$ and $\psi_i(q,\dot{q},t) = 0$, one could consider the trajectory requirement as [20]

$$\ddot{\phi} + \Sigma\dot{\phi} + K\phi = 0, \text{and } \dot{\psi} = -\Lambda\psi, \tag{22}$$

where ϕ and ψ are h- and $(m-h)$-vectors that contain the ϕ_i's and ψ_j's, respectively. The matrices Σ, K, and Λ can be chosen so that the solutions, ϕ and ψ, to the equations (22) tend to zero asymptotically as $t \to \infty$, so that the constraints $\phi_i = 0$ and $\psi_i = 0$ are ultimately satisfied. Equations (22) lead to trajectory requirements which can again be stated in the form of Eq. (12), and the control force is again given explicitly by Eq. (21)! The parameters that are used in the matrices Σ, K, and Λ will control the rate and nature of convergence of the trajectories of the dynamical system towards the manifold prescribed by the trajectory requirements, $\phi_i(q,t) = 0$ and $\psi_i(q,\dot{q},t) = 0$.

To illustrate the nature of this control force, let us go back to our toy problem, which we introduced in Sect. 1, of controlling a mass m in a time-varying gravity field so that it lies on the surface $\varphi(x,y,z,t) := x^2(t) + y^2(t) + z^2(t) - L^2 = 0$. The explicit control force that causes the mass to lie on this manifold asymptotically is given in the Appendix. We note that the control is nonlinear and no approximations related to the nonlinear nature of the "plant" are made. No a priori assumptions (such as a linear PD controller) are made about the controller either. Note that the control minimizes the control cost given in Eq. (20) at *each instant* of time for a user-specified weighting matrix N.

Flipping back to analytical dynamics, our closed-form equation given by (15) for the constrained motion of the system (10) presupposes that d'Alembert's prescription is valid for every mechanical system. What if it isn't? Before we take such a contrary stance, we need to inquire whether systems for which d'Alembert's principle doesn't hold do exist. Fortunately for us, it turns out that there are many such systems, the ones most commonly exemplified are those that may have sliding friction in them (for example, think of a mass sliding along a horizontal surface), because now the constraint force Q^C does perform work under virtual displacements [2]. Until recently, such systems have been a kind of eyesore for analytical dynamicists, because it has been difficult to include them within the general framework of Lagrangian mechanics. Constraint forces that do not obey d'Alembert's prescription are called nonideal, and often such systems are referred to as systems with nonideal constraints. Let us now turn to such systems and find out what insights they may provide, in our dual view, for the tracking control of nonlinear, nonautonomous systems.

3.3 Mechanical Systems with Nonideal Constraints and the Set of Controllers for Exact Trajectory Control

The difficulty of incorporating systems with nonideal constraints into the framework of Lagrangian mechanics—though such systems are fairly commonplace in the physical world—arises because of the following three main reasons:

1. We need to have the specification of constraints to be general enough so as to encompass problems of practical utility.
2. The specification must, in order to comply with physical observations, yield the accelerations of the constrained systems uniquely when using the math-ware of analytical dynamics that has been developed over the last 250 years.
3. When the constraint forces do no work, we must obtain the formalisms/equations that have thus far been obtained (e.g., Lagrange's equations, Gibbs-Appell equations, Poincare equations, Generalized Inverse equations, etc.).

It is for this reason that most texts and treatises on mechanics summarily dispatch these systems beyond their boundaries, early on in their treatments of analytical dynamics. For example, we find in Goldstein, at the beginning of his book, on page 17, the line [7]:

> This [D'Alembert's Principle] is no longer true for sliding friction and we must exclude such systems from our Lagrangian formulation.

In Pars's treatise on mechanics we again find on page 14, and we quote [13]:

> There are in fact systems for which this [D'Alembert's] Principle does not hold. But such systems will not be considered in this treatise.

Put differently, the main problem is how to modify and extend d'Alembert's principle so that the three difficulties mentioned above can be overcome. One way of doing this would be to extend d'Alembert's prescription to say that *at each instant of time, the work done by the force of constraint is prescribed for the specific system at hand*. Such a principle would then state that [25]:

For any virtual displacement $v(t)$ at time t, the work done by the force of constraint $W^C := v^T Q^C$ is prescribed to be equal to $v^T C(q, \dot{q}, t)$, (23)

where the n-vector $C(q, \dot{q}, t)$ is prescribed by the mechanician for the given, specific system being modeled. The prescription of C can be done through experimentation, and/or by analogy with other systems, or intuition, or otherwise. Since at any given instant of time t, W^C can be positive, negative, or zero; this allows the possibility that energy can be fed into the system at the constraint, or it can be removed at the constraint. We note that when $C \equiv 0$ ($W^C = 0$) for all time t, this extension of d'Alembert's principle reverts to d'Alembert's prescription.

For any sufficiently smooth C, one can find the explicit equation of motion for such a constrained system that has nonideal constraints and satisfies exactly the constraint requirements (8) and (9) (or alternately (12)). Dropping the arguments of the various quantities, the equation is [25]

$$M\,\ddot{q} = Q - M^{1/2}B^{+}\,(Aa - b) + M^{1/2}\left(I - B^{+}B\right)M^{-1/2}C := Q + Q^{C}, \tag{24}$$

where $B(q,\dot{q},t) = A(q,\dot{q},t)\,M(q,t)^{-1/2}$, and, as before, X^{+} denotes the Moore-Penrose inverse of the matrix X. We notice that the first two terms in the first equality on the right-hand side of Eq. (24) are identical to those on the right-hand side of Eq. (15), the nonideal nature of the constraint force having simply added an additional term on the right-hand side, for any given prescribed smooth function $C(q,\dot{q},t)$. Furthermore, when $C \equiv 0$ the equation reverts to the proper equation of motion (15) when d'Alembert's prescription is satisfied at all instants of time.

Using our dual view regarding the tracking control problem, what insights does this equation, which originates from analytical dynamics, provide for the constrained motion of a system with nonideal constraints? By choosing the Lipschitz continuous function $C(q,\dot{q},t)$ arbitrarily, Eq. (24) provides *all* the possible Lipschitz continuous controllers [25] that can make the uncontrolled system (7) *exactly* track the trajectory requirements specified by equations (8) and (9)!

The second and third members on the right-hand side in the first equality of Eq. (24) are M-orthogonal, and so

$$J(t) = \left\|B^{+}\,(Aa - b)\right\|^{2} + \left\|\left(I - B^{+}B\right)M^{-1/2}C\right\|^{2}. \tag{25}$$

The addition of the second term on the right-hand side increases the cost from its optimal value of $\|B^{+}(Aa - b)\|^{2}$ (see Eq. (19)) to that now provided by (25). We note that at those instants of time when $C = 0$, the control force becomes optimal and minimizes $J(t) = [Q^C]^T M^{-1} Q^C$.

As before, more generally, when using the weighting matrix N instead of M^{-1}, the explicit control that causes system (10) to exactly satisfy the trajectory requirements (8) and (9) is given in closed form by [21]

$$Q^{C} = -N(q,t)^{-1/2}A_N^{+}\,(Aa - b) + N^{-1/2}\left(I - A_N^{+}A_N\right)M^{-1/2}C, \tag{26}$$

for arbitrary continuous functions $C(q,\dot{q},t)$. When $C \equiv 0$, the control minimizes the general control cost given in (20) at each instant of time. The equation of motion of the controlled system is then

Table 2 Analogous concepts in analytical dynamics and control theory (detailed)

Analytical dynamics	Control theory
Unconstrained system $M(q,t)\ \ddot{q} = Q(q,\dot{q},t)$	Uncontrolled system or plant $M(q,t)\ \ddot{q} = Q(q,\dot{q},t)$
Constrained system $M(q,t)\ \ddot{q} = Q(q,\dot{q},t) + Q^C(q,\dot{q},t)$	Controlled system $M(q,t)\ \ddot{q} = Q(q,\dot{q},t) + Q^C(q,\dot{q},t)$
Constraints $\ddot{\phi} + \Sigma\dot{\phi} + K\phi = 0,$ $\dot{\psi} = -\Lambda\psi$	Trajectory requirements $\phi_i(q,t) = 0, i = 1, 2, \ldots, h$ $\psi_i(q,\dot{q},t) = 0, i = h+1, h+2, \ldots, m.$
Gauss's principle (GP) $J(t) = \left[Q^C(q,\dot{q},t)\right]^T M^{-1}(q,t)\ Q^C(q,\dot{q},t)$	Control cost $\int_0^T \left[Q^C(q,\dot{q},t)\right]^T M^{-1}(q,t)\ Q^C(q,\dot{q},t)\,dt$
Constraint force with GP $Q^C = -M^{1/2}(AM^{-1/2})^+(Aa-b)$	Control force or control
Optimal at each instant of time	Optimal over the interval of time [0,T]
Generalized Gauss's principle $J(t) = \left[Q^C(q,\dot{q},t)\right]^T N(q,t)\ Q^C(q,\dot{q},t),$ where $N > 0$	$\int_0^T \left[Q^C(q,\dot{q},t)\right]^T N(q,t)\ Q^C(q,\dot{q},t)\,dt,$ where $N > 0$
Equations of motion for nonideal constraints $M\ddot{q} = Q - N(q,t)^{-1/2} A_N^+ (Aa-b) + N^{-1/2}\left(I - A_N^+ A_N\right) M^{-1/2} C(q,\dot{q},t)$	Full set of continuous controllers that satisfy trajectory requirements for arbitrary continuous $C(q,\dot{q},t)$

$$M\ \ddot{q} = Q - N^{-1/2} A_N^+ (Aa-b) + N^{-1/2}\left(I - A_N^+ A_N\right) M^{-1/2} C := Q + Q^C. \tag{27}$$

The second and third members in the first equality on the right-hand side are N-orthogonal, and in the presence of the third member the control cost increases and is given by

$$J(t) = \left[Q^C\right]^T N Q^C = \left\| A_N^+ (Aa-b) \right\|^2 + \left\| \left(I - A_N^+ A_N\right) M^{-1/2} C \right\|^2. \tag{28}$$

We can now expand Table 1 to expose the various analogous concepts that we have developed (see Table 2).

4 Examples

We provide three examples in this section that utilize the connections that we have developed between constrained motion and control of nonlinear systems. The first deals with motion synchronization of three different highly nonlinear dynamical systems, the second deals with the optimal design of Lyapunov stable controllers for nonlinear systems, and the third deals with the energy control of nonlinear chain-type systems.

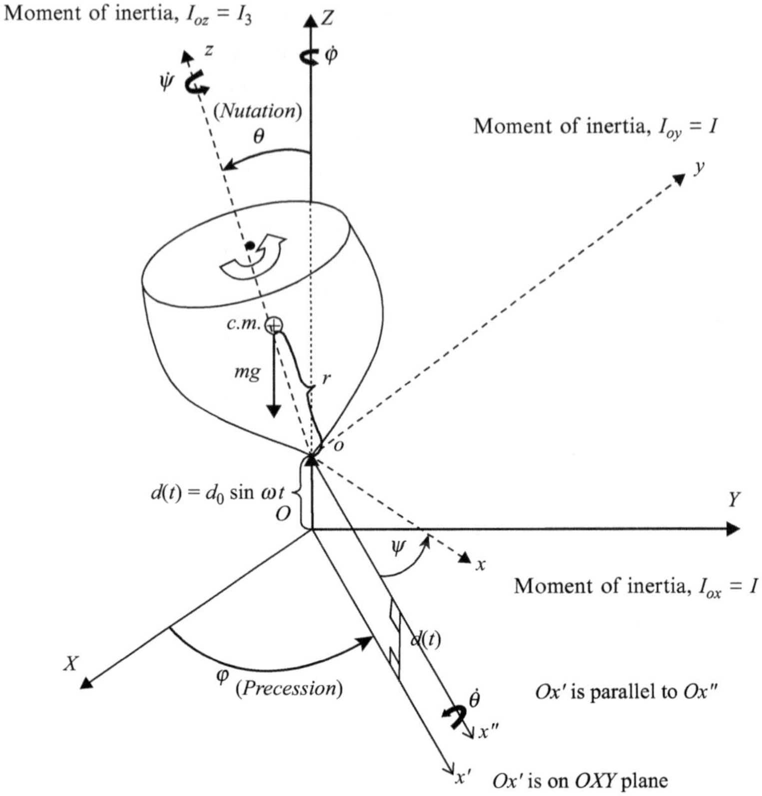

Fig. 2 Symmetric gyroscope with vertical support excitation $d(t) = d_0 \sin(\omega t)$

4.1 Determination of Control for Synchronization of Gyroscopes

Consider three gyroscopes mounted on a horizontal surface (that oscillates vertically) with each gyroscope having different dynamical characteristics. Our aim is to synchronize these three nonlinear systems so that two of the gyros—called "slave gyros"—track the motion of the third gyro called the "master gyro". While synchronization of nonlinear systems wherein each system is a copy of the other has been done before, here we look at the synchronization of different nonlinear systems [3, 17, 18]. Boccaletti et al. [3] gives an excellent review of synchronization of nonlinear systems.

A typical gyroscope (gyro) is shown in Fig. 2. We denote by m the mass of this typical gyro, $I_1 = I_2$ is the principal equatorial moment of inertia through the center of mass of the gyro, $I := I_1 + mr^2$, and I_3 is the polar moment of inertia about its symmetry axis. The point of support of the gyro is denoted by o, so that the moments of inertia about the axes ox and oy each equal I. The quantity r denotes the distance along the polar axis of the center of mass of the gyro from its point of support, and $d(t) = d_0 \sin \omega t$ is the time-varying amplitude of the vertical support motion that has frequency ω [22].

Using the Euler angles θ (nutation), ϕ (precession), and ψ (spin) (see Fig. 2), the Lagrangian for the system is given by

$$L = \frac{1}{2} I \left(\dot{\theta}^2 + \dot{\phi}^2 \sin^2 \theta \right) + \frac{1}{2} I_3 \left(\dot{\psi} + \dot{\phi} \cos \theta \right)^2 - mr\dot{d}\dot{\theta} \sin \theta - mgr \cos \theta. \quad (29)$$

The dots in Eq. (29) refer to differentiation with respect to time, t. Since ϕ and ψ are cyclic coordinates, the corresponding angular momenta $p_\psi = I_3 \left(\dot{\psi} + \dot{\phi} \cos \theta \right)$ and $p_\phi = I\dot{\phi}\sin^2\theta + p_\psi \cos \theta$ are conserved. The angular velocities $\dot{\phi}$ and $\dot{\psi}$ can be eliminated by using the Routhian [7]

$$R\left(\theta, \dot{\theta}, t\right) = L - p_\phi \dot{\phi}\left(p_\phi, p_\psi, \theta\right) - p_\psi \dot{\psi}\left(p_\phi, p_\psi, \theta\right). \quad (30)$$

The equation of motion, which is given by $\frac{d}{dt}\left(\frac{\partial R}{\partial \dot{\theta}}\right) - \frac{\partial R}{\partial \theta} = F_d$, then reduces to

$$I\ddot{\theta} + \frac{\left(p_\phi - p_\psi \cos \theta\right)\left(p_\psi - p_\phi \cos \theta\right)}{I \sin^3 \theta} - mgr \sin \theta - mr \sin \theta \, \ddot{d}(t) = F_d, \quad (31)$$

where F_d is the nonconservative force of damping, which we take here to be of linear-plus-cubic type, so that $F_d = -\widehat{c}\dot{\theta} - \widehat{e}\dot{\theta}^3$. Were we to further assume that $p_\phi = p_\psi = \overline{p}$ (which permits the gyro to be in the so-called "sleeping position") Eq. (31) can be further simplified to

$$\ddot{\theta} = -\gamma \sin \theta \sin \omega t - \alpha^2 \frac{(1 - \cos \theta)^2}{\sin^3 \theta} - c\dot{\theta} - e\dot{\theta}^3 + \beta \sin \theta := f\left(\theta, \dot{\theta}, t; P\right). \quad (32)$$

Under this assumption, Eq. (32) then is the differential equation that describes the motion of the symmetric gyro, where we have denoted $\alpha = \overline{p}/I$, $c = \widehat{c}/I$, $e = \widehat{e}/I$, $\beta = mgr/I$, and $\gamma = \omega^2 mrd_0/I$. The parameter set $P = \{\alpha, \beta, c, e, \gamma, \omega\}$ in Eq. (32) specifies the physical characteristics of the gyro and the harmonic vertical motion of the base on which it is supported. It may be pointed out that no assumption on the magnitude of the vertical displacement d_0 of the base has been made in arriving at this equation. We note in passing that no singularity arises in Eq. (32) due to the $\sin \theta$ term in the denominator. For various values of the parameters in the set P, a gyro, such as the one described above, can have motions that span the entire spectrum from regular periodic motion to period doubling and chaotic motion.

Consider three such gyros, each described by Eq. (32), that need to be synchronized so that two of them follow the motion of, say, the first (master) gyro. We shall take these three gyro systems to be *different* from each other, described by the parameter sets $P_i = \{\alpha_i, \beta_i, c_i, e_i, \gamma_i, \omega_i\}$, $i = 1, 2, 3$, and their dynamics will be investigated for the initial condition sets $IC_i = \left\{\theta_i^0, \dot{\theta}_i^0\right\}$, $i = 1, 2, 3$, given by

$$P_1 = \{10,\ 1,\ 0.5,\ 0.03,\ 35.8,\ 2.05\};\quad IC_1 = \left\{\ \theta_1^0 = -0.5, \dot{\theta}_1^0 = 1\right\}; \tag{33}$$

$$P_2 = \{10,\ 1,\ 0.5,\ 0.05,\ 35.5,\ 2\};\quad IC_2 = \left\{\theta_2^0 = 0.5, \dot{\theta}_2^0 = 1\right\};\ \text{and} \tag{34}$$

$$P_3 = \{10.5,\ 1,\ 0.5,\ 0.04,\ 38.5,\ 2.1\};\quad IC_3 = \left\{\theta_3^0 = 1, \dot{\theta}_3^0 = -0.5\right\}. \tag{35}$$

The Lyapunov exponents for each of the three uncontrolled gyros that have parameters described by the sets (33–35) are computed over a time span of 1,000 s by using the method described in Udwadia and von Bremen [26]. The integration for determining these exponents is performed using MATLAB's ode45 using a relative error tolerance of 10^{-9} and an absolute error tolerance of 10^{-13}. The Lyapunov exponent sets, l_i, of the three different gyro systems are computed to be $l_1 \approx \{0.211, -0.896, 0\}$, $l_2 \approx \{0.216, -1.001, 0\}$, and $l_3 \approx \{0.208, -0.936, 0\}$, respectively. The positive value of the largest Lyapunov exponent in each set indicates that the motions are chaotic for each of these gyros. Furthermore, the chaotic attractors for each system are different.

Figure 3 shows plots of $(\theta_i, \dot{\theta}_i)$, $i = 1, 2, 3$, for $50 \leq t \leq 100$ for the three uncontrolled gyros along with a figure (lower right corner) in which all three plots are superposed. The integration of the equations of motion throughout the following study is carried out using MATLAB's ode45 with a relative error tolerance of 10^{-9} and an absolute error tolerance of 10^{-12}. The differences in the responses between the three uncontrolled gyros,

$$h_{ij}(t) = \theta_i(t) - \theta_j(t), \qquad i \neq j, \tag{36}$$

are shown in Fig. 4. The three gyros can now be thought of as comprising *one* single, uncontrolled dynamical system. The uncontrolled equation of motion of this system is given by

$$a := \begin{bmatrix} \ddot{\theta}_1 \\ \ddot{\theta}_2 \\ \ddot{\theta}_3 \end{bmatrix} = \begin{bmatrix} f_1(\theta_1, \dot{\theta}_1, t;\, P_1) \\ f_2(\theta_2, \dot{\theta}_2, t;\, P_2) \\ f_3(\theta_3, \dot{\theta}_3, t;\, P_3) \end{bmatrix}, \tag{37}$$

where the function $f_i(\theta_i, \dot{\theta}_i, t; P_i)$ corresponds to the ith gyro and is obtained from Eq. (32) by using the parameter set P_i in it. We denote, as before, the uncontrolled acceleration of this dynamical system by a. Our task is to find the control force so that the last two gyros ($i = 2, 3$) track the motions of the master ($i = 1$) gyro. In order to do this, we apply a tracking control force Q^C so that the controlled dynamical system is described by the equations

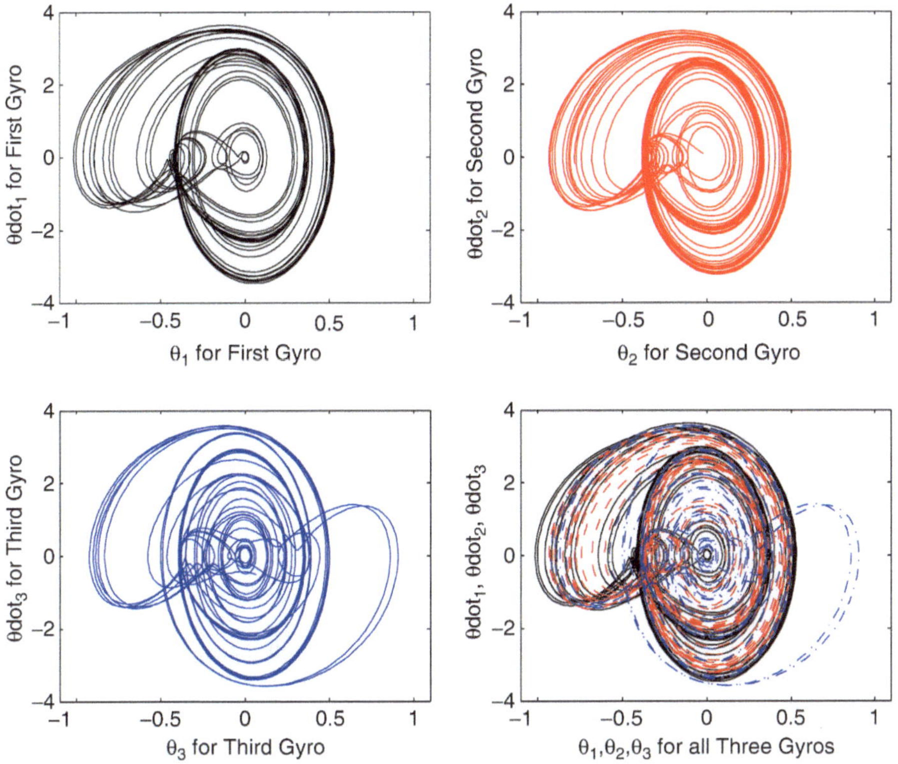

Fig. 3 $(\theta_i, \dot{\theta}_i)$ plots showing the dynamics of the three uncontrolled gyros for $50 \leq t \leq 100$. The *lower right* corner shows these plots superposed on one another; the first gyro is shown with a *solid line*, the second with a *dashed line*, and the third with a *dash-dot line*

$$\ddot{q} := \begin{bmatrix} \ddot{\theta}_1 \\ \ddot{\theta}_2 \\ \ddot{\theta}_3 \end{bmatrix} = \begin{bmatrix} f_1\left(\theta_1, \dot{\theta}_1, t; P_1\right) \\ f_2\left(\theta_2, \dot{\theta}_2, t; P_2\right) \\ f_3\left(\theta_3, \dot{\theta}_3, t; P_3\right) \end{bmatrix} + \begin{bmatrix} Q_1^C \\ Q_2^C \\ Q_3^C \end{bmatrix}. \tag{38}$$

The trajectory requirements (constraints) for synchronization are given by

1. $\phi_1(t) := \theta_1(t) - \theta_2(t) = 0.$
2. $\phi_2(t) := \theta_1(t) - \theta_3(t) = 0.$ (39)
3. The control force, Q_1^C, on the first gyro is zero for all time.

The last requirement above causes the motions of the master gyro to remain undisturbed, while the first two requirements call for the motions of the two "slave" gyros to track those of the master (first) gyro. Notice that the gyros do not have the

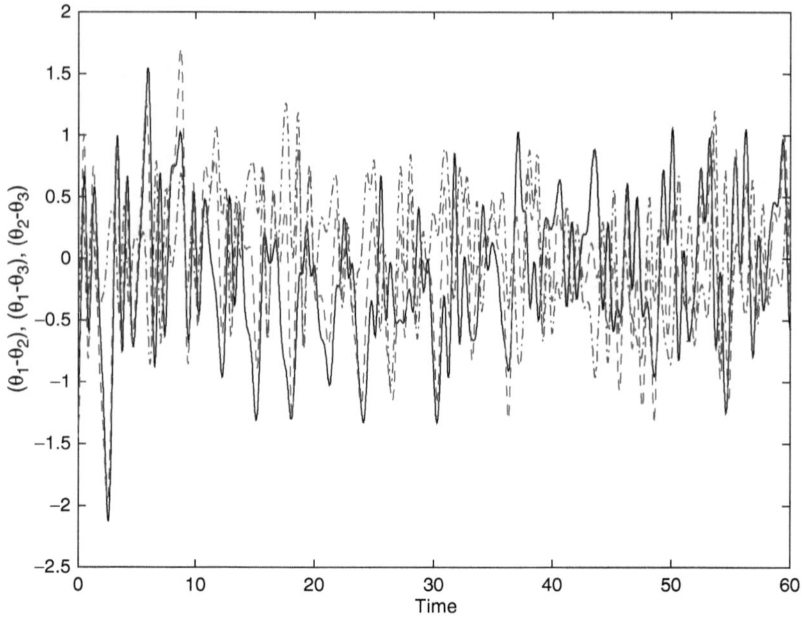

Fig. 4 The differences in the responses between the three uncontrolled gyros shown for a duration of 60 s. $h_{12}(t) = \theta_1(t) - \theta_2(t)$ is shown by the *solid line*, $h_{13}(t) = \theta_1(t) - \theta_3(t)$ is shown by the *dashed line*, and $h_{23}(t) = \theta_2(t) - \theta_3(t)$ is shown by the *dash-dot line*

same initial conditions. So we shall use the first of the two equations of (22) with $\varsigma = \varsigma_i = 1$, $\kappa = \kappa_i = 2$, $i = 1, 2$. The abovementioned constraint set (39) can then be expressed as

$$\ddot{\theta}_1 - \ddot{\theta}_2 = -\varsigma\left(\dot{\theta}_1 - \dot{\theta}_2\right) - \kappa\left(\theta_1 - \theta_2\right), \tag{40}$$

$$\ddot{\theta}_1 - \ddot{\theta}_3 = -\varsigma\left(\dot{\theta}_1 - \dot{\theta}_3\right) - \kappa\left(\theta_1 - \theta_3\right), \text{and} \tag{41}$$

$$\ddot{\theta}_1 = f_1\left(\theta_1, \dot{\theta}_1, t; P_1\right). \tag{42}$$

Expressing these trajectory requirements in the form of Eq. (12), we have

$$A = \begin{bmatrix} 1 & -1 & 0 \\ 1 & 0 & -1 \\ 1 & 0 & 0 \end{bmatrix}, \quad \text{and,} \tag{43}$$

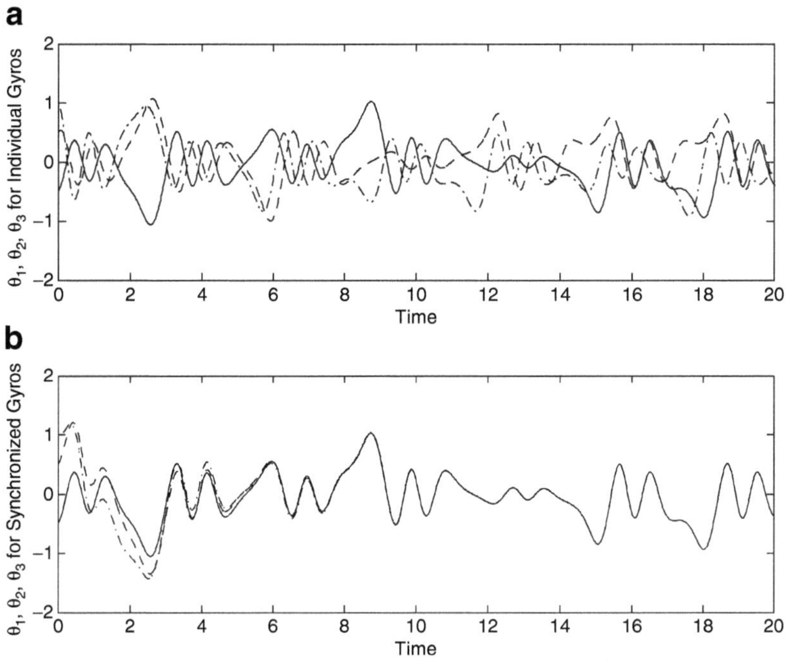

Fig. 5 (**a**) First 20 s of the response of the uncontrolled gyros with the master gyro shown with a *solid line*, the second gyro shown with a *dashed line*, and the third gyro shown with a *dash-dot line*; (**b**) Synchronization of the gyros by the application of control forces showing the slave gyros following the master (*solid line*) as required by the constraint Eqs. (40) and (41) with $\varsigma = 1$ and $\kappa = 2$. The motion of the master gyro remains unchanged, as required by (42)

$$b=\left[-\varsigma\left(\dot{\theta}_1-\dot{\theta}_2\right)-\kappa\left(\theta_1-\theta_2\right)\quad -\varsigma\left(\dot{\theta}_1-\dot{\theta}_3\right)-\kappa\left(\theta_1-\theta_3\right)\quad f_1\left(\theta_1,\dot{\theta}_1,t;P_1\right)\right]^T \tag{44}$$

The tracking control forces to be applied on this dynamical system are explicitly obtained, at each instant of time t, by simply using relation (14). This tracking control force minimizes the cost given in (13) at each instant of time. Figure 5a shows the time responses for the first 20 s of the three uncontrolled gyros, and Fig. 5b shows their controlled (synchronized) response, where the latter two gyros ($i = 2, 3$) are now slaved to the master ($i = 1$) gyro. We observe that the error between the responses gradually reduces to zero as required by Eqs. (40) and (41).

The plots in the $\left(\theta_i, \dot{\theta}_i\right)$-plane, $i = 1$, 2, 3, superposed on one another for all three gyros are shown in Fig. 6, indicating synchronization of the two slave gyros with the chaotic motion of the master gyro. The plots are made using the response of each of the gyros over a 50 s interval of time, starting at 50 s. We note that in this figure there are *three* separate plots that are superimposed on top of one another. As seen, the plots fall exactly on top of each other. The differences in the responses,

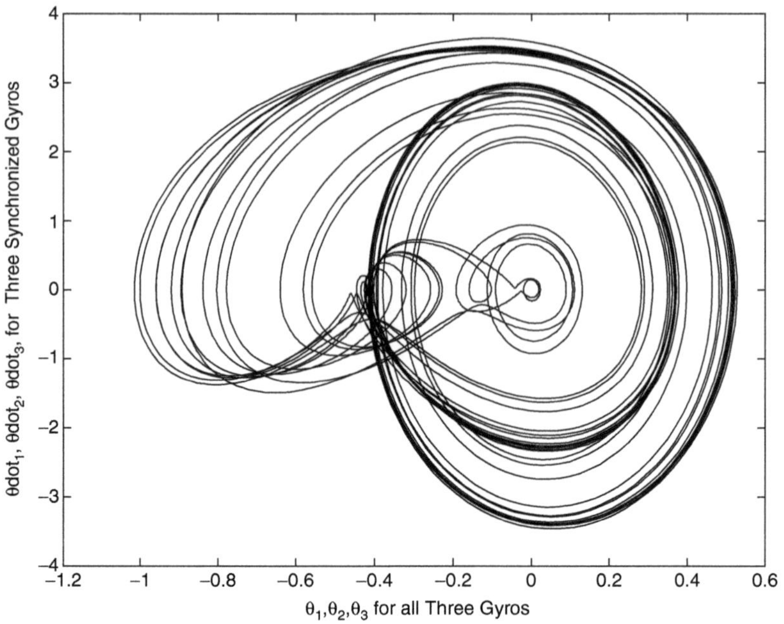

Fig. 6 Superimposed plots of $\left(\theta_i, \dot{\theta}_i\right)$, $i = 1, 2, 3$, of the three synchronized gyros for $50 \le t \le 100$. The master gyro is a chaotic system and its Lyapunov exponents [26] are $l_1 \approx \{0.211, -0.896, 0\}$. Each of the gyros executes the entire motion shown in the plot

$h_{ij}(t) = \theta_i(t) - \theta_j(t)$, $50 \le t \le 100$, between the motions of the three synchronized gyros are shown in Fig. 7. We notice that this error soon becomes of the same order of magnitude as the error tolerance (10^{-12}) used to numerically integrate the equations. The exponential convergence of $h_{ij}(t)$ towards zero as demanded by relations (40) and (41) is obvious.

4.2 Stable Controller Design

When trying to find a controller for a nonlinear nonautonomous system, one often postulates a controller based on experience, and/or intuition, and/or heuristics, and then analyzes its stability. Usually, Lyapunov's second method is used, and stability is checked by searching for a suitable Lyapunov function V and ensuring that its time derivative is nonpositive along the dynamical trajectory [9]. Though there are some standard methods that one can get guidance from in the search for a suitable Lyapunov function, when handling complex, nonlinear, and high-dimensional dynamical systems, this can become a difficult and time-consuming process, which may at times not be fruitful. When one is unable to find such a function, the stability of the postulated control is left uncertain.

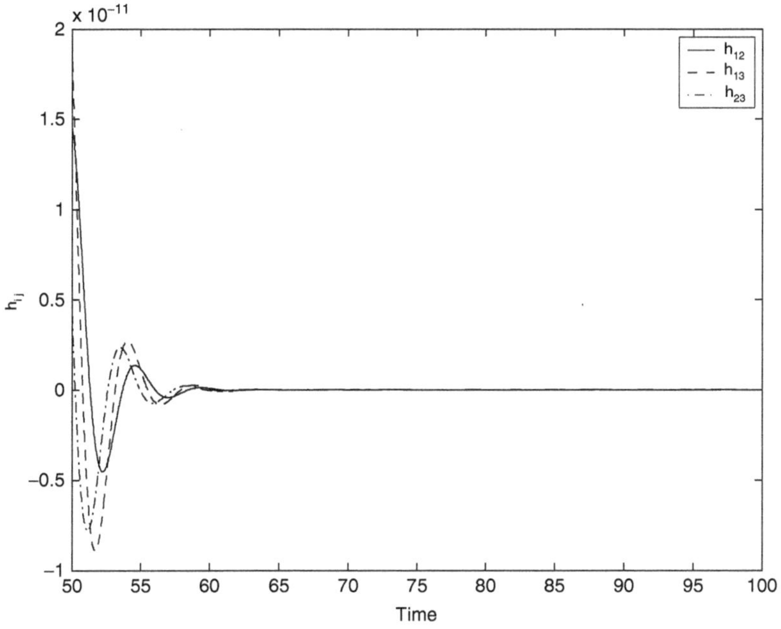

Fig. 7 $h_{12}(t) = \theta_1(t) - \theta_2(t)$ (*solid line*), $h_{13}(t) = \theta_1(t) - \theta_3(t)$ (*dashed line*), $h_{23}(t) = \theta_2(t) - \theta_3(t)$ (*dash-dot line*) for $50 \leq t \leq 100$. Note the vertical scale, which indicates that the error in synchronization is of the order of the numerical integration error tolerance, 10^{-12}

Using our knowledge of the connection between constrained motion and control, we could, so to speak, turn this problem on its head by first choosing a candidate Lyapunov function, V—namely, a positive definite function—and placing the constraint on the dynamical system that its derivative, $\dot{V}$, is nonpositive, thereby *making the candidate function a Lyapunov function* for the controlled system, thus ensuring stability. In fact, we can do this while *simultaneously* minimizing the control cost.

Consider the nonlinear, nonautonomous system (plant)

$$M\,\ddot{q} = -Kq + Lq^{(3)} + H := Q(q, \dot{q}, t)\,, \tag{45}$$

where

$$M(t) = \begin{bmatrix} m_1 \frac{(t+1)}{(t+2)} & 0 \\ 0 & m_2 \frac{(t+3)}{(t+2)} \end{bmatrix}, \qquad K = \begin{bmatrix} k_1 & -k_1 \\ -k_1 & k_1 + k_2 \end{bmatrix},$$
$$L = \begin{bmatrix} l_1 & 0 \\ 0 & l_2 \end{bmatrix}, \qquad H = q_1 q_2 \begin{bmatrix} c_1 \dot{q}_1 \\ c_2 \dot{q}_2 \end{bmatrix}, \tag{46}$$

and the 2-vector $q^{(3)} = [q_1^3 \; q_2^3]^T$. Our aim is to find a stable controller for this dynamical system that brings it to the fixed point $q_i = \dot{q}_i = 0$. We choose the Lyapunov candidate (positive definite) function

$$V(q, \dot{q}) = \frac{1}{2} a_1 \, q^T q + \frac{1}{2} a_2 \, \dot{q}^T \dot{q} + a_{12} \, \dot{q}^T q, \tag{47}$$

with the constants $a_1, a_2 > 0, a_1 a_2 > a_{12}^2$. In order to make this function a Lyapunov function for the controlled dynamical system, we place the constraint that

$$\dot{V} = \frac{dV(q, \dot{q})}{dt} := \frac{\partial V}{\partial \dot{q}} \ddot{q} + \frac{\partial V}{\partial q} \dot{q} = -\alpha V, \tag{48}$$

which can be rewritten in the form (12) (i.e., $A\ddot{q} = b$) with

$$A(q, \dot{q}) = a_2 \dot{q}^T + a_{12} q^T = [a_2 \dot{q}_1 + a_{12} q_1, a_2 \dot{q}_2 + a_{12} q_2] \text{, and} \tag{49}$$

$$b(q, \dot{q}) = -\alpha \left[\frac{1}{2} a_1 q^T q + \frac{1}{2} a_2 \dot{q}^T \dot{q} + a_{12} \dot{q}^T q \right] - a_1 q^T \dot{q} - a_{12} \dot{q}^T \dot{q}. \tag{50}$$

It is important that the constraint (12) be consistent, and when $A = 0$, we must have $b = 0$. We now know from (14) that the control force given by

$$Q^C = -M^{1/2} \left(A M^{-1/2} \right)^+ (Aa - b) \tag{51}$$

will cause the controlled system to not only exactly satisfy relation (48) but also simultaneously minimize the control cost $J(t) = [Q^C]^T M^{-1} Q^C$ at each instant of time. The equation of motion of the controlled system is then

$$M \, \ddot{q} = -Kq + Lq^{(3)} + H + Q^C \tag{52}$$

where Q^C is given in (51). Using the numerical values $m_1 = 1, m_2 = 2$, $k_1 = 100, k_2 = 100$, $l_1 = l_2 = 4$, and $c_1 = c_2 = 1$, in the relations shown in Eq. (46), and the initial conditions $q(0) = [1, -2]^T$ and $\dot{q}(0) = [-2, 3]^T$, the uncontrolled system given by Eq. (45) for these parameter values is *unstable*. We shall use the candidate Lyapunov function V given in Eq. (47) with $a_1 = 1, a_2 = 4$, and $a_{12} = 1$, and $\alpha = 1/2$. These parameter values ensure consistency of the constraint Eq. (12).

We obtain a simulation in the MATLAB environment of the controlled system given by Eq. (52) (using Q^C explicitly obtained from Eq. (51)). Numerical integration of the ode's is done using ode15s with a relative error tolerance of 10^{-8} and an absolute error tolerance of 10^{-12}. Figures 8 and 9 show the displacement and velocity response, respectively, of the controlled system as a function of time showing its asymptotic convergence to the fixed point $q = \dot{q} = 0$. This example illustrates the power that can be invoked through the recognition of the connection between constrained motion and the corresponding control problem.

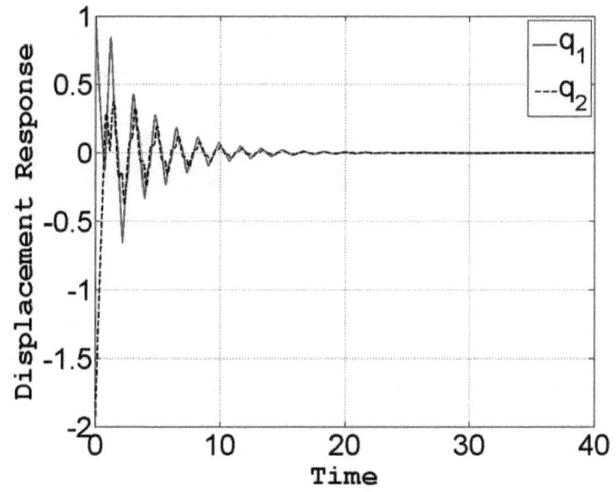

Fig. 8 Displacement response of the controlled nonlinear system

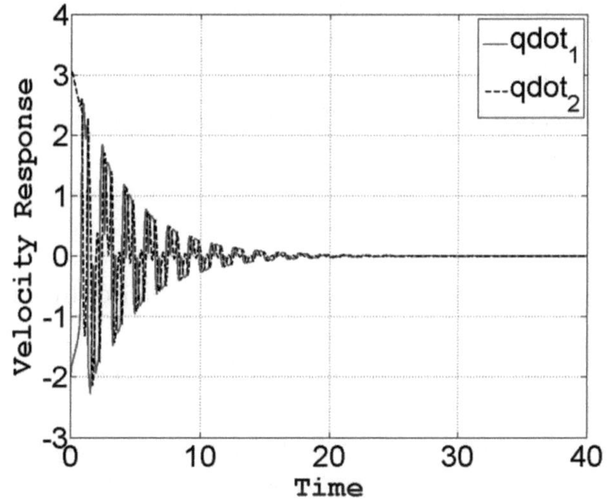

Fig. 9 Velocity response of the controlled nonlinear system

4.3 Energy Control of a 3-DOF Toda Oscillator

Energy control of nonlinear mechanical systems has become important nowadays and various energy harvesting schemes are being developed. We consider here the problem of energy control of a highly nonlinear mechanical system by using the so-called fundamental equation of mechanics to obtain the explicit nonlinear control force required to achieve the desired energy control.

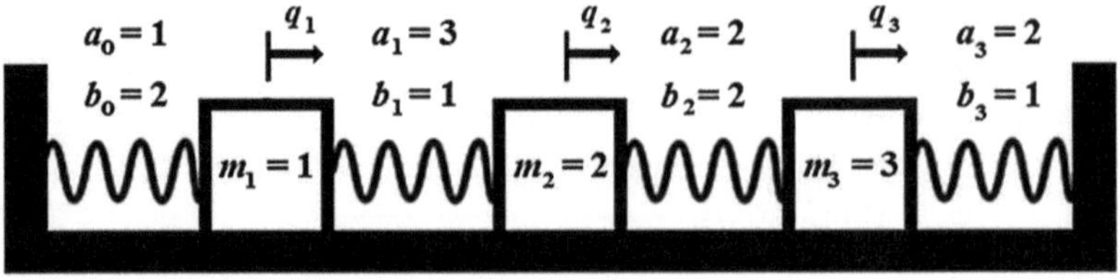

Fig. 10 A 3-DOF fixed-fixed Toda chain

We consider a 3-DOF fixed-fixed Toda chain [19] as shown in Fig. 10. The mass, displacement, and velocity of the ith mass ($i = 1, 2, 3$) in the chain are described by m_i, q_i, and $\dot{q}_i$, respectively. Given any nonzero initial energy state, H_0 of the chain, our aim is to stabilize the chain at a different nonzero desired energy level, H^*. And to achieve this, control can be applied to one or more of these three masses. In the present example, we control the energy of the chain by actuating the first mass, m_1, alone (see Fig. 10). We shall impose the requirement that the energy of the system be increased to the desired value H^* as a constraint on the mechanical system, and the constraint force that will cause this constraint to be satisfied will then be the requisite control force that would need to be applied to mass m_1. We begin with a description of the Toda potential.

(i) *Toda Potential and Spring Force*: The expression for the nonlinear potential of the Toda spring [19] is given by

$$u(q) = \frac{a}{b}\, e^{b\,q} - a\,q - \frac{a}{b}, \qquad a > 0, b > 0 \tag{53}$$

whereas its exponential spring force $F_s(q)$ can be derived from its potential as

$$F_s(q) \;= -F_{restoring}(q) = \frac{\partial u(q)}{\partial q} \;=\; a\,\left(e^{b\,q} -\; 1\right). \tag{54}$$

A plot of the Toda spring potential and the Toda spring force is shown in Figs. 11 and 12, respectively. For sufficiently small displacement, the spring force is approximately linear. However, the nonlinearity of the force gains prominence as the displacement increases. As can be inferred from Fig. 12, a larger force is required to stretch the spring by a unit distance than is required to compress it. Hence, the Toda chain considered possesses spring elements that are stronger in tension than in compression. Such systems arise frequently in structural subsystems such as the stringers in suspension bridges.

(ii) *Unconstrained System*: Consider the 3-DOF fixed-fixed Toda chain as shown in Fig. 10. The total energy of the chain can be written down as

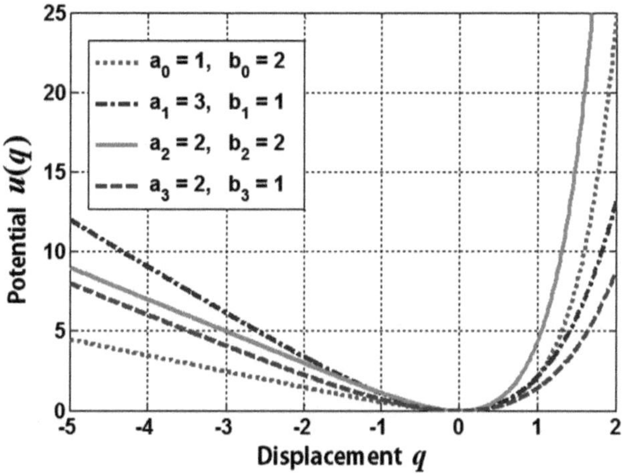

Fig. 11 Toda spring potential

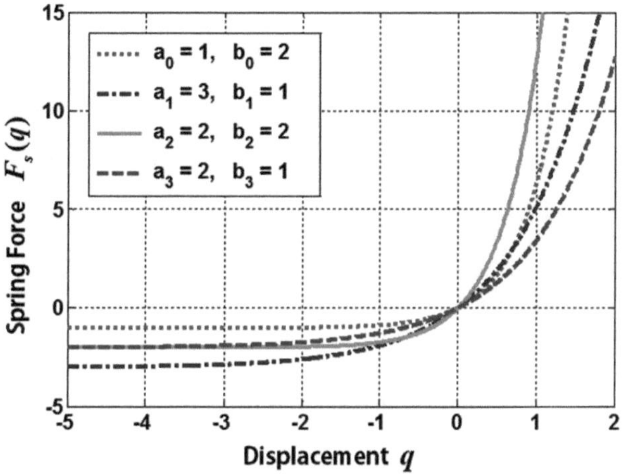

Fig. 12 Toda spring force

$$H(q,\dot{q}) = \sum_{i=1}^{3}\left[\frac{1}{2}\, m_i\, \dot{q}_i^2\right] + \sum_{i=0}^{3}\left[\frac{a_i}{b_i}\, e^{b_i(q_{i+1}-\, q_i)} - a_i\,(q_{i+1}-q_i) - \frac{a_i}{b_i}\right], \tag{55}$$

where $q_o \equiv q_4 \equiv 0$ describe the boundary conditions of the fixed-fixed chain. The equations of motion of the unconstrained (uncontrolled) system can be written down in matrix form as $M\,\ddot{q} = Q$ or more explicitly as

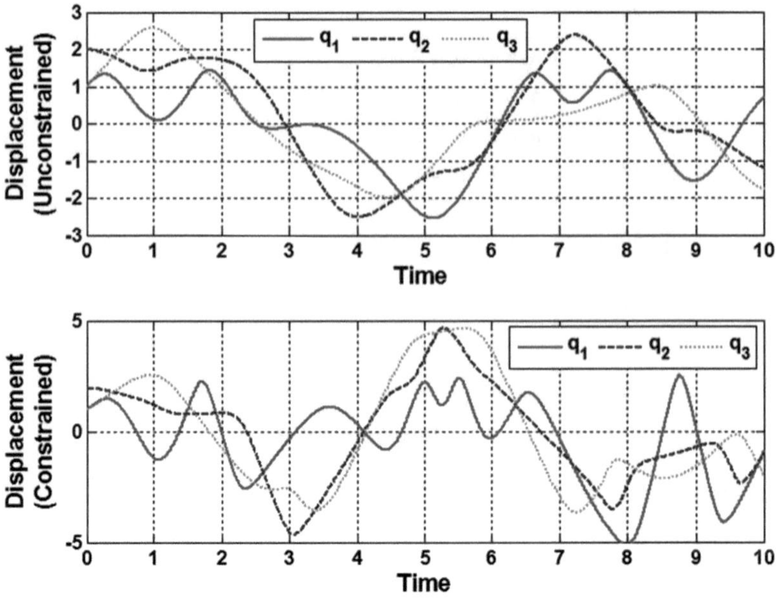

Fig. 13 Time history of displacements for the unconstrained system (*top*) and constrained system (*bottom*)

$$\begin{bmatrix} m_1 & 0 & 0 \\ 0 & m_2 & 0 \\ 0 & 0 & m_3 \end{bmatrix} \begin{bmatrix} \ddot{q}_1 \\ \ddot{q}_2 \\ \ddot{q}_3 \end{bmatrix} = \begin{bmatrix} a_1\left(e^{b_1(q_2-q_1)}-1\right) & - & a_0\left(e^{b_0(q_1)}-1\right) \\ a_2\left(e^{b_2(q_3-q_2)}-1\right) & - & a_1\left(e^{b_1(q_2-q_1)}-1\right) \\ a_3\left(e^{b_3(-q_3)}-1\right) & - & a_2\left(e^{b_2(q_3-q_2)}-1\right) \end{bmatrix}. \quad (56)$$

We take, for example, the initial conditions of this Toda chain to be

$$q_1(0) = 1, \dot{q}_1(0) = 2, q_2(0) = 2, \dot{q}_2(0) = 0, q_3(0) = 1, \dot{q}_3(0) = 2. \quad (57)$$

Figure 10 shows the parameter values of the masses ($m_i, i = 1, 2, 3$) used as well as the parameter values $a_i, b_i, i = 0, 1, 2, 3$ that characterize the four different Toda springs. Using these parameter values and the initial conditions given in Eq. (57), the unconstrained equations of motion given in Eq. (56) can now be numerically integrated. We note that for all the simulations presented in this subsection, the equations of motion have been integrated using the "ode45" scheme in the MATLAB environment with a relative integration error tolerance of 10^{-10} and an absolute error tolerance of 10^{-13}. Figure 13 (top) shows a plot of the displacements of the three masses from $t = 0$ to $t = 10$ time units for the unconstrained (uncontrolled) system.

The unconstrained Toda chain is a conservative system and the energy, being an integral of motion, remains constant throughout the duration of the

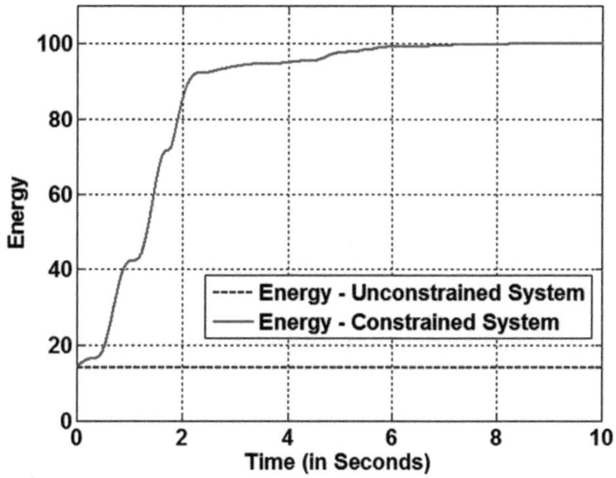

Fig. 14 Time history of energy of the 3-DOF Toda chain

simulation as can be inferred from Fig. 14. For the parameter values chosen, the energy level of the chain is H_0 =14.22 units. Our aim is to increase the energy of the chain to a new value.

(iii) *Constraints*: We shall assume that we want the nonlinear Toda chain described by Eq. (56) (with the parameter values as shown in Fig. 10) to have an energy level $H^* = 100$ units by controlling *only* mass m_1. In order to achieve this control objective, we impose the following two different types of constraints on the unconstrained system. The first deals with our objective to change the energy of the system to its desired value, H^*; the second deals with the fact that we want to achieve this by actuating just a single mass from amongst the three masses in the chain, namely, only mass m_1 (see Fig. 10).

1. *Energy Control Constraint*: The energy control constraint is given by

$$\frac{d}{dt}\left(H(q,\dot{q}) - H^*\right) + \beta\left(H(q,\dot{q}) - H^*\right) = 0, \tag{58}$$

where $\beta > 0$. The solution to this differential equation shows that as $t \to \infty$, $H(q,\dot{q}) \to H^*$. Notice that this constraint allows the 3-DOF Toda chain to be started from any arbitrary initial energy state H_0 (see Eq. 22) so that it reaches its desired energy state, H^*, as $t \to \infty$.

2. *No Control Constraints*: Since no control force is to be applied to masses m_2 and m_3 of the Toda chain, the second and third equations in the equation set (56) must remain unchanged in the controlled system. Therefore, the unconstrained equations of motion of masses m_2 and m_3 are themselves the constraints that guarantee that no control is applied to either of these two masses! Thus, in addition to the energy constraint given by Eq. (58),

the unconstrained system (Eq. 56) is also subjected to the following two constraints.

$$\begin{bmatrix} m_2 & 0 \\ 0 & m_3 \end{bmatrix} \begin{bmatrix} \ddot{q}_2 \\ \ddot{q}_3 \end{bmatrix} = \begin{bmatrix} a_2 \left(e^{b_2(q_3-q_2)} - 1\right) - a_1 \left(e^{b_1(q_2-q_1)} - 1\right) \\ a_3 \left(e^{b_3(-q_3)} - 1\right) - a_2 \left(e^{b_2(q_3-q_2)} - 1\right) \end{bmatrix} \tag{59}$$

When this set of constraints (Eqs. (58) and (59)) are expressed in the general form of Eq. (12), we obtain $A\ddot{q} = b$ or more explicitly

$$\begin{bmatrix} m_1\dot{q}_1 & m_2\dot{q}_2 & m_3\dot{q}_3 \\ 0 & m_2 & 0 \\ 0 & 0 & m_3 \end{bmatrix} \begin{bmatrix} \ddot{q}_1 \\ \ddot{q}_2 \\ \ddot{q}_3 \end{bmatrix} = \begin{bmatrix} \dot{q}^T Q - \beta\,(H - H^*) \\ a_2 \left(e^{b_2(q_3-q_2)} - 1\right) - a_1 \left(e^{b_1(q_2-q_1)} - 1\right) \\ a_3 \left(e^{b_3(-q_3)} - 1\right) - a_2 \left(e^{b_2(q_3-q_2)} - 1\right) \end{bmatrix}. \tag{60}$$

(iv) *Explicit Control Force*: With the matrices M, Q, A, and b at our disposal, the control force Q^C can be calculated using Eq. (14) and is given by

$$Q^C(q, \dot{q}) = \begin{bmatrix} -\xi_o\ (H - H^*)\ m_1\ \dot{q}_1 \\ 0 \\ 0 \end{bmatrix}, \tag{61}$$

where the value of $\beta = \xi_o\ m_1\ \dot{q}_1^2$ has been chosen to avoid any singularities in the control force, which might arise when the actuated mass m_1 has zero velocity. In the present example for illustration, the positive constant ξ_o has been chosen to be 0.03. The control force obtained (Eq. 61) is optimal and it minimizes the control cost given by Eq. (20) at each instant of time, with $N = M^{-1}$. Notice from Eq. (61) that the control force acting on the first mass appears to make it move like a self-excited oscillator!

(v) *Constrained System*: The equations of motion of the constrained (controlled) Toda chain can now be written down using Eq. (10) where M and Q are given by Eq. (56), and Q^C is given by Eq. (61). A plot of the displacements of the three masses of the controlled system (using the parameters shown in Fig. 10) is shown in Fig. 13 (bottom) from $t = 0$ to $t = 10$ time units. A plot of the time history of the energy is depicted in Fig. 14 for constrained system. As can be inferred from the figure, the application of the control force has resulted in an increase of the energy of the 3-DOF Toda chain from an initial energy level of $H_0 = 14.22$ units to the desired energy level of $H^* = 100$ units. Figure 15 shows a plot of the time history of the nonlinear control force acting on the first mass to achieve the desired transition. Once the desired energy level is attained, the control force automatically becomes zero, and we make use of the conservative nature of the chain to remain at the desired energy level for all future time.

It can be shown with some effort that the nonhomogeneous Toda chain that we have considered is controllable using control on just mass m_1 in the sense that the

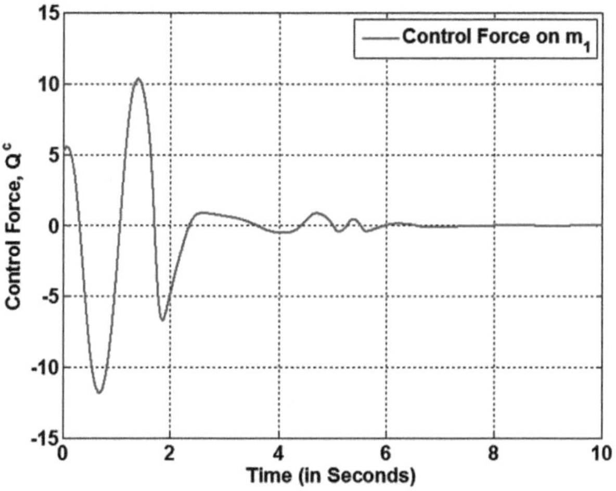

Fig. 15 Time history of control forces acting on the 3-DOF Toda chain

system can be "moved" from any arbitrary energy state $H_0 \neq 0$ to any other energy state $H^* \neq 0$ using the control described in Eq. (61). We don't prove that here, since it will take us too far afield from the central theme of this paper.

5 Conclusions and Open Problems

In this paper we have tried to establish a connection between the problem of constrained motion and the problem of trajectory tracking of nonlinear mechanical systems. This is its primary goal. The solution to the problem of constrained motion when d'Alembert's principle is valid, we see, provides the closed-form controller that will exactly cause the uncontrolled nonlinear mechanical system to track the trajectory requirements. No approximations and/or linearizations are made, and the control minimizes the control cost $J(t) = [Q^C]^T M^{-1} Q^C$ *at each instant of time*. This cost is the norm of the control force weighted by the inverse of the mass matrix of the mechanical system. For a general weighting matrix N, one can obtain the explicit control force given in Eq. (21). The simplicity of the controller allows for its application in real time for the control of complex mechanical systems.

When d'Alembert's principle is not valid, the constrained equations of motion provide, in closed form, the entire set of continuous controllers that would cause the trajectory requirements to be satisfied. Furthermore, when the weighting matrix in the control cost is a user-specified positive definite matrix N other than the matrix M^{-1} (which Nature uses) we obtain the explicit exact tracking control for the nonlinear system that minimizes this control cost at each instant of time. This is then further extended to get the full set of controllers that can provide exact tracking.

This duality between the constrained motion problem and the tracking control problem of nonlinear nonautonomous systems appears to be just the beginning of our explorations into the Janus-faced field of mechanics as stated in the prescient definition that Newton gave on the subject—the study of the motions of systems subjected to known forces *and* the study of the forces required to be applied to systems to generate known motions.

Examples provided in this paper deal with (i) the motion synchronization of dissimilar chaotic gyroscopes, (ii) the design of optimal Lyapunov stable controllers for nonlinear, nonautonomous systems, and (iii) the energy control of a nonhomogeneous Toda chain. These diverse examples exploit the connection between control theory and constrained motion theory; the closed-form control forces that are obtained are optimal and minimize the control cost at each instant of time; they provide exact trajectory tracking. The examples show the ease, simplicity, and efficacy with which control can be achieved. The full nonlinear "plant" is considered, without the need for any linearizations/approximations.

As mentioned earlier, the developments outlined herein form just the beginnings of a new path to our understanding of the synthesis of analytical dynamics and control. There are numerous open questions that remain unanswered, such as robustness of control, extensions to multi-body dynamics and the dynamics of continua, and applications to robotics, space systems, and fluid mechanical systems.

Appendix

We shall call the column 3-vector $q = [x, y, z]^T$. The plant we want to control corresponds to the unconstrained system (see Table 2) whose equation of motion is given in Eq. (2) as

$$M\, \ddot{q}(t) := \begin{bmatrix} m & 0 & 0 \\ 0 & m & 0 \\ 0 & 0 & m \end{bmatrix} \begin{bmatrix} \ddot{x} \\ \ddot{y} \\ \ddot{z} \end{bmatrix} = \begin{bmatrix} 0 \\ mg(x, y, z, t) \\ 0 \end{bmatrix} := Q \tag{62}$$

The trajectory requirement (constraint) is

$$\varphi(x, y, z, t) := x^2(t) + y^2(t) + z^2(t) - L^2 = q(t)^T q(t) - L^2 = 0, \tag{63}$$

and since we may not start on this manifold initially, we consider instead the constraint

$$\ddot{\varphi} + c\dot{\varphi} + k\varphi = 0, c > 0, k > 0 \tag{64}$$

whose solution as $t \to \infty$ is $\varphi = 0$. Thus we get asymptotic convergence to the trajectory requirement given in Eq. (63). Differentiating the constraint (63), Eq. (64) can be rewritten as

$$A\ddot{q} = [x \quad y \quad z] \begin{bmatrix} \ddot{x} \\ \ddot{y} \\ \ddot{z} \end{bmatrix} = -\dot{q}^T \dot{q} - c\dot{q}^T q - (k/2)\left(q^T q - L^2\right) := b \tag{65}$$

so that $A = [x\, y\, z]$ and $b = -\dot{q}^T \dot{q} - c\dot{q}^T q - k\left(q^T q - L^2\right)/2$. The control force Q^C that minimizes at each instant of time the control cost

$$J(t) = \left[Q^C\right]^T N Q^C \tag{66}$$

where N is a user-specified positive definite 3 by 3 matrix is given by Eq. (21) as

$$Q^C = -N^{-1} \begin{bmatrix} x/m \\ y/m \\ z/m \end{bmatrix} \left[A(MNM)^{-1}A^T\right]^+ \{gy + \dot{q}^T\dot{q} + c\dot{q}^T q + \tfrac{k}{2}(q^T q - L^2)\}$$

$$= -\frac{mN^{-1}}{\left(AN^{-1}A^T\right)} \begin{bmatrix} x \\ y \\ z \end{bmatrix} \left\{gy + \dot{q}^T\dot{q} + c\dot{q}^T q + \tfrac{k}{2}\left(q^T q - L^2\right)\right\}. \tag{67}$$

References

1. Appell, P.: Su rune Forme Generale des Equations de la Dynamique. C. R. Acad. Sci. Paris **129**, 459–460 (1899)
2. Awrejcewicz, J., Olejnik, P.: Analysis of dynamic systems with various friction laws. Appl. Mech. Rev. Trans. ASME **58**(6), 389–411 (2005)
3. Boccaletti, S., Kurths, J., Osipov, G., Valladares, D., Zhou, C.: The synchronization of chaotic oscillators. Phys. Rep. **366**, 1–101 (2002)
4. Dirac, P.A.M.: Lecture on Quantum Mechanics. Yeshiva University Press, New York (1964)
5. Gauss, C.F.: Uber ein neues allgemeines grundgesetz der mechanik. J. Reine Angew. Mathematik **4**, 232–235 (1829)
6. Gibbs, J.W.: On the fundamental formulae of dynamics. Am. J. Math. **2**, 49–64 (1879)
7. Goldstein, H.: Classical Mechanics. Addison-Wesley, New York (1976)
8. Jacobi, C.G.J.: Vorlesung uber Dynamik. G. Reimer, Berlin (1884)
9. Khalil, H.K.: Nonlinear Systems. Prentice Hall, New Jersey (2002)
10. Lagrange, JL.: Mechanique Analytique. Mme Ve Coureier, Paris, France (1811)
11. Moore, E.H.: On the reciprocal of the general algebraic matrix. Bull. Am. Math. Soc. **26**, 394–395 (1920)
12. Newton, I.: Mathematical Principles of Natural Philosophy, Vol. 1, Motte's translation of F. Cajori, University of California Press, Berkeley (1962)
13. Pars, L.A.: A Treatise on Analytical Dynamics. Oxbow Press, Woodbridge (1972)

14. Penrose, R.: A generalized inverse of matrices. Proc. Cambridge Phil. Soc. **51**, 406–413 (1955)
15. Peters, J., Mistry, M., Udwadia, F.E., Nakanishi, J., Schaal, S.: A Unifying framework for robot control with redundant DOFs. Auton. Rob. **24**, 1–12 (2008)
16. Poincare, H.: Su une Forme Nouvelle des Equations de la Mechanique. C. R. Acad. Sci. Paris **132**, 369–371 (1901)
17. Pyras, K.: Weak and strong synchronization of chaos. Phys. Rev. E **54**, 4508–4511 (1996)
18. Rulkov, N., Sushchik, M., Tsimring, L., Abarbanel, H.: Generalized synchronization of chaos in directionally coupled chaotic systems. Phys. Rev. **51**, 980–994 (1995)
19. Toda, M.: Theory of Nonlinear Lattices. Springer, New York (1989)
20. Udwadia, F.E.: A New perspective on the tracking control of nonlinear structural and mechanical systems. Proc. R. Soc. Lond. Ser. A **459**, 1783–1800 (2003)
21. Udwadia, F.E.: Optimal tracking control of nonlinear dynamical systems. Proc. R. Soc. Lond. Ser. A **464**, 2341–2363 (2008)
22. Udwadia, F.E., Han, B.: Synchronization of multiple chaotic gyroscopes using the fundamental equation of mechanics. J. Appl. Mech. 2011, **75**(2), 021011-1–021011-10 (2008)
23. Udwadia, F.E., Kalaba, R.E.: A new perspective on constrained motion. Proc. R. Soc. Lond. Ser. A **439**, 407–410 (1992)
24. Udwadia, F.E., Kalaba, R.E.: Analytical Dynamics: A New Approach. Cambridge University Press, Cambridge (1996)
25. Udwadia, F.E., Kalaba, R.E.: What is the general form of the explicit equations of motion for constrained mechanical systems? J. Appl. Mech. **69**, 335–339 (2002)
26. Udwadia, F.E., von Bremen, H.: An efficient and stable approach for computation of Lyapunov characteristic exponents of continuous dynamical systems. Appl. Math. Comput. **121**, 219–259 (2001)

The General Conception of the Intellectual Investigation of the Regular and Chaotic Behavior of the Dynamical System Hamiltonian Structure

Constantin Ruchkin

Abstract In this article the general conception of computer research of the intellectual method of the regular and chaotic behavior of the dynamical system of special Hamiltonian structure is presented. In accordance with the main idea of the proposed conception, graphic images of two-dimensional and three-dimensional Poincare's sections of the phase space of dynamical systems by methods artificial and computational intellect are investigated. The new approach generalizes existing classical methods of the recognition of graphic images for badly formalized graphic objects on Poincare's sections such as "a closed curve" and "a two-dimensional area." We propose to use the generalized Hough transform for identification of regular cases, and also methods of deformable active contours and deformable parametric models for the recognition of chaotic cases. On the basis of the given technique, the computer program of the intellectual research of the regular and chaotic behavior of the mechanical systems, showing serviceability and efficiency of the general conception, was developed.

1 Introduction

At the present time, the investigations of the nonlinear dynamical systems and the dynamical systems with special Hamiltonian structure are closely related to the classical methods of analytical dynamics such as KAM theory, Lyapunov's stability theory, and more [1, 9, 12, 16]. Currently, for the analysis of the nonlinear dynamical systems, new computer-based theories are increasingly used: ergodic theory, statistical forecasting methods, numerical methods of research of high accuracy and performance, computer cognitive research methods, and methods of computational and artificial intelligence [3, 5, 6, 10, 13, 14, 17, 19]. A new approach gives better results if they combine analytical, numerical, and algorithmic idea simultaneously. Recent investigations show that the new hybrid methods can be

C. Ruchkin (✉)
Donetsk National Technical University, FCSaT, str. Artema., 58., 83001 Donetsk, Ukraine
e-mail: c_ruchkin@mail.ru

J. Awrejcewicz (ed.), *Applied Non-Linear Dynamical Systems*, Springer Proceedings in Mathematics & Statistics 93, DOI 10.1007/978-3-319-08266-0_17

used to produce not only approximate solutions but also analytically exact solutions, which remain as the prerogative of analytical dynamics.

In this article, in addition to previous work [17], the general conception of the intellectual investigation of the nonlinear dynamical systems with the Hamiltonian structure will be presented. This concept is based on the methods and algorithms for the investigation of the regular and the chaotic behavior of nonlinear dynamical systems and obtains analytically exact solutions. Principles of the KAM theory about reorganizations (creation and destruction) of trajectory toruses in phase space at the perturbation of the regular behavior of Hamiltonian systems are the mathematical basis of the conception of the computer research by methods artificial and computational intellect.

2 Related Research and the Analysis of the Current State of the Problem

Fundamentals of the analytical study of the integrable dynamical systems with special Hamiltonian structure are based on KAM theory. However, in the practical computing, algorithmic problems related to the complexity of the behavior classification of the nonlinear dynamical systems of the highest order are arise. Of [4, 6, 10, 11, 13–15, 17] are devoted to the solution of these problems and the development of algorithms and intelligent methods for the study of dynamical systems. Consider this work in more detail.

For the first time, mathematical models of pattern recognition, models of an object, and pattern recognition training in the article of Nejmark, Yu. I. are discussed. In their works [13, 14], authors propose to use the pattern recognition methods to study mathematical models of the dynamical system. The possibilities of using the pattern recognition methods to study mathematical models with a large number of parameters are discussed. The principal point is to study models by constructing phase and parametric portraits. This allows one to solve the problems of predicting the states of the object described by the mathematical model in hand and controlling the object and analyzing and studying problems that follow from the particular content of the model. Examples of three mathematical models are given to illustrate this problem.

In [10], it developed and implemented a methodology and data processing algorithms for interactive study of the behavior of the nonlinear dynamical systems based on mathematical modeling, taking into account the possible degeneracy of the singular points, and technological constraints. The behavior of the nonlinear dynamical system near degenerate critical points is well studied and classified. Along with this, the classical analysis is ambiguous in those cases when a singular point is degenerate, such as Kolmogorov. To analyze the behavior of the system in such a case, the algorithm is based on the idea of algorithmic search dividing lines in the state space.

The main important feature of the dynamic behavior of the system is the chain-recurrent set—the set of pseudo-periodic trajectories. It brings a lot of periodic orbits, and its location is the initial step of many methods of investigation of dynamical systems. The localization of these sets (except for the simplest-fixed points) is difficult when using classical methods of numerical analysis, as they provide some information on the asymptotic behavior of the system. To solve this problem, it uses the idea of partitioning the phase space into a finite set of cells and modeling the behavior of the system in accordance with the transformation of these cells by the action of the system. Refinement of the phase portrait occurs in sequential subdivision of cells covering. The practical implementation of such methods faced with a rapid increase in the number of cells, which increases the complexity of the design and implementation of workable methods. To study and simulate the dynamical systems in [15], a method of symbolic image is investigated. This method allows you to receive important characteristics of the original dynamical system using the algorithms on a directed graph whose vertices are the elements of the phase space, and the edges are defined by the behavior of the system on the elements.

The localization of the chain-recurrent sets is achieved by allocating a symbolic image of the graph of strongly connected component. This method depends strongly on the method of the construction and of the recognition of the symbolic images. The literature describes the most simple method of constructing the image of the cell (we will call it point method), which examines the behavior of the system through the system's behavior on a fixed set of points [4, 15].

In the article [11], the methods of state recognitions of the dynamical systems are considered. It is supposed that the characteristics of the dynamical systems, in a random way, change in time and are described by some nonstationary time series. Spectral expansion of nonstationary random process in Fourier series is applied. In the procedure of the recognition of states of system as initial attributes, the set of the random variables received from the description of features of spectrograms is used.

3 Basic Statement of the General Conception of the Intellectual Investigation

The computer solution of the prediction problem of the regular and chaotic behavior of the nonlinear dynamical systems can be obtained by several stages. At the first stage, the process of the discretization of the system's solutions by means of the numerical integration is used. At the second stage, the process of graphics and geometric modeling of the results is applicable. If the dimension of the system is less than three, the result of this integration is conveniently represented in graphical form on an image as a series of points forming a curve in a space and characterizing the state of the system at any time At the third stage, the system's behavior by the kind of a phase trajectory is analyzed. In the case of the systems of the third and higher orders, the result of numerical simulation of the system can be represented in graphical form in three-dimensional space only by special images (projections and sections of Poincare). However, the computer simulation results in the fact that some information on the nature of the system is lost and the results are incorrect. For more information about the qualitative behavior of the trajectory space of the system we construct sections of space curves. The general algorithm can be represented as:

1. Submission of the dynamical system in the Hamiltonian form
2. The initial conditions and parameters of integration
3. Finding numerical solutions for a limited time period
4. Construction of the phase portrait of the dynamical system
5. Calculation and visualization of the Poincare section of the phase space
6. Intelligent analysis of graphic images and finding critical points, periodic, quasiperiodic solutions, and other regular or chaotic solutions
7. The perturbation solutions of the system and reanalysis

4 General Proposition of KAM Approach and the Integrable Hamiltonian System

The system with the Hamiltonian is called integrable if there is a canonical transformation to a new variable angle effect. The existence of n integrals of motion *J1, J2, ..., Jn* means that a 2n-dimensional phase space of an integrable Hamiltonian path belongs to the n-dimensional set, which by analogy with the circle in the case of a harmonic oscillator has the topology of the n-dimensional torus. The trajectory is given by the winding. If the ratio of the frequencies along the meridians parallels rationally $\omega_1 : \omega_2 = m : s$, where m and s are relatively prime, then the trajectory is closed. If the frequency ratio is irrational, then the trajectory of a dense way fills the surface n-dimensional torus. In this case the motion is ergodic.

According to the KAM theory, torus on which the winding "not irrational" ($|\omega_1 : \omega_2 - m : s| > K(\varepsilon) : S^{2,5}$) become unstable and will be destroyed. In the last cases torus will be destroyed with $\omega_1 : \omega_2 = \left(\sqrt{5} - 1\right)/2$. Also according to the KAM theory, smaller tori are born along with the destruction of some tori, which in the Poincare section correspond to elliptic fixed points interspersed with hyperbolic fixed points. The process of destruction of some tori and the birth of other smaller breeds continues the self-similar distribution of elliptic and hyperbolic fixed points in the Poincare section (Fig. 1).

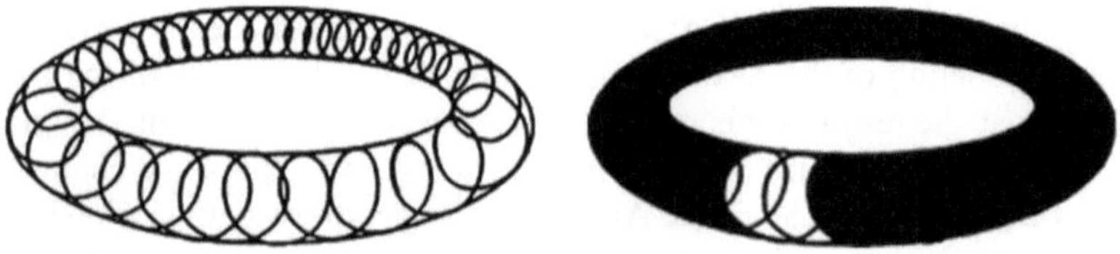

Fig. 1 Closed and unclosed trajectories on the torus

5 The Construction of Poincare's Sections

The numerical researches of phase space, which consist of a set of phase trajectories, will be conducted by means of Poincare's sections. Poincare's sections have dimensionality on unit is less than dimensionality of researched dynamical system. The exceptional interest of the dynamical systems of the third and fourth order is represented. The result of these researches can be displayed graphically on the computer monitor. Poincare's sections (local or global sections) for such systems will represent certain graphic images on a plane or in space accordingly. If points of a phase flow form a curve, it is possible to speak about the regular behavior (periodic or multiperiodic) of Hamiltonian systems. The cloud of points appearing in the section of Poincare of a phase flow will testify to the approach of a "chaotic" behavior of the system.

The particular interest is represented by dynamical systems for which there is a possibility of reconstruction and investigation of global Poincare's section, and for them there is a possibility of the creation of all possible motion. Though the creation of the section of Poincare of phase space happens approximately by means of numerical integration methods on the fixed interval of time, it appears enough what to understand in the overall picture of the behavior of Hamiltonian system.

The received phase portraits of two-dimensional and three-dimensional sections of Poincare of Hamiltonian systems can be researched by means of statistical or determined pattern recognition techniques. The complexity of the application of classical methods of recognition consist that main graphic images of two-dimensional and three-dimensional Poincare's sections are consisted concept "a closed curve" and "a two-dimensional area".

6 Recognition's Problem of Integrable Cases

In the work [18] the global spherical Poincare sections are investigated with help the computer program "Modeler". This section builds a three-dimensional sphere and shows a set of points (point cloud). In regular cases, these sets of points form a three-dimensional "closed curve" with self-crossing or without self-crossing. A particular case of this curve is a circle, which lies on the surface of a sphere. The center of this circle can coincides with the center of the sphere. If the center of the circle pass through the center of the sphere, then the radius of the sphere and the circle are the same, otherwise the radius of the circle is smaller than the radius of the sphere. Therefore, the problem of the detection and recognition of three-dimensional convex closed analytic curves constructed on the Poisson sphere, is an important and actual problem which can be solved by means of the generalized spherical Hough transform. In the work [18] it is shown how to solve the problem of the detection and recognition of the circles formed by the points lying on the surface of arbitrary sphere. To solve this problem, it is proposed to use a method, which generalizes the classical Hough transform to three dimensions on a spherical and generalized Hough transform. In the proposed algorithm we used an accumulator array increased dimension, which coincides with the number of unknown parameters in the equation of the family of the required curves. Computational complexity of the voting process carried out in the Hough extended accumulator space depends on the sampling of the parameter space. The main limitation on the use of the method—noisy source image—must be overcome in the preliminary stage.

7 Recognition's Problem of Chaotic Cases

So, the chaotic behavior of the system can be detect in the case when a phase portrait of a dynamical system is include closed areas. In this paper, to solve this problem, we use the methods of deformable active contours and deformable parametric models [7].

The method of active contour means to building variable circuit which is composed of n points in two-dimensional spaces. Each point of this contour is the approximated border of the object. For each point is the energy E_i: $E_i = aE_{int}(v_i) + bE_{ext}(v_i)$, where $E_{int}(v_i)$ is the energy component, which depends on the shape of the contour; $E_{ext}(v_i)$ is the energy component, which depends on image properties, such as gradient; a, b are the weighting coefficients to ensure each energy contribution to the overall equation criterion; and E_i, E_{int}, E_{ext} is the square matrix. The value at the center of each matrix corresponds to the energy in the energy point v_i (i—the vertex contour). The remaining values in the matrices correspond to the energy in each point in the environment v_i. Each vertex v_i has the potential to move to any point v_i, corresponding to the minimum value of the energy E_i, where p_{jk} (v_i) is the point (x, y), which corresponds to points on the image in the energy matrix. If the energy function is chosen optimally, the top loop V consistently moves in and stays close to the borders of the object. In the cases of symmetric dynamical systems for solving the problem of random selection of areas on the bitmap is used the method of parametric deformable model. Deformable models are called parametric when the shape of the detected object is determined only by the parameters of the system. Conformity assessment model configuration shown in the image data is also produced by the energy of the model. This model depends on the material parameters; the sum of the internal energy, which expresses the value of the configuration of the model limits set by the developer; and the external energy, which measures the goodness of fit model and the data in the image. The adaptation process of the model is similar of process of searching parameter vector and of reaching the global maximum (minimum) of the energy model. The specific model is characterized by the method of defining the form of the simulated object and function that calculates the energy model. The kind of the models is defined by a set of algebraic curves of the second, third, or higher order with certain restrictions imposed on their possible configurations. Internal energy models set additional limits on the desired configuration, bringing the weight penalty function to the total energy model for unwanted deformations. The calculation of the external energy is the basis of image features that are found in the allocated object. The external energy model increases when the model tends to take the position of the image where the pixels form a structure similar to a recognizable object. Energy optimization models were made using a gradient search method local minimum.

The proposed by the author the model and the algorithm for detection chaotic regions or algebraic curve uses the methods of deformable active contours and deformable parametric models.

8 Intellectual Research Problem of the Motion of a Rigid Body with a Fixed Point

The problem of the motion of a rigid body with a fixed point is considered in the classical setting. This problem refers to the direct problems of mechanics. For a given static, kinematic, and dynamic (structural) parameters of a rigid body with a fixed point and the initial traffic conditions necessary to determine its trajectory in space at any given time, to determine the type of trajectories, and to establish the nature of the movement of the body [6, 17].

A mathematical model of the motion of a free rigid body in a mobile basis is a system of six ordinary differential Euler equations:

$$J\dot{\omega} = J\omega \times \omega + r \times \nu, \quad \dot{\nu} = \nu \times \omega, \tag{1}$$

where $J = Diag\ (A, B, C)$ is the inertia tensor, $\omega = (p, q, r)$ is the angular velocity of the body in the projection on the movable axle, $\nu = (\nu_1, \nu_2, \nu_3)$ is the vertical unit vector, and $r = (r_1, r_2, r_3)$ is the vector from a fixed point to the center of the mass of the body.

Equation (1) to determine in the phase space $R^{12}\ (\omega, \nu, J, r) = R^6\ (\omega, \nu) \times R^6$ (J, r) family of possible (regular and chaotic) trajectories of the dynamical system. The initial conditions are $\omega_0 = \omega(0)$, $\nu_0 = \nu(0)$.

The regular case is characterized by "a good behavior" of a dynamical system, sustainable trajectory with respect to certain perturbations of the initial conditions, and the ability to determine the position of the system at a sufficient period of time. Chaotic motion has shown substantial dependence of solutions on the initial data, and the trajectories are exponentially unstable. In the chaotic case, even a small perturbation to predict the behavior of the system is possible only in a finite interval of time, called the time horizon.

Thus, the aim of the work [17] is to develop, design, and create an interactive computer system which can be used to find the solution of system (1), determine its character, and visualize the motion of a rigid body with a fixed point. If system (1) can be integrated and obtains the explicit dependence of the phase variables of time, it could be easily distinguished from the regular case of a chaotic event.

It is known [6] that the solution of (1) can be reduced to quadrature if four integrals are found. However, for arbitrary values of the design parameters, there are only three of the first algebraic integrals: $H\ (\omega,\ \nu)$, the energy; $G\ (\omega,\ \nu)$, the kinetic; and $I\ (\nu)$, the geometric:

$$H\ (\omega, \nu) = \frac{1}{2} J\omega \cdot \omega - r \times \nu = h, G\ (\omega, \nu) = J\omega \cdot \nu = g, I\ (\nu, \nu) = \nu \cdot \nu = 1. \tag{2}$$

The fourth integral exists only in the three most common cases, Euler–Poinsot, the Lagrange–Poisson, and Kowalewski, and some special cases. All of these cases are characterized by a specific set of design parameters, and the existence of four integrals, which allow reducing the system of differential equations (1) to a system of algebraic equations. The features of the integrable cases are the deterministic phase space trajectories and the increasingly stable trajectory of the body.

In the other integrable cases, the solution of a system of differential equations (1) can be received by the Runge-Kutta method of fifth order. Determine the nature of the family of movements and the opportunity to build the structure by analyzing the trajectory of the phase space of the dynamical system under study. For a fixed point of $R^6(J,\ r)$, the first integrals (2) are isolated in the phase space $R^6(\omega,\ \nu)$, a compact three-dimensional manifold $Q^3{}_{h,g}$ invariant with respect to the flow. The arrangement and relationship $Q^3{}_{h,g}$ in $R^3(\omega)$ and $R^3(\nu)$ give a fairly general idea of changing the physical quantities of the (1).

The basis for further qualitative analysis of the orbital structure of the phase space is the method of phase Poincaré sections and the theory of invariant curves. The onset of the chaotic regime occurs when the Poincaré section of the phase flow to form a cloud of points - as a two-dimensional region. If the phase flow lies on the curve, then we can talk about the regular behavior of the system. To investigation the topology of $Q^3{}_{h,g}$ by constructing an integrable case for local and global sections $P^2 \subset Q^3{}_{h,g}$ for the non-integrable case on the Poisson sphere $S^2(\nu) = 1$. Global section of P^2 is in $Q^3{}_{h,g}$ compact two-dimensional manifold. The trajectory, is filled of $Q^3{}_{h,g}$, or cross the face of P^2. P^2 are defined curves along which the phase flow is not transversal to the selected section, and the track points back at the material section for the final time interval.

On the basis of the given concept, the computer program of intellectual research of the regular and chaotic behavior of several mechanical systems, showing serviceability and efficiency of the given approach, was developed. An example of solving the problem of recognition of regular and chaotic cases on 3d Poincare's section is shown in Fig. 2. This Poincaré section is calculated by Runge-Kutta method for (1) and is reconstructed on the Poisson sphere using equations (2).

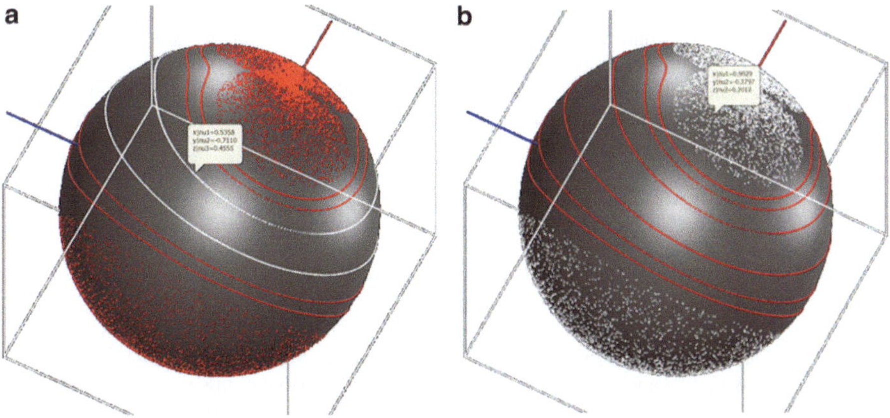

Fig. 2 Recognition of the regular (**a**) and chaotic (**b**) cases on 3d Poincare's section

9 Conclusions and Future Tasks

In this article the general concept of the intellectual investigation of the nonlinear dynamical systems with a special Hamiltonian structure will be presented. The concept is based on the methods and algorithms for the study of the regular and chaotic behavior of the nonlinear dynamical system. The new approaches generalizing existing classical methods of recognition for badly formalized concept "a closed curve" and "a two-dimensional area" for recognition are considered. We propose to use the generalized Hough transform to identification of regular cases, and also methods of deformable active contours and deformable parametric models for the recognition of chaotic cases. With its help we can solve the problem of detecting the regular or chaotic behavior of dynamical systems of special Hamiltonian structure. The result of the computer program of the intellectual research of the regular and chaotic behavior of a mechanical system is shown.

Future investigations will address the following problems: the choice of method of integration, the choice of the kind and form of the Poincare section, the specific analysis of graphical images and pattern recognition methods in space, and full automation of the research of the dynamical system.

References

1. Arnold, V.I.: Small denominators and problems of stability of motion in classical and celestial mechanics. Russ. Math. Surv. **18**, 85 (1963)
2. Ballard, D.H.: Generalizing the Hough Transform to Detect Arbitrary Shapes, p. 122. Computer Science Department, University of Rochester, New York (1980)
3. Bolsinov, A.V., Fomenko, A.T.: Integrable Hamiltonian system. In: Geometry, Topology, Classification, vol. 2, p. 448. Udmun-ton, Izhevsk (1999)

4. Fundinger, D.: Investigating dynamics by multilevel phase space discretization. Ph.D. thesis, Institut fur Parallele und Verteilte Systeme, Abteilung Bildverstehen, 2006. http://elib.uni-stuttgart.de/opus/volltexte/2006/2614
5. Galgani, L., Giorgilli, A., Strelcyn, J.-M.: Chaotic motions and transition to stochasticity in the classical problem of the heavy rigid body with a fixed point. Nuovo Cimento **61B**(1), 1–20 (1981)
6. Gashenenko, I.N., Richter, P.H.: Enveloping surfaces and admissible velocities of heavy rigid bodies. Int. J. Bifurcat. Chaos **14**(8), 2525–2553 (2004)
7. Kass, M., Witkin, A., Terzopoulos, D.: Snakes: active contour models. Int. J. Comput. Vis. **1**(4), 321–331 (1988)
8. Khoshelham, K.: Extending generalized Hough transform to detect 3D objects in laser range data. IAPRS **XXXVI**(Part 3), 206–210 (2007)
9. Kolmogorov, A.N.: On the conservation of conditionally periodic motions under small perturbations of the Hamiltonian. Dokl. Akad. Nauk SSSR, **98**, 527–530 (1954)
10. Kononova, A.I., Troyanovskiy, V.M.: Isolation zones and special points in the state space of a nonlinear dynamical system. Rev. Appl. Ind. Math. **16**(4), 669–672 (2009)
11. Kuk, J.V.: Intellectual of recognition systems of states a dynamical objects with nonstationary characteristics. In: Kuk, J.V., Lavrikova, E.I. (eds.) Artificial Intelligence, vol. 4, pp. 763–773. NAS Ukraine, Donetsk (2006)
12. Moser, J.: KAM theory and problems of stability: Per./Per. In: Treshchev, D.V. (ed.) Regular and Chaotic Dynamics, p. 448. Institute of Computer Researches, Moscow-Izhevsk (2001)
13. Neimark Yu, I., Teklina, L.G.: On possibilities of using pattern recognition methods to study mathematical models. Pattern Recogn. Image Anal. **22**(1), 144–149 (2012)
14. Nejmark, Yu. I.: New the approach to numerical research of specific dynamic systems by pattern recognition and statistical modeling methods. In: Nejmark, Yu. I., Kotelnikov, I.V., Teklina, L.G. (eds.) Application-Oriented Nonlinear Dynamics, vol. 18(2), pp. 3–15s. News High Schools (2010)
15. Osipenko, G.: Dynamical Systems, Graphs, and Algorithms. Lecture Notes in Mathematics, vol. 1889, p. 288. Springer, Berlin (2007)
16. Poincare, A.: New Methods of Celestial Mechanics: Fav Works in 3 Volumes, p. 999. Nauka, Moscow (1972)
17. Ruchkin, K.A.: Development of computer system for analysis of Poincare sections. In: Artificial intelligence, vol. 1, pp. 83–87. NAS Ukraine, Donetsk (2009)
18. Ruchkin, K.A.: Application of Hough transform to the recognitions problem of regular solutions of dynamical systems. In Artificial intelligence, **1**, 112–119. NAS Ukraine, Donetsk (2013)
19. Schwartzman, S.: Global cross sections of compact dynamical systems. Proc. Natl. Acad. Sci **48**, 786–791 (1962)

Dynamic Properties of Two-Axle Freight Wagon with UIC Double-Link Suspension as a Non-smooth System with Dry Friction

Jan Matej, Jarosław Seńko, and Jan Awrejcewicz

Abstract The influence of chosen parameters on the lateral dynamic behavior of the two-axle freight wagon with UIC double-link suspension is presented. This type of suspension uses the advantages of the studied non-smooth mechanical system, since dry friction is used to damp the system vibrations. Mathematical models of this suspension with and without lateral bump-stop are derived owing to non-smooth mechanics assumptions being based on the Coulomb law regarding friction and implemented into the MBS program. Numerical simulations of dynamics of the analyzed system are performed on a straight track followed by the methods appropriate for predicting the dynamic stability of railway vehicles. A dynamic reply of the vehicle to the railway track excitation, in the form of the initial condition, is monitored and studied. The carried out analysis mainly concerns the investigation of limit cycle dynamics exhibited by the system elements in terms of safety of the freight wagon. In addition, an influence of the friction coefficient coupling the interacting elements of suspension and being responsible for damping properties of our non-smooth system is analyzed. Furthermore, an occurrence of a low critical speed is illustrated and its physical meaning is explained, and the usefulness of applying the lateral bump-stop in the structure of UIC double-link suspension is justified.

1 Introduction

Many of freight wagons in Europe are equipped with the UIC link suspension in which damping is provided only by dry friction in pivoted joints of linkages. The structure of this suspension is quite simple, but its elements have strongly nonlinear

J. Matej (✉) • J. Seńko
Department of Automobile and Heavy Machines, Warsaw University of Technology, 84 Narbutta Street, 02-524 Warsaw, Poland
e-mail: jmt@simr.pw.edu.pl; jsenko@wp.pl

J. Awrejcewicz
Department of Automation and Biomechanics, Lodz University of Technology, Stefanowski Street, 90-924 Lodz, Poland
e-mail: awrejcew@p.lodz.pl

J. Awrejcewicz (ed.), *Applied Non-Linear Dynamical Systems*, Springer Proceedings in Mathematics & Statistics 93, DOI 10.1007/978-3-319-08266-0_18

characteristics. Such a type of suspension was examined by many researchers [1, 2]. Nonlinear simulation model for freight wagons with UIC double-link suspension in two-axle freight wagon was developed, for instance, by Hoffmann in [1], but without the lateral bump-stop limiting lateral displacements of the lower link. The purpose of the presented article is to compare the dynamic properties of the two-axle freight wagon equipped with standard UIC double-link suspension with and without lateral bump-stop. The mathematical model of the UIC double-link suspension, prepared according to non-smooth mechanics assumptions, was implemented into the MBS program [6]. The numerical examinations were conducted on the straight track. They included different values of coefficient of friction in joints of the UIC double-link suspension without and with lateral bump-stop, as well as the state of loading the car body (empty/fully loaded). In the case of the leaf spring, a piecewise linear and progressive characteristic in vertical direction was taken into account. The utilized MBS program enabled to use pre-calculated and tabulated wheel–rail contact parameters generated for different lateral positions for S1002 and UIC60 real, nonlinear profiles of the wheel and rail. We recognize the wagon movement as stable, if dynamic reply of the vehicle to the railway track excitation has the form of decaying lateral oscillations of wheel sets, returning to the center line of the track. At sufficiently high speed of the vehicle, the oscillations following an external disturbance grow and lead to a limit cycle. In general, freight wagons have low critical speeds. Fully developed limit cycle oscillation is usually termed "hunting." The lowest vehicle speed at which sustained oscillations appear is named the critical speed.

2 Elements of the Freight Wagon with the UIC Double-Link Suspension

The considered two-axle freight wagon with UIC double-link suspension consists of the body and two wheel sets are shown in Fig. 1. In the structure of the wagon the leaf spring ***2*** is located between the axle box ***4*** and the car body ***1*** (Fig. 2). Wheel sets are guided in lateral and longitudinal direction, utilizing the links ***3***. Longitudinal displacements of wheel set are blocked by guiding fork ***6***. Brackets ***5*** enable the transfer of vertical load from the car body to the axle boxes and wheels. Each of the wheel sets can move relatively to the car body in the range of lateral and longitudinal clearances, equal to 20 and 22,5 mm accordingly.

Upper and lower links ***3***, tumbling blocks ***7***, pins ***8***, C-washers ***9*****,** and links ***10***, shown in Fig. 3 and in Fig. 4, are the main elements of the UIC double-link suspension. Each tumbling block rests on the pin and is secured in the lateral direction by the C-washer. When the wheel set moves relative to the car body, the upper and lower links roll or slide over the tumbling blocks. Rolling in the joint is possible, because the radii of the links' cross sections are smaller than the radii of the half hole in the tumbling blocks.

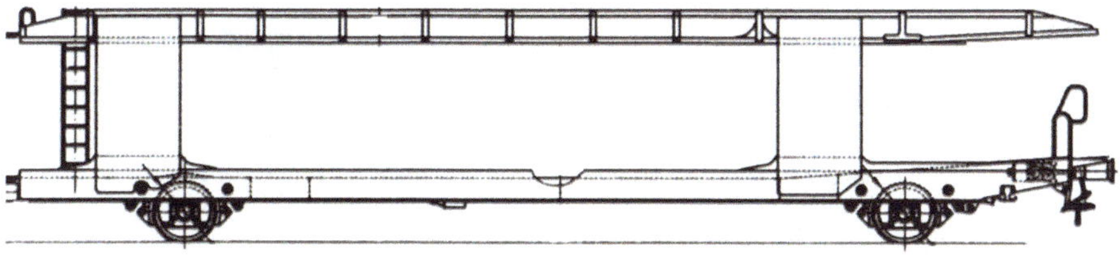

Fig. 1 Side view of the two-axle freight wagon adapted to transport of cars [4]

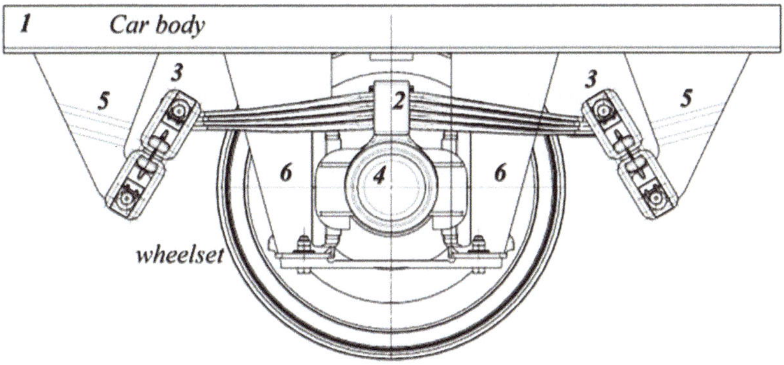

Fig. 2 Sketch of wheel set guiding with the UIC double-link suspension

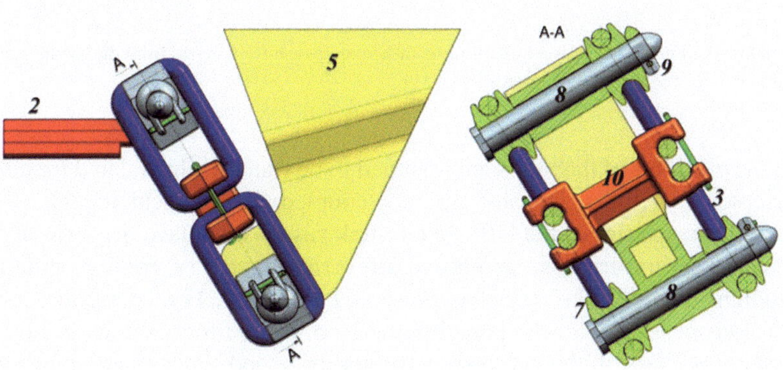

Fig. 3 Elements of UIC double-link suspension without the lateral bump-stop

3 Mathematical Model of the UIC Double-Link Suspension

Authors of this article used the mathematical models of standard UIC double-link suspension proposed by Piotrowski [5]. The main assumptions, according to these models, were the following: contacting elements of joints are cylindrical; the Coulomb law of dry friction is applied for the description of friction in the

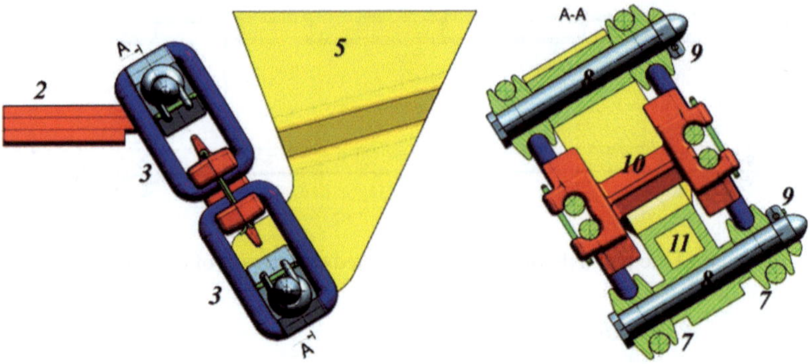

Fig. 4 Elements of UIC double-link suspension with the lateral bump-stop

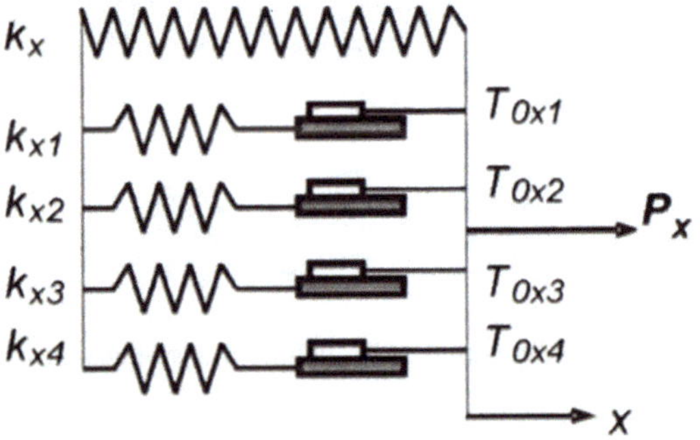

Fig. 5 Rheological model of the UIC double-link suspension for longitudinal direction in the case without and with the lateral bump-stop [5]

joints; the elements of the joints are assumed to be rigid; the lateral and longitudinal displacements of the suspension are not coupled. Rheological models, used to describe the properties of the UIC double-link suspension, have the form of elastic elements with dry friction, composed of springs and dry friction sliders. For longitudinal direction the model is presented in Fig. 5. For lateral direction the model of suspension with the lateral bump-stop has the form shown in Fig. 6 [4]. When there is a lack of the bump-stop, the spring k_2 and slider u_0 are removed.

In case of the UIC suspension with the lateral bump-stop, the two links are active, if the lateral displacement of the leaf spring pivots $Y < u_0$. When $Y > u_0$, only the upper link is active and the stiffness of the suspension is equal to $k + k_2$.

The break force T_{0y} and spring with stiffness of k_1, describing the elastic element with dry friction, may be replaced by the differential equation for the force T (Fig. 7). Then the continuity condition for the slider and spring has a form

$$\dot{T}/k_1 + v_S = \dot{\varsigma} - \dot{Y} \tag{1}$$

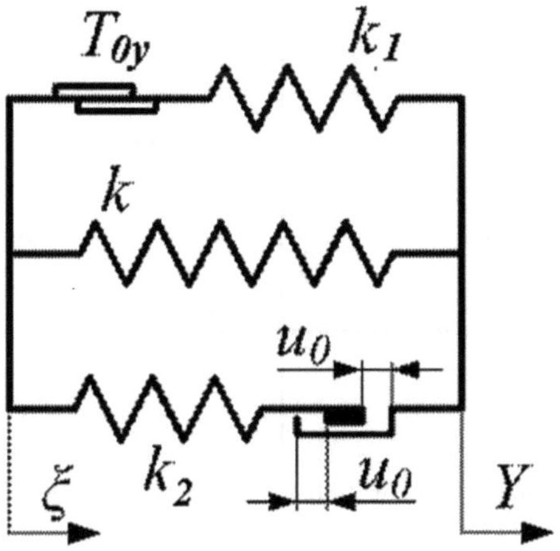

Fig. 6 Rheological models of the UIC double-link suspension with lateral bump-stop for lateral direction [4]

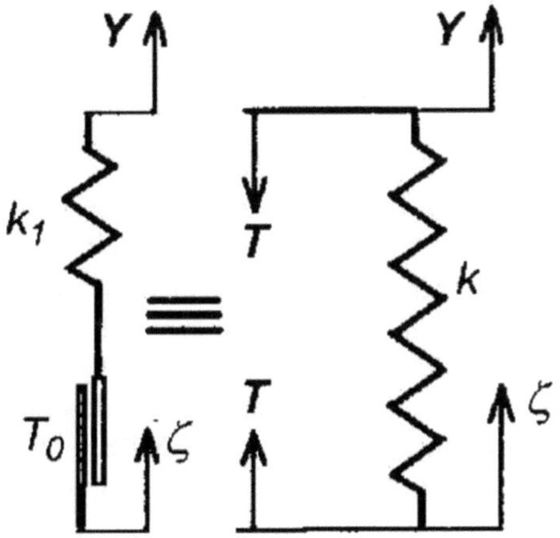

Fig. 7 The slider and the spring replaced by friction force T [5]

where v_s is the velocity of sliding and T_0 is the break force (friction force). The non-smooth relations, basing on the Coulomb law of dry friction and describing the characteristics of friction slider, are as follows [5]:

$$\Omega : \quad T \in [-T_0, +T_0], \tag{2}$$

$$v_S \in -K\left(\dot{T}, D\Omega\right). \tag{3}$$

The cone **K** is described by the velocity of sliding:

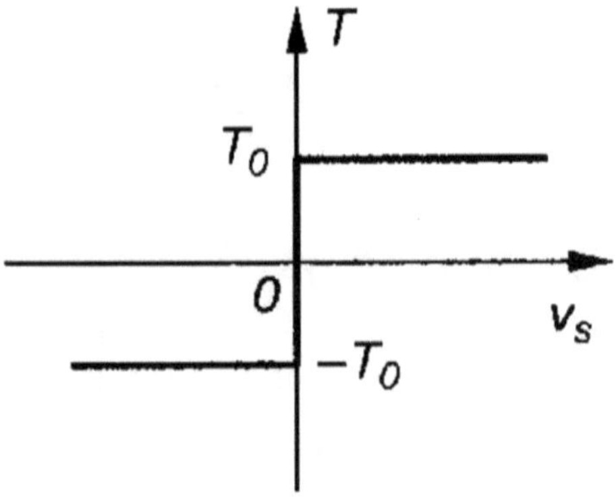

Fig. 8 Relation between the velocity of sliding and the force of friction

$$v_s \in \begin{cases} \{0\} & if \quad |T| < T_0 \\ R^+ & if \quad T = +T_0 \\ R^- & if \quad T = -T_0 \end{cases},$$

where the function $[.]^+$ is defined as

$$[u]^+ = \begin{cases} u & if \quad u \geq 0 \\ 0 & if \quad u < 0 \end{cases}. \tag{4}$$

When $|T_0| = T_0$, it is necessary to consider the differential succession of non-smooth relation (4). The characteristics shown in Fig. 8 are non-smooth, multi-valued, and non-differentiable. It was applied to the vehicle simulation model. In that way the models of the suspension have been described by the differential equations of the first order, implemented in the MBS model of the 2-axle freight wagon in the following manner, for the lateral and longitudinal direction:

$$\dot{T}_y = \begin{cases} k_{y1}(-\dot{Y}) & if \quad |T_y| < T_{0y}, \\ -[-k_{y1}(-\dot{Y})]^+ & if \quad T_y = +T_{0y}, \\ [k_{y1}(-\dot{Y})]^+ & if \quad T_y = -T_{0y}, \end{cases} \tag{5}$$

$$\dot{T}_{xi} = \begin{cases} k_{xi}(-\dot{X}) & if \quad |T_{xi}| < T_{0xi} \\ -[-k_{xi}(-\dot{X})]^+ & if \quad T_{xi} = +T_{0xi}, \\ [k_{xi}(-\dot{X})]^+ & if \quad T_{xi} = -T_{0xi} \end{cases} \quad (i = 1, 4). \tag{6}$$

In formulas (5) and (6), the lateral deflection is described by Y, while X indicates the longitudinal deflection of the suspension. The restoring forces P_y, P_x of the suspension in lateral and longitudinal directions are defined as

Table 1 Parameters of the UIC double-link suspension for the lateral direction without the bump-stop

f	T_{0y}/Q	k_y/Q [1/m]	k_{y1}/Q [1/m]
0,20	0,36518E-01	0,34534E+01	0,10519E+02
0,50	0,83620E-01	0,34718E+01	0,10155E+02

Table 2 Parameters of the UIC double-link suspension for the lateral direction with the bump-stop

f	T_{0y}/Q	k_y/Q [1/m]	k_{y1}/Q [1/m]	k_{y2}/Q [1/m]
0,20	0,36518E-01	0,34534E+01	0,10519 IE+02	0,34534E+01
0,50	0,83620E-01	0,34718E+01	0,10155E+02	0,34658E+01

Table 3 Parameters of the UIC double-link suspension model for the longitudinal direction

f	T_{0x1}/Q	T_{0x2}/Q	T_{0x3}/Q	T_{0x4}/Q	k_x/Q	k_{x1}/Q	k_{x2}/Q	k_{x3}/Q	k_{x4}/Q
0,2	0,750E−2	0,536E−2	0,69E−3	0,32E−2	0,712E+1	0,428E+1	0,245E+1	0,308E+0	0,107E+1
0,5	0,183E −1	0,117E −1	0,21E −2	0,89E −2	0,661E +1	0,448E +1	0,229E +1	0,415E +0	0,126E +1

$$P_y = k_y Y + T_y, \qquad P_x = k_x X + \sum_{i=1}^{i=4} T_{xi}. \tag{7}$$

The parameters of the model of the UIC double-link suspension without and with lateral bump-stop have been prepared for dimensions located in the middle of the tolerance field, given by the technical specification of DIN 5545 Standards [4]. Four values of the coefficient of friction between rolling/sliding elements of joints have been assumed: $f = 0{,}2$; 0,3; 0;4, 0,5. Values of stiffness and dry friction force for lateral/longitudinal directions were scaled with the vertical load Q of the suspension in the following manner: k/Q, k_1/Q, k_2/Q, k_{xi}/Q, T_{0y}/Q, T_{0xi}/Q, $(i = 1, 4)$. Chosen unit parameters are presented for $f = 0{,}2$ and $f = 0{,}5$ in Tables 1–3.

4 The MBS Model of the Two-Axle Freight Wagon with the UIC Double-Link Suspension

A multi-body simulation model of the two-axle freight wagon (18° of freedom) was prepared in the MBS program [6]. The UIC double-link suspension was modeled as nonlinear, massless element. Mathematical models of this suspension were involved into multi-body model in the form of the differential equations (5) and (6), for each of wheel set. These differential equations have been integrated with the wheel set template in the MBS program. For the nominal UIC60 profile for rails and nominal S1002 profile for wheels with nominal radii equal to 0,42 m, the one-point contact model (based on Kalker's rolling contact theory [3]) was assumed. The nonlinear wheel–rail geometry and pre-tabulated contact functions according to the

Table 4 Mass and inertial parameters for two-axle freight wagon with UIC double-link suspension

		Moments of inertia (kg m^2)		
Part	Mass (kg)	I_x	I_y	I_z
Empty car body	13,450	282,000	318	321,000
Fully loaded car body	27,450	595,140	618,140	596,970
Wheel set	990	535	74	535

FASTSIM algorithm of Kalker [6] were used. The coefficient of friction between wheel and rail, accepted equal to 0,4. The track model was treated as a uniform structure with the mechanical properties described by the parameters given in [6]. The inertial parameters used in simulations for the empty and loaded freight wagon are presented in Table 4. The other parameters are as follows: the base of the wagon equal to 10 m, track gauge equal to 1,435 m, rail inclination equal to 1:40, and the border displacement of the leaf spring pivoting in lateral direction $u_0 = 10$ mm. The distance of the wagon center of gravity from the rail level was accepted as 1,527 and 2,162 m for empty and fully loaded car body appropriately.

5 Results of Numerical Simulation

The simulations have been performed for two cases: without and with the lateral bump-stop in UIC double-link suspension. Lateral vibrations were caused by the initial condition introduced in the form of lateral displacement of leading wheel set, equal to 0,005 m, relative to the track center line. As a response we can observe the hunting motion of wheel sets. We stated that critical speed of empty and fully loaded wagon is very small and close to 10 m/s. Freight wagons are operated usually with the maximum velocity smaller or equal to 120 km/h. In order to judge the properties of the suspension, further investigations were done for velocity bigger than critical speed of the wagon. As a criterion of evaluation, the lateral displacements of leading wheel set were accepted. In the range of velocity from 10 to 20 m/s, an advantage of UIC double-link suspension without lateral bump-stop, with the coefficient of friction between rolling/sliding elements equal to 0,2 was stated. But for velocity close to 35 m/s, lateral displacements of leading wheel set of empty wagon violently grow and undesirable wheel flange contact between wheels and rails is observed (Fig. 9). Simultaneously, the lateral amplitudes of leading wheel set of empty wagon with the UIC double-link suspension equipped with the lateral bump-stop achieve the values close to 5 mm for velocity equal to 15 m/s (Fig. 10). The limit cycle of this wheel set is presented in Fig. 11. For velocity from 20 to 40 m/s, lateral amplitudes of leading wheel set for this case are smaller than for the compared type of suspension without the bump-stop. In addition they are smaller than 1 mm. It means that there is no flange contact between wheels and rails.

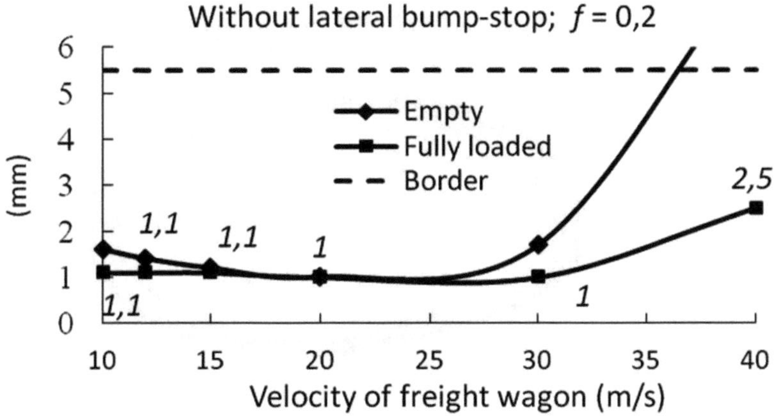

Fig. 9 Amplitudes of lateral displacements of leading wheel set for wagon with the suspension without the lateral bump-stop and coefficient of friction between rolling/sliding elements of joints equal to 0,2, in the function of wagon velocity

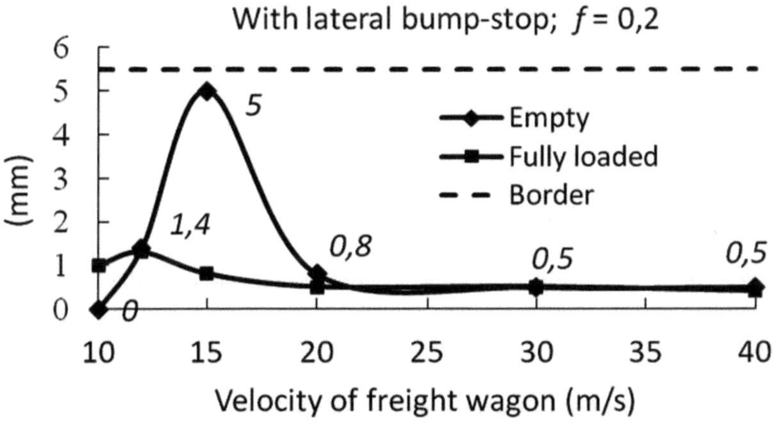

Fig. 10 Amplitudes of lateral displacements of leading wheel set for wagon with the suspension with the lateral bump-stop and coefficient of friction between rolling/sliding elements of joints equal to 0,2, in the function of wagon velocity

In the context of previous conclusion according to the UIC double-link suspension with lateral bump-stop, we need to explain the cause of the large lateral amplitude of leading wheel set, close to 0,005 m, examining the case presented in Fig. 10 for empty wagon. Therefore an analysis of a resonance between the lateral excitation frequency of the wheel sets f_{w} and the yaw eigenfrequency of car body f_{yaw}, as well as the roll eigenfrequency f_{roll} of the car body motion, was conducted. These frequencies can be approximated in the following manner [1]:

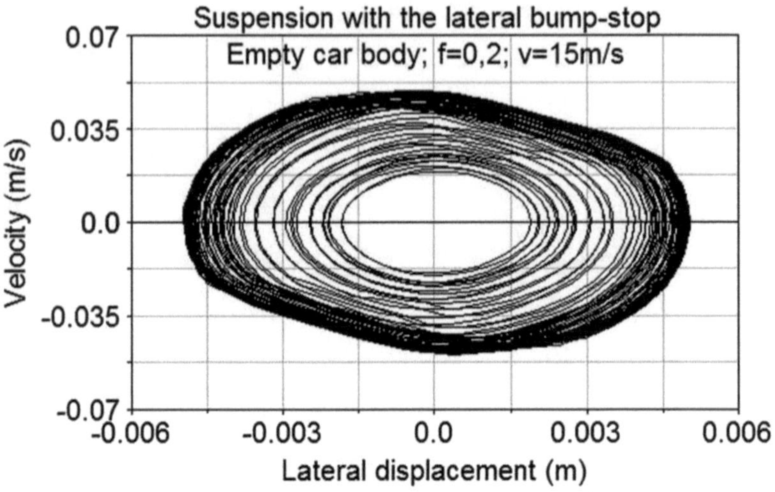

Fig. 11 Limit cycle of leading wheel set of empty wagon with the UIC double-link suspension equipped with the lateral bump-stop, for $f = 0{,}2$ and $v = 15$ m/s

$$\begin{aligned} f_w &= v/2\pi\sqrt{\lambda/b_0 r_0}, \\ f_{yaw} &= 1/2\pi\sqrt{4a^2k_y/I_z}, \\ f_{roll} &= 1/2\pi\sqrt{4\left(L_1^2k_z + L_2^2k_y\right)/I_x}, \end{aligned} \tag{8}$$

where a is the half wheelbase, k_y and k_z denote appropriately the lateral stiffness of suspension and the vertical stiffness of the leaf spring, I_z and I_x are the yaw and the roll moments of inertia of the car body, v is the speed of vehicle, λ denotes the wheel conicity, $2b_0$ and r_0 are the distance between the nominal rolling circles and the nominal rolling radius of wheel, and L_1 and L_2 indicate the lateral and vertical distances between the center of car body mass and the UIC suspension, respectively.

Values of parameters in formulas (8) were assumed to be known. The value of wheel conicity parameter was accepted as $\lambda = 0{,}13$, which corresponds to lateral displacements of wheel set in the scope up to 6 mm. Using Klingel's formula, we got from the equations (8) the value of lateral excitation frequency of the wheel set f_w equal to 1,53 Hz. Applying the spectrum analysis (fast Fourier transform) of lateral oscillations of wheel sets we obtained more probable value of f_w equal to **1,5** Hz. In the next step the hysteresis loop presented in Fig. 12 and according to lateral direction of the UIC double-link suspension with the bump-stop was analyzed. The following values of k_y stiffness were estimated: $k_{y1} = 113470$ N/m, $k_{y2} = 225860$ N/m, and $k_{y3} = 560760$ N/m. Next, the yaw and roll eigenfrequencies of the empty car body were calculated: $f_{yaw1} = 1{,}05$ Hz, $f_{yaw2} = 1{,}49$ Hz, $f_{yaw3} = 2{,}35$ Hz, $f_{roll1} = 0{,}93$ Hz, $f_{roll\,2} = 1{,}16$ Hz, and $f_{roll\,3} = 1{,}68$ Hz. We noticed that the values of the lateral excitation frequency of the wheel set $f_w = 1{,}5$ and yaw

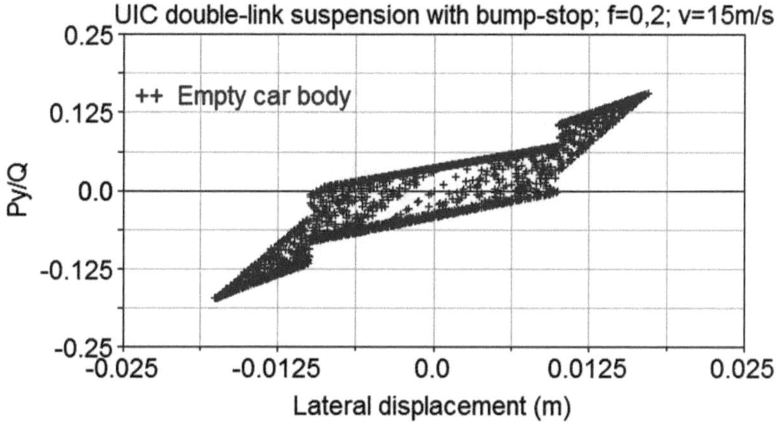

Fig. 12 Hysteresis loops for empty wagon equipped with the UIC double-link suspension with the bump-stop, for velocity equal to 15 m/s and coefficient of friction between rolling/sliding elements equal to 0,2

eigenfrequency $f_{\mathrm{yaw2}} = 1{,}49$ Hz are almost identical. It means that lateral oscillations of wheel set and car body yaw motion are in the resonance. That is the direct cause of the high leading wheel set amplitude of examined freight wagon, equipped with UIC double-link suspension with the lateral bump-stop. Similar analysis for empty wagon running with velocity greater than 20 m/s and for fully loaded wagon in the range of velocities from 10 m/s to 40 m/s showed that instability is caused by small damping of the lateral displacements of wheel set. Hysteresis loop for fully loaded wagon equipped with the UIC double-link suspension with the bump-stop, for velocity equal to 15 m/s and coefficient of friction between rolling/sliding elements equal to 0,2, is shown in Fig. 13. Similarly, but more worse situation according to hunting motion of leading wheel sets we can observe for coefficients of friction between rolling/sliding elements of joints greater than 0,2, especially for $f = 0{,}5$ (Figs. 14 and 15). Also in these cases the high lateral displacements of leading wheel sets are caused first through a resonance between the lateral excitation frequency of the wheel sets f_{w} and the yaw eigenfrequency f_{yaw} of the empty and fully loaded car body.

When the speed of wagon increases, we still can observe the limit cycles of wheel sets caused by small damping properties of the suspension. According to earlier assumptions, the lateral and longitudinal directions of the UIC double-link suspension should not be coupled. In practice, wheel sets have the possibilities to yaw and in that way the longitudinal displacement of the leaf spring occurs. In the end, suspension works in two directions; however the longitudinal restoring forces P_x are almost four times smaller than lateral restoring forces P_y (Fig. 16).

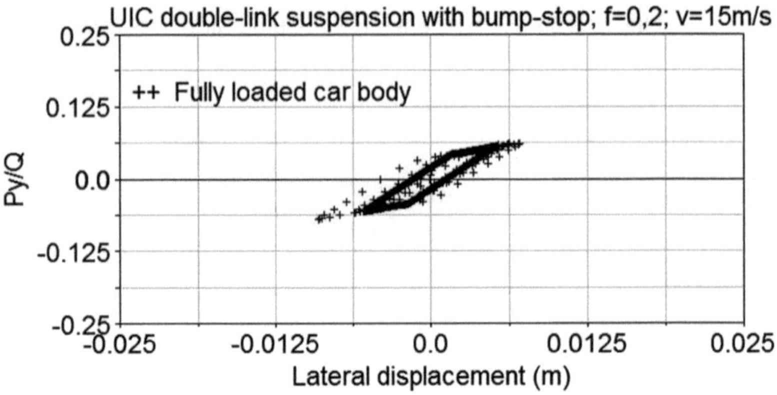

Fig. 13 Hysteresis loop for fully loaded wagon equipped with the UIC double-link suspension with the bump-stop, for velocity equal to 15 m/s and coefficient of friction between rolling/sliding elements equal to 0,2

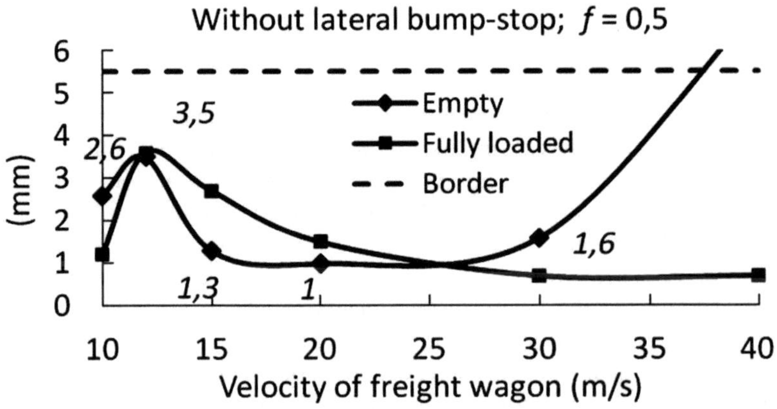

Fig. 14 Amplitudes of lateral displacements of leading wheel set for wagon with the suspension without the lateral bump-stop and coefficient of friction between rolling/sliding elements of joints equal to 0,5, in the function of wagon velocity

6 Conclusions

In the presented article the influence of chosen parameters on the lateral dynamic behavior of the two-axle freight wagon with the UIC double-link suspension was showed. Mathematical models of this suspension with and without lateral bump-stop were derived owing to non-smooth mechanics assumptions being based on the Coulomb law regarding friction.

Very small and close to 10 m/s, critical speed of empty and fully loaded freight wagon with the UIC double-link suspension was stated.

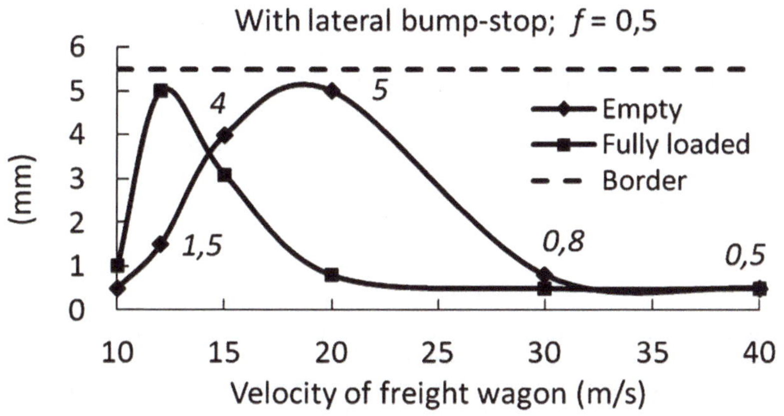

Fig. 15 Amplitudes of lateral displacements of leading wheel set for wagon with the suspension with the lateral bump-stop and coefficient of friction between rolling/sliding elements of joints equal to 0,2, in the function of wagon velocity

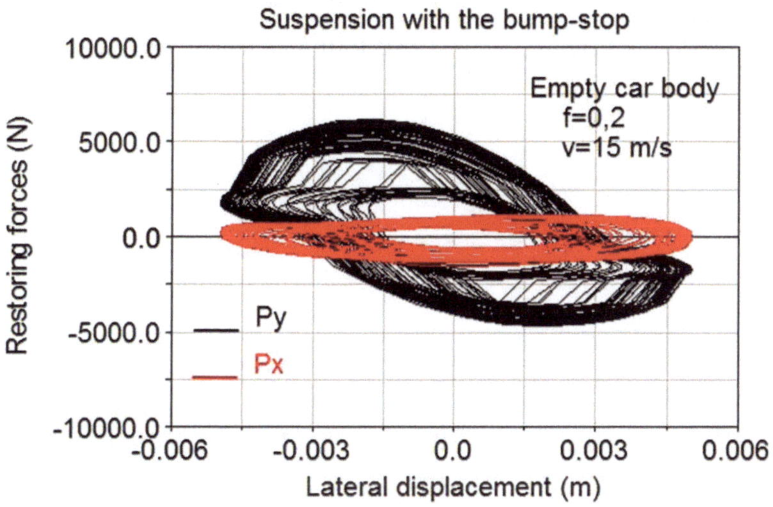

Fig. 16 Restoring forces in the function of lateral displacement of leading wheel set for empty wagon equipped with the UIC double-link suspension with the bump-stop, for velocity equal to 15 m/s and coefficient of friction between rolling/sliding elements equal to 0,2

An advantage of the UIC double-link suspension without lateral bump-stop, for the coefficient of friction between rolling/sliding elements equal to 0,2, was showed for velocity smaller than 30 m/s.

The appearance of high lateral amplitudes of wheel sets, of the empty or fully loaded freight wagon with the UIC double-link suspension with or without lateral bump-stop, was interpreted as a resonance between the lateral excitation frequency of the wheel sets and the yaw eigenfrequency of the car body.

It was also demonstrated, that if the speed of two-axle freight wagon increases, we still can observe the limit cycles of wheel sets, caused by small damping properties of the suspension, for two analyzed cases of suspension.

References

1. Hoffmann, M., True, H.: The dynamics of European two-axle railway freight wagons with UIC standard suspension. Vehicle Syst. Dynam. **46**, 225–236 (2008)
2. Jönsson P.: Modelling and laboratory investigations on freight wagon link suspensions, with respect to vehicle-track dynamic interaction, Ph. D. thesis, Stockholm, Sweden (2004)
3. Kalker, J.J.: Three-dimensional elastic bodies in rolling contact. Kluwer Academic Publishers, Dordrecht/Boston/London (1990)
4. Matej, J., Piotrowski, J., Seńko, J.: Analysis of ride dynamics of L02A Car Wagon unit based on numerical simulations. Internal Report, Warsaw (2010)
5. Piotrowski, J.: Model of the UIC link suspension for freight wagons. Archive of Appl. Mech. **73**, 517–532 (2003)
6. Wheel–rail element. Reference guide, ADAMS RAIL users manual. 2005 (MSC Software).

Dynamics of Mechanical Systems with Mecanum Wheels

Klaus Zimmermann, Igor Zeidis, and Mohamed Abdelrahman

Abstract The kinematics and dynamics of a mechanical system with mecanum wheels is studied. A mecanum wheel is a wheel with rollers attached to its circumference. Each roller rotates about an axis that forms an angle with the plane of the disk (for the omni-wheels, the axes of the rollers lie in the plane of the wheel and in an ideal case are tangent to the outer circumference of the wheel). Such a design provides additional kinematic advantages for the mecanum wheels in comparison with the conventional wheels. Within the framework of non-holonomic mechanics, the equations of motion are derived for the case of an arbitrary angle at which the rollers are attached (usually, this angle is assumed to be equal to 45°). In robotics, a simplified approach, in which the equations of non-holonomic kinematic constraints are solved approximately by means of a pseudo-inverse matrix, is frequently applied. Such an approximate approach leads to "holonomization" of the system and allows Lagrange's equations of the second kind to be used. In the present paper, the equations of motion obtained on the basis of the principles of non-holonomic mechanics are compared with the approximate equations. It is shown that for translational motions and for the rotation of the system about its center of mass, both these approaches lead to the same result.

1 Introduction

Mobility is an important property for people. Thus, its generation, conservation, and, if necessary, recovery are an important goal and a great challenge for life scientists and engineers. Some tasks encompass fields from the design of faster and—at the same time—energy-saving transportation systems in the air to the

K. Zimmermann (✉) • I. Zeidis • M. Abdelrahman
Department of Technical Mechanics, Technische Universität Ilmenau, Max-Planck-Ring 12, 98693 Ilmenau, Germany
e-mail: klaus.zimmermann@tu-ilmenau.de; igor.zeidis@tu-ilmenau.de; mohamed.abdelrahman@tu-ilmenau.de

J. Awrejcewicz (ed.), *Applied Non-Linear Dynamical Systems*, Springer Proceedings in Mathematics & Statistics 93, DOI 10.1007/978-3-319-08266-0_19

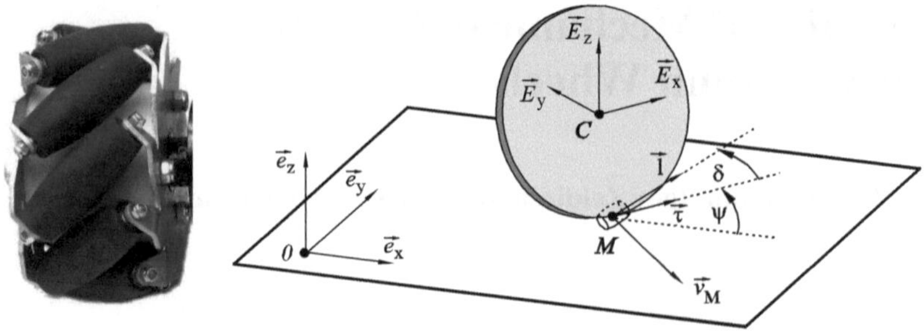

Fig. 1 A mecanum wheel (*left*) and the mechanical model (*right*)

recovery of upright walking through exoskeletons. For service robots and equipment for physically disabled persons, mobile systems with a high maneuverability play an important role. The conventional wheel moves back into the focus of interest. Seemingly, its basic mechanical function "rolling" is easy to understand and all mechanics are adequately described. But, there exist some new developments, which connect the advantages of wheels and legs in a whole system (e.g., "whegs"). Furthermore, the so-called omnidirectional wheels (e.g., "mecanum wheels"), generating constraints different from the conventional wheel, lead to investigations, based on the mechanics and control of non-holonomic systems. At the present time, vehicles with mecanum wheels (Fig. 1, left) are gaining ground for different applications. The wheels of such vehicles have rollers that are arranged along the rim at an angle to the wheel plane [5]. As a rule, this angle is equal to 45°. Such wheels have additional kinematical possibilities in comparison with conventional wheels. Due to these possibilities, a vehicle with mecanum wheels can move forward-backward and leftward-rightward and rotate in an arbitrary way. Usually a mecanum-wheeled vehicle has four wheels. By varying the rate and the direction of rotation of each wheel, one can implement, for example, a translational motion of the vehicle in any direction, as well as arbitrary turns and rotations on the spot. The kinematics of wheeled systems, including those with mecanum wheels, are reviewed in [2]. The issues of kinematics, dynamics, and control of systems with mecanum wheels in a non-holonomic treatment are considered in [6, 10, 11] for a number of particular cases. There are a great number of studies on robotics, in which the kinematics and dynamics of robots with four mecanum wheels is approximately treated in terms of holonomic mechanics (see, e.g., [3, 4, 7–9]). In these studies, pseudo-inverse matrices are used to resolve the kinematic constraint relations.

In the present study, the kinematics of a vehicle with four mecanum wheels are considered in terms of non-holonomic mechanics for an arbitrary orientation of the rollers of the mecanum wheels. The equations obtained are compared with the equations implied by an approximate holonomic treatment, and the types of motion for which both treatments coincide are found.

2 Non-holonomic Model of a Mechanical System with Mecanum Wheels

Using a non-holonomic model of a mecanum wheel, we will obtain the kinematic relations and the equation of motion for a mechanical system with such wheels.

2.1 *Kinematics of a Mecanum Wheel*

For a conventional wheel, the contact between the wheel and the supporting plane is characterized by the condition that the wheel is rolling without slip. This means that the velocity of the point by which the wheel contacts the plane at each current instant is equal to zero. Then the projections of the velocity of the contact point onto the direction lying in the wheel plane, as well as onto the direction perpendicular to this plane, are equal to zero. For a mecanum wheel, there is only one direction the projection onto which of the velocity of the point of contact of the wheel with the supporting plane vanishes. This direction can be arbitrary, but it is fixed relative to the wheel.

As a model of a mecanum wheel we will consider the rolling of a disk of radius R centered at the point C on a horizontal plane. The plane of the disk is always vertical. Let $\vec{l}$ denote the unit vector along the axis of attachment of the rollers, $\vec{\tau}$ the unit vector lying in the wheel plane tangent to the rim at the point of contact, and δ the angle between the plane of the wheel and the roller axis (between the vectors $\vec{\tau}$ and $\vec{l}$). The angle δ is constant. The kinematic constraint relation for a mecanum wheel implies that the vector of velocity $\vec{v}_M$ of the point M of contact of the wheel with the plane points along the line perpendicular to the axis of the roller, i.e., the projection of the velocity of the point M onto the roller axis is equal to zero (see Fig. 1, right).

The kinematic constraint relation has the form

$$\vec{v}_M \cdot \vec{l} = 0. \tag{1}$$

The velocity $\vec{v}_M$ is defined by the equation

$$\vec{v}_M = \vec{v}_C + \vec{\omega} \times \overrightarrow{CM} \tag{2}$$

where $\vec{v}_C$ is the velocity of the center C and $\vec{\omega}$ is the angular velocity of the disk.

Let $\left\{O, \vec{e}_x, \vec{e}_y, \vec{e}_z\right\}$ be a fixed reference frame (inertial system) and let C be the origin of a movable reference frame (body fixed frame) $\left\{C, \vec{E}_x, \vec{E}_y, \vec{E}_z\right\}$. The unit vectors $\vec{E}_x$ and $\vec{E}_y$ are parallel to the horizontal plane, the vector $\vec{E}_z$

lying in the disk plane and the vector $\vec{E}_y$ being perpendicular to this plane. Let φ be the angle of rotation of the disk about the axis passing through the point C perpendicular to the plane of the disk, and ψ the angle formed by the disk plane with a line parallel to the vector $\vec{E}_x$ (the angle between the vector $\vec{\tau}$ and the vector $\vec{e}_x$, see Fig. 1, right). The vectors $\vec{\omega}$ and $\overrightarrow{CM}$ are defined by

$$\vec{\omega} = \dot{\varphi}\,\vec{E}_y + \dot{\psi}\,\vec{E}_z,\ \overrightarrow{CM} = -R\,\vec{E}_z. \tag{3}$$

Let x_C, y_C, R be the coordinates of the point C in the reference frame $\left\{O, \vec{e}_x, \vec{e}_y, \vec{e}_z\right\}$. Then

$$\begin{aligned}\vec{v}_C &= (\dot{x}_C \cos\psi + \dot{y}_C \sin\psi)\,\vec{E}_x + (-\dot{x}_C \sin\psi + \dot{y}_C \cos\psi)\,\vec{E}_y,\\ \vec{\omega}\times\overrightarrow{CM} &= -R\dot{\varphi}\,\vec{E}_x.\end{aligned} \tag{4}$$

The vector $\vec{l}$ is expressed by

$$\vec{l} = \cos\delta\,\vec{E}_x + \sin\delta\,\vec{E}_y. \tag{5}$$

Finally, the kinematic relation (1) becomes

$$\dot{x}_C \cos(\psi + \delta) + \dot{y}_C \sin(\psi + \delta) = R\dot{\varphi}\cos\delta. \tag{6}$$

On the basis of the analysis of the kinematic constraints of type (6) it is shown [1] that if a mechanical system is based on n mecanum wheels in such a way that (a) $n \geq 3$; (b) not all vectors $\vec{l}_i$ are parallel to each other; and (c) the points of contact of the wheels with the plane do not lie on one line, then it is always possible to find control functions φ_i $(i = 1, \ldots, n)$ that implement any prescribed motion of the system's center of mass.

2.2 *Kinematics of a Vehicle with Four Mecanum Wheels*

Consider a model of a four-wheeled vehicle with mecanum wheels (Fig. 2).

Let now C be the center of mass of the system (the body together with the wheels). The coordinates of the point C in the reference frame $\left\{O, \vec{e}_x, \vec{e}_y, \vec{e}_z\right\}$ are x_C, y_C, and R; the quantities $\left|\overrightarrow{CO_1}\right| = \rho_1$ and $\left|\overrightarrow{CO_2}\right| = \rho_2$ are the distances from the center of mass to the axes of the respective wheel pairs, and $2\,l$ is the distance between the centers of the wheels of one axis.

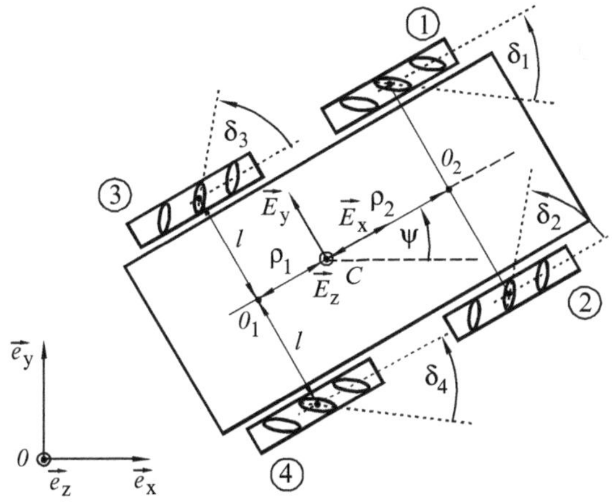

Fig. 2 A vehicle with four mecanum wheels

The corresponding kinematic relations are obtained on the basis of Eq. (6), in which δ should be replaced by $-\delta_1$ and $-\delta_4$ for wheels 1 and 4 and by δ_2 and δ_3 for wheels 2 and 3, respectively. These relations have the form

$$\dot{x}_C \cos(\psi - \delta_1) + \dot{y}_C \sin(\psi - \delta_1) - \dot{\psi}\,(l \cos\delta_1 + \rho_2 \sin\delta_1) = R\dot{\varphi}_1 \cos\delta_1, \quad (7)$$

$$\dot{x}_C \cos(\psi + \delta_2) + \dot{y}_C \sin(\psi + \delta_2) + \dot{\psi}\,(l \cos\delta_2 + \rho_2 \sin\delta_2) = R\dot{\varphi}_2 \cos\delta_2, \quad (8)$$

$$\dot{x}_C \cos(\psi + \delta_3) + \dot{y}_C \sin(\psi + \delta_3) - \dot{\psi}\,(l \cos\delta_3 + \rho_1 \sin\delta_3) = R\dot{\varphi}_3 \cos\delta_3, \quad (9)$$

$$\dot{x}_C \cos(\psi - \delta_4) + \dot{y}_C \sin(\psi - \delta_4) + \dot{\psi}\,(l \cos\delta_4 + \rho_1 \sin\delta_4) = R\dot{\varphi}_4 \cos\delta_4, \quad (10)$$

where φ_i is the angle of rotation of the disk about the rotation axis and δ_i is the angle between the wheel plane and the roller axis ($i = 1, \ldots, 4$).

Equations (7)–(10) define four non-holonomic constraints. Note that if only translational motions ($\dot{\psi} = 0$) or only rotations about the center of mass ($\dot{x}_C = \dot{y}_C = 0$) are allowed, then the constraints become holonomic.

2.3 *Dynamics of a Vehicle with Mecanum Wheels*

The configuration of the mechanical system under consideration is defined by seven parameters, which involve the coordinates $q^1 = x_C$, $q^2 = y_C$ of the system's center of mass C, the angle $q^3 = \psi$ between the abscissa axis of the fixed reference frame and a straight line perpendicular to the axis of the wheel pair, and the angles of rotation of the wheels $q^{i+3} = \varphi_i$ ($i = 1, \ldots, 4$).

To describe the motion of the system we will use Lagrange's equations with undetermined multipliers corresponding to the non-holonomic constraints:

$$\frac{d}{dt}\frac{\partial T}{\partial \dot{x}_C} - \frac{\partial T}{\partial x_C} - \lambda_1 \cos(\psi - \delta_1) - \lambda_2 \cos(\psi + \delta_2) - \lambda_3 \cos(\psi + \delta_3)$$
$$- \lambda_4 \cos(\psi - \delta_4) = 0, \tag{11}$$

$$\frac{d}{d\,t}\frac{\partial T}{\partial \dot{y}_C} - \frac{\partial T}{\partial y_C} - \lambda_1 \sin(\psi - \delta_1) - \lambda_2 \sin(\psi + \delta_2) - \lambda_3 \sin(\psi + \delta_3)$$
$$- \lambda_4 \sin(\psi - \delta_4) = 0, \tag{12}$$

$$\frac{d}{d\,t}\frac{\partial T}{\partial \dot{\psi}} - \frac{\partial T}{\partial \psi} + \lambda_1 (l \cos\delta_1 + \rho_2 \sin\delta_1) - \lambda_2 (l \cos\delta_2 + \rho_2 \sin\delta_2)$$
$$+ \lambda_3 (l \cos\delta_3 + \rho_1 \sin\delta_2) - \lambda_4 (l \cos\delta_4 + \rho_1 \sin\delta_4) = 0, \tag{13}$$

$$\frac{d}{dt}\frac{\partial T}{\partial \dot{\varphi}_i} - \frac{\partial T}{\partial \varphi_i} + \lambda_i R \cos\delta_i = M_i, i = 1, \ldots, 4. \tag{14}$$

Here, T is the kinetic energy of the system, M_i are the torques applied to the respective wheels, and λ_i are Lagrange's undetermined multipliers ($i = 1, \ldots, 4$).

The kinetic energy T is defined as the sum of the kinetic energies of the body and the wheels and is expressed as follows:

$$T = \frac{1}{2}m_0\left(\dot{x}_C^2 + \dot{y}_C^2\right) + \frac{1}{2}J_0\dot{\psi}^2 + \frac{1}{2}m_1\left[4\left(\dot{x}_C^2 + \dot{y}_C^2\right)\right.$$
$$+ 4\dot{x}_C\dot{y}_C(\rho_1 - \rho_2)\sin\psi - 4\dot{y}_C\dot{\psi}(\rho_1 - \rho_2)\cos\psi$$
$$\left. + 4\dot{y}_C\dot{\psi}(\rho_1 - \rho_2)\cos\psi + 2\dot{\psi}^2\left(\rho_1^2 + \rho_2^2 + 2l^2\right)\right]$$
$$+ 2J_2\dot{\psi}^2 + \frac{1}{2}J_1\left(\dot{\varphi}_1^2 + \dot{\varphi}_2^2 + \dot{\varphi}_3^2 + \dot{\varphi}_4^2\right). \tag{15}$$

Here m_0 is the mass of the body, m_1 is the mass of each of the wheels, J_0 is the mass moment of inertia of the body about the vertical axis passing through the center of mass, J_1 is the mass moment of inertia of each wheel about its axis of rotation, and J_2 is the moment of inertia of each wheel about the vertical axis passing through the center of the wheel.

Consider a practically important case, for which $\delta_i = \delta$ ($i = 1, \ldots, 4$). Then, eliminating the multipliers λ_i in Eqs. (11)–(14), we obtain

$$(m_0 + 4m_1)\ddot{x}_C + 2m_1\ddot{\psi}(\rho_1 - \rho_2)\sin\psi + 2m_1\dot{\psi}^2(\rho_1 - \rho_2)\cos\psi$$
$$= -\frac{2J_1}{R^2\cos^2\delta}\left[\left(\ddot{x}_C + \dot{y}_C\dot{\psi}\right)(1 + \cos 2\psi\cos 2\delta)\right.$$
$$\left. + \left(\ddot{y}_C - \dot{x}_C\dot{\psi}\right)\sin 2\psi\cos 2\delta + \ddot{\psi}(\rho_1 - \rho_2)\sin\psi\sin^2\delta\right]$$
$$+ \frac{1}{R\cos\delta}\left[(M_1 + M_4)\cos(\psi - \delta) + (M_2 + M_3)\cos(\psi + \delta)\right], \tag{16}$$

$$(m_0 + 4m_1)\,\ddot{y}_C - 2m_1\ddot{\psi}\,(\rho_1 - \rho_2)\cos\psi + 2m_1\dot{\psi}^{\,2}(\rho_1 - \rho_2)\sin\psi$$
$$= -\frac{2J_1}{R^2\cos^2\delta}\left[\left(\ddot{x}_C + \dot{y}_C\dot{\psi}\right)\sin 2\psi\cos 2\delta\right.$$
$$\left.+ \left(\ddot{y}_C - \dot{x}_C\dot{\psi}\right)(1 - \cos 2\psi\cos 2\delta) - \ddot{\psi}\,(\rho_1 - \rho_2)\cos\psi\sin^2\delta\right]$$
$$+ \frac{1}{R\cos\delta}\left[(M_1 + M_4)\sin(\psi - \delta) + (M_2 + M_3)\sin(\psi + \delta)\right], \qquad (17)$$

$$\left(J_0 + 4J_2 + 2m_1(\rho_1^2 + \rho_2^2 + 2l^2)\right)\ddot{\psi}$$
$$+ 2m_1\ddot{x}_c(\rho_1 - \rho_2)\sin\psi - 2m_1\ddot{y}_c(\rho_1 - \rho_2)\cos\psi$$
$$+ \frac{2J_1}{R^2\cos^2\delta}\left[\ddot{\psi}\,(l^2 + 0.5(\rho_1^2 + \rho_2^2) + (l^2 - 0.5(\rho_1^2 + \rho_2^2))\cos 2\delta\right.$$
$$+ l(\rho_1 + \rho_2)\sin 2\delta)$$
$$+ \ddot{x}_c(\rho_1 - \rho_2)\sin\psi\sin^2\delta - \ddot{y}_c(\rho_1 - \rho_2)\cos\psi\sin^2\delta$$
$$\left.+ \dot{x}_C\dot{\psi}\,(\rho_1 - \rho_2)\cos\psi\sin^2\delta + \dot{y}_C\dot{\psi}\,(\rho_1 - \rho_2)\sin\psi\sin^2\delta\right]$$
$$= \frac{1}{R\cos\delta}\left[l\,(M_2 - M_1 + M_4 - M_3)\cos\delta + (\rho_1\,(M_4 - M_3)\right.$$
$$\left.+ \rho_2\,(M_2 - M_1))\sin\delta\right]. \qquad (18)$$

As a result, for given torques M_i $(i = 1, \ldots, 4)$, we have a system of three equations for three variables, x_C, y_C, and ψ. Then the angles φ_i of rotation of the wheels can be found from Eqs. (7)–(10).

In the vehicles that are usually considered, the angle between the plane of the mecanum wheel and the roller axis is equal to 45° ($\delta = \pi/4$), and the center of mass of the system coincides with its geometric center ($\rho_1 = \rho_2 = \rho$). For this case, Eqs. (16)–(18) are simplified and become

$$\left(m + \frac{4J_1}{R^2}\right)\ddot{x}_C + \frac{4J_1}{R^2}\dot{y}_C\dot{\psi} = \frac{1}{R}\left[(M_1 + M_2 + M_3 + M_4)\cos\psi\right.$$
$$\left.+ (M_1 - M_2 - M_3 + M_4)\sin\psi\right], \qquad (19)$$

$$\left(m + \frac{4J_1}{R^2}\right)\ddot{y}_C - \frac{4J_1}{R^2}\dot{x}_C\dot{\psi} = \frac{1}{R}\left[(M_1 + M_2 + M_3 + M_4)\sin\psi\right.$$
$$\left.- (M_1 - M_2 - M_3 + M_4)\cos\psi\right], \qquad (20)$$

$$\left[J_C + \frac{4J_1}{R^2}(l + \rho)^2\right]\ddot{\psi} = -\frac{l + \rho}{R}(M_1 - M_2 + M_3 - M_4), \qquad (21)$$

where $m = m_0 + 4\ m_1$ is the total mass of the system and $J_C = J_0 + 4J_2 + 4m_1$ $(\rho^2 + l^2)$ is the moment of inertia of the entire system relative to the center of mass.

3 Approximate Model

Let v_x and v_y be the projections of the velocity of the center of mass of the vehicle onto the axes of the coordinate system $\left\{C, \vec{E}_x, \vec{E}_y, \vec{E}_z\right\}$ that is attached to the vehicle's body and is turned by the angle ψ relative to the fixed reference frame $\left\{O, \vec{e}_x, \vec{e}_y, \vec{e}_z\right\}$. Since

$$v_x = \dot{x}_C \cos\psi + \dot{y}_C \sin\psi, v_y = -\dot{x}_C \sin\psi + \dot{y}_C \cos\psi, \tag{22}$$

the kinematic constraint equations (7)–(10) become

$$v_x \cos\delta_1 \ - v_y \sin\delta_1 - \dot{\psi}\,(l \cos\delta_1 + \rho_2 \sin\delta_1) = R\dot{\varphi}_1 \cos\delta_1, \tag{23}$$

$$v_x \cos\delta_2 + v_y \sin\delta_2 + \dot{\psi}\,(l \cos\delta_2 + \rho_2 \sin\delta_2) = R\dot{\varphi}_2 \cos\delta_2, \tag{24}$$

$$v_x \cos\delta_3 + v_y \sin\delta_3 - \dot{\psi}\,(l \cos\delta_3 + \rho_1 \sin\delta_3) = R\dot{\varphi}_3 \cos\delta_3, \tag{25}$$

$$v_x \cos\delta_4 - v_y \sin\delta_4 + \dot{\psi}\,(l \cos\delta_4 + \rho_1 \sin\delta_4) = R\dot{\varphi}_4 \cos\delta_4. \tag{26}$$

In the matrix form, Eqs. (23)–(26) are represented as follows:

$$J \cdot \vec{v}_C = \vec{v}_W,\ \vec{v}_C = \left(v_x, v_y, \dot{\psi}\right)^T,\ \vec{v}_W = (R\dot{\varphi}_1, R\dot{\varphi}_2, R\dot{\varphi}_3, R\dot{\varphi}_4)^T, \tag{27}$$

$$J = \begin{pmatrix} \cos\delta_1 & -\sin\delta_1 & -(l\cos\delta_1 + \rho_2 \sin\delta_1) \\ \cos\delta_2 & \sin\delta_2 & l\cos\delta_2 + \rho_2 \sin\delta_2 \\ \cos\delta_3 & \sin\delta_3 & -(l\cos\delta_3 + \rho_1 \sin\delta_3) \\ \cos\delta_4 & -\sin\delta_4 & l\cos\delta_4 + \rho_1 \sin\delta_4 \end{pmatrix}. \tag{28}$$

Following [9], we solve the matrix equation (27) for the vector $\vec{v}_C$ using the pseudo-inverse matrix J^+ to obtain

$$\vec{v}_C = J^+ \cdot \vec{v}_W,\ J^+ = \left(J^T \cdot J\right)^{-1} \cdot J^T. \tag{29}$$

The pseudo-inverse matrix gives an approximate solution of the matrix equation (27) that minimizes the Euclidean norm $\left\| J \cdot \vec{v}_C - \vec{v}_W \right\|$ of the residual.

In the particular case, for which all δ_i $(i = 1, \ldots, 4)$ are equal to $\pi/4$ and $\rho_1 = \rho_2 = \rho$, the solution (29) becomes

$$v_x = \frac{R}{4}\left(\dot{\varphi}_1 + \dot{\varphi}_2 + \dot{\varphi}_3 + \dot{\varphi}_4\right), \tag{30}$$

$$v_y = -\ \frac{R}{4}\left(\dot{\varphi}_1 - \dot{\varphi}_2 - \dot{\varphi}_3 + \dot{\varphi}_4\right) \tag{31}$$

$$\dot{\psi} = -\ \frac{R}{4\,(l+\rho)}\left(\dot{\varphi}_1 - \dot{\varphi}_2 + \dot{\varphi}_3 - \dot{\varphi}_4\right). \tag{32}$$

The expression (15) for the kinetic energy becomes

$$T = \frac{1}{2}m\ \left(v_x^2 + v_y^2\right) + \frac{1}{2}J_C\,\dot{\psi}^2 + \frac{1}{2}J_1\left(\dot{\varphi}_1^2 + \dot{\varphi}_2^2 + \dot{\varphi}_3^2 + \dot{\varphi}_4^2\right). \tag{33}$$

Substituting relations (30)–(32) into expression (33) yields

$$\begin{aligned} T = {} & \frac{1}{2}\left[m\frac{R^2}{8} + \frac{R^2}{16(\rho+l)^2}J_C + J_1\right]\left(\dot{\varphi}_1^2 + \dot{\varphi}_2^2 + \dot{\varphi}_3^2 + \dot{\varphi}_4^2\right) \\ & + \left[m\frac{R^2}{8} - \frac{R^2}{16(\rho+l)^2}J_C\right]\left(\dot{\varphi}_1\,\dot{\varphi}_4 + \dot{\varphi}_2\dot{\varphi}_3\right) \\ & - \frac{R^2}{16(\rho+l)^2}J_C\left(\dot{\varphi}_1\dot{\varphi}_2 - \dot{\varphi}_1\,\dot{\varphi}_3 - \dot{\varphi}_2\dot{\varphi}_4 + \dot{\varphi}_3\,\dot{\varphi}_4\right). \end{aligned} \tag{34}$$

For this kinetic energy, write out Lagrange's equations of the second kind for 4 variables $q^i = \varphi_i$

$$\frac{d}{d\,t}\frac{\partial T}{\partial \dot{\varphi}_i} - \frac{\partial T}{\partial \varphi_i} = M_i,\ i = 1,\ldots,4. \tag{35}$$

Notice that this procedure is valid only for systems subject to holonomic constraints. By differentiating Eqs. (30)–(32) with respect to time and substituting into the resulting relations the expressions for $\ddot{\varphi}_i$ found from Eq. (35) we obtain the equations for the coordinates x_C and y_C of the center of mass and the angle ψ of rotation of the vehicle:

$$\begin{aligned} \left(m + \frac{4J_1}{R^2}\right)\left(\ddot{x}_C + \dot{y}_C\dot{\psi}\right) = {} & \frac{1}{R}\left[(M_1 + M_2 + M_3 + M_4)\cos\psi\right. \\ & \left. + (M_1 - M_2 - M_3 + M_4)\sin\psi\right], \end{aligned} \tag{36}$$

$$\begin{aligned} \left(m + \frac{4J_1}{R^2}\right)\left(\ddot{y}_C - \dot{x}_C\dot{\psi}\right) = {} & \frac{1}{R}\left[(M_1 + M_2 + M_3 + M_4)\sin\psi\right. \\ & \left. - (M_1 - M_2 - M_3 + M_4)\cos\psi\right], \end{aligned} \tag{37}$$

$$\left[\left(J_C + \frac{4J_1}{R^2}(l+\rho)^2\right]\ddot{\psi} = -\frac{l+\rho}{R}\left(M_1 - M_2 + M_3 - M_4\right). \tag{38}$$

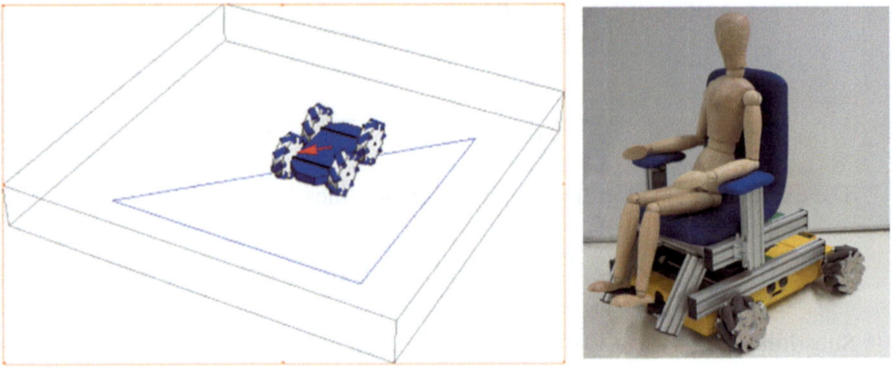

Fig. 3 Simulation of the vehicle motion using Mathematica® (realized by B. Adamov) (*left*) and the prototype of a vehicle with four mecanum wheels (*right*)

4 Conclusions

Equations of motion for a vehicle with four mecanum wheels were derived proceeding from a non-holonomic model for mecanum wheels. By comparing the non-holonomic model with an approximate model used in robotics we established that both models lead to the same result in the particular cases for which the vehicle either moves translationally or rotates about its center of mass. In both these cases, the constraints become holonomic. Based on the model considered in this study, an experimental verification of simulation results (Fig. 3 left) was realized, using a prototype of a mobile robot with four mecanum wheels (Fig. 3, right).

Acknowledgments The authors thank Boris Adamov from Moscow State Lomonosov University for the permission to publish the simulation results. The authors would also like to thank Siegfried Oberthür and Tobias Kästner for the practical realization of the prototype. This study was supported by the Development Bank of Thuringia and the Thuringian Ministry of Economic Affairs with funds of the European Social Fund (ESF) under grant 2011 FGR 0127.

References

1. Borisov, A.V., Kilin, A.A., Mamaev, I.S.: An omni-wheel vehicle on a plane and a sphere. Rus. J. Nonlin. Dyn. **7**(4), 785–801 (2011) (Russian)
2. Campion, G., Bastin, G., D'Andrea-Novel, B.: Structural properties and classification of kinematic and dynamic models of wheeled mobile robots. IEEE Trans. Robot. Automat. **12**(1), 47–62 (1996)
3. Dickerson, S.L., Lapin, B.D.: Control of an omni-directional robotic vehicle with mecanum wheels. In: National Telesystems Conference Proceedings, Atlanta, USA, pp. 323–328. March 26–27, 1991

4. Han, K.-L., Choi, O.K., Lee, I., Choi, S.: Design and control of an omni-directional mobile robot for mobile haptic interface. In: International Conference on Control, Automation and Systems, COEX, Seoul, pp. 1290–1295. Oct. 14–17, 2008
5. Ilon, B.E.: Wheels for a course stable selfpropelling vehicle movable in any desired direction on the ground or some other base. US Patent 3,876,255, 1975
6. Martynenko, Y.G.: Stability of steady motions of a mobile robot with roller-carrying wheels and a displaced centre of mass. J. Appl. Math. Mech. **74**(4), 436–442 (2010)
7. Muir, P.F., Neumann, C.P.: Kinematic Modeling for Feedback Control of an Omnidirectional Wheeled Mobile Robot, Autonomous Robot Vehicles, pp. 25–31. Springer, NewYork (1990)
8. Tsai, C.-C., Tai, F.-C., Lee, Y.-R.: Motion controller design and embedded realization for mecanum wheeled omnidirectional robots. In: Proceedings of the 8th World Congress on Intelligent Control and Automation, Taipei, Taiwan, pp. 546–551. June 21-25, 2011.
9. Viboonchaicheep, P., Shimanda, A., Kosaka, Y.: Position rectification control for mecanum wheeled omni-directional vehicles. In: 29th Annual Conference of IEEE Industrial Electronics Society, pp. 854–859, 2006
10. Wampfler, G., Salecker, M., Wittenburg, J.: Kinematics, dynamics, and control of omnidirectional vehicles with mecanum wheels. Mech. Base. Des. Struct. Mach. **17**(2), 165–177 (1989)
11. Zimmermann, K., Zeidis, I., Behn, C.: Mechanics of Terrestrial Locomotion, With a Focus on Nonpedal Motion Systems. Heidelberg, Springer (2009)

Reliability Analysis of the Dynamics of a Horizontal Drill-String

T.G. Ritto and Rubens Sampaio

Abstract Nowadays directional drilling, such as horizontal drilling, is very common. This chapter deals with the reliability analysis of a horizontal drill-string dynamics. A stochastic probabilistic computational model is considered, where a bar model (tension/compression) is used for the drill-string, and the system is discretized by means of the finite element method. The frictional forces between the column and the borehole are uncertain and relevant to the dynamic behavior of the structure. Therefore, a stochastic model is used for the frictional coefficient: a random field with exponential autocorrelation function. Because a probabilistic approach is employed, the reliability is measured in terms of the probability of the system not perform as it should, i.e., exceed stress limit, start a crack, or do not achieve a target efficiency. For the computation of the failure probability, importance sampling is applied to avoid too many stochastic simulations, and the stochastic system is analyzed for different scenarios.

1 Introduction

A drill-string is a slender structure used to drill rocks in search of oil and gas. Basically, it is composed of thin tubes (drill-pipes) and thicker tubes (drill-collars) with a bit fixed at the drilling end. In the beginning of oil exploration there were mainly vertical wells, but nowadays directional (horizontal in special) drilling is very common [1, 22]. Concerning the bit-rock interaction, Perneder et al. [12] proposed bit-rock interface laws in directional drilling, and, concerning a curved

T.G. Ritto (✉)
Federal University of Rio de Janeiro, Department of Mechanical Engineering, Centro de Tecnologia, Ilha do Fundão, 21945-970, Rio de Janeiro, Brazil
e-mail: tritto@mecanica.ufrj.br

R. Sampaio
PUB-Rio, Mechanical Engineering Department, Rua Marquês de São Vicente, 225, 22451-900, Rio de Janeiro, Brazil
e-mail: rsampaio@puc-rio.br

J. Awrejcewicz (ed.), *Applied Non-Linear Dynamical Systems*, Springer Proceedings in Mathematics & Statistics 93, DOI 10.1007/978-3-319-08266-0_20

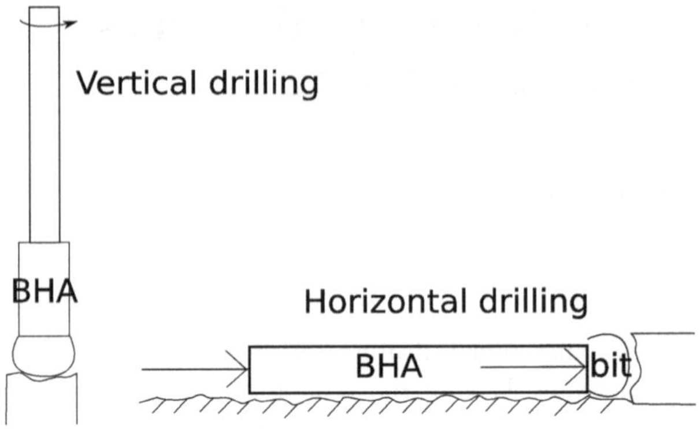

Fig. 1 Sketch of a vertical and a horizontal drilling

drill-string, Hu et al. [5] used curved-beam elements to model the drill-string directional drilling. Figure 1 sketches a vertical and a horizontal drilling, where BHA means Bottom Hole Assembly.

The safety of structures might be assessed if uncertainties are taken into account, and there are many ways to take into account uncertainties in a dynamical system, such as stochastic models and fuzzy randomness [9, 10]. This chapter describes stochastic tools that are used to assess the reliability of the structure [2, 6, 11, 19]. Since the number of Monte Carlo simulations to assess small failure probabilities might be big, importance sampling is applied to diminish the simulations required for a fixed precision.

There are many works in the literature concerned with the modeling of the drill-string dynamics in a vertical well [3, 4, 7, 13, 21], while some works have been devoted to the stochastic analysis of the drill-string dynamics [14, 16, 18, 20]. There are few articles in the literature dealing with the horizontal drill-string dynamics. Ritto et al. [17] use a simplified model for the system and propose a stochastic model for the frictional coefficient: a random field with exponential autocorrelation function. The present chapter starts from this computational model to propose a reliability analysis.

The drill-string is modeled using a bar model (tension/compression) and is discretized by means of the finite element method. An oscillatory force (due to mud motor) is imposed on the system and there is a bit-rock interaction as the column moves forward. A failure occurs if, for instance, the stress limit is exceeded or a crack is initiated. The present chapter considers a failure if the efficiency of the system (defined by output/input power ratio) is below a target value. This result can further be used for robust optimization [15].

The article is organized as follows. The deterministic and stochastic system are presented in Sects. 2 and 3. There, the equations of the system and of the stochastic model are presented. Then, the efficiency measure is defined in Sect. 4, where the measure of the efficiency relates output and input power. The proposed reliability

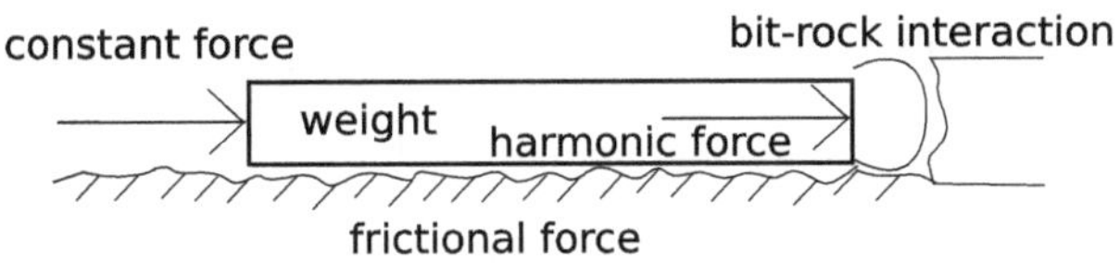

Fig. 2 Sketch of the system analyzed

analysis, which takes into account the efficiency of the system, is discussed in Sect. 5, and the numerical results are shown in Sect. 6. Finally, the concluding remarks are made in Sect. 7.

2 Deterministic System

The present model [17] considers a bar with axial vibration only. A sketch of the system analyzed is shown in Fig. 2. Weight, friction force, and bit-rock interaction force are some of the forces taken into account in the modeling.

Only a part of the horizontal drill-string, called Bottom Hole Assembly (BHA), is considered, and the equation of motion is given by

$$\rho A \frac{\partial^2 u(x,t)}{\partial t^2} - EA \frac{\partial^2 u(x,t)}{\partial x^2} = f_{\mathrm{sta}}(x,t) + f_{\mathrm{har}}(x,t) + f_{\mathrm{bit}}(\dot{u}(x,t)) + f_{\mathrm{fric}}(\dot{u}(x,t)), \tag{1}$$

where u is the axial displacement, ρ is the mass density of the material of the column, A is the cross-sectional area, and E is the elasticity modulus. The space and time variables are $x \in [0, L]$, $t \in [0, T]$, where L is the length of the structure and T is the duration of the time analysis. The right-hand side of the equation presents the forces per unit length acting on the system. f_{sta} is related to the static force imposed at $x = 0$, and f_{har} is the harmonic force imposed by the mud motor at $x = L$. The force related to the friction field (Coulomb friction model) is given by

$$f_{\mathrm{fric}}(\dot{u}(x,t)) = -\mu(x)(\rho A)g \; \mathrm{sgn}(\dot{u}(x,t)), \tag{2}$$

where μ is the friction coefficient and sgn is the sign function. And the force related to the bit-rock interaction f_{bit} is given by

$$\begin{cases} f_{\mathrm{bit}}(\dot{u}(x,t)) = [c_1 \exp(-c_2 \dot{u}(x,t)) - c_1]\, \delta(x-L) \\ f_{\mathrm{bit}}(\dot{u}(x,t)) = 0 \quad \text{for} \quad \dot{u}(L,t) \le 0, \end{cases} \tag{3}$$

where the first expression is used if $\dot{u}(L,t) > 0$, c_1 and c_2 are the two constants of the bit-rock interaction, and this force is applied at $x = L$. The discretized equations of the deterministic system, using linear shape functions, are written as

$$\mathbf{M\ddot{u}}(t) + \mathbf{C\dot{u}}(t) + \mathbf{Ku}(t) = \mathbf{f}_{\text{sta}} + \mathbf{f}_{\text{har}}(t) + \mathbf{f}_{\text{bit}}(\dot{\mathbf{u}}(t)) + \mathbf{f}_{\text{fric}}(\dot{\mathbf{u}}(t)), \tag{4}$$

with the appropriate boundary and initial conditions. In the above equation, $\mathbf{u}$ is the displacement vector, $\mathbf{M}$ is the mass matrix, and $\mathbf{K}$ is the stiffness matrix. The proportional damping matrix $\mathbf{C} = \alpha \mathbf{M}$ (α is a positive constant) is added a posteriori to the computational model. The force vectors $\mathbf{f}_{\text{sta}}$, $\mathbf{f}_{\text{har}}$, $\mathbf{f}_{\text{bit}}$, and $\mathbf{f}_{\text{fric}}$ are related to the static, harmonic, bit-rock interaction, and frictional forces, respectively. The force related to the bit mass is included in the matrix $\mathbf{M}$, since it depends linearly on the acceleration. More details of the model can be found in [17].

3 Stochastic Systems

It is assumed a random frictional coefficient field because of the difficulty to model the friction details. Thus, the random field $\{\nu(x) : x \in [0, L]\}$ is defined as a collection of real-valued random variables from a probability space $(\Omega, \mathcal{F}, P)$, where Ω is the sample space, $\mathcal{F}$ is the σ-algebra, and P is the probability measure.

It is also assumed that ν is a stationary truncated Gaussian random field on $[0, L]$ with exponential autocorrelation function

$$R(x_1, x_2) = \sigma^2 \exp\left(-\frac{|x_2 - x_1|}{b}\right), \tag{5}$$

where b is the correlation length, which measures the decay of the autocorrelation function. The stochastic field is truncated such that $0 \leq \nu(x) \leq 1$.

Finally, it is assumed that the truncated Gaussian field is expanded with Karhunen-Loève expansion [8] using standard Gaussian random variables. As long as the probability of having $\nu(x) \notin [0, 1]$ is very low, this approximation will be good (which is the case of the present analysis):

$$\nu(x, \xi) = \underline{\mu}(x) + \sum_{k=1}^{N} \sqrt{\lambda_k} Z_k(\xi) \phi_k(x), \tag{6}$$

where $\underline{\mu}$ is the mean value of the frictional coefficient, λ_k and ϕ_k are the kth eigenvalue and kth eigenvector of the autocorrelation function R, Z_k are independent standard Gaussian random variables, and N defines the precision of this representation (as N increases the representation gets better). The random frictional force (per unit length) is written as

$$F_{\text{fric}}(\dot{u}(x,t), \xi) = -\nu(x, \xi)(\rho A) g \operatorname{sgn}(\dot{u}(x,t)). \tag{7}$$

And the stochastic system is written as

$$\mathbf{M\ddot{U}}(t,\xi) + \mathbf{C\dot{U}}(t,\xi) + \mathbf{KU}(t,\xi) = \mathbf{f}_{\text{sta}} + \mathbf{f}_{\text{har}}(t) + \mathbf{f}_{\text{bit}}(\dot{\mathbf{U}}(t,\xi)) + \mathbf{F}_{\text{fric}}(\dot{\mathbf{U}}(t,\xi), \xi) \tag{8}$$

where $\mathbf{U}$ is the random response and $\mathbf{F}_{\text{fric}}$ is the random vector related to the frictional random field. This is a nonlinear system of equations because the bit-rock interaction is nonlinear and because the frictional force is discontinuous (if the speed at a point changes sign, the friction force of this point also changes sign).

4 Efficiency Measure

The efficiency of the horizontal drilling is analyzed in the following way. We establish a measure that is given by the ratio between output and input power; if this parameter increases, the efficiency of the process also increases. The input and output power are given by

$$p_{\text{in}}(t) = f_{\text{sta}}\dot{u}(0,t) + f_{\text{har}}(t)\dot{u}(L,t) \tag{9}$$

and

$$p_{\text{out}}(t) = f_{\text{bit}}(t)\dot{u}(L,t)\,. \tag{10}$$

And the average power is given by

$$\begin{aligned} \overline{p}_{\text{in}} &= \frac{1}{t_1 - t_0}\int_{t_0}^{t_1} (f_{\text{sta}}\dot{u}(0,t) + f_{\text{har}}(t)\dot{u}(L,t))\text{d}t \\ \overline{p}_{\text{out}} &= \frac{1}{t_1 - t_0}\int_{t_0}^{t_1} f_{\text{bit}}(t)\dot{u}(L,t)\text{d}t\,. \end{aligned} \tag{11}$$

Let us define the ratio between the output and input power in the time interval (t_0, t_1) by $y = \overline{p}_{\text{out}}/\overline{p}_{\text{in}}$. When the stochastic system is considered, this ratio becomes a random variable Y.

5 Reliability Analysis

The reliability analysis aims to evaluate the probability of failure, i.e., the probability that the system response does not satisfy a performance criterion. The functional relationship between the performance criterion and the random variables can be expressed as a performance function as follows:

$$Z = g(\mathbf{X})\,, \tag{12}$$

where $\mathbf{X}$ is a vector of random variables. In the present analysis, the elements of $\mathbf{X}$ are the random variables related to the discretized random field presented in Sect. 3.

The failure surface or the limit state is defined when $Z = 0$, and the failure occurs when $Z < 0$. The probability of failure is given by (Schueller et al., 2004)

$$PF = \int_{g(\mathbf{X})<0} f_{\mathbf{X}}(\mathbf{x})d\mathbf{x}\,, \tag{13}$$

in which $f_{\mathbf{X}}$ is the joint probability density function of $\mathbf{X}$. The result of the integration shown in Eq. 13 is the probability of failure of the system under analysis. In general, the integration above is not easily obtained. The standard Monte Carlo method is a non-intrusive and well-suitable method to calculate the failure probability, where the failure probability is calculated from the sum of all the realizations for which a failure occurs among N runs, as follows:

$$\widehat{PF} = \frac{1}{N}\sum_{i=1}^{N}\left(\mathbf{1}_{g(\mathbf{X})<0}\right)\,, \tag{14}$$

where $\widehat{PF}$ is an unbiased estimator of PF, Eq. 13, and $\mathbf{1}_{g(\mathbf{X})<0}$ is an indicator that is equal to 1 when $g(\mathbf{X}) < 0$ and 0 otherwise. Eq. 14 means that the probability of failure can be estimated as a ratio of the number of runs that lead to failure and the number of the total runs performed, $\widehat{PF} = n_f/n_s$. By the Central Limit Theorem, $\widehat{PF}$ has approximately a normal distribution $Normal(PF, N^{-1}Var(PF))$ for large N, where $Var(PF)$ can be estimated via the sample variance unbiased estimator, $S^2 = (1/(N-1))\sum_{i=1}^{N}(PF_i - \widehat{PF})$, with which we can estimate a confidence interval. When PF is small, a large number of Monte Carlo simulations is necessary for convergence. For instance, if $PF = 10^{-3}$ an average of 1000 simulations is necessary for a failure to happen.

An alternative to Monte Carlo is a technique called importance sampling, known as the fundamental variance reduction techniques. It involves choosing a sampling distribution $f^*_{\mathbf{X}}$ that favors important samples, improving the Monte Carlo convergence. Equation 14 is modified as follows:

$$\begin{aligned} PF &= \int_{g(\mathbf{X})<0} f_{\mathbf{X}}(\mathbf{x})d\mathbf{x} = \\ &= \int_{g(\mathbf{X})<0} \frac{f_{\mathbf{X}}(\mathbf{x})}{f^*_{\mathbf{X}}(\mathbf{x})} f^*_{\mathbf{X}}(\mathbf{x})d\mathbf{x} = \mathbf{E}_{f^*}\{W_{\mathbf{X}}\}\,, \end{aligned} \tag{15}$$

where $W_{\mathbf{X}} = f_{\mathbf{X}}/f^*_{\mathbf{X}}$ is called the likelihood ratio. The optimal sampling distribution can be calculated, but it is necessary to know a priori what we want to estimate. The probability of failure defined in Eq. 15 can be estimated using the standard Monte Carlo sampling, but weighting each failure run with $W_{\mathbf{X}}$, as follows:

$$\widehat{PF} = \frac{1}{N}\sum_{i=1}^{N}\left(W_{\mathbf{X}}\,.\,\mathbf{1}_{g(\mathbf{X})<0}\right)\,. \tag{16}$$

6 Numerical Results

The values of the parameters used in the simulations are the following: $E = 210$ GPa, $\rho = 7850$ km/m^3, $g = 9.81$ m/s^2, $D_\mathrm{i} = 0.10$ m (inner diameter), $D_\mathrm{o} = 0.15$ m (outer diameter), L/D_o=400, $m_\mathrm{bit} = 20$ kg, $\alpha = 0.2$, $c_1 = 1.4\times 10^3$ N, $c_2 = 400$, $\mu = 0.1$, $\sigma = 0.1 \times \mu$, $b = 10$ and $\omega_f = 100 \times 2\pi/60$ rad/s, $t \in [0, 10]$s, $\Delta t = 0.0001$s, $N = 100$, $f_\mathrm{sta} = 5500$N, and $F_o = 550$N where m_bit is the mass attached at $x = L$ to model the bit. Free-free boundary conditions are considered, such that the column can rigidly move to the right. The system is discretized with 40 finite elements and the system of differential equations is integrated applying the Newmark integration scheme assuming zero initial condition.

First, the deterministic responses are analyzed, to give an idea of the system dynamics. Figure 3 shows the bit speed evolving in time. The mean speed is around 20 meters per hour, but it oscillates from 4 to 34 meters per hour. Figure 4 shows the input power divided by the output power; this ratio also oscillates in time. For the present case, the efficiency is about 15%. Since there are losses not considered in the model, such as lateral and torsional vibrations, this value can be thought of as an upper bound for the efficiency.

Now, the stochastic analysis. Figure 5 shows an approximation of the probability density of the output/input power ratio, called Y. It presents a bimodal distribution, where the higher peek is related to situations where the rock is so easy to drill that the bit speed increases with time.

Figure 6 shows 1,000 Monte Carlo (MC) realizations of Y and a target efficiency of 10%. It can be observed that few realizations are under the target value, which means that a lot of Monte Carlo simulations will be needed to estimate the probability that Y is under 10%. This probability is the failure probability, or the risk of the operation.

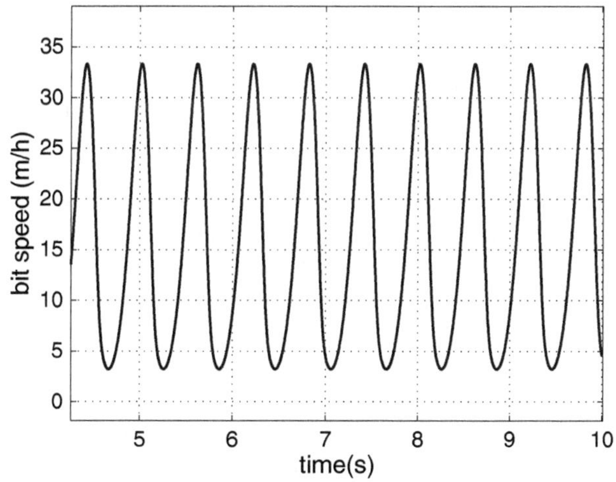

Fig. 3 Bit speed

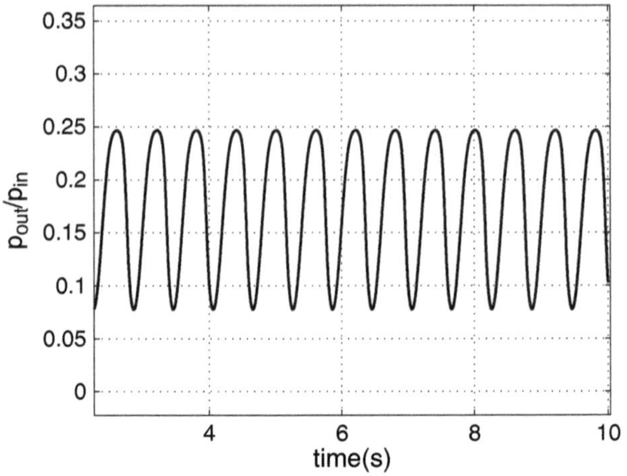

Fig. 4 Output power divided by the input power

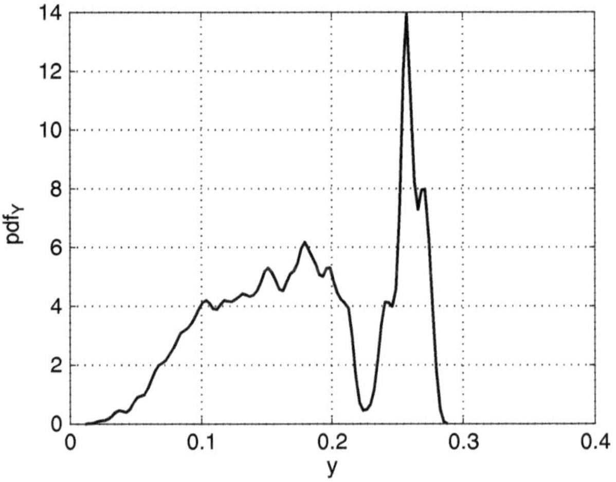

Fig. 5 Approximation of the probability density function of random variable Y

Since the present application has only one random field with constant mean taken into account, a simple way to perform importance sampling is to use as sampling probability density function $f^*_{\mathbf{X}}$ the original distribution with the mean shifted, such that failures occur more often, say around 50%. A quick test with few Monte Carlo simulations can be done to define the size of the shift. Figure 7 shows 1,000 importance sampling (IS) realizations, where the proposed probability density function of the random field has a mean value multiplied by 1.1 and same covariance function. It is noted that many more points are under 10% now, improving the convergence of the estimate. Figure 7 shows the convergence of the failure probability for different target values, using Monte Carlo simulations.

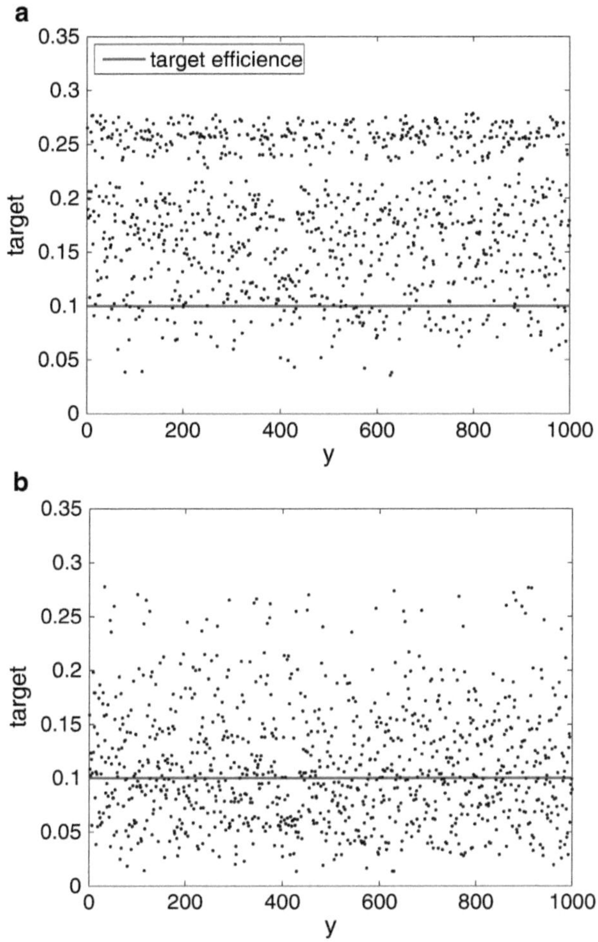

Fig. 6 Samples of the MC strategy (**a**) and of the IS strategy (**b**)

For target values of 15% or 20%, the convergence is reasonable (using 1000 simulations), but for target value of 5%, the convergence is unacceptable.

To avoid too many simulations, the failure probability is estimated using conveniently MC or IS, such that the convergence is reasonable for 1000 realizations. For lower target values, the proposed IS probability density function of the random field has mean value multiplied by 1.2, to guarantee a reasonable convergence of the estimate.

Finally, we want to propose the construction of a map that takes into account efficiency and probability of failure thresholds to define a safe region to operate. Figure 8 shows this map for two different cases. An upper bound of 5 % is assumed for the value of failure probability, and a lower bound of 5 % is assumed for the value

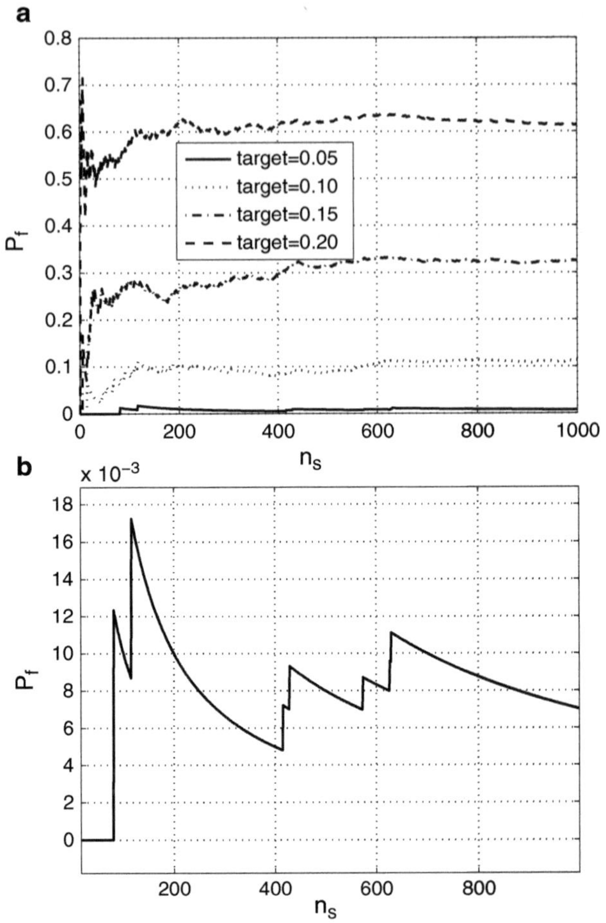

Fig. 7 Convergence of $\hat{P}_f$ for different target values (**a**) and zoom of the lower curve (**b**)

of the efficiency. These limits are in dashed line in Fig. 8. It should be noticed that when the target value increases, the failure probability also increases, meaning that if we impose a higher target efficiency value, the risk of not performed as desired increases.

Figure 8a shows a comparison of the system with 50RPM and 100RPM excitation frequency at the bit. It is noted that the response is almost the same. This is due to the fact that these excitation frequencies are far away from the natural frequencies of the structure and, also, the computational model does not consider the rotation of the bit. If torsional vibrations are included in the model, we would expect, in a regular operation, an increase in the efficiency of the system.

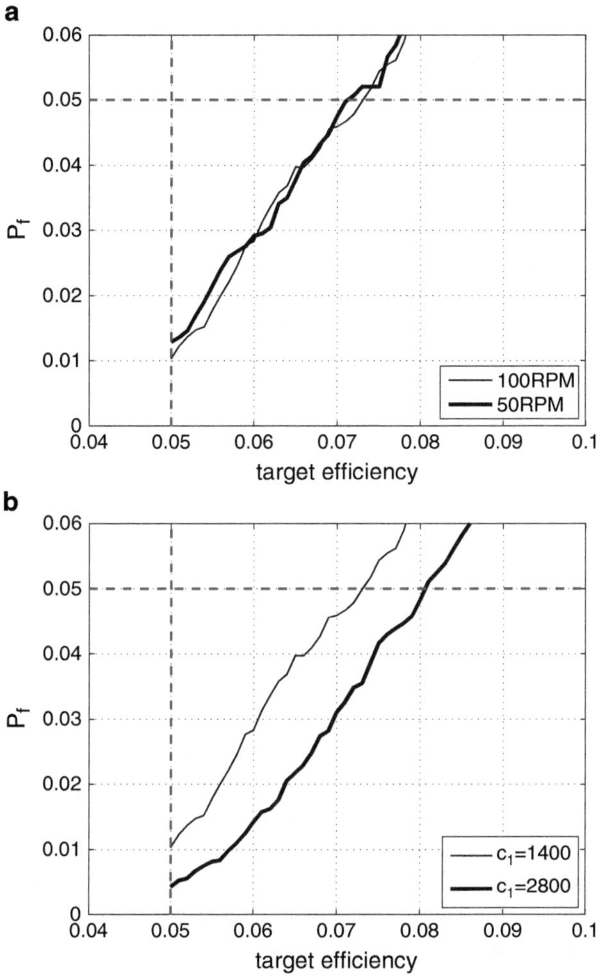

Fig. 8 Map with the safe region, taking into account efficiency and probability of failure thresholds. (**a**) Different frequency of excitation, (**b**) different value of c_1 (related to the bit-rock interaction)

Figure 8b shows the a comparison of the system with two different values of the bit-rock interaction constant $c_1 = 1.4 \times 10^3\,\text{N}$ and $c_1 = 2.8 \times 10^3\,\text{N}$. It is noted that the system performs better when c_1 increases. This is explained by the fact that the bit force is significantly bigger for the same bit speed, when c_1 is increased, as shown in Fig. 9.

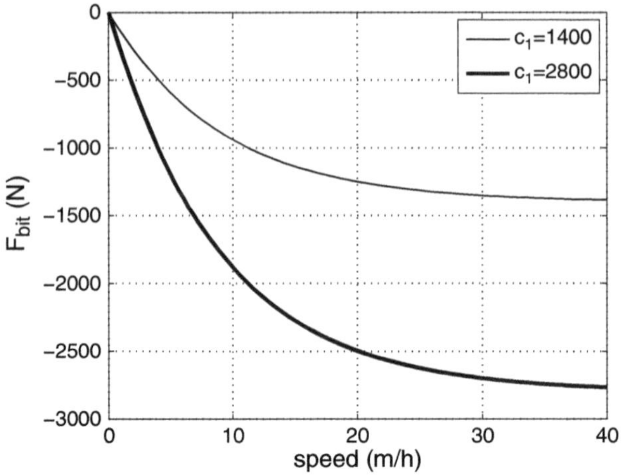

Fig. 9 Bit-rock interaction model for two values of c_1

7 Concluding Remarks

This chapter considered a simplified model to analyze the dynamics of a horizontal drill-string. The focus of the present analysis was the reliability of the operation, which was defined by the probability of not achieving a target efficiency. The efficiency measured was given by the mean of the input/output ratio.

To compute the probability of failure, importance sampling strategy was used together with Monte Carlo simulations, such that a reasonable convergence of the estimate is achieved with a small number of numerical simulations.

When the target values increase, the failure probability also increases, meaning that if we impose a higher target value, the risk of do not perform as desired increases. A map that takes into account efficiency and probability of failure thresholds was proposed to obtain a safe region to operate.

Acknowledgements The authors are grateful to the financial support from the Brazilian agencies: CAPES, CNPq, and FAPERJ.

References

1. Allouche, E., Ariaratnam, S., Lueke, J.: Horizontal directional drilling profile of an emerging industry. J. Const. Eng. Manag. ASCE **26**, 68–76 (2000)
2. Au, S.K., Beck, J.L.: Important sampling in high dimensions. Struct. Saf. **25**(2), 139–163 (2003)
3. Christoforou, A.P., Yigit, A.S.: Dynamic modeling of rotating drillstrings with borehole interactions. J. Sound Vib. **206**(2), 243–260 (1997)

4. Divenyi S., Savi M.A., Wiercigroch M., Pavlovskaia E.: Drill-string vibration analysis using non-smooth dynamics approach. Nonlinear Dynam. **70**, 1017–1035 (2012)
5. Hu Y., Di Q., Zhu W., Chen Z., Wang W.: Dynamics characteristics analysis of drillstring in the ultra-deep well with spacial curved beam finite element. J. Petroleum Sci. Eng. **82–83** 166–173 (2012)
6. Jensen, H.A.: Reliability-based optimization of uncertain systems in structural dynamics. AIAA J. **40**(4), 731–738 (2002)
7. Khulief, Y.A., Al-Naser, H.: Finite element dynamic analysis of drillstrings. Finite Elem. Anal. Des. **41**, 1270–1288 (2005)
8. Loève, M.: Probability theory. Graduate Texts in Mathematics, Springer, USA (1977)
9. Moller, B., Graf, W., Beer, M.: Safety assessment of structures in view of fuzzy randomness. Comput. Struct. **81**(15), 1567–1582 (2003)
10. Moller, B., Beer, M.: Engineering computation under uncertainty-Capabilities of non-traditional models. Comput. Struct. **86**(10), 1024–1041 (2008)
11. Patelli, E., Pradlwarter, H.J., Schueller, G.I.: On multinormal integrals by Importance Sampling for parallel system reliability. Struct. Saf. **33**(1), 1–7 (2011)
12. Perneder, L., Detournay, E. Downton G.: Bit/rock interface laws in directional drilling. Int. J. Rock Mech. Min. Sci. **51**, 81–90 (2012)
13. Richard T., Germay C., Detournay, E.: A simplified model to explore the root cause of stick-slip vibrations in drilling systems with drag bits. J. Sound Vib. **305**, 43–456 (2007)
14. Ritto T.G., Soize C., Sampaio R.: Nonlinear dynamics of a drill-string with uncertain model of the bit-rock interaction. Inter. J. Non-Linear Mech. **44**(8), 865–876 (2009)
15. Ritto, T.G., Soize, C. Sampaio, R.: Robust optimization of the rate of penetration of a drill-string using a stochastic nonlinear dynamical model. Computat. Mech. **45**(5), 415–427 (2010)
16. Ritto, T.G., Sampaio, R.: Stochastic drill-string dynamics with uncertainty on the imposed speed and on the bit-rock interaction parameters. Int. J. Uncert. Quant. **2**(2), 111–124 (2012)
17. Ritto, T.G., Escalante, M.R., Sampaio, R., Rosales, M.B.: Drill-string horizontal dynamics with uncertainty on the frictional force. J. Sound Vib. **32**(1) 145–153 (2013)
18. Ritto, T.G., Sampaio, R.: Measuring the efficiency of vertical drill-strings: A vibration perspective. Mech. Res. Comm. **52**, 32–39 (2013)
19. Schueller, G.I., Pradlwarter, H.J., Koutsourelakis, P.S.: A critical appraisal of reliability estimation procedures for high dimensions. Probabilist Eng. Mech. **19**(4), 463–474 (2004)
20. Spanos, P.D., Chevallier, A.M., Politis, N.P.: Nonlinear stochastic drill-string vibrations. J. Vib. Acoust., Trans. ASME **124**(4), 512–518 (2002)
21. Tucker, R.W., Wang, C.: An integrated model for drill-string dynamics. J. Sound Vib. **224**(1), 123–165 (1999)
22. Willoughby D.A.: Horizontal Directional Drilling, Utility and Pipeline applications. Civil Engineering Series, McGraw-Hill, New York, USA (2005)

Transition in Oscillatory Behavior in Mouse Oocyte and Mouse Embryo Through Oscillatory Spherical Net Model of Mouse Zona Pellucida

Andjelka N. Hedrih

Abstract Zona pellucida, a 3D matrix that surrounds mammalian oocyte, dynamically changes its elasticity during the maturation and fertilization process. We consider fertilization as a biomechanical oscillatory phenomenon and hypothesized that mature oocyte and embryo are in different oscillatory states. Using an oscillatory spherical net model of mouse ZP [4], eigen circular frequencies of mouse oocyte and mouse embryo were calculated. Frequency analysis of circular frequencies of ZP under periodical external excitation force in the form of sperm cell impact was done for both states. To determine the conditions for dynamical absorption under impact of sperm cells on ZP of mouse oocyte and embryo, numerical analyses were done.

1 Introduction

Zona pellucida (ZP), the outermost surface of the oocyte, dynamically changes its elasticity during the maturation and fertilization process [1,6,7]. The purpose of this structure is to protect the oocyte, to work as high selective structure for high-quality spermatozoa, to select the "right one," to ensure polyspermy block, and to guide the embryo through the oviduct. The final fertilization process occurs at ZP. Considering fertilization process as a biomechanical phenomenon we hypothesized that mature oocyte [3,5] and embryo are in different oscillatory states. If they are in different oscillatory states, what are oscillatory properties of mouse embryo that do not allow penetration of other sperm cells through ZP in fertilized oocyte?

If there is only an initial perturbation by kinetic and potential energy given to oscillatory structures, only free vibrations of ZP appear. In this case material particles at the initial moment obtain the initial displacement measured from their equilibrium positions and initial velocities. In order for free oscillations to appear, it is enough that only one mass particle position is perturbed from its equilibrium

A.N. Hedrih (✉)
Department for Bio-medical science, State University of Novi Pazar,
Vuka Karadzica bb, 36 300 Novi Pazar, Serbia
e-mail: handjelka@hm.co.rs

J. Awrejcewicz (ed.), *Applied Non-Linear Dynamical Systems*, Springer Proceedings in Mathematics & Statistics 93, DOI 10.1007/978-3-319-08266-0_21

position or that only one mass particle at its equilibrium position obtains initial velocity. If we apply one or multifrequency external excitation forces to a ZP discrete net, it oscillates in multifrequency forced regime oscillations.

The aim of our study was to compare eigen circular frequencies of mouse oocyte and mouse embryo ZP as a property of the systems and to determine the conditions for dynamical absorption under impact of sperm cells on ZP of mouse oocyte and embryo through spherical surface net model of mZP [4].

2 Determination of the Eigen Circular Frequencies

Using spherical surface net model of mZP [4] and a method of discrete continuum [2] we considered that the system of ZP oscillatory net oscillates in a free regime after ovulation without the presence of spermatozoa. In a spherical surface net model of mZP [4] ZP is modeled as a one-layer network that envelops the oocyte/embryo that we supposed that is solid, elastic, and rigid. The network consists of orthogonal chains of material particles interconnect with elastic massless springs on a specific manner. Material particles represent the ZP glycoproteins. See Fig. 1a. To do a frequency analysis of a part of spherical net model and to determine a particular set of the eigen circular frequencies of mZP, we use the smallest part of the mZP oscillatory spherical surface model (see Fig. 1b) that still preserves the molar ratio of the mZP glycoproteins (ZP1:ZP2–ZP3 is 1:5).

We applied one or multifrequency external excitation forces to a one knot molecule *ZP* discrete net (third in the chain or ninth in this chain using symmetry

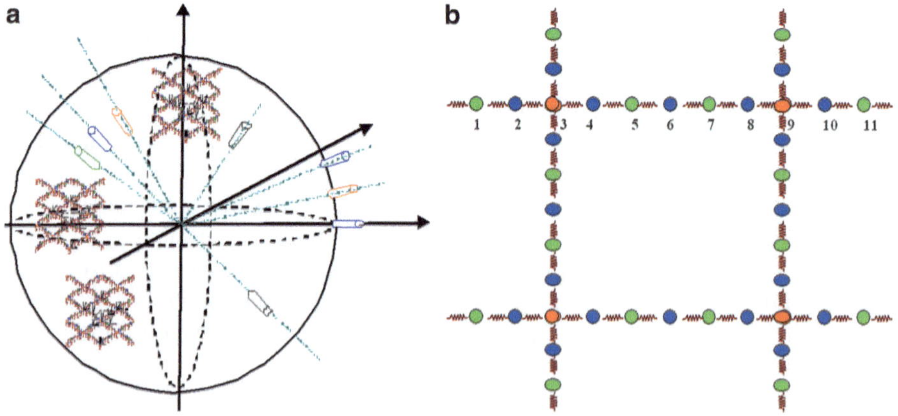

Fig. 1 (**a**) Model of *ZP* spherical surface that shows a radial direction of axis of constructive elements of the model—*ZP* proteins. *Axis* shows directions of movements of *ZP* proteins. Each *ZP* protein is connected to the sphere with elastic springs that can oscillate in radial direction. (**b**) Part of the *ZP* network on a part of the sphere (oocyte). *Orange* (*ZP1*), *blue* (*ZP2*), and *green* (*ZP3*) represent *ZP* proteins. The network is identical in both circular and meridian direction

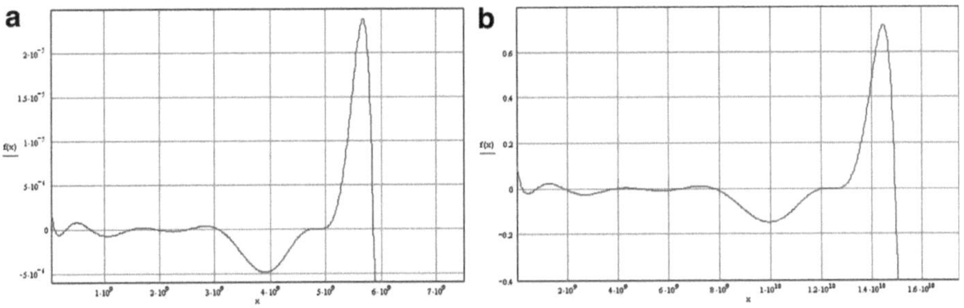

Fig. 2 Presentation of two characteristic branches of the eleven degree polynomial function graphs along square of circular frequencies ω^2: (**a**) before and (**b**) after fertilization of ZP for different material parameters (module of elasticity and masses particles)

in order of eleven mass particles in chain). Frequency equation that describes the oscillations of the mZP glycoproteins is in the form of chain with eleven material particles. The frequency equation is an eleven degree polynomial function along square of possible circular frequencies ω^2 and is in the form

$$f\left(\omega^2\right) = \left|\mathbf{C} - \omega^2\mathbf{A}\right| = 0. \tag{1}$$

C is matrix of coefficient of elasticity, and **A** is matrix of coefficient of mass inertia. For obtaining eigen circular frequencies we used graphical method and Mathcad software. See Fig. 2a, b. Relative molecular masses of the three mZP glycoproteins, Mr*ZP*1, 200,000 Da; Mr*ZP*2, 120,000 Da; and Mr*ZP*3, 83,000 Da transformed in kg, were used. The zeros (roots) in the frequency equations are squares of eigen circular frequencies ω_s^2. For considered nonhomogenous chains there are 11 squares of eigen circular frequencies: ω_s^2, $s = 1, 2, 3, 4, 5, 6, 7, 8, 9, 10, 11$.

For our numerical experiment we approximate that coefficient of elasticity is equal for all material particles and calculated it from the experimental data of Sun et al. [9] according to the formula

$$c = \frac{E\left(R^2 - r^2\right)\pi}{2R} \tag{2}$$

E is Young module of elasticity, R is half diameter of the mouse oocyte/embryo, and r is internal radius of oocyte or is half diameter of the oocyte/embryo minus approximate thickness of mZP.

For an oocyte: $E_o = 17.5$kPa; $2R_o = 56.2$ μm, average diameter of an oocyte from [9]; and $\delta_o = 4.8$ μm, approximate ZP thickness of the oocyte was used for calculating coefficient of elasticity. Its value is $c_o = 0.253$ N/m.

For mouse embryo: $E_e = 42.2$kPa; $2R_e = 61$ μm, average diameter of an embryo from [9]; and $\delta_e = 5.34$ μm, approximate ZP thickness of the embryo was used for calculating coefficient of elasticity. Its value is $c_e = 0.646$ N/m.

For obtaining eigen circular frequencies we used graphical method and Mathcad software tool and correction factor 10^6.

Using Matcad software tool and graphical method (See Fig. 2a) the obtained eigen circular frequencies of mouse oocyte ω_{os}, $s = 1, 2, 3, 4, 5, 6, 7, 8, 9, 10, 11$ for corresponding material parameters (module of elasticity and masses particles) are: ω_s^2, $s = 1, 2, 3, 4, 5, 6, 7, 8, 9, 10, 11$ before fertilization for corresponding material parameters (module of elasticity and masses particles) are: $\omega_{o1} = 9.318 \times 10^9$rad/s, $\omega_{o2} = 1.72 \times 10^{10}$ rad/s, $\omega_{o3} = 2.707 \times 10^{10}$ rad/s, $\omega_{o4} = 3.856 \times 10^{10}$ rad/s, $\omega_{o5} = 4.379 \times 10^{10}$rad/s, $\omega_{o6} = 5.011 \times 10^{10}$rad/s, $\omega_{o7} = 5.522 \times 10^{10}$rad/s, $\omega_{o8} = 6.887 \times 10^{10}$ rad/s, $\omega_{o9} = 6.952 \times 10^{10}$ rad/s, $\omega_{o10} = 7.024 \times 10^{10}$ rad/s, and $\omega_{o11} = 7.645 \times 10^{10}$ rad/s.

Using Matcad software tool and graphical method (See Fig. 2b) the obtained eigen circular frequencies of mouse embryo ω_{es}, $s = 1, 2, 3, 4, 5, 6, 7, 8, 9, 10, 11$ for corresponding material parameters (module of elasticity and masses particles) are: $s=1, 2, 3, 4, 5, 6, 7, 8, 9, 10, 11$ for corresponding material parameters (module of elasticity and masses particles) are: $\omega_{e1} = 1.484 \times 10^{10}$rad, $\omega_{e2} = 2.725 \times 10^{10}$ rad/s, $\omega_{e3} = 4.324 \times 10^{10}$ rad/s, $\omega_{e4} = 6.205 \times 10^{10}$ rad/s, $\omega_{e5} = 6.977 \times 10^{10}$rad/s, $\omega_{e6} = 8.005 \times 10^{10}$rad/s, $\omega_{e7} = 8.806 \times 10^{10}$rad/s, $\omega_{e8} = 1.096 \times 10^{11}$ rad/s, $\omega_{e9} = 1.193 \times 10^{11}$ rad/s, $\omega_{e10} = 1.127 \times 10^{11}$ rad/s, and $\omega_{e11} = 1.221\text{x}\ 10^{11}$ rad/s.

These two obtained sets of square of eigen circular frequencies of mZP ω_s^2, $s = 1, 2, 3, 4, 5, 6, 7, 8, 9, 10, 11$ before and after fertilization for corresponding material parameters (module of elasticity and masses particles) are also sets of resonant frequency squares of corresponding one frequency external excitation square of frequencies $\Omega_{rez,s}^2 = \omega_s^2$, $s = 1, 2, 3, 4, 5, 6, 7, 8, 9, 10, 11$.

3 Forced Oscillations

The system of ordinary differential equations of mass particles forced dynamics is in the following matrix form:

$$\mathbf{A}\{\ddot{x}\} + \mathbf{C}\{x\} = \{Q\} = \{F_0\}\cos\Omega t \tag{3}$$

Particular solution for kth mass particle in the chains, when one frequency external excitation is applied to the third mass particle in chain, is in the form

$$x_k\left(t, \Omega_3^2\right) = C_{(3)k}\left(\Omega_3^2\right)\cos\Omega_3 t \tag{4}$$

Amplitude $C_{(3)k}(\Omega_3^2)$ of particular solution for kth mass particle in chain forced vibration displacement under the one frequency external excitation applied to the third mass particle in chain is in the form

$$C_{(3)k}\left(\Omega_3^2\right) = \frac{\Delta_{(3)k}\left(\Omega_3^2\right)}{\Delta\left(\Omega_3^2\right)} \tag{5}$$

for $\Omega_3^2 \neq \omega_s^2$, $s = 1, 2, 3,, 4, 5, 6, 7, 8, 9, 10, 11$, and where determinant of the previous system is in the form

$$\Delta\left(\Omega_3^2\right) = \left|\mathbf{C} - \Omega_3^2\mathbf{A}\right| \neq 0 \tag{6}$$

and must be different from zero. Then the condition $\Omega_3^2 \neq \omega_s^2$, $s = 1, 2, 3, 4, 5, 6, 7, 8, 9, 10, 11$ have to be fulfilled.

For obtaining amplitudes $C_{(3)k}(\Omega_3^2)$ of particular solutions, it is necessary to find sub-determinant $\Delta_{(3)}(\Omega_3^2)$. Corresponding sub determinants $\Delta_{(3)}(\Omega_3^2)$ is possible to obtain by substituting corresponding 3-rd column in determinant of the system, by the column containing amplitude of the external excitation force F_{03}.

4 Conditions for Dynamical Absorption

Dynamical absorption is present in the systems with multiple degrees of freedom (in our case chain is system with 11 degrees of freedom) when external periodical force is applied to the system. Depending on the material properties and parameters of the oscillatory system structure (coefficient of elasticity, mass of material particles, external excitation frequencies, amplitudes) it is possible that under the influence of periodical external force one or more material particles don't oscillate in forced mode with external excitation frequency and that other mass particles are in this forced oscillatory mode. In the theory of oscillation [8] this phenomenon is known as dynamical absorption.

External excitation frequencies at which dynamical absorption occurs of mass particles on certain positions were read out from amplitude-frequency graphs for each mass particle in chain. For determine the amplitude of forced vibrations $C_k(\Omega_3^2)$, $k = 1, 2, 3, 4, 5, 6, 7, 8, 9, 10, 11$ for each mass particle, when third mass particle are loaded by external one frequency excitation with frequency Ω_3 and external periodical force $F_3 = F_{03}\cos\Omega_3 t$, the required condition is:

$$C_k\left(\Omega_3^2\right) = 0, \textit{for some of } k = 1, 2, 3, 4, 5, 6, 7, 8, 9, 10, 11 \textit{ when } \Delta\left(\Omega_3^2\right) \neq 0 \tag{7}$$

Graphs of amplitude-frequency stationary forced regimes for forced vibration of $C_3(\Omega_3^2)$ third, $C_6(\Omega_6^2)$ sixth, and $C_9(\Omega_3^2)$ ninth material particle in chain excited by external excitation $F_3 = F_{03}\cos\Omega_3 t$ force with amplitude F_{03} and frequency Ω_3 applied to third mass particle in chain in mZP net model of oocyte and embryo are given in Figs. 3, 4, and 5, respectively. When external excitation $F_3 = F_{03}\cos\Omega_3 t$ force with amplitude F_{03} and frequency Ω_3 is applied to third mass particle in chain, frequencies under which dynamical absorption occurs in forced oscillatory regimes in the third material particle oscillatory displacement in the chain are:

For mouse oocyte: 1.233×10^{10}rad/s, 2.49×10^{10}rad/s, 4.605×10^{10}rad/s, and 5.38×10^{10}rad/s.

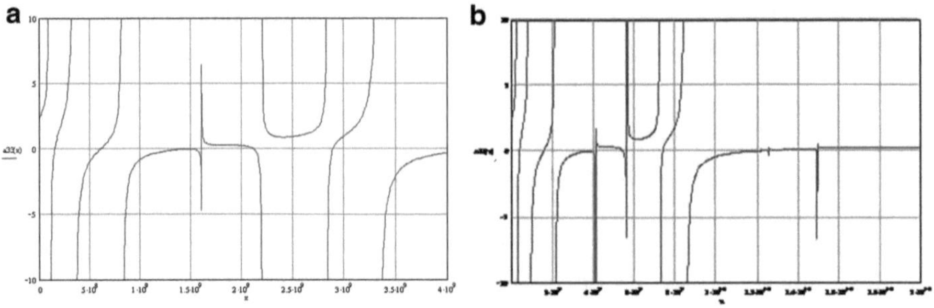

Fig. 3 Amplitude-frequency stationary forced regimes for forced oscillations of $C_3(\Omega_3^2)$ third material particle in chain excited by external excitation $F_3 = F_{03}\cos\Omega_3 t$ force with amplitude F_{03} and frequency Ω_3 applied to third mass particle in chain; $\mathrm{x} = \Omega^2$. for (**a**) oocyte, (**b**) embryo

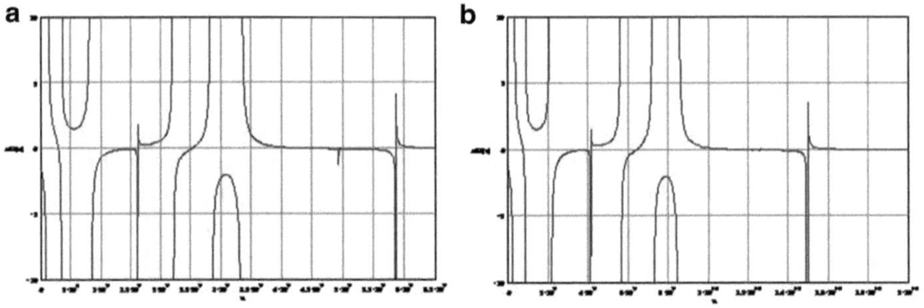

Fig. 4 Amplitude-frequency stationary forced regimes for forced oscillations of $C_3(\Omega_3^2)$ sixth material particle in chain excited by external excitation $F_3 = F_{03}\cos\Omega_3 t$ force with amplitude F_{03} and frequency Ω_3 applied to third mass particle in chain; $\mathrm{x} = \Omega^2$.for (**a**) oocyte, (**b**) embryo

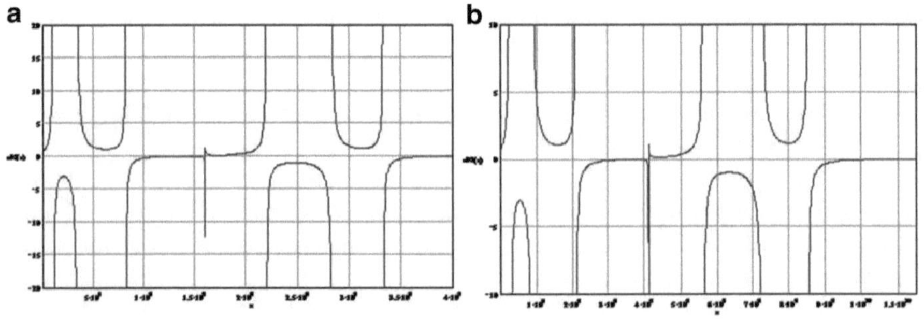

Fig. 5 Amplitude-frequency stationary forced regimes for forced oscillations of $C_3(\Omega_3^2)$ ninth material particle in chain excited by external excitation $F_3 = F_{03}\cos\Omega_3 t$ force with amplitude F_{03} and frequency Ω_3 applied to third mass particle in chain; $\mathrm{x} = \Omega^2$. for (**a**) oocyte, (**b**) embryo. It is visible that there is no dynamical absorption

And for the embryo: 2.145×10^{10}rad/s, 3.95×10^{10}rad/s, 7.335×10^{10}rad/s, 8.602×10^{10} rad/s, 1.22×10^{11}rad/s, and 1.118×10^{11}rad/s.

We can see that in the forced regime one mode of forced oscillatory displacement of this third mass particle does not exist at a number of external excitation frequencies—frequencies of dynamical absorptions upon that mass particle are loaded. Other mass particles are not directly loaded and are in forced oscillatory regimes at these frequencies of dynamical absorption of third loaded mass particle. This is a dynamical paradox that this mass particle is loaded by external single frequency force and that loaded mass particle is not in the forced oscillatory regime and that other mass particles in chain oscillate in forced regimes at these frequencies. Amplitude of external excitation has no influence on the appearance of dynamical absorption. Only external excitation frequency in relation with chain material properties (system structure properties like eigen circular frequencies, coefficients of elasticity/rigidity, and masses of material particles) influenced the appearance of dynamical absorption.

When external excitation $F_3 = F_{03} \cos \Omega_3 t$ force with amplitude F_{03} and frequency Ω_3 is applied to third mass particle in chain, frequencies under which dynamical absorption occurs for the sixth material particle in the chain are:

For mouse oocyte: 1.672×10^{10} rad/s and 1.596×10^{10} rad/s
For the embryo: 2.683×10^{10} rad/s and 8.655×10^{10} rad/s
Under these conditions there is no dynamical absorption on the ninth material particle in the chain either in oocyte or in embryo.

5 Conclusions

Due to limited length of the paper we did not analyze the forced oscillations of knot molecules in ZP net; we analyzed only the phenomenon of dynamical absorption for certain molecules in the model. We did the analysis on a part of spherical surface net of mZP in the system with finite number of degree of freedom. In the real system of ZP there are a lot of molecules and almost indefinite eigen circular frequencies.

Young module of elasticity of mouse ZP in mature oocyte (before fertilization) has 2.5 times higher value compared to ZP of mouse embryo (after fertilization) [9]. Sperm cells could not penetrate the ZP of embryo. From a biological point of view, after fertilization occurs polyspermy block, a phenomenon where new sperm cells could not attach to the ZP and the already attached have been rejected. This process is receptor and enzymatic mediated and includes repulsive electric charge between ZP and sperm cells.

We tried to perceive these events from the aspect of theory of oscillations. Sperm cells have impact on a ZP in the form of multifrequency external excitation forces. We applied external excitation to the one knot molecule *ZP* discrete net (third in the chain). According to the spherical surface net model of mouse ZP, our numerical analysis shows that eigen circular frequencies of mouse embryo ZP have higher values compared to eigen circular frequencies of mouse oocyte ZP. This indicates that for potential penetration through mouse embryo, sperm cells will require more energy. In the living system this is not possible owing to sperm energy loss with time. After fertilization there are no sperm cells capable to respond to these increased demands.

On amplitude-frequency graphs zeros correspond to dynamical absorptions in forced oscillatory regime. On these frequencies certain material particle is not oscillating and external force has no effect on it. For the same molecules, frequencies when dynamical absorption occurs are higher for ZP of mouse embryo then for ZP of mouse oocyte. For sixth molecule in the chain in case of embryo there are six frequencies where dynamical absorption is possible and for the case of the oocyte there four. All these results confirm that oocyte and embryo are in different oscillatory states and that these oscillatory states have biological purpose. As in biological systems this transition is an irreversible process; further investigations of conditions under which this irreversible process is possible should be done. If small nonlinearities are included in the system it is possible to determine the dependence of frequencies on initial conditions as well as resonant jumps in the system.

Acknowledgments The author would like to thank Prof. Katica (Stevanović) Hedrih from the Mathematical Institute of SASA, Belgrade, Serbia, for the help with useful consultations. Parts of this research were supported by the Ministry of Education, Science and Technological Development of the Republic of Serbia through Grant ON174001 and State University of Novi Pazar.

References

1. Familiari, G., Heyn, R., Relucenti, M., Sathananthan, H.: Structural changes of the zona pellucida during fertilization and embryo development. Frontier Biosci. **13**, 6730–51 (2008)
2. Hedrih (Stevanović), K.: Discrete continuum method, symposium, recent advances in analytical dynamics control, stability and differential geometry. In: Proceedings of the Mathematical institute SANU Edited by Vladan Djordjević, p. 151, pp. 30–57, 2002. ISBN 86-80593-32-X. http://www.mi.sanu.ac.yu/publications.htm
3. Hedrih, A.: Frequency analysis of knot mass particles in oscillatory spherical net model of mouse zona pellucida. In: Lecture Session, Short Paper, Abstract book of 23rd International Congress of Theoretical and Applied Mechanics, (IUTAM ICTAM Beijing 2012), Beijing, China, SM01-049, p. 209, 19–24 Aug 2012. ISBN 978-988-16022-3-7
4. Hedrih, A.: Modeling oscillations of zona pellucida before and after fertilization. ENOC Young Scientist Prize Paper. *EUROMECH* Newsletter 40. Europian Mech. Soc. **40**, 6–14. 2011

5. Hedrih, A., Lazarevic, M., Mitrovic-Jovanovic, A.: Fertilization as a biomechanical oscillatory phenomenon in mammals. In: Proceedings of the Fourth Serbian (29th Yu) Congress on Theoretical and Applied Mechanics, Vrnjacka Banja, Serbia, pp. 579–584, 4–7 June 2013
6. Murayama, Y., Mizuno, J., Kamakura, H., Fueta, Y., Nakamura, H., Akaishi, K., Anzai, K., Watanabe, A., Inui, H., Omata, S.: Mouse zona pellucida dynamically changes its elasticity during oocyte maturation, fertilization and early embryo development. Hum. Cell **19**, 119–125 (2006)
7. Papi, M., Sylla, L., Parasassi, T., Brunelli, R., Monaci, M., Maulucci, G., Missori, M., Arcovito, G., Ursini, F., De Spirito, M.: Evidence of elastic to plastic transition in the zona pellucida of oocytes using atomic force spectroscopy. Appl. Phys. Lett. **94**, 153902 (2009)
8. Rašković, P.D.: Teorija oscilacija (Theory of Oscillations). Beograd, Naučna knjiga (1965)
9. Sun, Y., Wan, K.T., Roberts, K.P., Bischof, J.S., Nelson, B.J.: Mechanical property characterization of mouse zona pellucida. IEEE T Nanobiosci. **2**(4), 279–286 (2003)

Constrained n-Body Problems

Wojciech Szumiński and Maria Przybylska

Abstract We consider a problem of mass points interacting gravitationally whose motion is subjected to certain holonomic constraints. The motion of points is restricted to certain curves or surfaces. We illustrate the complicated behaviour of trajectories of these systems using Poincaré cross sections. For some models we prove the non-integrability analysing properties of the differential Galois group of variational equations along certain particular solutions of considered systems. Also some integrable cases are identified.

1 Introduction

Let us consider several point masses interacting mutually according to a certain low. This is just the n-body problem. For the classical gravitational or electrostatic interactions such problem with $n > 2$ is not integrable. Let us restrict the motion of points to certain surfaces or curves. These holonomic constrains modify interactions of points. In some cases these modifications lead to the non-integrability, and in others to the integrability. The described constrained classical n-body problems can be considered as a source of toy models for testing various methods and tools for studying dynamics of classical systems. In fact this paper arose from such investigations. Several simple examples show that, in fact, one can meet interesting and difficult problems investigating this kind of systems and moreover, such, let us say, academic investigations give unexpected results.

To describe them let us recall the anisotropic Kepler problem which appears in quantum mechanics of solid. It was thoroughly investigated by Guztwiller [5]. The rescaled Hamiltonian of the problem is given by

$$H = \frac{1}{2}\left(p_1^2 + p_2^2 + p_3^2\right) - \frac{1}{\sqrt{x^2 + \mu(y^2 + z^2)}}, \tag{1}$$

W. Szumiński (✉) • M. Przybylska
Institute of Physics University of Zielona Góra, 65-417 Zielona Góra, Poland
e-mail: uz88szuminski@gmail.com; M.Przybylska@if.uz.zgora.pl

J. Awrejcewicz (ed.), *Applied Non-Linear Dynamical Systems*, Springer Proceedings in Mathematics & Statistics 93, DOI 10.1007/978-3-319-08266-0_22

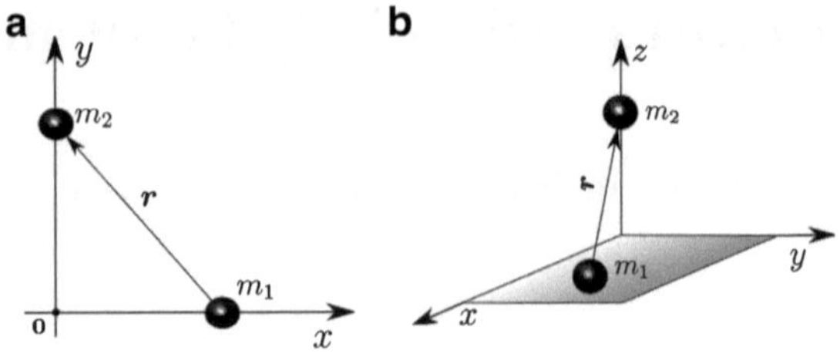

Fig. 1 Motion of two masses on: (**a**) perpendicular axes (geometry of model 1), and (**b**) plane and perpendicular axis (geometry of model 2)

where μ is a positive constant. For the two-degrees-of-freedom version of this problem the Hamiltonian reads

$$H = \frac{1}{2}\left(p_1^2 + p_2^2\right) - \frac{1}{\sqrt{x^2 + \mu y^2}}. \tag{2}$$

Unexpectedly, these systems can be considered as gravitational two-body problems with constraints. To see this, let us consider two masses, one mass moving along a line and the second mass moving along a perpendicular line; see Fig. 1a. The Hamiltonian of the system is the following:

$$H_1 = \frac{p_1^2}{2m_1} + \frac{p_2^2}{2m_2} - \frac{Gm_1m_2}{\sqrt{x^2 + y^2}}. \tag{3}$$

So, by a simple rescaling we obtain Hamiltonian (2). Similarly, let one mass moves along a line, and the other moves in a plane perpendicular to this line; see Fig. 1(b). The Hamiltonian has the form

$$H_2 = \frac{p_1^2}{2m_1} + \frac{1}{2m_2}(p_2^3 + p_3^2) - \frac{Gm_1m_2}{\sqrt{x^2 + y^2 + z^2}} \tag{4}$$

and again its simple rescaling gives (1).

As we can see the Hamiltonians (2) and (1) differ from the Hamiltonians of standard planar and spatial Kepler problem only in the parameter μ. For $\mu \neq 0$, contrary to the standard Kepler problem, the force is not radial. The dynamics of the anisotropic Kepler problem is dramatically different from that of the standard Kepler problem.

The chaotic behaviour of the anisotropic Kepler problems was investigated in numerous papers (see, e.g., [2,3,5]) and the non-integrability of the planar problem was proved in [4] and for planar and spatial problem in [1]. The non-integrability proof in [1] uses the differential Galois approach and the authors state that for $\mu \notin \{0, 1\}$ there are no meromorphic integrals besides the Hamiltonian itself. But there is no written about meromorphic functions of what variables authors say. If one considers meromorphic functions of coordinates and momenta, then already Hamiltonian is not a meromorphic function; thus the system trivially is not meromorphically integrable for all values of μ. Thus below we formulate these theorems in a more precise way.

Theorem 1. *Hamiltonian system defined by* (2) *is integrable in the Liouville sense with first integrals which are meromorphic in* (x, y, p_1, p_2, r), *where* $r = \sqrt{x^2 + \mu y^2}$, *if and only if* $\mu \in \{0, 1\}$.

In the case when $\mu = 1$ this system has two additional functionally independent additional first integrals:

$$I_1 = p_2 x - p_1 y, \qquad I_2 = p_2(p_1 y - p_2 x) + \frac{x}{\sqrt{x^2 + y^2}};$$

thus it is super-integrable.

The spatial anisotropic Kepler problem defined by (1) has an invariant subspace defined by $z = p_3 = 0$. In this subspace it coincides with the previous system. Thus, the necessary conditions of the integrability are the same as for the previous system.

Theorem 2. *Hamiltonian system defined by* (1) *is integrable in the Liouville sense with first integrals which are meromorphic in* $(x, y, z, p_1, p_2, p_3, r)$, *where* $r = \sqrt{x^2 + \mu(y^2 + z^2)}$, *if and only if* $\mu \in \{0, 1\}$.

In the case when $\mu = 1$ it coincides with the three-dimensional standard Kepler problem, and it has the following first integrals:

$$\boldsymbol{c} = \boldsymbol{r} \times \boldsymbol{p}, \qquad \boldsymbol{e} = \boldsymbol{p} \times \boldsymbol{c} - \frac{\boldsymbol{r}}{r},$$

where $\boldsymbol{r} = (x, y, z)$, $\boldsymbol{p} = (p_1, p_2, p_3)$, and $r = \sqrt{x^2 + y^2 + z^2}$. Among them one can find three functionally independent and pairwise commuting.

Hamiltonian (2) (and also (1)) because of the presence of square root r is not single-valued and meromorphic in coordinates and momenta. Thus, formally, in order to apply the differential Galois theory approach to such a Hamiltonian system, we have to extend it to the corresponding Poisson system introducing r as additional variable. However, in calculations one can work with the original Hamiltonian system, and the only trace of this extension is the fact that we study the integrability in the class of meromorphic functions of not only coordinates and momenta but also of r. This extension procedure as well as its application to a certain three-body problem was given in [7]. The similar trick is applied to all remaining Hamiltonian systems with algebraic potentials considered in this paper.

The above examples show that it is reasonable to examine similar classes of constrained n-body systems. In the next section we will give several examples of such systems with a few degrees of freedoms. In a case when the considered system reduces to a system with two degrees of freedom the Poincaré cross sections give us quickly insight into the dynamics of the systems. However, a challenging problem is to prove that they are non-integrable and to find values of parameters for that they become integrable. For some presented problems we prove their non-integrability using the so-called Morales–Ramis theory [8]. It is based on the analysis of differential Galois group of variational equations obtained by linearisation of equations of motion along a particular solution. The main theorem of this theory states that if the considered system is meromorphically integrable in the Liouville sense, then the identity component of the differential Galois group of the variational equations is Abelian. For a precise definition of the differential Galois group and differential Galois theory, see, e.g., [9].

2 Integrability Analysis of Several Restricted n-Body Problems

Model 3: Two Masses on Two Inclined Straight Lines

The direct generalisation of model 1 from Fig.1a is the following. Assume that mass m_1 moves along horizontal line $q_2 = 0$ and it has coordinates $(q_1, 0)$ and mass m_2 with coordinates $q_2(\cos\phi, \sin\phi)$ moves along a straight line inclined to the horizontal line: see Fig. 2. The Hamiltonian function is given by

$$H = \frac{p_1^2}{2m_1} + \frac{p_2^2}{2m_2} - \frac{Gm_1m_2}{\sqrt{q_1^2 + q_2^2 - 2q_1q_2\cos\phi}}. \tag{5}$$

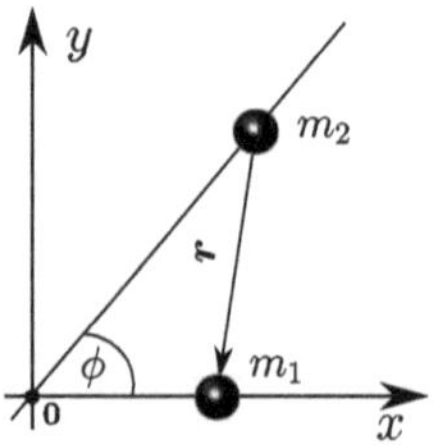

Fig. 2 Geometry of model 3

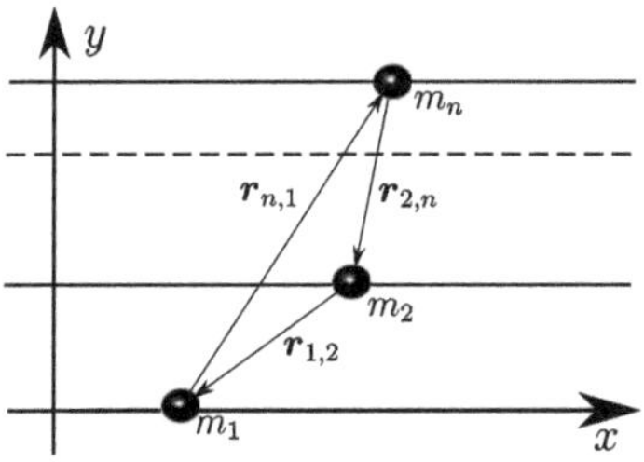

Fig. 3 Geometry of model 4

In the Appendix we will prove the following theorem:

Theorem 3. *The system governed by Hamiltonian* (5) *is integrable in the class of functions meromorphic in* (q_1, q_2, p_1, p_2, r) *where* $r = \sqrt{q_1^2 + q_2^2 - 2q_1q_2\cos\phi}$, *ff*

- *either* $\phi \in \{0, \pi\}$ *and* $m_1, m_2 \in \mathbb{R}$, *or*
- $\phi \in \{\pi/2, 3\pi/2\}$ *and* $m_2 = m_1$.

Model 4: Masses Moving on the Parallel Lines

Let us consider a problem of n masses moving in parallel lines; see Fig. 3. As a generalised coordinates we use the relative displacements $q_i = x_i - x_{i-1}$ along axis x, for $i = 2, \ldots, n$ and $q_1 = x_1$.

The Lagrange and the Hamiltonian functions do not depend on variable q_1, which is a cyclic variable and its corresponding momentum $p_1 = c$ becomes a parameter. Thus, we obtain the reduced system with $n - 1$ degrees of freedom. Model of $n = 2$ masses is integrable. The reduced system with $n = 3$ masses has two degrees of freedom and it is described by the following Hamiltonian:

$$H=\frac{1}{2}\left(\frac{(c-p_2)^2}{m_1}+\frac{(p_2-p_3)^2}{m_2}+\frac{p_3^2}{m_3}-\frac{2Gm_2m_3}{\sqrt{(a-b)^2+q_3^2}}+m_1\left(-\frac{2Gm_2}{\sqrt{a^2+q_2^2}}-\frac{2Gm_3}{\sqrt{b^2+(q_2+q_3)^2}}\right)\right). \tag{6}$$

We assumed that masses m_2 and m_3 move along horizontal curves $y = a$ and $y = b$, respectively. Figure 4 shows the Poincaré cross sections related to (6). Clearly, the system is generally non-integrable. However, a proof of this fact is an open question.

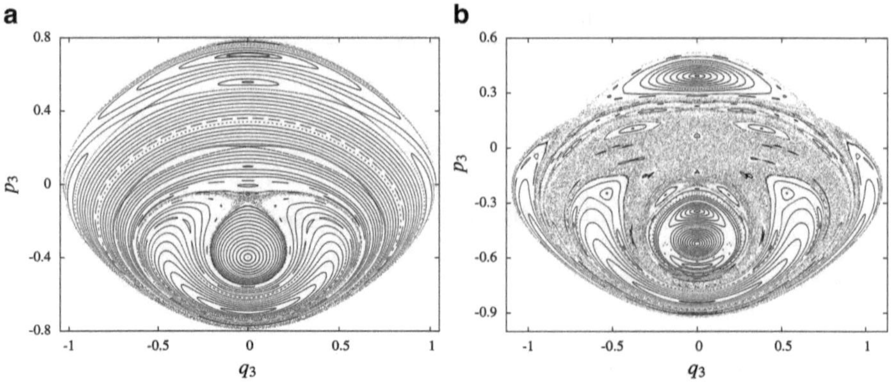

Fig. 4 Poincaré sections for model 4. Parameters: $m_1 = 1,\ m_2 = 2,\ m_3 = 1,\ G = 1,\ a = 3,$ $b = 1,\ c = 0,$ cross-plain $q_2 = 0,\ p_2 > 0$. (**a**) $E = -2.25$, (**b**) $E = -2.1$

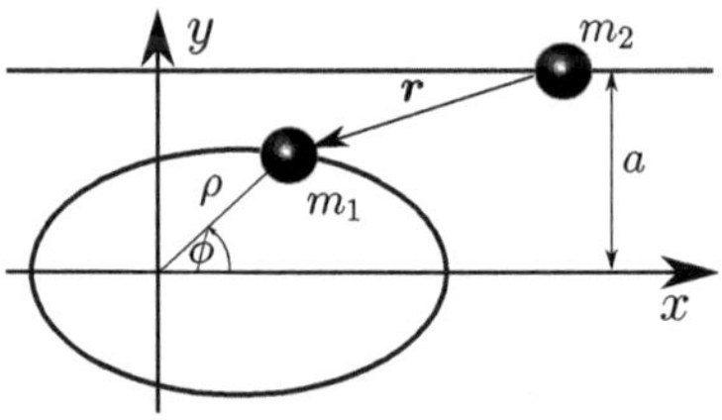

Fig. 5 Geometry of model 5

Model 5: Two Masses Moving on an Ellipse and a Straight Line Parallel to the Main Axis of the Ellipse

In Fig. 5 the geometry of the system is shown. Now we assume that the mass m_1 moves on the ellipse with coordinates $\rho(\cos\phi, \sin\phi)$, where $\rho = c/(1 + e\cos\phi)$, and mass m_2 moves along a straight line parallel to the main axis of ellipse with coordinates (x, a). The Hamiltonian function is given by

$$H = \frac{1}{2}\left(\frac{p_x^2}{m_2} + \frac{p_\phi^2\,(1 + e\cos\phi)^4}{c^2 m_1\left(1 + e^2 + 2e\cos\phi\right)} - \frac{2Gm_1m_2}{\sqrt{\left(\frac{c\cos\phi}{1+e\cos\phi} - x\right)^2 + \left(\frac{c\sin\phi}{1+e\cos\phi} - a\right)^2}}\right). \tag{7}$$

Figure 6 shows the Poincaré cross sections. They present that for certain fixed values of parameters, the system is not integrable. In fact we can prove the following theorem:

Theorem 4. *If* $a = 0$ *and* $(m_1 - m_2)m_1m_2 \neq 0$ *then the system governed by Hamiltonian* (7) *is not completely integrable with first integrals which are*

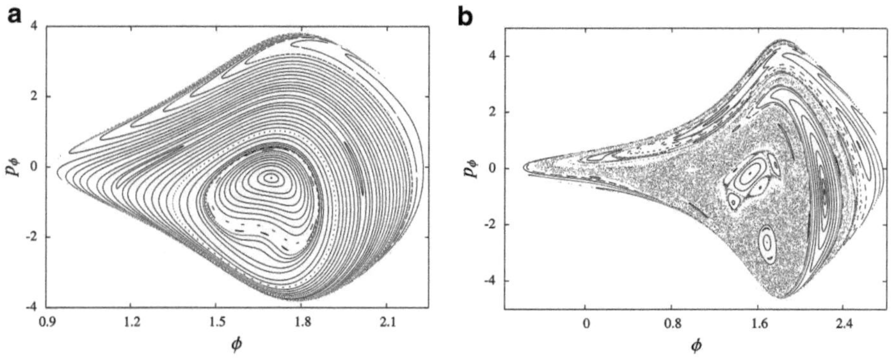

Fig. 6 Poincaré sections for model 5. Parameters: $m_1 = 1$, $m_2 = 2$, $G = 1$, $a = 3$, $c = 2$, $e = 0.5$, cross-plain $x = 0$, $p_x > 0$. (**a**) $E = -1$, (**b**) $E = -0.5$

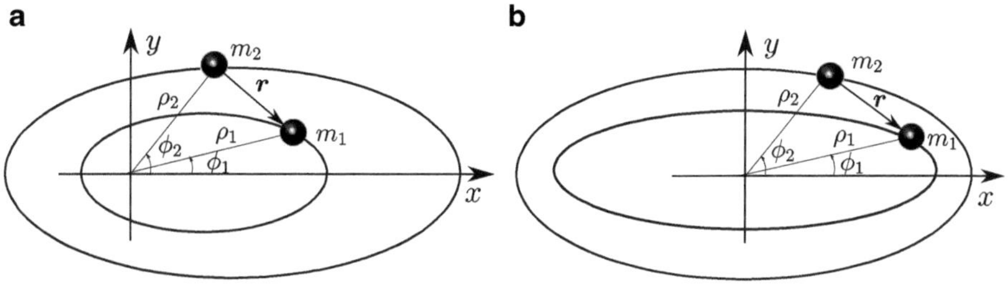

Fig. 7 Motion of two masses on: (**a**) two confocal ellipses (geometry of the model 6), and (**b**) two concentric ellipses with parallel main axes (geometry of the model 7)

meromorphic in (x, ϕ, p_1, p_2, r), *where*

$$r = \sqrt{\left(\frac{c\cos\phi}{1 + e\cos\phi} - x\right)^2 + \left(\frac{c\sin\phi}{1 + e\cos\phi}\right)^2}.$$

This theorem is in particular true for the circle when $e = 0$ and $c = \rho$.

Model 6: Two Mass Points Moving in Two Conics

In Fig. 7a the geometry of the system is presented. In this case, masses m_1 and m_2 move along two confocal ellipses with coordinates $\rho_1(\cos\phi_1, \sin\phi_1)$ and $\rho_2(\cos\phi_2, \sin\phi_2)$, where

$$\rho_1 = \frac{c_1}{1 + e_1\cos\phi_1}, \qquad \rho_2 = \frac{c_2}{1 + e_2\cos\phi_2},$$

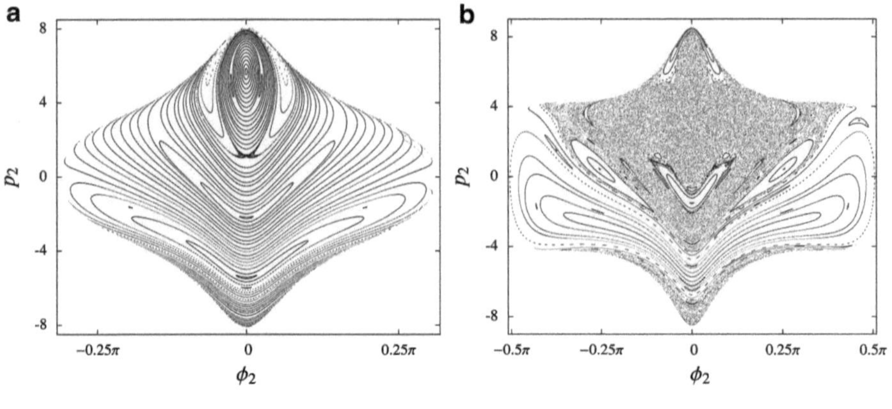

Fig. 8 Poincaré sections for model 6. Parameters: $m_1 = 2$, $m_2 = 2$, $G = 1$, $c_1 = 1$, $c_2 = 2$, $e_1 = \frac{1}{2}$, $e_2 = \frac{3}{2}$, cross-plain $\phi_1 = 0$, $p_1 > 0$. (**a**) $E = -4.4$, (**b**) $E = -1.8$

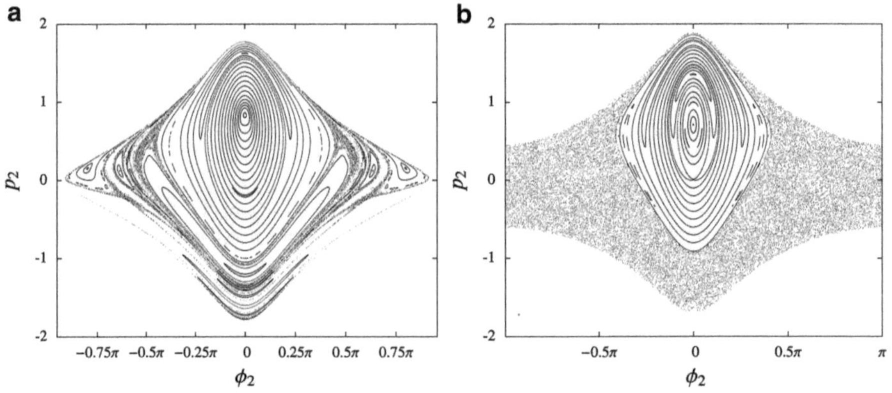

Fig. 9 Poincaré sections for model 6. Parameters: $m_1 = 2$, $m_2 = 1$, $G = 1$, $c_1 = 3$, $c_2 = 1$, $e_1 = \frac{3}{5}$, $e_2 = 0$, cross-plain $\phi_1 = 0$, $p_1 > 0$. (**a**) $E = -0.7$, (**b**) $E = -0.5$

and interact gravitationally. Hamiltonian function takes the form

$$H = \frac{(1 + e_1 \cos\phi_1)^4 p_1^2}{2c_1^2 m_1 \left(2e_1 \cos\phi_1 + e_1^2 + 1\right)} + \frac{(1 + e_2 \cos\phi_2)^4 p_2^2}{2c_2^2 m_2 \left(2e_2 \cos\phi_2 + e_2^2 + 1\right)} - \frac{G m_1 m_2}{B},$$

$$B = \sqrt{\left(\frac{c_1 \cos\phi_1}{1 + e_1 \cos\phi_1} - \frac{c_2 \cos\phi_2}{1 + e_2 \cos\phi_2}\right)^2 + \left(\frac{c_1 \sin\phi_1}{1 + e_1 \cos\phi_1} - \frac{c_2 \sin\phi_2}{1 + e_2 \cos\phi_2}\right)^2}. \tag{8}$$

To present the dynamics of considered system we make several Poincaré cross sections; see Figs. 8–9.

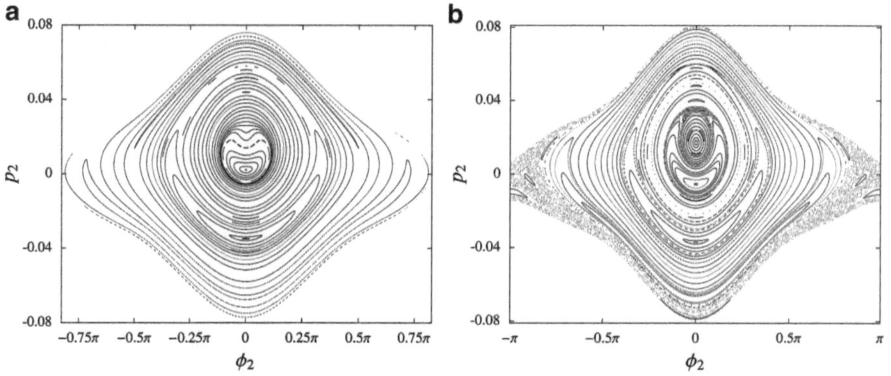

Fig. 10 Poincaré sections related to model 7. Parameters: $m_1 = 1$, $G = 1$, $a_1 = 0.8$, $a_2 = 1.1$, $b_1 = 1$, $b_2 = \frac{a_2 b_1}{a_1} = 1.4$, $m_2 = \frac{m_1 a_1}{a_2} = 0.73$, cross-plain $\phi_1 = 0$, $p_1 > 0$ (**a**) $E = -0.82$, (**b**) $E = -0.8$

Model 7: Two Masses Moving in Concentric Ellipses with Parallel Main Axes

The geometry of the system is shown in Fig. 7b. In this case, masses m_1 and m_2 move in two ellipses which have common centres and parallel main axes. Using the standard trigonometric parametrisations of points on ellipses $(a_i \cos\phi_i, b_i \sin\phi_i)$ for $i = 1, 2$, we can derive the Hamiltonian:

$$H = \frac{1}{2}\left(\frac{p_1^2}{a_1^2 m_1 \cos\phi_1^2 + a_2^2 m_1 \sin\phi_1^2} + \frac{p_2^2}{b_1^2 m_2 \cos\phi_2^2 + b_2^2 m_2 \sin\phi_2^2}\right.$$
$$\left. - \frac{2Gm_1 m_2}{\sqrt{(a_2\cos\phi_1 - b_2\cos\phi_2)^2 + (a_1\sin\phi_1 - b_1\sin\phi_2)^2}}\right), \tag{9}$$

where a_1, a_2 and b_1, b_2 are major and minor semi-axes of both ellipses. The Poincaré cross sections are shown in Fig. 10.

Model 8: n-Masses Moving in the Circles

Let us consider the motion of n-masses moving on the concentric circles which interact gravitationally. As a generalised coordinate we use the relative angles θ_i; see Fig. 11. Similarly to the fourth model the Hamiltonian function has one cyclic variable θ_1 and its corresponding momentum $p_1 = c$ is a first integral of the system. Thus, we get the reduced system with $n-1$ degrees of freedom. Case of two masses is of course integrable, but the model of $n = 3$ has much more complex dynamics.

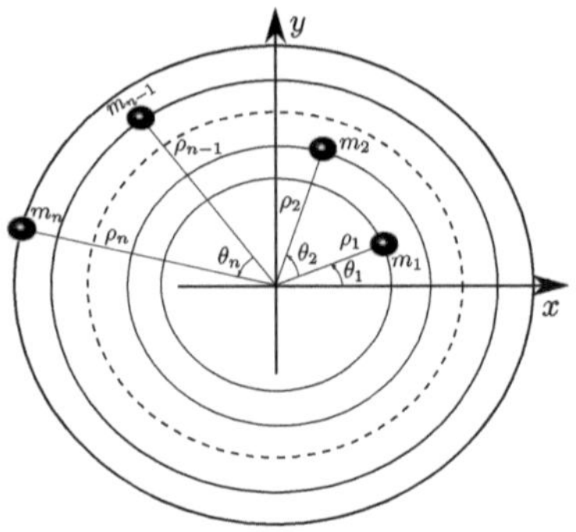

Fig. 11 Geometry of model 8

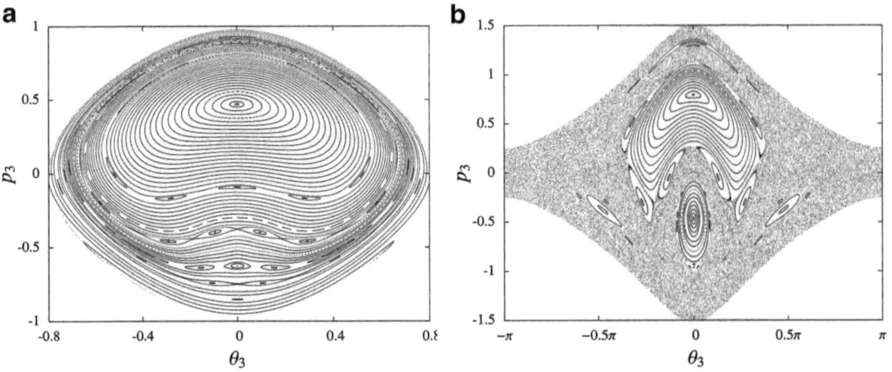

Fig. 12 Poincaré sections for model 8. Parameters: $m_1 = 1$, $m_2 = 2$, $m_3 = 1$, $G = 1$, $\rho_1 = 2$, $\rho_2 = 3$, $\rho_3 = 1$, $c = 0$, cross-plain $\theta_2 = 0$, $p_2 > 0$. (**a**) $E = -3.5$, (**b**) $E = -2.8$

To present this complexity we make several Poincaré sections; see Fig. 12. The Hamiltonian of this reduced system has the form

$$H = \frac{1}{2}\left(\frac{(c-p_2)^2}{\rho_1^2 m_1} + \frac{(p_2-p_3)^2}{\rho_2^2 m_2} + \frac{p_3^2}{\rho_3^2 m_3} - \frac{2Gm_1m_2}{\sqrt{\rho_1^2+\rho_2^2-2\rho_1\rho_2\cos\theta_2}}\right. \\ \left. - \frac{2Gm_2m_3}{\sqrt{\rho_2^2+\rho_3^2-2\rho_2\rho_3\cos\theta_3}} - \frac{2Gm_1m_3}{\sqrt{\rho_1^2+\rho_3^2-2\rho_1\rho_3\cos(\theta_2+\theta_3)}}\right). \tag{10}$$

Acknowledgements The authors are very grateful to Andrzej J. Maciejewski for many helpful comments and suggestions. This research has been supported by grant No. DEC-2011/02/A/ST1/00208 of the National Science Centre of Poland.

Appendix: Proof of Theorem 3

Hamilton equations for Hamiltonian (5) have the form

$$\dot{q}_1 = \frac{p_1}{m_1},\ \dot{q}_2 = \frac{p_2}{m_2},\ \dot{p}_1 = \frac{G m_1 m_2 (\alpha q_2 - q_1)}{\left(q_1^2 + q_2^2 - 2\alpha q_1 q_2\right)^{3/2}},\ \dot{p}_2 = \frac{G m_1 m_2 (\alpha q_1 - q_2)}{\left(q_1^2 + q_2^2 - 2\alpha q_1 q_2\right)^{3/2}}, \tag{11}$$

where $\alpha := \cos\phi$. In order to simplify the calculations, we make the following non-canonical transformation:

$$\begin{bmatrix} q_1 \\ q_2 \\ p_1 \\ p_2 \end{bmatrix} = \begin{bmatrix} \frac{1}{\sqrt{\beta^2} - \mu_2} & \frac{\sqrt{\beta^2-\mu_2^2}(\mu_1+\mu_2)}{\left(\mu_2-\sqrt{\beta^2}\right)\sqrt{\mu_1^2-\mu_2^2)}} & 0 & 0 \\ 0 & 1 & 0 & 0 \\ 0 & 0 & \frac{\mu_1-\mu_2}{2\sqrt{\beta^2}-2\mu_2} & \frac{\sqrt{\beta^2-\mu_2^2}\sqrt{\mu_1^2-\mu_2^2}}{\left(\mu_2-\sqrt{\beta^2}\right)(\mu_1+\mu_2)} \\ 0 & 0 & 0 & 1 \end{bmatrix} \begin{bmatrix} x_1 \\ x_2 \\ y_1 \\ y_2 \end{bmatrix},$$

$$\mu_1 = m_1 + m_2, \quad \mu_2 = m_2 - m_1, \quad \beta = \sqrt{(m_1 - m_2)^2 + 4 m_1 m_2 \alpha^2}. \tag{12}$$

System (11) after this transformation takes the form

$$\dot{x}_1 = y_1,\ \dot{y}_1 = \frac{G x_1 (\mu_1 - \beta)(\mu_2 - \beta)^3}{2\left(-2x_2 x_1 (\beta + \mu_1)\sqrt{\frac{\beta^2-\mu_2^2}{\mu_1^2-\mu_2^2}} + \frac{2\beta x_2^2(\beta+\mu_1)(\beta-\mu_2)}{\mu_1-\mu_2} + x_1^2\right)^{3/2}},$$

$$\dot{x}_2 = \frac{2y_2}{\mu_1 + \mu_2},\ \dot{y}_2 = \frac{G(\beta-\mu_2)^2\left(x_1\sqrt{\left(\mu_1^2-\mu_2^2\right)\left(\beta^2-\mu_2^2\right)} - (\mu_1+\mu_2)x_2(\beta+\mu_1)(\beta-\mu_2)\right)}{4\left(-2x_2x_1(\beta+\mu_1)\sqrt{\frac{\beta^2-\mu_2^2}{\mu_1^2-\mu_2^2}} + \frac{2\beta x_2^2(\beta+\mu_1)(\beta-\mu_2)}{\mu_1-\mu_2} + x_1^2\right)^{3/2}}. \tag{13}$$

It has invariant manifold $\mathscr{N} = \left\{(x_1, x_2, y_1, y_2) \in \mathbb{C}^4 \mid x_1 = y_1 = 0\right\}$ and its restriction to $\mathscr{N}$ is

$$\dot{x}_2 = \frac{2y_2}{\mu_1 + \mu_2}, \qquad \dot{y}_2 = -\frac{G(\mu_1 + \mu_2)\left[(\mu_1 - \mu_2)(\beta - \mu_2)\right]^{3/2}}{8\sqrt{2}x_2^2\sqrt{\beta^3}(\beta + \mu_1)}. \tag{14}$$

Let the particular solution of (13) be defined by (14), and $\boldsymbol{Z} = [X_1, X_2, Y_1, Y_2]^T$ denotes the variations of $[x_1, x_2, y_1, y_2]^T$. Then, the variational equations along this particular solution have the form $\dot{\boldsymbol{Z}} = A\boldsymbol{Z}$, where

$$A = \begin{bmatrix} 0 & 0 & 1 & 0 \\ 0 & 0 & 0 & \frac{2}{\mu_1+\mu_2} \\ \frac{G(\mu_1-\beta)(\mu_1-\mu_2)^{3/2}(\mu_2-\beta)^3}{4\sqrt{2}\beta^{3/2}x_2^3(\beta+\mu_1)^{3/2}(\beta-\mu_2)^{3/2}} & 0 & 0 & 0 \\ -\frac{G(\beta+3\mu_1)(\beta-\mu_2)(\mu_1-\mu_2)^2\sqrt{\beta+\mu_2}\sqrt{\mu_1+\mu_2}}{16\sqrt{2}\beta^{5/2}x_2^3(\beta+\mu_1)^{3/2}} & \frac{G(\beta-\mu_2)^{3/2}(\mu_1-\mu_2)^{3/2}(\mu_1+\mu_2)}{4\sqrt{2}\beta^{3/2}x_2^3\sqrt{\beta+\mu_1}} & 0 & 0 \end{bmatrix}.$$

Equations for X_1 and Y_1 form a subsystem of normal variational equations and can be rewritten as a one second-order differential equation for variable $X \equiv X_1$

$$\ddot{X} + \left(-\frac{G\,(\mu_1-\mu_2)^{3/2}\,(\beta-\mu_1)\,(\beta-\mu_2)^{3/2}}{4\sqrt{2}\beta^{3/2}x_2^3\,(\beta+\mu_1)^{3/2}} \right) X = 0. \tag{15}$$

We transform this equation using the following change of independent variable:

$$t \longrightarrow z = -\frac{4\sqrt{2}E\sqrt{\frac{\beta(\beta+\mu_1)}{(\mu_1-\mu_2)(\beta-\mu_2)}}}{G\left(\mu_1^2-\mu_2^2\right)} x_2(t), \tag{16}$$

where E is a level of Hamiltonian transformed by means of (12) and restricted to $\mathscr{N}$. Then normal variational equation (15) takes the form

$$X'' + pX' + qX = 0, \quad p = -\frac{1}{2z} + \frac{1}{2(z-1)}, \quad q = \frac{-\beta+\mu_1}{2(\beta+\mu_1)z^2} + \frac{\beta-\mu_1}{2(\beta+\mu_1)(z-1)z}, \tag{17}$$

where $' \equiv \frac{\mathrm{d}}{\mathrm{d}z}$. We recognise that this equation is a Riemann P equation (see, e.g., [6, 8])

$$\frac{\mathrm{d}^2X}{\mathrm{d}z^2} + \left(\frac{1-a-a'}{z} + \frac{1-c-c'}{z-1}\right)\frac{\mathrm{d}X}{\mathrm{d}z} + \left(\frac{aa'}{z^2} + \frac{cc'}{(z-1)^2} + \frac{bb'-aa'-cc'}{z(z-1)}\right)X = 0, \tag{18}$$

with exponents

$$a = \tfrac{1}{4}\left(3+\sqrt{1+\tfrac{16\beta}{\beta+\mu_1}}\right),\ a' = \tfrac{1}{4}\left(3-\sqrt{1+\tfrac{16\beta}{\beta+\mu_1}}\right),\ b = c' = 0,\ \ b' = -1,\ \ c = \tfrac{1}{2}. \tag{19}$$

The differences of exponents are given by

$$\lambda = a - a' = \frac{1}{2}\sqrt{\frac{17\gamma+1}{\gamma+1}}, \qquad \sigma = b - b' = 1, \qquad \nu = c - c' = \frac{1}{2}, \tag{20}$$

where $\gamma = \beta/\mu_1$. The Riemann P equation is solvable iff one of the four numbers $\lambda+\sigma+\nu$, $-\lambda+\sigma+\nu$, $\lambda-\sigma+\nu$, and $\lambda+\sigma-\nu$ is an odd integer or λ or $-\lambda$

and σ or $-\sigma$ and ν or $-\nu$ belong (in an arbitrary order) to the so-called Schwarz table [6, 8]. Conditions $\pm\lambda + \sigma + \nu = 2p + 1$, where $p \in \mathbb{Z}$, give the following expression for γ

$$\gamma = \frac{-3 + 5p - 2p^2}{1 - 5p + 2p^2},$$

that takes only two non-negative values 0 and 1. Similarly, conditions $\lambda - \sigma + \nu = 2p + 1$ and $\lambda + \sigma - \nu = 2p - 1$, where $p \in \mathbb{Z}$, give

$$\gamma = \frac{p - 2p^2}{-2 - p + 2p^2}, \quad \gamma = \frac{-1 + 3p - 2p^2}{-1 - 3p + 2p^2},$$

respectively, that only take two non-negative values 0 and 1.

Since two differences of exponents are equal to 1/2 and 1, only the first case in the Schwarz table is admissible that leads to the condition $\lambda = 1/2 + p$, where $p \in \mathbb{Z}$. It gives

$$\gamma = -\frac{p + p^2}{-4 + p + p^2},$$

and this expression takes only two non-negative values 0 and 1. Value $\gamma = 1$ gives $\alpha = \cos\phi = \pm 1$, and that implies $\phi \in \{0, \pi\}$. Parameter γ vanishes only when $\beta = 0$ that gives $m_2 = m_1$ and simultaneously $\alpha = \cos\phi = 0$. These are the only cases when the identity component of differential Galois group of Riemann P equation (18) with exponents (19) is solvable that is necessary for its Abelianity and the integrability of the system.

References

1. Arribas, M., Elipe, A., Riaguas, A.: Non-integrability of anisotropic quasi-homogeneous Hamiltonian systems. Mech. Res. Comm. **30**(3), 209–216 (2003)
2. Casasayas J., Llibre J.: Qualitative Analysis of the Anisotropic Kepler Problem. Memoirs of Amer. Math. Soc. **52**(312), 1–115 (1984)
3. Devaney R. L.: Blowing up Singularities in Classical Mechanical Systems. Amer. Math. Monthly. **89**, 535–552 (1982)
4. Gutzwiller M.C.: Bernoulli sequences and trajectories in the anisotropic Kepler problem. J. Math. Phys. **18**(4), 806–823 (1977)
5. Gutzwiller M.C.: Chaos in Classical and Quantum Mechanics. Springer, New York (1990)
6. Kimura T.: On Riemann's equations which are solvable by quadratures. Funkcial. Ekvac. **12**, 269–281 (1970)
7. Maciejewski A.J., Przybylska M., Simpson L., Szumiński W.: Non-integrability of the dumbbell and point mass problem. Celestial Mech. Dynam. Astronom. **117**, 315–330 (2013)
8. Morales Ruiz J.J.: Differential Galois Theory and Non-Integrability of Hamiltonian Systems. Birkhäuser, Basel (1999)
9. Van der Put M, Singer M.F.: Galois Theory of Linear Differential Equations. Springer, Berlin (2003)

Hunting French Ducks in Population Dynamics

Ferdinand Verhulst

Abstract Equations with periodic coefficients for singularly perturbed growth can be analysed by using fast and slow timescales in the framework of Fenichel geometric singular perturbation theory and its extensions. The analysis is restricted to one-dimensional time-periodic ordinary differential equations and shows the presence of slow manifolds, canards and the dynamical exchanges between several slow manifolds. There exist permanent (or periodic) canards and periodic solutions containing canards.

1 Introduction

In this note we consider systems with slow-fast motion in a singularly perturbed setting; the slow motion is characterised by the exponential closeness of solutions to slow manifolds. In the case that the solution moves along a stable slow invariant manifold and at some point the slow manifold becomes unstable, we have the possibility of 'exponential sticking' or canard (French duck) behaviour. In this case, the solution continues for an $O(1)$ time along the slow invariant manifold that has become unstable and jumps after that away, for instance, to the neighbourhood of another invariant set. Following Pontrjagin (see [10]), one also calls this 'delay of stability loss'.

This delay or sticking process is closely connected to the so-called *canard* phenomenon for differential equations that can be described as follows: *Canard solutions are bounded solutions of a singularly perturbed system that, starting near an normally hyperbolic attracting slow manifold, cross a singularity of the system of differential equations and follow for an* $O(1)$ *time a normally hyperbolic repelling slow manifold.*

F. Verhulst (✉)
Mathematisch Instituut, University of Utrecht, PO Box 80.010,
3508 TA Utrecht, The Netherlands
e-mail: f.verhulst@uu.nl

J. Awrejcewicz (ed.), *Applied Non-Linear Dynamical Systems*, Springer Proceedings in Mathematics & Statistics 93, DOI 10.1007/978-3-319-08266-0_23

The canard behaviour will depend on the dimension of the problem and the nature of the singularity. An example of canard behaviour was found by the Strassbourg group working in non-standard analysis for a perturbed van der Pol equation; see [2] for details and references. In this example, the singularity crossed is a fold point. The analysis of this problem is quite technical.

Canards arising at transcritical bifurcations have been described in [6] and [9]. The purpose of the present note is to study such phenomena in growth equations that can be handled explicitly, both analytically and numerically; this may increase our understanding. In Sects. 2.4 and 4, we consider simple illustrations of exchange of stability between slow manifolds. The equations in Sect. 5 model growth phenomena with daily or seasonal fluctuations as appear in models of mathematical biology and economics. They are a natural extension of the logistic model introduced in [19].

We shall also meet *permanent canards*, solutions that are attracted to a slow manifold, travel across a singularity resulting in instability of the slow manifold, but are remaining near the slow manifold without leaving the unstable part.

The numerics which we used to obtain illustrations is based on the package CONTENT; see [8].

2 Fenichel Theory

In the period 1950–1960, many results were obtained for the asymptotic analysis of singularly perturbed ordinary differential equations; see for instance [15]. Later, these results were supplemented by essential qualitative insight based on developments of invariant manifold theory. This slow manifold analysis, as it is usually called, is often named after one of its main inventors, N. Fenichel. The canard aspects are an extension of the theory. We will summarise some important aspects.

2.1 The Slow Manifold: Fenichel's Results

Approximation theorems like Tikhonov's theorem ([15]) are concerned with the attraction, at least for some time, to the regular expansion that corresponds with a stable critical point of a boundary layer equation. The theory is quite general and deals with non-autonomous equations.

In many problems, it is possible to associate with the regular expansion a manifold in phase or solution space and to consider the attraction properties of the flow near this manifold. This raises the question of whether these manifolds really exist or whether they are just a phantom phenomenon. Such questions were addressed and answered in a number of papers by Fenichel (1971–1979) and other authors; for references and extensive introductions the reader is referred to the survey papers [3, 9] and [5]; see also [16].

Consider for $t \geq 0$ the $(n+m)$-dimensional non-autonomous system:

$$\dot{x} = f(x, y, t) + \varepsilon \cdots, \quad x \in D \subset \mathbb{R}^n,$$
$$\varepsilon \dot{y} = g(x, y, t) + \varepsilon \cdots, \quad y \in G \subset \mathbb{R}^m.$$

A corresponding theory can be developed for the system:

$$x' = \varepsilon f(x, y, \varepsilon t) + \varepsilon^2 \cdots, \quad y' = g(x, y, \varepsilon t) + \varepsilon \cdots,$$

where the prime denotes differentiation with respect to $\tau = t/\varepsilon$.

We will call y the fast variable and x the slow variable. The zero set of $g(x, y, t)$ is given by $y = \phi(x, t)$, which represents a first-order approximation M_0 of the slow manifold M_ε. The flow on M_ε is to a first approximation described by $\dot{x} = f(x, \phi(x, t), t)$.

In the Fenichel theory we assume

$$\text{ReSp } g_y(x, \phi(x, t)) \neq 0, x \in D.$$

That is, all the eigenvalues of the linearised flow near M_0, derived from the equation for y, have nonzero real parts.

A manifold is called hyperbolic if the local linearisation is structurally stable (real parts of eigenvalues all nonzero), and it is normally hyperbolic if in addition the expansion or contraction near the manifold in the transversal direction is larger than in the tangential direction (the slow drift along the slow manifold).

Note that this perspective on dynamics allows for interesting phenomena. One might approach M_ε for instance by a stable branch, stay for some time near M_ε, and then leave again a neighbourhood of the slow manifold by an unstable branch. This produces solutions indicated as 'pulse-like', 'multibump solutions', etc. This type of exchanges of the flow near M_ε is what one often looks for in geometric singular perturbation theory.

2.2 *Existence of the Slow Manifold*

The question of whether the slow manifold M_ε, approximated by $y = \phi(x, t)$, persists for $\varepsilon > 0$ was answered by Fenichel. The main result is as follows. If M_0 is a compact manifold that is normally hyperbolic, it persists for $\varepsilon > 0$ (i.e., there exists for sufficiently small, positive ε a smooth manifold M_ε close to M_0). Corresponding with the signs of the real parts of the eigenvalues, there exist stable and unstable manifolds of M_ε, smooth continuations of the corresponding manifolds of M_0, on which the flow is fast.

There are some differences between the cases where M_0 has a boundary or not. For details, see [3, 4] and the original papers by Fenichel.

2.3 The Compactness Property

Note that the assumption of compactness of D and G is essential for the uniqueness of the slow manifold. In many examples and applications, M_0, the approximation of the slow manifold obtained from the fast equation, is not bounded. This can be remedied, admittedly in an artificial way, by applying a suitable cutoff of the vector field far away from the domain of interest. In this way, compact domains arise that coincide locally with D and G. However, this may cause some problems with the uniqueness of the slow manifold. Consider for instance the following example from [16]:

Example. Consider the system

$$\begin{cases} \dot{x} & = 1, x(0) = x_0 > 0, \\ \varepsilon \dot{y} & = -\frac{y}{x^2}, y(0) = y_0 \geq 0. \end{cases} \tag{1}$$

Putting $\varepsilon = 0$ produces $y = 0$, which corresponds with M_0. We can obtain a compact domain for x by putting $x_0 \leq x \leq L$ with x_0 and L positive constants independent of ε. However, the limiting behaviour of the solutions depends on the initial condition and L. Integration of the equations yields

$$y(x) = y_0 \exp\left(\frac{1}{\varepsilon}(\frac{1}{t + x_0} - \frac{1}{x_0})\right).$$

As t tends to infinity, the solution for $y(t)$ tends to

$$y_0 \exp\left(-\frac{1}{\varepsilon x_0}\right),$$

so the solutions are, after an initial fast transition, all exponentially close to $y = 0$. However, there are an infinite number of slow manifolds dependent on x_0 and L, all tunnelling into an exponentially small neighbourhood of M_0 given by $y = 0$; see Fig. 1.

It is easy to modify the example to keep the variables on a compact domain; the slow manifold will then be unique. Consider for instance:

Example.

$$\begin{cases} \dot{x} & = \cos t, x(0) = x_0 > 1, \\ \varepsilon \dot{y} & = -\frac{y}{x^2}, y(0) = y_0 \geq 0. \end{cases} \tag{2}$$

We have $x(t) = x_0 + \sin t$ and $(x_0 - 1)^2 \leq x^2(t) \leq (x_0 + 1)^2$. It is easy to see that we have the estimate:

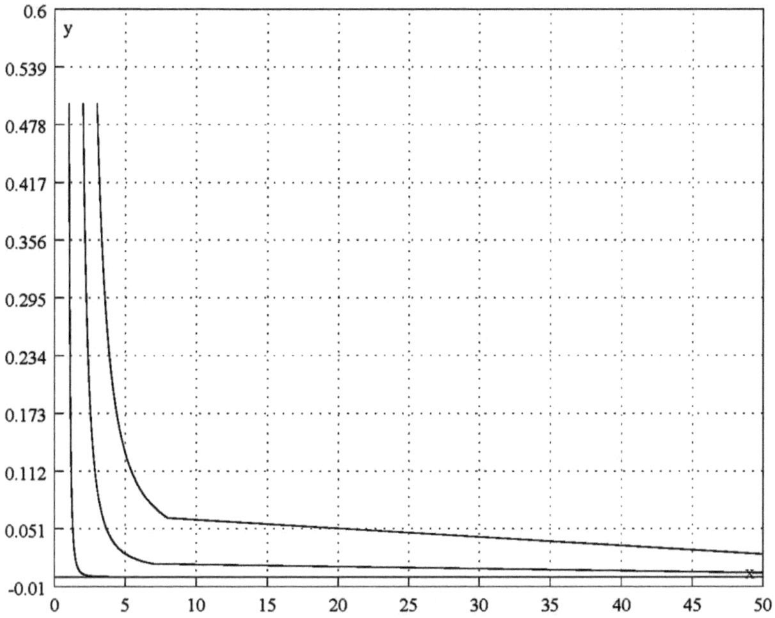

Fig. 1 Slow manifold behaviour of system (1) with initial conditions $x_0 = 1, 2, 3$, $y_0 = 0.5$, $\varepsilon = 0.1$. The limiting behaviour depends on the initial condition and the time interval

$$y_0 e^{-\frac{1}{\varepsilon}\frac{t}{(x_0-1)^2}} \leq y(t) \leq y_0 e^{-\frac{1}{\varepsilon}\frac{t}{(x_0+1)^2}}.$$

The solutions decrease towards $y = 0$.

One might wonder about the practical use of exponential closeness as such solutions are difficult to distinguish numerically. The phenomenon is important and of practical use when there is a change of stability, a bifurcation of the slow manifold. As we will see in a number of examples, exponentially close orbits may trigger different canard phenomena.

2.4 *Permanent or Periodic Canards*

Consider an example discussed in [16] (example 8.13):

$$\varepsilon \dot{y} = -x(t)y(1-y), \;\; y(0) = y_0, \;\; 0 \leq y_0 \leq 1.$$

The continuous function $x(t)$ may be explicitly given or can be derived from a coupled oscillator equation; ε is a small positive parameter. The slow manifolds are $y = 0$ and $y = 1$; one expects that if $x(t)$ changes sign periodically, the solutions will move very fast and periodically from one slow manifold to the other. This turns out not to be the case.

Solving the initial value problem we find

$$y(t) = \frac{y_0 e^{-\frac{1}{\varepsilon}\int_0^t x(s)ds}}{1 - y_0 + y_0 e^{-\frac{1}{\varepsilon}\int_0^t x(s)ds}}.$$

The solutions $y = 0$ and $y = 1$ are the (exact) slow manifolds of the dynamical system. If $x(t) > 0, y = 0$ is stable, $y = 1$ unstable; if $x(t) < 0, y = 1$ is stable, $y = 0$ unstable.

Assume that $x(t)$ is T-periodic with alternating positive and negative values. We can express $x(t)$ as

$$x(t) = a + f(t),\ a = \frac{1}{T}\int_0^T x(t)dt,\ \int_0^T f(t)dt = 0.$$

The solution can then be written as

$$y(t) = \frac{y_0 e^{-\frac{a}{\varepsilon}t} E(t)}{1 - y_0 + y_0 e^{-\frac{a}{\varepsilon}t} E(t)},\ E(t) = e^{-\frac{1}{\varepsilon}\int_0^t f(s)ds}. \tag{3}$$

The function $E(t)$ is positive and bounded. We conclude:

1. The condition for periodicity $y(0) = y(T)$ applied to eq. (3) produces the requirement $a = 0$. If $a = 0$, all solutions with $0 < y_0 < 1$ are T-periodic.
2. If $a > 0$, $\lim_{t\to\infty} y(t) = 0$; although $x(t)$ takes alternating positive and negative values, the solution stays near $y = 0$; we have a permanent canard. See Fig. 2 where $x(t) = a + \sin t$. Increasing n does not change the picture qualitatively.

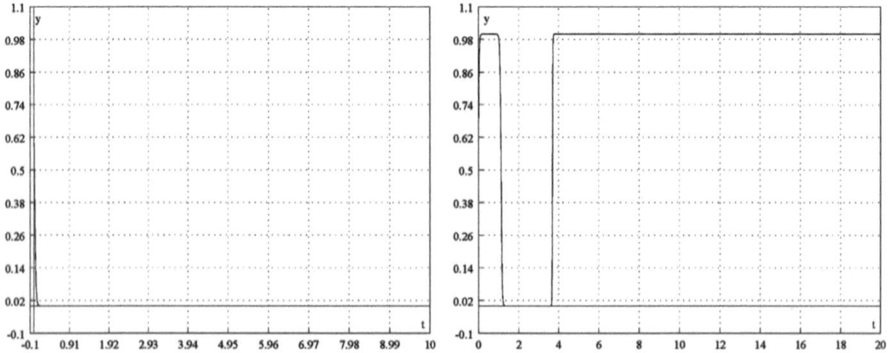

Fig. 2 Permanent canards resulting from the dynamics of $\varepsilon\dot{y} = -(a + \sin nt)y(1 - y)$, $y_0 = 0.6$, $\varepsilon = 0.01$. *Left* $a = 0.5$, $n = 1$, *right* $a = -0.5$, $n = 1$. Although the stability of the slow manifolds changes periodically, the solutions tend after transient behaviour to $y = 0$ and $y = 1$, respectively, to remain in their neighbourhood

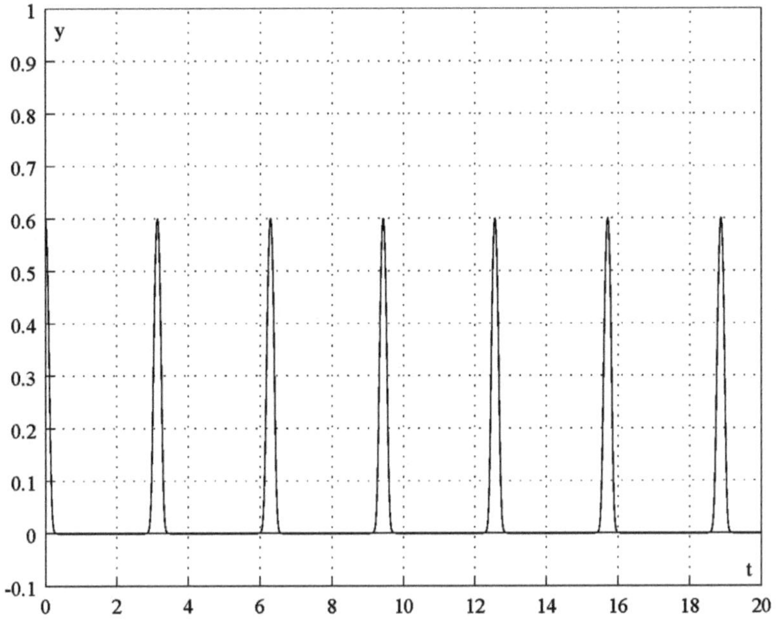

Fig. 3 Spike solutions from the equation $\varepsilon \dot{y} = -\sin(2t)y(1-y)$, $y_0 = 0.6$

3. If $a < 0$, $\lim_{t\to\infty} y(t) = 1$; the slow manifold $y = 1$ shows permanent canard behaviour after a transient excursion to $y = 0$. See Fig. 2. Increasing n does not change the picture qualitatively.
4. If $a = 0$, we have even more interesting canard behaviour with the possibility of spikes for $y(t)$; see Fig. 3 where we have near-homoclinic behaviour in the sense that the spikes return periodically; between the spikes the solutions are exponentially close to the slow manifold $y = 0$. In the case of Fig. 3 we have

$$E(t) = e^{\frac{1}{2\varepsilon}(\cos 2t - 1)}.$$

The spikes arise whenever $\cos 2t - 1 = O(\varepsilon)$; the spike width in this case is $O(\sqrt{\varepsilon})$.

Increasing n for $x(t) = \sin nt$, the spikes diminish in height; this can also be directly concluded from eq. (3).

In [16] it is indicated that if $x(t)$ is derived from a chaotic oscillator, we may have chaotic jumping between the slow manifolds.

3 Periodic Solutions

In a slow-fast system of the form

$$\dot{x} = f(x, y, t), \ \varepsilon\dot{y} = g(x, y, t),$$

with a normally hyperbolic slow manifold (so without critical points), periodic solutions can be found by localising to the slow manifold and applying standard theory for periodic solutions. The idea is based on the rigorous theory of slow manifolds. This is not the main subject of this note but see for instance [17].

Transcritical bifurcations play a part in [6] and [12]. In the last reference a prey-predator system is analysed with a Holling II interaction term. This produces the intersection of two slow manifolds with a periodic solution involving a canard when the intersection is crossed. In the case of a constant carrying capacity, this model is also considered in [18]. As we shall see, a modified logistic equation can be considered as a standard form for this behaviour; an additional advantage of the one-dimensional equations is that often they can be analysed in more analytic detail.

Although we can solve a number of equations explicitly, it is convenient to have general theorems. From [11] we have the one-dimensional equation:

$$\dot{x} = f(x, t), \tag{4}$$

with $f(x, t)$ defined for $x \in \mathbb{R}$ and $t \in [0, \infty)$, continuous and satisfying the uniqueness condition; $f(x, t)$ is T-periodic in t.

- The solutions $x(t)$ of the eq. (4) are either periodic or monotonic in the sense that $x(0) < x(T) < x(2T) < \ldots < x(nT) < \ldots$ (or $x(0) > x(T) > \ldots$ etc.); see Theorem 9.1 in [11].
- If an isolated periodic solution of eq. (4) is Lyapunov-stable, then it is asymptotically stable; see Theorem 9.2 in [11].

4 A Modified Logistic Equation

Consider the non-autonomous equation for $y \geq 0$:

$$\varepsilon\dot{y} = x(t)y - y^2, \tag{5}$$

in which the growth rate $x(t)$ is sufficiently smooth and can take positive and negative values. Values of t for which $x(t)$ vanishes correspond with a transcritical bifurcation. The equation can model daily or seasonal changes of the growth rate. We solve the equation for general continuous $x(t)$ and $y(0) = y_0 > 0$. Putting

$$\Phi(t) = \int_0^t x(s)ds,$$

we find for the solution of eq. (5):

$$y(t) = \frac{e^{\frac{1}{\varepsilon}\Phi(t)}}{\frac{1}{y_0} + \frac{1}{\varepsilon}\int_0^t e^{\frac{1}{\varepsilon}\Phi(s)}ds}. \tag{6}$$

If $x(t)$ is T-periodic, we can write

$$x(t) = a + f(t)$$

and

$$\Phi(t) = at + F(t),\ F(t) = \int_0^t f(s)ds, \tag{7}$$

with a a real constant and $f(t)$ a zero average T-periodic function; $F(t)$ is bounded and $F(T) = 0$. In the cases that $x(t)$ is quasi- or almost-periodic, we can write similar expressions; see for an example Fig. 4.

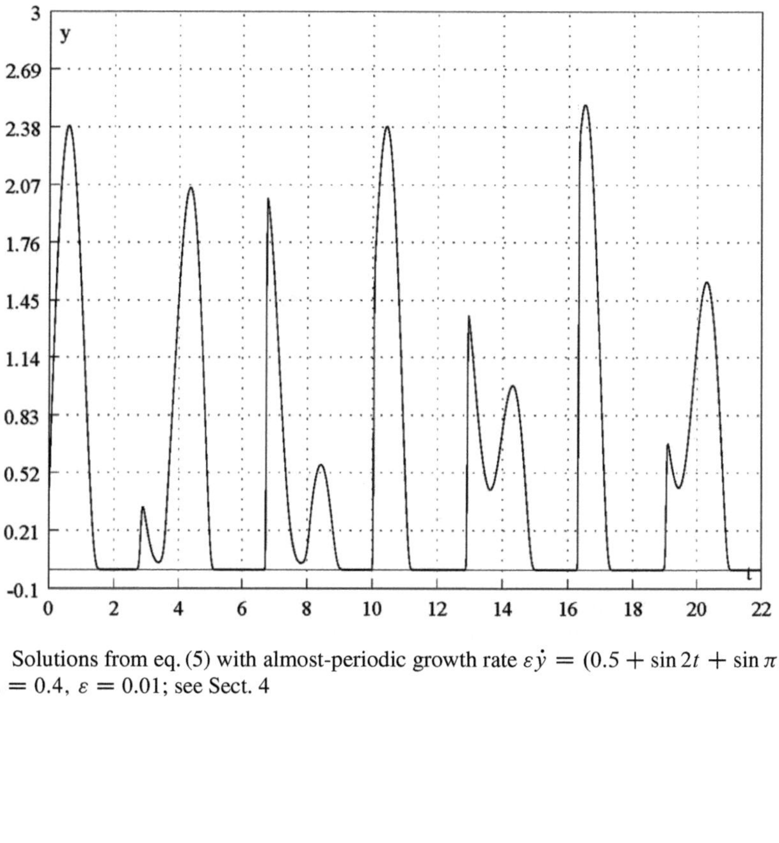

Fig. 4 Solutions from eq. (5) with almost-periodic growth rate $\varepsilon\dot{y} = (0.5 + \sin 2t + \sin \pi t)y - y^2$, $y_0 = 0.4$, $\varepsilon = 0.01$; see Sect. 4

From now on, we assume that $x(t)$ is continuous and T-periodic and δ is a small positive constant, but independent of ε; we have the following cases:

1. If the growth rate is negative, $a + f(t) \leq -\delta < 0$ for $0 \leq t \leq T$, we state that $\lim_{t\to\infty} y(t) = 0$. This can be deduced from the equation with a simple estimate from $\varepsilon\dot{y} \leq -\delta < 0$ for all time, but also from the explicit solution (6). Clearly $a < 0$ and the proof is simple: multiplying with $\exp(-at/\varepsilon)$ produces a bounded numerator and a monotonically increasing denominator. There is no periodic solution with $y_0 > 0$.
2. Less trivial is the case $a = 0$. In this case $\Phi(t)$ is bounded. The solution $y(t)$, given by expression (6), decreases to zero as $t \to \infty$ as the integral takes over a positive function. Again, there is no periodic solution with $y_0 > 0$. In Fig. 5 we choose $x(t) = \sin t$ for an illustration; $y = 0$ is associated with a permanent canard.
3. $a > 0$ with $a + \min_{0\leq t\leq T} f(t) < -\delta < 0$. In this case a periodic solution exists. Several proofs are possible, but a simple one runs as follows:

   Assuming the periodicity condition $y(0) = y(T) = y_0 (> 0)$, we obtain from the solution (6) the expression

$$y_0 = \frac{e^{\frac{aT}{\varepsilon}} - 1}{\frac{1}{\varepsilon}\int_0^T e^{\frac{1}{\varepsilon}\Phi(s)} ds}.$$

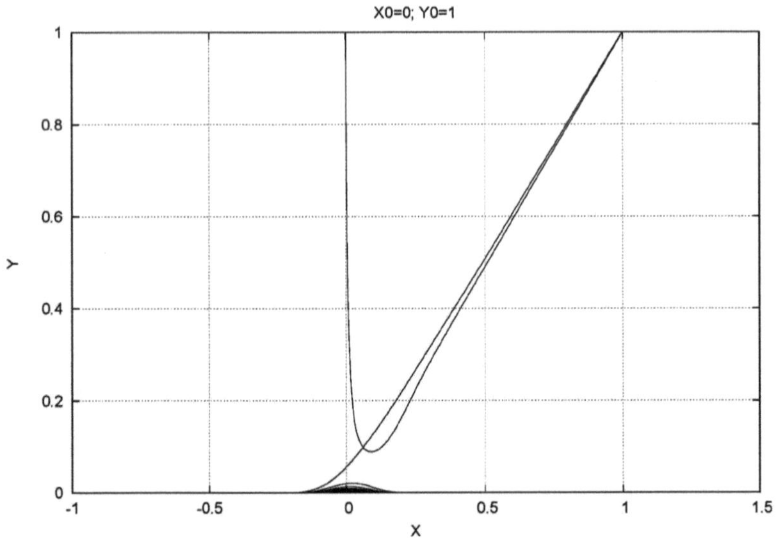

Fig. 5 Solution approaching the periodic solution in $y = 0$ of the modified logistic equation (5) with $x(t) = \sin t, x(0) = 0, y(0) = 1, \varepsilon = 0.01$. To the right of the y-axis, $y = 0$ corresponds with an unstable slow manifold

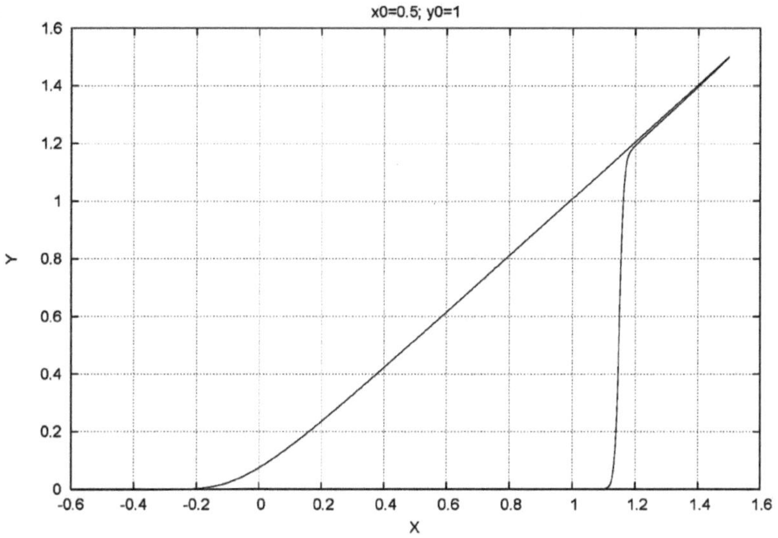

Fig. 6 Canard-like periodic solution of the modified logistic equation (5) with $x(t) = 0.5 + \sin t$, $\varepsilon = 0.01$. To the right of the y-axis, $y = 0$ is an unstable slow manifold; $x = 0$ corresponds with a transcritical bifurcation

With the right-hand side being positive, we have constructed a positive solution for y_0 and so a positive periodic solution.

For more explicit choices of $x(t)$ we can show that y_0 can both be exponentially small and can be $O(1)$. See Fig. 6. The periodic solution follows closely the slow manifold $y = 0$, produces a canard and increases quickly at a positive value of $x(t)$; this value depends on a (in this case 0.5) and the shape of $x(t)$ as it depends on the time interval of exponential attraction during the stable slow manifold phase. For instance, taking for simplicity the non-periodic transition $x(t) = -a + bt$, $a, b > 0$, we have $\Phi(t) = -at + \frac{1}{2}bt^2$. It is easy to show that at the time of transition of the y-axis, $t = a/b$, we have $y(a/b) = O(exp.(-a^2/(2b\varepsilon)$. For $t > a/b$ the solution remains exponentially close to $y = 0$ (canard) until the solution in an $O(\varepsilon)$-neighbourhood of $t = 2a/b$ jumps off to the slow manifold $y = x(t)$. In the case of the example displayed in Fig. 6, this jump off (end of the canard) is delayed somewhat by the shape of $\Phi(t) = 0.5t + 1 - \cos t$ in this example.

The transcritical bifurcation takes place also periodically by the descent of $x(t)$ following the slow manifold $y = x(t)$. Starting in an $O(\varepsilon)$-neighbourhood of the periodic solution on the slow manifold at $x = y = 1$ and choosing for the transition the simple example $x(t) = 1 - bt$, $b > 0$, we find

$$y(\frac{1}{b}) = \frac{e^{\frac{1}{\varepsilon 2b}}}{1 + \frac{1}{\varepsilon}\int_0^{\frac{1}{b}} e^{\frac{1}{\varepsilon}(t - \frac{1}{2}bt^2)} ds} = O(\sqrt{\varepsilon}).$$

Assuming that we can non-trivially linearise the slow manifold and the solutions on it near the transcritical bifurcation, this estimate is typical.

Solutions near the slow manifold $y = 0$ correspond with near-extinction, followed by a sudden explosive increase of the y-variable. The stability question is more delicate in this case; we will consider this in Sect. 5.

4. If $a < 0$, we find a negative expression for y_0 and no periodic solution, but a permanent canard associated with $y = 0$.
5. If the growth rate is periodic but always positive, $a + f(t) \geq \delta > 0$, we find a periodic solution in an ε-neighbourhood of the slow manifold $y = x(t)$. This is an example of the theory mentioned earlier; see [17].

5 The Periodic P.F. Verhulst Model

Consider the classical logistic equation of [19], but now with periodically varying growth rate $r(t)$ and carrying capacity $K(t)$, period T. In standard notation for the population size $N(t)$ with *positive growth rate* $r(t)$, the equation is

$$\varepsilon \dot{N} = r(t)N\left(1 - \frac{N}{K(t)}\right), \quad N(0) = N_0 > 0. \tag{8}$$

We have $K(t) > m > 0$. Without the fast growth perspective, the equation was studied in [1, 14] and [13].

It is a natural assumption, at least for limited intervals of time, that $r(t)$ can take negative values. For such cases we modify the logistic equation to

$$\varepsilon \dot{N} = r(t)N - \frac{N^2}{R(t)}, \quad N(0) = N_0 > 0 \tag{9}$$

with $R(t) > 0$ and T-periodic. Without this modification, a negative growth rate would be accompanied by a positive nonlinear term; there is no rationale for this.

Equations (8) and (9) describe the dynamics if interaction with other populations is negligible.

5.1 Positive Growth Rate

The slow manifolds that exist for eq. (8) are approximated by the limit cases $N(t) = 0$ and $N(t) = K(t)$. Consider the stability of the slow manifolds. The 'eigenvalues' are respectively $r(t)$ and $-r(t)$ so that the slow manifold $N(t) = K(t)$ is stable if $r(t) > \delta > 0$. Replacing $x(t)$ by $r(t)$ we can repeat the calculation of Sect. 4 to obtain

$$N(t) = \frac{e^{\frac{1}{\varepsilon}\Phi(t)}}{\frac{1}{N_0} + \frac{1}{\varepsilon}\int_0^t \frac{r(s)}{K(s)} e^{\frac{1}{\varepsilon}\Phi(s)} ds}. \tag{10}$$

As before, we put $r(t) = a + f(t)$ with a a constant and $f(t)$ a T-periodic function with average zero. From eq. (10) we find with the periodicity condition $N_0 = N(T) > 0$:

$$N_0 = \frac{e^{\frac{aT}{\varepsilon}} - 1}{\frac{1}{\varepsilon}\int_0^T \frac{r(s)}{K(s)} e^{\frac{1}{\varepsilon}\Phi(s)} ds}. \tag{11}$$

As $a > 0$, a periodic solution exists. We conclude: Assume that the growth rate is T-periodic with $r(t) \geq \delta$ with $\delta > 0, 0 \leq t \leq T$, so we have stability of the slow manifold $N(t) = K(t)$; N_0 from eq. (11) is positive. For the periodic solution, we have $N(t) = K(t) + O(\varepsilon)$ for all time. This is not immediately clear from the exact solution (10).

5.2 Positive and/or Negative Growth Rate

The solution of eq. (9) is easy to find:

$$N(t) = \frac{e^{\frac{1}{\varepsilon}\Phi(t)}}{\frac{1}{N_0} + \frac{1}{\varepsilon}\int_0^t \frac{1}{R(s)} e^{\frac{1}{\varepsilon}\Phi(s)} ds}. \tag{12}$$

For intervals of time when $r(t) \geq \delta > 0$, a slow manifold is given by $N(t) = r(t)R(t)$. On such intervals, this slow manifold is stable. When $r(t) \neq 0$, $N(t) = 0$ is a slow manifold.

For a periodic solution to exist we apply the periodicity condition $N(T) = N_0$. This produces

$$N_0 = \frac{e^{\frac{aT}{\varepsilon}} - 1}{\frac{1}{\varepsilon}\int_0^T \frac{1}{R(s)} e^{\frac{1}{\varepsilon}\Phi(s)} ds}. \tag{13}$$

As $r(t)$ changes sign during a period, we have the possibility of periodic transitions between the approximate slow manifold $N(t) = r(t)R(t)$ for $r(t) > 0$ and a neighbourhood of the slow manifold $N(t) = 0$ if $r(t) < 0$ We draw a number of conclusions:

1. Assume stability of the trivial solution ($r(t) \leq -\delta < 0$ for $0 \leq t \leq T$). As expected, it follows from solution (12) that $N(t) \to 0$ for $t \to \infty$, so the population will become extinct. No periodic solution exists.

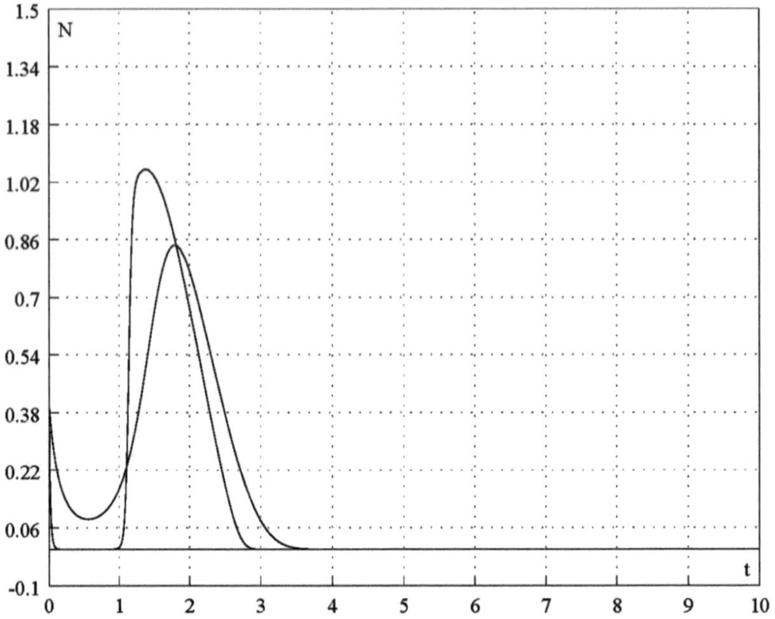

Fig. 7 The case of eq. (9) with sign changing $r(t) = -0.5 + \sin t$, $R(t) = 2 + \cos t$, $N_0 = 0.4$. The first interval where $r(t) > 0$ is $\pi/6 < t < 5\pi/6 (\pi/6 = 0.52\ldots, 5\pi/6 = 2.62\ldots)$. At $t = 5\pi/6$, the trivial solution becomes stable; at $t = 13\pi/6 = 6.80\ldots$, the trivial solution becomes unstable again. If $\varepsilon = 0.1$, a canard develops after $r(t)$ becomes negative (at $t = 2.62\ldots$) that puts the stable slow manifold $N = r(t)R(t)$ that exists for positive values of $r(t)$ outside of reach. If $\varepsilon = 0.01$, a canard develops earlier, at $t = 0.52\ldots$, the transient maximum is larger, a permanent canard develops near $t = 2.62\ldots$

2. Assume that $r(t)$ changes sign with $a < 0$ (or $\int_0^T r(t)dt < 0$). In the periodicity condition (13), the numerator is negative and the denominator is positive, so no periodic solution exists. All solutions of eq. (9) are monotonic. This can also be seen as follows. The integral in the denominator of solution (12) is positive, so we have the estimate:

$$N(t) \leq N_0 e^{\frac{1}{\varepsilon}(at+F(t)}.$$

In the case $a < 0$, this is a strong estimate of exponential decay, but how do we conciliate this with the stability of the slow manifold $N = r(t)R(t)$ for intervals of time when $r(t) > 0$? In Fig. 7 we have $r(t) = -0.5 + \sin t$ so we start with negative values of $r(t)$ for $0 \leq t < \pi/6 = 0.52\ldots$; $r(t)$ is positive for $\pi/6 < t < 5\pi/6 = 2.62\ldots$. After $t = 13\pi/6 = 6.80\ldots$, a canard develops that keeps the solution close to the trivial solution $N(t) = 0$. The solution cannot make the jump transition back to the slow manifold $N = r(t)R(t)$. Taking $\varepsilon = 0.01$, a

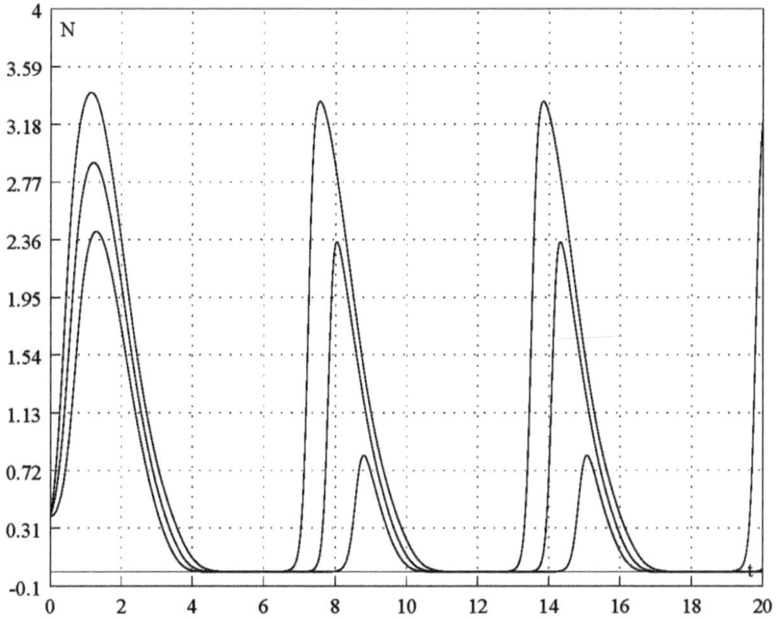

Fig. 8 Three solutions of eq. (9) approaching periodic solutions in the cases $r(t) = a + \sin t$, $R(t) = 2 + \cos t$ with $a = 0.5$ (largest amplitude), $a = 0.3, 0.1$; $N(0) = 0.4, \varepsilon = 0.1$. Canards develop at the transitions from positive to negative values of $r(t)$

canard develops earlier, between $t = \pi/6 = 0.52$ and 1; but near $t = 1$, the solution grows again after which a permanent canard emerges near $t = 2.62$.

3. Assume that $r(t)$ changes sign with $a > 0$ (or $\int_0^T r(t)dt > 0$). In the periodicity condition (13), the numerator is positive, so a unique periodic solution exists. In Fig. 8 we present three solutions approaching periodic solutions if $r(t) = a + \sin t$ with $a = 0.5, 0.3, 0.1$; $R(t) = 2 + \cos t$. Canards develop at the sign transitions of $r(t)$. For instance, if $a = 0.5$, the first transition is at $t = 7\pi/6 = 3.67\ldots$, but the canard delays the transition until a little above $t = 4$. If $a > 0$ is diminished, the corresponding canard increases in size, as expected. In all cases, each cycle faces the possibility of near-extinction.
4. Assume that we have the boundary case when $r(t)$ changes sign with $a = 0$ or $\int_0^T r(t)dt = 0$. In this case the periodicity condition (13) cannot be satisfied. In solution (12), the numerator varies periodically; the integral in the denominator produces a positive, constant contribution at each cycle. The maximum population density will decrease algebraically with time; see Fig. 9.

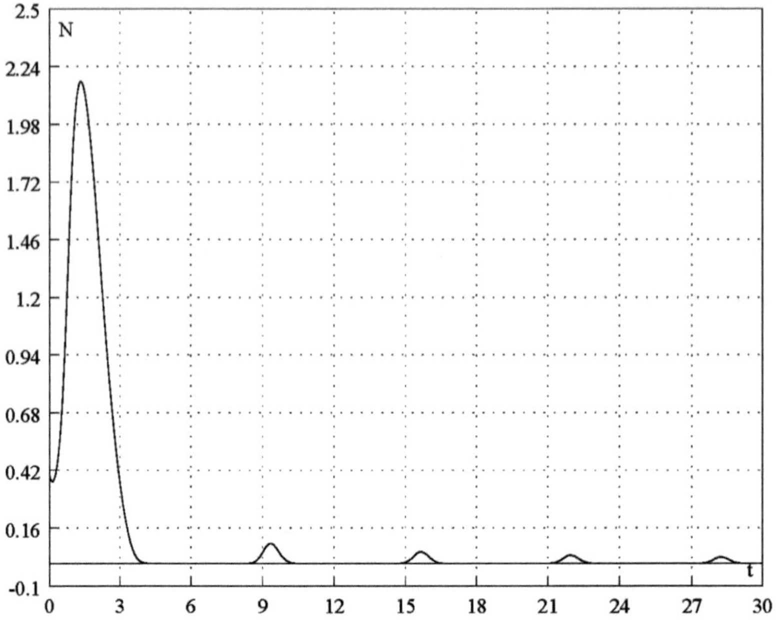

Fig. 9 Solution of eq. (9) in the case $r(t) = \sin t$, $R(t) = 2 + \cos t$, $N(0) = 0.4, \varepsilon = 0.1$. The solution decreases as predicted; for smaller values of ε the decrease is faster

6 Conclusion

It is remarkable that we can analyse the dynamics of a number of one-dimensional growth models in some analytic detail. The analysis may serve as an inspiration for problems with more complicated exchange of stability of slow manifolds.

Our models can be extended to involve one or more predators and to involve spatial diffusion. This will be a subject of a forthcoming paper.

Acknowledgements Taoufik Bakri kindly introduced me to the use of CONTENT. Comments by Odo Diekmann on the interpretation of the time-varying P.F. Verhulst model are gratefully acknowledged.

References

1. Bernadete, D.M., Noonburg, V.W., Pollina, B.: Qualitative tools for studying periodic solutions and bifurcations as applied to the periodically harvested logistic equation. MAA Monthly **115**, 202–219 (2008)
2. Eckhaus, W.: Relaxation oscillations including a standard chase on French ducks. Asymptotic Analysis II. Lect. Notes Math.**985**, 449–494 (1983)

3. Jones, C.K.R.T.: Geometric singular perturbation theory. Dynamical Systems, Lect. Notes Math. Montecatini Terme **1609**, 44–118 (1994)
4. Kaper, T.J.: An introduction to geometric methods and dynamical systems theory for singular perturbation problems. Proc. Symp. Appl. Math. **56** (1999)
5. Kaper, T.J., Jones, C.K.R.T.: A primer on the exchange lemma for fast-slow systems. IMA Vols. Math. Appl. **122** 65–88 (2000)
6. Krupa, M., Szmolyan, P.: Extending slow manifolds near transcritical and pitchfork singularities. Nonlinearity **14**, 1473–1491 (2001a)
7. Krupa, M., Szmolyan, P.: Relaxation oscillation and canard explosion. J. Diff. Equat.**174**, 312–368 (2001b)
8. Kuznetsov, Y.A., Levitin, V.V.: A multiplatform environment for analysing dynamical systems. Dynamical Systems Laboratory, Centrum voor Wiskunde en Informatica, Amsterdam, (1997) (http://www.math.uu.nl/people/kuznet/)
9. Kuznetsov, Y.A., Muratori, S., Rinaldi, S.: Homoclinic bifurcations in slow-fast second order systems. Nonlinear Anal. Theory, Methods Appl. **25**, 747–762 (1995)
10. Neihstadt, A.I.: Asymptotic investigation of the loss of stability by an equilibrium as a pair of imaginary eigenvalues slowly cross the imaginary axis. Usp. Mat. Nauk **40**, 190–191 (1985)
11. Pliss, V.A. Nonlocal problems of the theory of oscillations. Academic Press, NewYork (1966)
12. Rinaldi, S., Muratori, S., Kuznetsov, Y.A.: Multiple attractors, catastrophes and chaos in seasonally perturbed predator-prey communities. Bull. Math. Biol. **55**, 15–35 (1993)
13. Rogovchenko, S.P., Rogovchenko, Y.V.: Effect of periodic environmental fluctuations on the Pearl-Verhulst model. Chaos, Sol. Fract. **39**, 1169–1181 (2009)
14. Sonneveld, P., van Kan, J.: On a conjecture about the periodic solution of the logistic equation. J. Math. Biol. **8**, 285–289 (1979)
15. Tikhonov, A.N.: Systems of differential equations containing a small parameter multiplying the derivative. Mat. Sb. **31**(73), 575–586 (1952)
16. Verhulst, F.: Methods and Applications of Singular Perturbations. Springer, NewYork (2005)
17. Verhulst, F.: Periodic solutions and slow manifolds. Int. J. Bifur. Chaos. **17**(8) 2533–2540 (2007)
18. Verhulst, F., Bakri, T.: The dynamics of slow manifolds. J. Indones. Math. Soc. (MIHMI) **13**, 73–90 (2007)
19. Verhulst, P.F.: Notice sur la loi que la population poursuit dans son accroissement. Correspondance Mathématique et Physique **10**, 113–121 (1838)

Model of Nonlinear Fractal Oscillator in Nanosystem

Valeriy S. Abramov

Abstract Based on the theory of fractional calculus and the concept of fractal the model of nonlinear fractal oscillator is proposed. Hamilton operators in the representation of occupation numbers for different structural states are obtained. The features of the behavior of energy spectrum, deformation, and stress fields for different structural states of the fractal dislocation in a model nanosystem are investigated.

1 Introduction

Serge Haroche and David Wineland, 2012 Nobel Laureates in Physics, proposed experimental methods that made it real to measure individual quantum systems and govern them [8, 10]. The experimental studies of the features of the statistical properties of individual quantum systems in neutron spin measurements [9], with the observation of Bose-Einstein condensation [13], showed the presence of correlations in the measured values. Near singular points (Dirac points) Dirac fermions in molecular graphene show quantum and statistical features of behavior [11]. Actuality of fundamental research of individual quantum systems is related to the possible use of them in quantum information technology. Using the principles of quantum optics [15], the recording and subsequent reading of quantum information (the coded in the polarization states of photons) is carried out on the quantum states of single atoms or collective quantum states of atomic ensemble. It is noted the main difficulty in the processing of quantum information: the results of photodetection are random and reveal the presence of quantum chaos [17], and there is the phenomenon of entanglement states of photons [7]. Physical properties of these quantum systems (nanosystems) are essentially nonlinear. The methods of nonlinear dynamics have been applied to the theoretical description of the chaos in structural mechanics [6], the analysis of nonlinear chaotic models [16], and rare attractors and nonlinear oscillators [18]. Fractal dislocation is one of the nonclassical structural objects in

V.S. Abramov (✉)
National Academy of Sciences of Ukraine, Donetsk, Ukraine
e-mail: vsabramov@mail.ru

J. Awrejcewicz (ed.), *Applied Non-Linear Dynamical Systems*, Springer Proceedings in Mathematics & Statistics 93, DOI 10.1007/978-3-319-08266-0_24

nanostructured materials [1–5]. In order to describe possible correlation effects and statistical properties of the deformation field of fractal dislocation of mixed states a two-point model was proposed [3].

This model is based on the theory of fractional calculus [14] and the concept of a fractal [12]. Accounting for the effect of bifurcation of solutions of nonlinear equations [2] leads to the appearance of the four branches of the lattice nodes displacement function in model nanosystem. The aim of this paper is to describe the nonlinear fractal oscillator in a model of fractal dislocation. The investigation of features of behavior correlation and statistical properties of mixed states of the fractal dislocation in the two-point model.

2 Fractal Oscillator in a Model of Fractal Dislocation

At the construction of model of fractal dislocation in the [1–5] was used the Hamilton operator $\widehat{H}_2$ from [4] for the energy spectrum of fractal dislocation

$$\widehat{H}_2=\varepsilon_1\widehat{n}_1+\varepsilon_2\widehat{n}_2+\varepsilon_3\widehat{n}'_3, \quad \widehat{n}_1=\widehat{a}_1^+\widehat{a}_1, \quad \widehat{n}_2=\widehat{a}_2^+\widehat{a}_2, \quad \widehat{n}'_3=\widehat{a}_3\widehat{a}^+{}_3, \quad \widehat{n}_3=\widehat{a}^+{}_3\widehat{a}_3. \tag{1}$$

Here $\widehat{n}_1, \widehat{n}_2, \widehat{n}_3$ are the operators of occupation numbers states of the dislocation with nondimensional own energies $\varepsilon_1 = \varepsilon_2$, ε_3. Relations between the new $\widehat{a}_1^+, \widehat{a}_2^+, \widehat{a}_3$ and old $\widehat{\psi}_1, \widehat{c}^+, \widehat{c}$ operators are defined by expressions

$$\widehat{a}_1^+=t_{11}\widehat{\psi}_1+t_{21}\widehat{c}^+ + t_{31}\widehat{c}, \quad \widehat{a}_2^+=t_{12}\widehat{\psi}_1+t_{22}\widehat{c}^+ +t_{32}\widehat{c}, \quad \widehat{a}_3=t_{23}\widehat{c}^+ +t_{33}\widehat{c}, \quad \widehat{\psi}_1=D_z^\alpha. \tag{2}$$

In expressions (2) the elements t_{ij} of the matrix $\widehat{T}$ are defined by the relations

$$\begin{aligned} &t_{11} = k', \quad t_{12} = -k, \quad t_{13} = 0, \quad t_{21} = k\ cn\,(u_\alpha, k)\,, \quad t_{22} = k'\ cn\,(u_\alpha, k)\,, \\ &t_{23} = -sn\,(u_\alpha, k)\,, \quad t_{31} = k\ sn\,(u_\alpha, k)\,, \quad t_{32} = k' sn\,(u_\alpha, k)\,, \quad t_{33} = cn\,(u_\alpha, k)\,. \end{aligned} \tag{3}$$

Here k and $u_\alpha = F(\varphi_\alpha, k)$ are the module and the argument of the Jacobi elliptic functions $sn(u_\alpha, k)$, $cn(u_\alpha, k)$; $(k')^2 = 1 - k^2$; F is an incomplete elliptic integral of the first kind; φ_α is the polar angle. Using (2), we find the commutation relations for the new operators

$$\begin{aligned} &\left[\widehat{a}_1, \widehat{a}_1^+\right] = \widehat{n}'_1 - \widehat{n}_1 = -2kk's_{21}\widehat{b}_{\alpha 2} + k^2 b_0 \widehat{b}_{\alpha 3}, \\ &\widehat{b}_{\alpha 2} = \left[\widehat{\psi}_1, \widehat{z}\right] = \alpha I_z^{1-\alpha}, \quad \widehat{n}'_1 = \widehat{a}_1\widehat{a}_1^+, \\ &\widehat{n}'_2 = \widehat{a}_2\widehat{a}_2^+, \quad \left[\widehat{a}_2, \widehat{a}_2^+\right] = \widehat{n}'_2 - \widehat{n}_2 = -2kk's_{21}\widehat{b}_{\alpha 2} - (k')^2 b_0 \widehat{b}_{\alpha 3}, \quad \widehat{\psi}_2 = D_z^{1-\alpha}, \end{aligned}$$

$$s_{21} = (cn\,(u_\alpha, k) + sn(u_\alpha, k))\,/\sqrt{2}, \quad \left[\widehat{a}_3, \widehat{a}_3^+\right] = \widehat{n}_3' - \widehat{n}_3 = b_0 \widehat{b}_{\alpha 3}, \quad b_0 = 1 - 2n_{30},$$
$$n_{30} = sn^2\,(u_\alpha, k), \quad \widehat{b}_{\alpha 3} = \left[\widehat{\psi}_2, \widehat{z}\right] = (1-\alpha)\, I_z^\alpha = \left[\widehat{c}, \widehat{c}^+\right], \tag{4}$$

where $\widehat{z}$ is the coordinate operator. Structure operators fractional partial derivative (integral) of the Riemann-Liouville D_z^α (I_z^α) on dimensionless coordinate z with the index order α is defined as

$$D_z^\alpha \Phi = \partial_z \int_{z_0}^{z} \Phi\,(\xi)\, |z-\xi|^{-\alpha} d\xi / \Gamma\,(1-\alpha), \quad I_z^\alpha \Phi = \int_{z_0}^{z} \Phi\,(\xi)\, |z-\xi|^{\alpha-1} d\xi / \Gamma\,(\alpha), \tag{5}$$

where ∂_z is operator ordinary partial derivative on z and Γ is gamma function. Indices $\alpha, 1-\alpha$ have the meaning of fractal dimensions along the axis Oz. Taking into account (2) and (4) we obtain the relations

$$2\widehat{c}\widehat{c}^+ = \widehat{z}^2 + \left[\widehat{\psi}_2, \widehat{z}\right] - \widehat{\psi}_2^2, \quad 2\widehat{c}^+\widehat{c} = \widehat{z}^2 - \left[\widehat{\psi}_2, \widehat{z}\right] - \widehat{\psi}_2^2, \quad \left\{\widehat{c}, \widehat{c}^+\right\} = \widehat{c}\widehat{c}^+ + \widehat{c}^+\widehat{c} = \widehat{z}^2 - \widehat{\psi}_2^2. \tag{6}$$

Acting by the operator $\left\{\widehat{c}, \widehat{c}^+\right\}$ to the function $\Phi_{\alpha c}$, we obtain the equation of a fractal oscillator

$$\left(\widehat{z}^2 - \widehat{\psi}_2^2\right) \Phi_{\alpha c} = \left(\widehat{z}^2 - D_z^{1-\alpha} D_z^{1-\alpha}\right) \Phi_{\alpha c} = \left(\widehat{c}\widehat{c}^+ + \widehat{c}^+\widehat{c}\right) \Phi_{\alpha c} = (2n_{\alpha c} + 1)\, \Phi_{\alpha c}. \tag{7}$$

Here $n_{\alpha c}, \Phi_{\alpha c}$ are the eigenvalues (generally fractional values may depend on z, α) and eigenfunctions of the fractal oscillator. Acting by the operator $\widehat{b}_{\alpha 3}$ to the function $\Phi_{\alpha c}$, we obtain the equation

$$\widehat{b}_{\alpha 3} \Phi_{\alpha c} = \left[\widehat{c}, \widehat{c}^+\right] \Phi_{\alpha c} = (1-\alpha)\, I_z^\alpha \Phi_{\alpha c} = \chi_{\alpha c} \Phi_{\alpha c}. \tag{8}$$

Function $\chi_{\alpha c}$ in the general case may depend on α and z. Acting by the operator D_z^α on the left of (8), we obtain the nonlinear equation

$$D_z^\alpha\,(\chi_{\alpha c} \Phi_{\alpha c}) = D_z^\alpha\,(1-\alpha)\, I_z^\alpha \Phi_{\alpha c} = (1-\alpha)\, \Phi_{\alpha c} = D_z^\alpha \left[\widehat{c}, \widehat{c}^+\right] \Phi_{\alpha c} = D_z^\alpha \widehat{b}_{\alpha 3} \Phi_{\alpha c}. \tag{9}$$

To find the eigenvalues $n_{\alpha c}$ and eigenfunctions $\Phi_{\alpha c}$, equations (7) and (9) must be solved together. These equations are fundamental to describe the nonlinear fractional oscillator. Note that if $\alpha = 0$ from (7) and (8) follow the equation for the well-known linear quantum oscillator with positive integer eigenvalues

$n_c = 0, 1, 2,..$; eigenfunctions Φ_c; boson commutation relations for the operators $\widehat{c}, \widehat{c}^+$; and parameter $\chi_{\alpha c} = 1$

$$\left(\widehat{z}^2 - D_z^1 D_z^1\right)\Phi_c = \left(\widehat{z}^2 - \frac{d^2}{dz^2}\right)\Phi_c = \left(\widehat{c}\widehat{c}^+ + \widehat{c}^+\widehat{c}\right)\Phi_c = (2n_c + 1)\,\Phi_c, \quad \left[\widehat{c}, \widehat{c}^+\right]\Phi_c = \Phi_c. \tag{10}$$

On the basis of the commutation relations (4) we obtain the Hamiltonians in the representation of occupation numbers for other structural states of fractal dislocation:

$$\begin{aligned}
&\widehat{H}_1 = \varepsilon_2(\widehat{n}_1 + \widehat{n}_2) + \varepsilon_3\widehat{n}_3, \quad \widehat{H}_2 = \varepsilon_2(\widehat{n}_1 + \widehat{n}_2) + \varepsilon_3\widehat{n}_3' = \widehat{H}_1 + \varepsilon_3 b_0\widehat{b}_{\alpha 3},\\
&\widehat{H}_3 = \varepsilon_1\widehat{n}_1 + \varepsilon_2\widehat{n}_2' + \varepsilon_3\widehat{n}_3 = \widehat{H}_1 - \varepsilon_2\widehat{A}_{\alpha 2}, \quad \widehat{A}_{\alpha 2} = 2kk's_{21}\widehat{b}_{\alpha 2} + (k')^2 b_0\widehat{b}_{\alpha 3},\\
&\widehat{H}_4 = \varepsilon_1\widehat{n}_1' + \varepsilon_2\widehat{n}_2 + \varepsilon_3\widehat{n}_3 = \widehat{H}_1 + \varepsilon_2\widehat{A}_{\alpha 1}, \quad \widehat{A}_{\alpha 1} = -2kk's_{21}\widehat{b}_{\alpha 2} + k^2 b_0\widehat{b}_{\alpha 3},\\
&\widehat{H}_5 = \varepsilon_1\widehat{n}_1 + \varepsilon_2\widehat{n}_2' + \varepsilon_3\widehat{n}_3' = \widehat{H}_1 + 2\varepsilon_2 kk's_{21}\widehat{b}_{\alpha 2} + \left(\varepsilon_3 + (k')^2\varepsilon_2\right) b_0\widehat{b}_{\alpha 3},\\
&\widehat{H}_6 = \varepsilon_1\widehat{n}_1' + \varepsilon_2\widehat{n}_2 + \varepsilon_3\widehat{n}_3' = \widehat{H}_1 - 2\varepsilon_2 kk's_{21}\widehat{b}_{\alpha 2} + \left(\varepsilon_3 - k^2\varepsilon_2\right) b_0\widehat{b}_{\alpha 3},\\
&\widehat{H}_7 = \varepsilon_1\widehat{n}_1' + \varepsilon_2\widehat{n}_2' + \varepsilon_3\widehat{n}_3 = \widehat{H}_1 + \varepsilon_2 b_0\widehat{b}_{\alpha 3},\\
&\widehat{H}_8 = \varepsilon_1\widehat{n}_1' + \varepsilon_2\widehat{n}_2' + \varepsilon_3\widehat{n}_3' = \widehat{H}_1 + (\varepsilon_2 + \varepsilon_3)\, b_0\widehat{b}_{\alpha 3}.
\end{aligned} \tag{11}$$

Dimensionless displacement u points of fractal dislocation (deformation field) are connected with a parameter λ_α (stress field) by model relations (Hooke's law):

$$u = \lambda_\alpha/\lambda_0 = F(\varphi, k), \quad u_\alpha = u - u_0, \tag{12}$$

where λ_0 is the normalization parameter and u_0 is the constant (critical) displacement.

3 The Two-Point Model for Mixed States Nanosystem

In this two-point model [3] based on the Hamiltonian $\widehat{H}_2$ (1) the deformation field of the fractal linear dislocation [1, 2] in a model sample of finite nanosize with volumetric discrete lattice $N_1 \times N_2 \times N_3$ is investigated. Deviations of the lattice nodes from the state of equilibrium in a separate plane $N_1 \times N_2$ for two different points of $z_1(j')$ and $z_2(j')$ are described by non-Hermitian displacement operators $\widehat{u}(z_1)$ and $\widehat{u}(z_2)$, corresponding to the rectangular matrix with dimensions $N_1 \times N_2, j \in [1, N_3]$. The original rectangular matrix displacements $\widehat{u}(z_1)$ and $\widehat{u}(z_2)$ with elements $u_{nm}(z_1) = u_{\varepsilon 1}(z_1)$, $u_{nm}(z_2) = u_{\varepsilon 1}(z_2)$ in bulk lattice $N_1 \times N_2 \times N_3 = 30 \times 40 \times 67$ were obtained by the method of iterations on an index m for the first and second branches of the dimensionless complex displacement functions $u(z) = u_{\varepsilon 1}(z)$ and $u(z) = u_{\varepsilon 2}(z)$ by the formulas in [3], respectively,

$$u(z) = u_{\varepsilon 1}(z) = (g_1 - g_2 + g_4)/2, \quad u(z) = u_{\varepsilon 2}(z) = (g_1 - g_2 - g_4)/2. \quad (13)$$

Functions g_1, g_2, g_3 by analogy with [3] are modeled by expressions

$$\begin{aligned} g_1(u,\alpha) &= (1-\alpha)\left(1 - 2sn^2\left(u - u_0, k\right)\right)/\left(p_0 - p_1 n - p_2 m - p_3 j\right), \\ g_2(z,\alpha) &= 2^{-2\alpha}3^{3\alpha-1/2}\,|z - z_c|^{-\alpha}\Gamma\left(\alpha+1/3\right)\Gamma\left(\alpha+2/3\right)/\sqrt{\pi}\,\Gamma\left(\alpha+1/2\right), \\ g_3(z,\alpha) &= 3^{3\alpha-1/2}\,2\,|z - z_c|^{-2\alpha}\Gamma\left(\alpha+1/3\right)\Gamma\left(\alpha+2/3\right)/\pi, \\ g_4 &= \left[(g_1+g_2)^2 - g_3\right]^{1/2}. \end{aligned} \quad (14)$$

Here p_0, p_1, p_2, p_3 are some governing parameters. The initial parameters were the following: $\alpha = 0.5$; $k = 0.5$; $u_0 = 29.537$; $p_0 = 0.01$; $p_1 = 0.00075$; $p_2 = p_3 = 0$; $z_c = 2.813$. In the calculations it should be : $z_1 = 0.053 + h_z(j' + 33)$; $z_2 = 6.653 - h_z(j' + 33)$; $j' \in [-33; 33]$; $h_z = 0.1$, which corresponds to the forward and backward waves of displacements $u_{nm}(z_1)$, $u_{nm}(z_2)$; $n = \overline{1, 30}$; $m = \overline{1, 40}$. The value of j' plays the role of dimensionless current discrete time. For $j' = 0$ we have $z_1 = z_2 = 3.353$. The choice of calculation scheme for the forward and backward waves, an example of the behavior of real parts $u'_{\varepsilon 1} = \mathrm{Re}u_{\varepsilon 1}$ and $u'_{\varepsilon 2} = \mathrm{Re}u_{\varepsilon 2}$ for the first and second branches of the direct wave of complex displacement functions from (12) are shown in Fig. 1. The values of functions $|z_1 - z_c|$, $|z_2 - z_c|$ at the

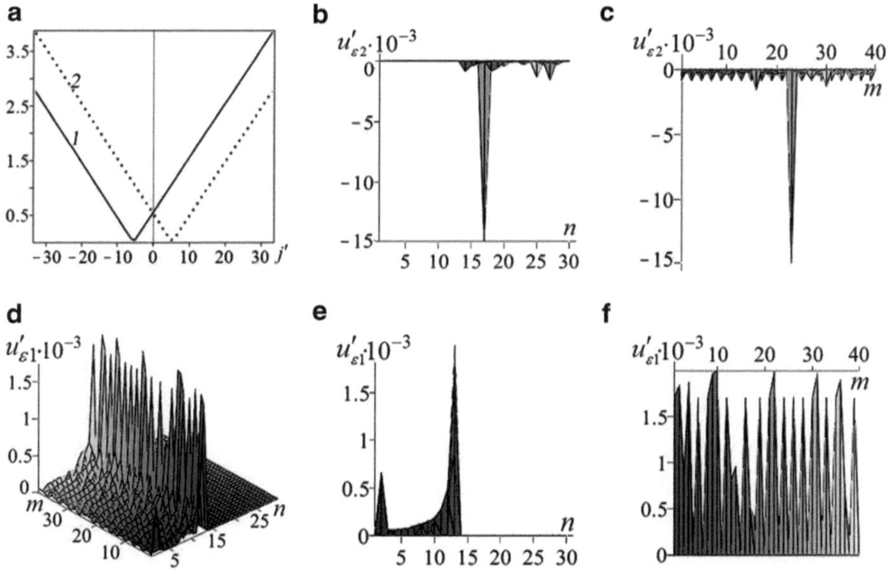

Fig. 1 Dependencies of $|z_1 - z_c|$ (curve 1), $|z_2 - z_c|$ (curve 2) on index j' at $z_c = 2.813$ for the forward and backward waves (**a**); behavior $\mathrm{Re}u_{\varepsilon 1}(z_1)$ (**d**), its projections (**e**, **f**); projections $\mathrm{Re}u_{\varepsilon 2}(z_1)$ (**b**, **c**) on the lattice index n, m for the forward wave at $z_1 = 0.053$

minimum (equal to 0.04) are achieved at $j' = -5$, $j' = 5$, respectively. The values of these functions at the intersection point ($j' = 0$) are equally 0.54 (Fig. 1a). The choice of the governing parameters p_0, p_1, p_2, p_3 determines the state of a discrete rectangular sublattice $N_1 \times N_2$ with fractal dislocation, localized within this region nOm parallel to the axis Om at the moment time $j' = -33$ for the forward wave (Fig. 1b–f). The core of the linear dislocation for the first branch $u'_{\varepsilon 1}$ is localized at $n = 14$ (Fig. 1e). Deviation nodes of the core along the axis Om are pronounced stochastic behavior (Fig. 1f). At $n < 14$ there is a pronounced stochastic behavior of the whole plane nOm, and at $n > 14$ these deviations are practically zero. The core of the linear dislocation for the second branch $u'_{\varepsilon 2}$ is localized at $n = 17$ (Fig. 1b). In this case, on the background of the stochastic deviations nodes of the core along the axis Om, a pronounced peak (at $m = 23$) is observed (Fig. 1c). The appearance of this peak indicates about the presence of nonlinear fractional oscillator with special properties. At $n > 15$ there is a pronounced stochastic behavior of the whole plane nOm, and at $n < 15$ these deviations are practically zero. The results in Fig. 1 make it possible to interpret the linear fractal dislocation as a set of coupled nonlinear fractal oscillators.

For the description of mixed states the effective composite operators of displacements are introduced:

$$\widehat{u}_1 = \widehat{\rho}_{12}\widehat{u}^+(z_1), \quad \widehat{u}_3 = \widehat{\rho}_{12}\widehat{u}^+(z_2), \quad \widehat{u}_5 = \widehat{u}(z_1)\widehat{\rho}_{12}^T, \quad \widehat{u}_7 = \widehat{u}(z_2)\widehat{\rho}_{12}^T, \tag{15}$$

$$\widehat{u}_2 = \widehat{\rho}_{21}\widehat{u}(z_1), \quad \widehat{u}_4 = \widehat{\rho}_{21}\widehat{u}(z_2), \quad \widehat{u}_6 = \widehat{u}^+(z_1)\widehat{\rho}_{21}^T, \quad \widehat{u}_8 = \widehat{u}^+(z_2)\widehat{\rho}_{21}^T. \tag{16}$$

For the states $p = 1, 2, \ldots 8$, respectively. Here the symbols "+" and "T" mean the operation of Hermitian conjugation and transposition. The square matrices with sizes $N_1 \times N_1$ for $p = 1, 3, 5, 7$ (squeezed states) and $N_2 \times N_2$ for $p = 2, 4, 6, 8$ (stretched states) correspond to the introduced operators $\widehat{u}_p$, so that $\widehat{u}_5 = \widehat{u}_1^+$, $\widehat{u}_7 = \widehat{u}_3^+$, $\widehat{u}_6 = \widehat{u}_2^+$, $\widehat{u}_8 = \widehat{u}_4^+$. The density state operators $\widehat{\rho}_{12}$, $\widehat{\rho}_{12}^T$, $\widehat{\rho}_{21}$, $\widehat{\rho}_{21}^T$ are represented by

$$\widehat{\rho}_{12} = \widehat{\xi}_{N1}^T\widehat{\xi}_{N2}/N_1N_2, \quad \widehat{\rho}_{12}^T = \widehat{\xi}_{N2}^T\widehat{\xi}_{N1}/N_1N_2, \quad \widehat{\rho}_{21} = \widehat{\rho}_{12}^T, \quad \widehat{\rho}_{21}^T = \widehat{\rho}_{12}, \tag{17}$$

where $\widehat{\xi}_{N1}$, $\widehat{\xi}_{N2}$ are row vectors of dimensions $1 \times N_1$, $1 \times N_2$, with elements equal to one. The rectangular matrices $\widehat{\rho}_{12}$, $\widehat{\rho}_{21}$ have dimensions $N_1 \times N_2$, $N_2 \times N_1$. For the operators in (17) the normalization conditions are fulfilled $\widehat{\xi}_{N1}\widehat{\rho}_{12}\widehat{\xi}_{N2}^T = 1$; $\widehat{\xi}_{N2}\widehat{\rho}_{21}\widehat{\xi}_{N1}^T = 1$. Having performed an averaging over the index nodes n, m by calculating trace Sp of square matrices (15), (16), the averaged functions u_p, $|u_p|$, $tg\varphi_p$ for $p = 1, 2, \ldots 8$ are obtained:

$$u_p = Sp\widehat{u}_p = u'_p + iu''_p = |u_p|\exp(i\varphi_p), \quad u_p^* = Sp\widehat{u}_p^+, \quad tg\varphi_p = u''_p/u'_p, \tag{18}$$

where $u_p^{'} = \mathrm{Re}u_p$, $u_p^{''} = \mathrm{Im}u_p$, the symbol "$_*$" means the operation of complex conjugation, and $|u_p|$, φ_p are a module, a phase of the complex averaged functions u_p. Here the averaging across indexes j, j' is not made. Then we find the correlation function of the first order K_{pq} for $p, q \in [1, 8]$

$$\begin{gathered} K_{pq} = S_{pq} - H_{pq} = K'_{pq} + iK''_{pq} = |K_{pq}| \exp(i\theta_{pq}), \quad |H_{pq}| = |u_p| \cdot |u_q|, \\ S_{pq} = Sp\widehat{S}_{pq} = S'_{pq} + iS''_{pq} = |S_{pq}| \exp(i\psi_{pq}), \quad \widehat{S}_{pq} = \widehat{u}_p\widehat{u}_q^{+}, \quad \widehat{S}_{pq}^{+} \neq \widehat{S}_{pq}, \\ H_{pq} = (Sp\widehat{u}_p)\left(Sp\widehat{u}_q^{+}\right) = u_p u_q^{*} = H'_{pq} + iH''_{pq} = |H_{pq}| \exp(i\delta_{pq}), \quad \delta_{pq} = \varphi_p - \varphi_q. \end{gathered} \tag{19}$$

Having done the normalization of the above functions, we obtain the distribution function of mixed states of Bose-Einstein type f_{pp} (ground state) and $f_{pp}^{'}$ (excited state), Fermi-Dirac type F_{pp} (ground state) and $F_{pp}^{'}$ (excited state), and inversion parameters d_{pp} in the form

$$f'_{pp} - f_{pp} = 1, \quad f'_{pp} = S_{pp}/H_{pp}, \quad f_{pp} = K_{pp}/H_{pp}, \tag{20}$$

$$F'_{pp} + F_{pp} = 1, \quad F_{pp} = H_{pp}/S_{pp}, \quad F'_{pp} = K_{pp}/S_{pp}, \quad d_{pp} = 1 - 2F_{pp}. \tag{21}$$

Distribution functions of the Fermi-Dirac type carry information about the deformation field. The distribution functions of Bose-Einstein type carry information about the stress field. The behavior of the distribution functions of the Fermi-Dirac type, Bose-Einstein type, and inversion parameters on the dimensionless time j' for squeezed and stretched states is shown in Figs. 2 and 3. From these figuresa significant difference in the behavior of these functions of the dimensionless time j' for the first $u_{\varepsilon 1}$ and second $u_{\varepsilon 2}$ branches of the complex displacement function u is seen.

Note that the behavior of these functions for the first branch $u_{\varepsilon 1}$, as a function of the dimensionless coordinates z, discussed in detail in [3]. For the second branch $u_{\varepsilon 2}$ is characterized by the presence of peaks up to the dependencies f_{11},f_{22} on j' at $j' = 17$. For a squeezed state (Fig. 2b, d), this peak is formed from the ground state with $F_{11} = 0.228$ ($d_{11} = 0.544$), and for a stretched state (Fig. 3b, d), this peak is formed from an inverted state with $d_{22} = -0.402$ at $F_{22} = 0.701$.

The appearance of such peaks indicates the presence of nonlinear fractal oscillators with special properties. These results can be interpreted by analogy in terms of transient effects such as induction, avalanche, self-induced transparency, and echo, known from quantum optics [15]. At $p \neq q$ from (19) it follows that the function K_{pq} is complex. For some values p, q this function has a sense of cross-correlated function (for a pair of different points z_1, z_2).

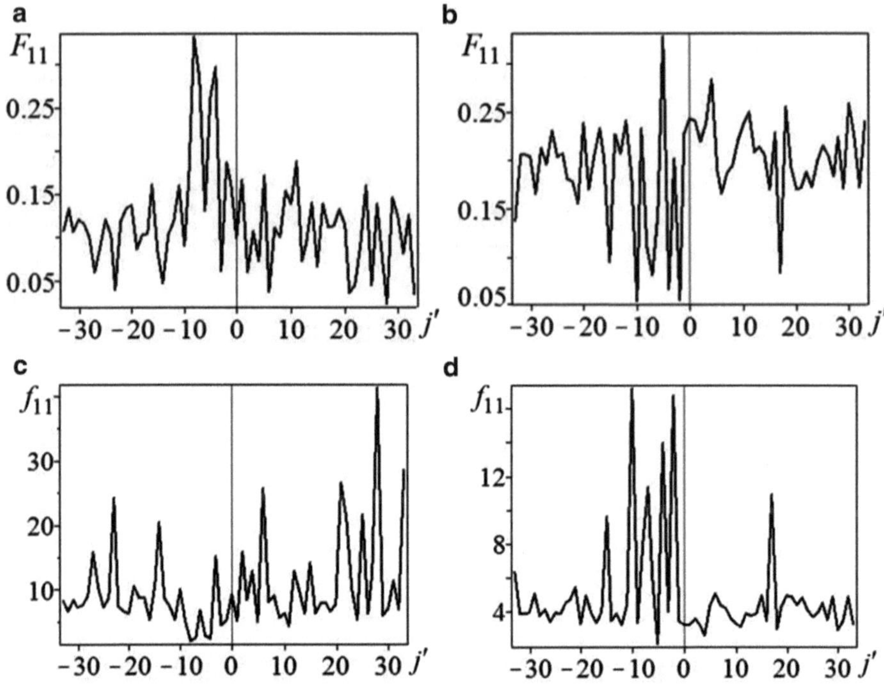

Fig. 2 Dependencies of the distribution functions of the Fermi-Dirac type (**a, b**) and Bose-Einstein type (**c, d**) on j' for squeezed states $(p, p) = (1, 1)$ of branches $u_{\varepsilon 1}$ (**a, c**), $u_{\varepsilon 2}$ (**b, d**)

4 The Characteristic Functions for Effective Pair Operators of Displacements

To investigate the correlation between the effective pair operators of displacements $\widehat{S}_{pq}, \widehat{S}^{+}_{pq}$, there is a need for second-order correlation functions G_{pq}

$$G_{pq}=V_{pq}-W_{pq}, \quad V_{pq}=Sp\widehat{V}_{pq}, \quad \widehat{V}_{pq}=\widehat{S}_{pq}\widehat{S}^{+}_{pq}, \quad \widehat{V}^{+}_{pq}=\widehat{V}_{pq},$$
$$W_{pq}=\left(Sp\widehat{S}_{pq}\right)\left(Sp\widehat{S}^{+}_{pq}\right)=S_{pq}S^{*}_{pq}=|S_{pq}|^{2}, \quad h'_{pq}+h_{pq}=1, \quad h'_{pq}=W_{pq}/V_{pq},$$
$$h_{pq}=G_{pq}/V_{pq}, \quad D_{pq}=1-2h'_{pq}, \quad C'_{pq}=1/h'_{pq}, \quad C'_{pq}-C_{pq}=1. \tag{22}$$

Dependencies from (23) of the inversion parameters D_{57} (Fig. 4a, b) and D_{68} (Fig. 5a, b), the distribution functions of Bose-Einstein type C'_{57} (Fig. 4c, d) and C'_{68} (Fig. 5c, d) on the dimensionless time j' for the squeezed and stretched states (p, q) show significant differences in the behavior deformation and stress fields for the two branches of the complex displacement functions $u_{\varepsilon 1}, u_{\varepsilon 2}$.

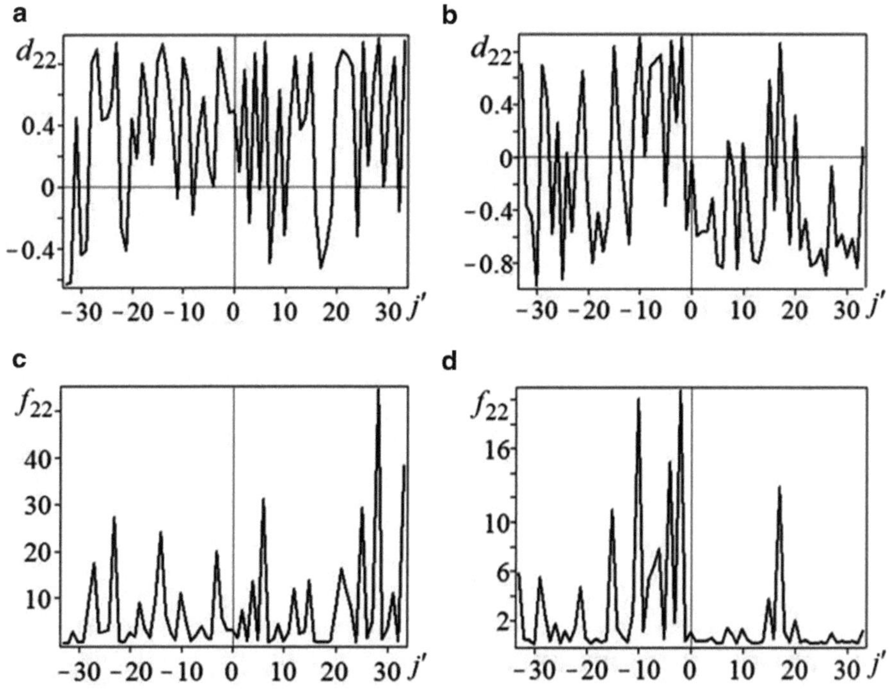

Fig. 3 Dependencies of the inversion parameters (**a**, **b**) and the distribution functions of Bose-Einstein type (**c**, **d**) on j' for stretched states $(p,p)=(2,2)$ of branches $u_{\varepsilon 1}$ (**a**, **c**), $u_{\varepsilon 2}$ (**b**, **d**)

Thus functions h_{57}, h_{68} are positive and can be interpreted as the distribution functions of Fermi-Dirac type for effective pair operators of displacements. Using (19) we find a representation for

$$\left|S_{pq}\right|^2 = \left(\left|K_{pq}\right| - \left|u_p\right| \cdot \left|u_q\right|\right)^2 + 2\left|u_p\right| \cdot \left|u_q\right| \cdot \left|K_{pq}\right| \left(1 + \cos \Phi_{pq}\right), \tag{23}$$

where $\Phi_{pq} = \delta_{pq} - \theta_{pq}$. The analysis of the influence of correlation effects for the phase Φ_{pq} is conveniently carried out in terms of characteristic functions R_{pq}, r_{pq}, μ^2_{pq}, which are given by

$$R'_{pq} + R_{pq} = 1, \quad R'_{pq} = 1 + \cos \Phi_{pq}, \quad R_{pq} = -\cos \Phi_{pq}, \tag{24}$$

$$r'_{pq} - r_{pq} = 1, \quad r'_{pq} = 2/\left(1 - R_{pq}\right), \quad r_{pq} = \left(1 + R_{pq}\right)/\left(1 - R_{pq}\right), \tag{25}$$

$$\left(\mu'_{pq}\right)^2 + \mu^2_{pq} = 1, \quad \left(\mu'_{pq}\right)^2 = \cos^2\left(\Phi_{pq}/2\right) = 1/r'_{pq}, \quad \mu^2_{pq} = \sin^2\left(\Phi_{pq}/2\right). \tag{26}$$

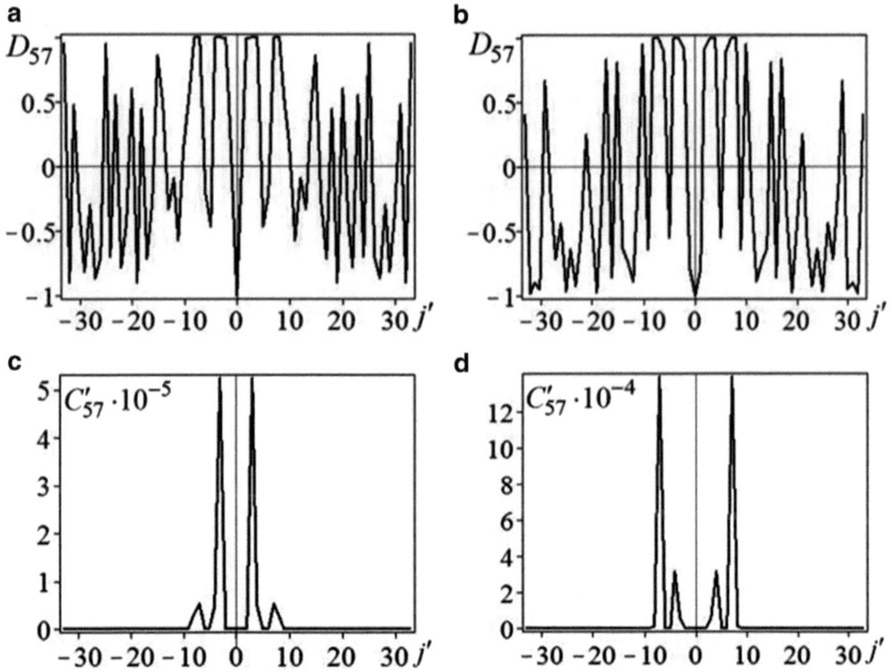

Fig. 4 Dependencies of the inversion parameters (**a**, **b**) and the distribution functions of Bose-Einstein type (**c**, **d**) on j' for squeezed states $(p, q) = (5, 7)$ of branches $u_{\varepsilon 1}$ (**a**, **c**), $u_{\varepsilon 2}$ (**b**, **d**)

Functions R_{pq} from (24) have a sense of the normalized parameter describing the population inversion between the states p and q taking into account correlations (Figs. 6a, b, 7a, b).

Functions r_{pq}, r'_{pq} from (25) make sense of the characteristic distribution functions of Bose-Einstein type in ground, excited states, respectively (Figs. 6c, d, 7c, d). Functions $\mu^2_{pq}, (\mu'_{pq})^2$ from (26) make sense of the characteristic distribution functions of Fermi-Dirac type in ground, excited states, respectively. Functions $\mu^2_{pq}, (\mu'_{pq})^2$ also can be interpreted as the parameters of the polarization degree, the entanglement degree of states [7]. Dependencies of the characteristic phase functions R_{13} (Fig. 6a, b) and R_{57} for squeezed states (30×30) coincide.

By analogy dependencies of the characteristic phase functions R_{24} (Fig. 7 a, b) and R_{68} for stretched states (40×40) coincide. Detailed behavior of the functions h_{13}, h_{24} is shown in Fig. 8: observed shapingpulsesof complex form. Thus on the whole interval $j' \in [-33; 33]$ functions h_{13}, h_{24} are of variable sign and oscillate around zero, but the behavior of these functions is different. Changing the sign of the functions h_{13}, h_{24} is related to the change of sign of the second-order correlation functions, which indicates a change of the nature of relations between the pair operators of displacement $\widehat{S}_{pq}$.

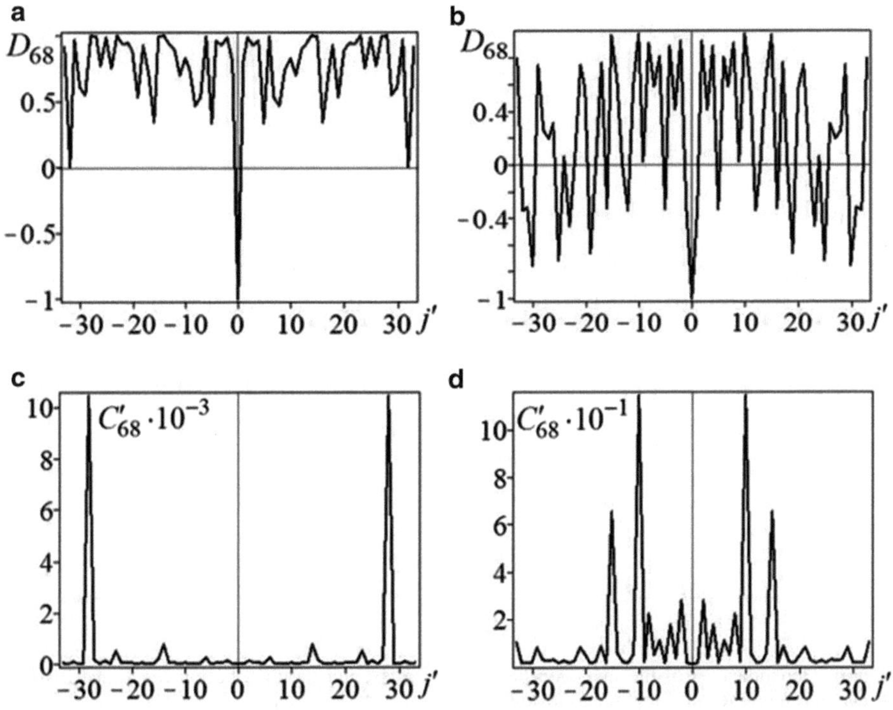

Fig. 5 Dependencies of the inversion parameters (**a**, **b**) and the distribution functions of Bose-Einstein type (**c**, **d**) on j' for stretched states $(p, q) = (6, 8)$ of branches $u_{\varepsilon 1}$ (**a**, **c**), $u_{\varepsilon 2}$ (**b**, **d**)

On the other hand, the interpretation to change the statistics for the pair operators of displacement is allowed $\widehat{S}_{pq}$: from the statistic of Fermi-Dirac type (positive values of the distribution functions h_{13}, h_{24}) to the statistic of Bose-Einstein type (negative values h_{13}, h_{24}). However, dependencies of distribution functions h_{13}, h_{57} for squeezed states (30x30) differ sharply. By analogy the dependencies of distribution functions h_{24}, h_{68} for stretched states (40x40) also differ sharply. In this case, the function h_{13}, h_{24} changes sign near zero, and there is a change in the type of the interaction: repulsion or attraction for pair operators of displacement (similar change in the type of interaction between biphonons).

5 Conclusions

Within the model of the fractal dislocation it is proposed to describe the fractal nonlinear oscillator. We obtain a system of coupled nonlinear equations, which is based on the theory of fractional calculus and the concepts of fractal. For a description of other possible structural states of fractal dislocation corresponding

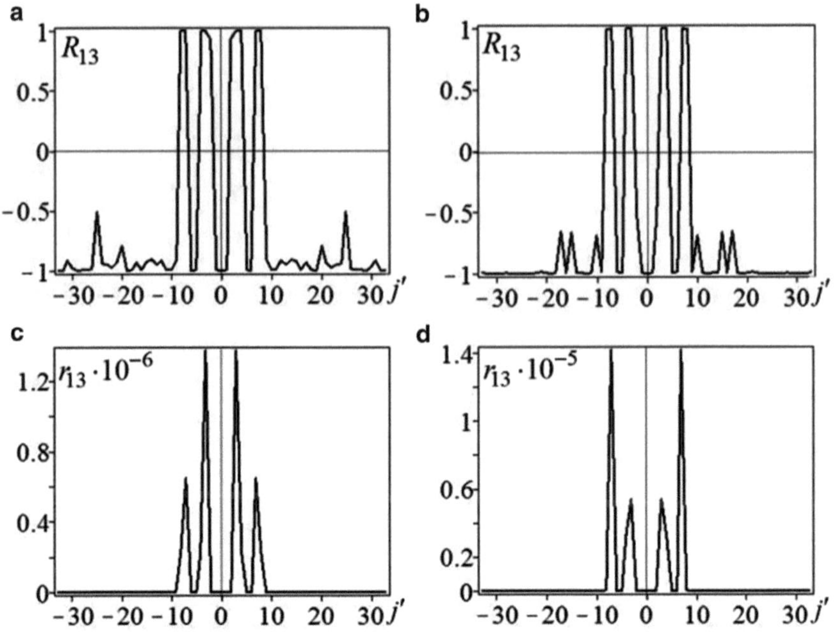

Fig. 6 Dependencies of the inversion parameters (**a**, **b**) and the distribution functions of Bose-Einstein type (**c**, **d**) on j' for squeezed states $(p,q)=(1,3)$ of branches $u_{\varepsilon1}$ (**a**, **c**), $u_{\varepsilon2}$ (**b**, **d**)

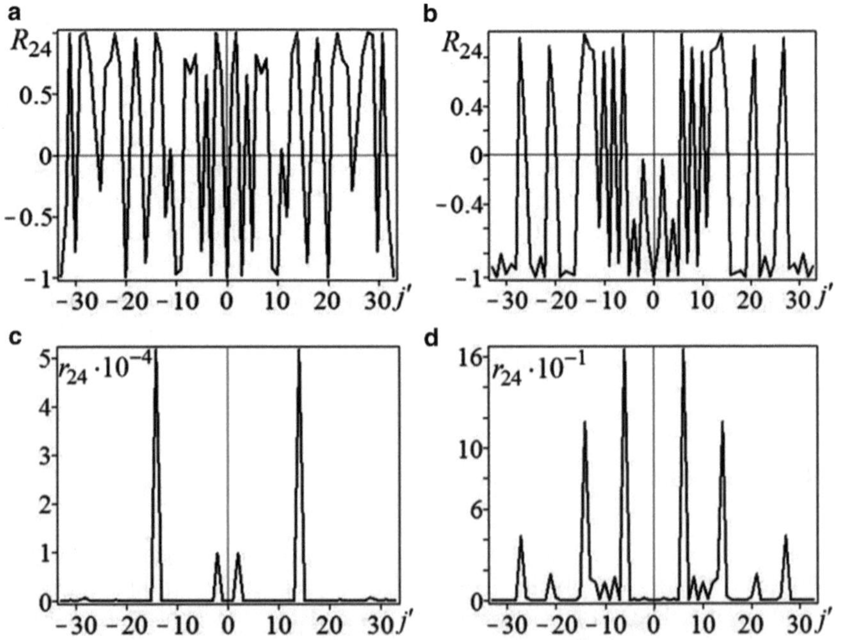

Fig. 7 Dependencies of the inversion parameters (**a**, **b**) and the distribution functions of Bose-Einstein type (**c**, **d**) on j' for stretched states $(p,q)=(2,4)$ of branches $u_{\varepsilon1}$ (**a**, **c**), $u_{\varepsilon2}$ (**b**, **d**)

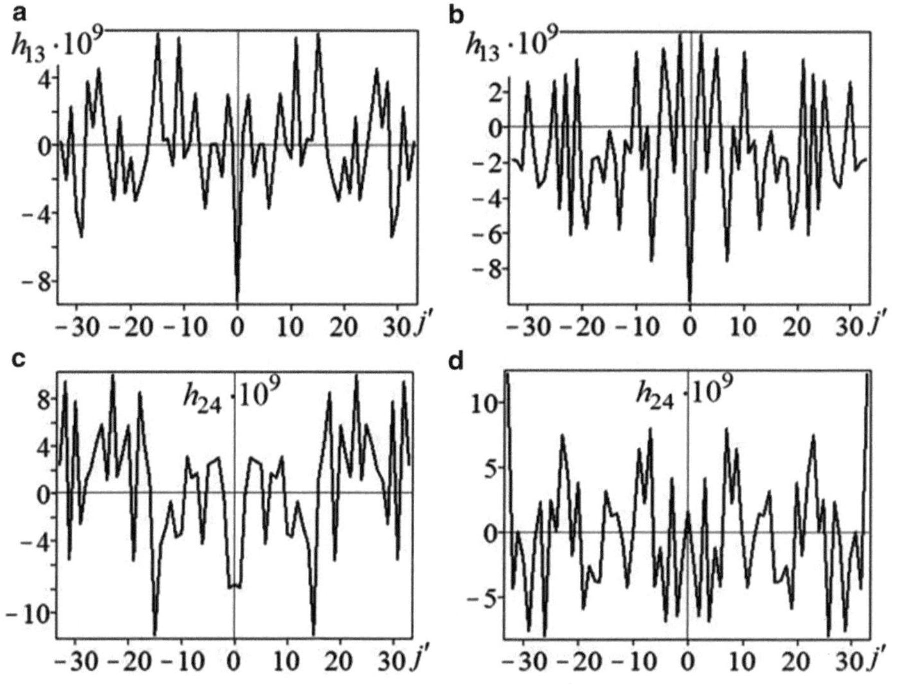

Fig. 8 Dependencies of the distribution functions on j' for squeezed $(p, q) = (1, 3)$ (**a**, **b**) and stretched (**c**, **d**) states of branches $u_{\varepsilon 1}$ (**a**, **c**), $u_{\varepsilon 2}$ (**b**, **d**)

Hamilton operators are obtained. Within the two-point model for mixed states (squeezed and stretched) the features of correlation and statistical properties of deformations and stress fields of fractal dislocation are investigated. By the method of numerical simulation it is shown that the fractal dislocation can be represented as a set of coupled fractal oscillators. The appearance of the individual peaks in the core of fractal dislocation (for the first and second branches of the displacement functions) indicates the presence of separate fractal nonlinear oscillators with special properties. It is shown that for the pair displacement operators there is the change in the type of correlation connection and statistics and there is the effect of the presence of entanglement states.

References

1. Abramov, V.S.: Fractal dislocation as one of non-classical structural objects in the nano-dimensional systems. Metallofiz. i Noveishie Tekhnologii. **33**(2), 247–251 (2011)
2. Abramov, V.S.: Behavior of the deformation field of fractal dislocation in the presence of bifurcations. Bul. Donetsk Nat. Univers. Ser. A. **2**, 23–29 (2011)

3. Abramov, V.S.: Ordering and statistical properties of the deformation field of fractal dislocation in a model nanosystem. Bul. Donetsk Nat. Univers. **1**, 90–96 (2013)
4. Abramova, O.P., Abramov, S.V.: The influence of fractal fields of deformation and stress on the energy spectrum of dislocation. Bul. Donetsk Nat. Univers. **1**(1), 119–123 (2005)
5. Abramova, O.P., Abramov, S.V.: Features of behaviour of the deformation field of fractal quasi-two-dimensional structures. Bul. Donetsk Nat. Univers. **1**, 9–17 (2013)
6. Awrejcewicz, J., Krysko, V.A.: Chaos in Structural Mechanics. Springer, Berlin, Heidelberg (2008)
7. Chekhova, M.V., Fedorov, M.V.: The Schmidt modes of biphoton qutrits: Poincare-sphere representation. J. Phys. B At. Mol. Opt. Phys. **46**(095502), 1–10 (2013)
8. Chou, C.W., Hume, D.B., Rosenband, T., Wineland, D.J.: Optical clocks and relativity. Science **329**, 1630–1633 (2010)
9. Erhart, J., Sponar, S., Sulyok, G.: Experimental demonstration of a universally valid error-disturbance uncertainty relation in spin measurements. Nat. Phys. **8**, 185–189 (2012)
10. Gleyzes, S., Kuhr, S., Guerlin, C., et al.: Quantum jumps of light recording the birth and death of a photon in a cavity. Nature **446**, 297–300 (2007)
11. Gomes, K.K., Mar, W., Ko, W., et al.: Designer Dirac fermions and topological phases in molecular graphene. Nature **483**(7389), 306–310 (2012)
12. Mandelbrot, B.B.: The Fractal Geometry of Nature. Freeman, New York (1982)
13. Perrin, A., Bücker, R., Manz, S.: Hanbury Brown and Twiss correlations across the Bose-Einstein condensation threshold. Nat. Phys. **8**, 195–198 (2012)
14. Samko, S.G., Kilbas, A., Marichev, O.: Fractional Integrals and Derivatives. Theory and Applications. Gordon and Breach Science Publication, New York (1990)
15. Scully, M.O., Zubairy, M.S.: Quantum Optics. Cambridge University Press, Cambridge (1997)
16. Skiadas, C.H., Skiadas, C.: Chaotic Modeling and Simulation: Analysis of Chaotic Models, Attractors and Forms. Taylor and Francis/CRC, London (2009)
17. Stockmann, H.-J.: Quantum Chaos. An Introduction. Cambridge University Press, Cambridge (1999)
18. Zakrzhevsky, M.V., Smirnova, R.S., Schukin, I.T., et al.: Nonlinear Dynamics and Chaos. Bifurcation Groups and Rare Attractors. RTU Publishing House, Riga (2012)

Dynamical Statement Networks

Wojciech Cholewa

Abstract This paper addresses dynamics in diagnostic expert systems. The introduced dynamic statement network may be used not only in static diagnosing systems but also in dynamic monitoring systems. Within these networks, statements consist of contents and values. Statement values are based on a concept introduced in intuitionistic fuzzy sets, i.e. they contain independent belief about validity and invalidity of information presented by statement contents. The relationships between statements are modelled in the form of a set of necessary as well as sufficient conditions. The author compared requirements placed against static and dynamic statement networks and devised a manner in which a static network may be transformed into a dynamic one which, in turn, facilitates non-monotonic reasoning required for monitoring systems.

Keywords Expert system • Intuitionistic network • Knowledge base • Diagnostic reasoning

1 Introduction

Systems facilitating evaluation of a technical state of objects are of crucial practical significance. A recognition of the actual state or a change in this state frequently provides argumentation for decision-making in reference to further operation of objects in question. Such systems may be developed and devised as diagnostic expert systems to support a recognition of the current state of an object [3].

Monitoring systems facilitate observations of objects in the course of time. They may include diagnostic modules that would allow for recognition of the present state on an object during its changes in time. The purpose of the considered systems may refer to objects of different classes such as machines, industrial and economic processes, patients at intensive care wards, meteorological processes, and many more.

W. Cholewa (✉)
Faculty of Mechanical Engineering, Silesian University of Technology,
Konarskiego 18a, 44-100 Gliwice, Poland
e-mail: wojciech.cholewa@polsl.pl

J. Awrejcewicz (ed.), *Applied Non-Linear Dynamical Systems*, Springer Proceedings in Mathematics & Statistics 93, DOI 10.1007/978-3-319-08266-0_25

One finds a great variety of the currently available expert systems and their detailed overview would go beyond the purpose of the presented paper. Nevertheless, it is essential to note that before research into new classes of such systems can proceed, the corresponding needs must be meticulously analysed. Such an analysis should indicate those features of the available systems that require further modifications.

The majority of expert systems that enable inference in conditions of inaccurate and incomplete data require a considering of the complete knowledge base during their development. Devising such systems by combination of independently developed and independently verified component subsystems constitutes a challenging, or impossible, task.

Practical applications of expert systems in diagnostics involve an assumption that we frequently do not have the complete data which are required to enable the inference process. Some types of expert systems such as bayesian network-based systems allow for substitution of incomplete data with default ones. A number of methods were developed that enable arbitrary acknowledgment or acquisition of such default data, e.g. on the basis of available historical data. The results of inference related to application of default data are communicated to the system user who is to take a decision concerning further operation of the object upon analysis of received information.

While taking a decision, the user should know whether the result of the inference process is conditioned by direct results of both observation and measurements or whether it originates from the default data whose occurrence is expected for an average object, however, not necessarily for the analysed object in question. The decisions should base on a belief that there are indications of their validity, i.e. plausibility, advisability, and legality. Systems basing on intuitionistic logic are considered to be an interesting example of inference systems that enable collection of judgments. In this logic, propositional calculus allows for a syntax transformation that ensures its justification. This differentiates intuitionistic logic from classical logic whose propositional calculus maintains the truthfulness of transformed syntax.

The following section of the paper presents essential operation activities of expert systems in which knowledge base is saved as intuitionistic statement networks. Such systems were introduced in order to limit any inconveniences occurring during development and applications of discussed systems. This, in particular, applies to difficulties with such systems resulting from:

- A need of investigation into a complete knowledge base during system development
- No possibility of differentiation between conclusions formulated on the basis of default data that substitute missing (unknown) data as well as conclusions conditioned on a complete set of data
- No possibility of integration with a knowledge base of detailed explanations and sets of source information that justify proper operation of the system and facilitate interpretation

2 Statements

Diagnostic systems recognize the object state basing on the available diagnostic knowledge directly pertaining to the object family to which the examined object belongs as well as on available data determined as values of features (attributes) of the given object. These features may directly concern the impact of the object onto the surrounding, the impact of the surrounding onto the object, and may define the structure of the object, its features, etc. The features may manifest ratio and interval quantitative values as well as ordinal and nominal qualitative values. Strings of signs come as a particular kind of feature values. Recording of such data in the form of OA matrix (object-attribute matrix) whose elements are values of features is a commonly applied representation method. The OA matrix is also referred to as Aristotelian table [8] or an informational system [7]. The columns of the OA matrix correspond to features (attributes), whereas rows reflect different instances of the object, instances being various objects belonging to the studied object class, or an object of the selected class at different time intervals.

A hypothesis stating that for the object class in question the OA matrix contains unfailing regularities allowing for inference process, constitutes the basic assumption related to diagnostic applications of the OA matrix.

Knowledge acquisition for the needs of diagnostic systems may be realized as a result of applied processes of machine learning [9] as well as a result of cooperation with domain experts. In each case, a need to define a degree of generality of considered data arises. Not only a considerably high but also significantly low degree of generality of data may hinder or even block entirely a possible definition of a sustainable solution. An optimization should result in assuming appropriate granularity of information represented by the data. One expects that the optimal granularity should allow for identification and generalization of dependencies occurring between the data. Furthermore, it is expected that the correct interpretation of obtained information granules will be maintained.

A number of formalisms relating to possible methods of information granulation is currently well known [2]. Statements, most suitable for practical uses, constitute a kind of informational granules that facilitate development and construction of knowledge models in a form of statement networks [4].

The noun *statement* has several meanings. Here, the statement is information on recognition of proposition resulting from observed facts or representing an opinion. The statement s consists of a content $c(s)$ and value $b(s)$

$$s = \langle\, c(s), b(s)\, \rangle. \tag{1}$$

The content $c(s)$ of a statement s is presented as an indicative sentence, i.e. a sentence which may be either entirely true or entirely false. On the basis of the content only, one may not conclude whether the statement has been recognized or not. The information on recognition of the statement is represented as a value $b(s)$ of the statement s, independently of its content $c(s)$.

The values of statements may be analysed as elements of OS matrix (object-statement matrix) which is analogically constructed as the OA matrix. The columns of the OS matrix correspond to statements, whereas rows refer to object instances. The values of statements may be determined on the basis of the values of features contained in the OA matrix. It may be assumed that the OA matrix includes all data required for definition or recognition of the statement value. This results in the OA matrix being transformable into the OS matrix. Having considered this occurrence it appears that the data on the object may be represented in a form of statements.

It is very important that not only data on the object design and working conditions but also results of inferring on the object state may appear in a form of statements. This, in turn, allows for an inference process carried out in a set of statements in which diagnostic knowledge is represented by relations occurring between the statement values. It should be assumed that

- the set of considered statements is a finite set in which contents of statements do not change,
- the set of considered relations between statements is known and fixed for the studied class of objects.

In inference processes realized by expert systems, conclusions are either acknowledged or rejected on the basis of formerly recognized premisses. Sets of statements studied as sets of premisses and conclusions are fixed and finite ones. They are determined during system development and construction. During the system operation no new conclusions, appearing as new statements, are generated or discovered. This, subsequently, means that inference processes in diagnostic expert systems that are realized with the use of OS matrices are processes conducted in so-called closed worlds.

3 Intuitionistic Statement Value

A value of a statement constitutes a carrier of information on recognition or the lack of recognition of the statement content. The boolean type is the fundamental type of a statement value. It consists of logical values *true* and *false* which are frequently encoded as numerical values, 1 and 0, respectively. If the value of a statement is not equal to one of these values, then it must be equal to the other value. It means the Boolean type complies with a so-called law of excluded middle.

The boolean type of statement values facilitates construction of accurate inference systems. Due to restricted quality of available knowledge and available information, a use of approximate inference systems is frequently required. A degree of belief about the truth of a statement content is an example of a value of statements considered as nodes of belief networks which are also referred to as bayesian networks [6]. Not only a certainty factor or subjective probability of a statement truthfulness or degree of belief in the truth of a statement [5] but also a degree of belonging to a fuzzy or rough set of statements recognized as true, and a large number of similar concepts, may become a value of a statement. A common feature that is shared by the enumerated types of statement values is that these values are represented in a form of a point value by a one number only.

In practical applications of expert systems, one needs to determine statement values in a manner that allows differentiation between the following groups of statements:

- statements that are not recognized as either true or false,
- statements for which the numbers of for and against arguments on their recognition are equal

Additionally, these systems require that conclusions be formulated independent on the basis of a belief about the truth of selected statements as well as on the basis of belief about the falsity of other statements.

The expected manner of representing a statement value is hard to introduce if a statement value is written down as a single number. For this purpose it is essential to consider a statement value defined in a form of pairs of numbers. An example of such a value may include a pair consisting of a necessity degree and possibility degree originating from modal logic. Another concept of a statement value which is worthy of attention is a definition of this value (2) derived from the theory of intuitionistic fuzzy sets (IFS) [1].

An intuitionistic value $b(s)$ of statement s, also referred to as the statement value, is represented by the ordered pair

$$b(s) = \langle\, p(s), n(s)\, \rangle \quad \text{for} \quad p(s), n(s) \in [\,0\,,1\,] \tag{2}$$

where $p(s)$ is a degree of validity (justification, truthfulness) and $n(s)$ is a degree of invalidity (lack of justification, lack of truthfulness) of statement s. These values are also referred to as evaluations of positive and negative information. Within IFS it is assumed that

$$p(s) + n(s) \in [\,0\,,1\,]\,, \tag{3}$$

and it is not assumed that

$$p(s) + n(s) = 1.0\,. \tag{4}$$

Considering (2), the value $b(s)$ of an unrecognized statement s, i.e. a value of the statement not recognized as true or false, takes the following form:

$$b(s) = \langle 0, 0 \rangle \tag{5}$$

Values $p(s)$ and $n(s)$ allow for defining a hesitation margin $h(s)$ for statement s which is also known as a degree of nondetermination

$$h(s) = 1 - p(s) - n(s) \tag{6}$$

where (3) results in

$$h(s) \in [0, 1] \tag{7}$$

The literature on IFS uses notation $\mu(s)$, $\nu(s)$, $\pi(s)$ instead of $p(s)$, $n(s)$, $h(s)$.

The statement values (2) may be determined directly with respect to the assumption (3). They may also result from inference on the basis of other statements in accordance with an assumed knowledge model. A potential imperfection of the applied knowledge model and inconsistency of the studied set of statement values may lead to unexpected results of the inference process manifested as disagreements and contradictions in the set of selected statement values which may not satisfy restrictions (3). Such possibility leads to an extension of the definition (6) of hesitation margin

$$h(s) = \begin{cases} 1 - p(s) - n(s) & \text{if } p(s) + n(s) \leq 1 \\ 0 & \text{if } p(s) + n(s) > 1 \end{cases} \tag{8}$$

In addition a disagreement level $d(s)$ for values $p(s)$ and $n(s)$ of statement s not satisfying restriction (3) is also introduced:

$$d(s) = \begin{cases} 0 & \text{if } p(s) + n(s) \leq 1 \\ p(s) + n(s) - 1 & \text{if } p(s) + n(s) > 1 \end{cases} \tag{9}$$

where

$$d(s) \in [0, 1]. \tag{10}$$

Occurrence of positive values $d(s)$ implies contradiction in either the applied knowledge model or in the assumed statement values.

4 Statement Network

Elements which statement networks consist of are statements $s \in S$ considered as pairs (1) of both contents $c(s)$ and values $b(s)$ of these statements. The state of statement networks is determined by a set of statement values

$$b(S) = \{b(s) : s \in S\} \tag{11}$$

The essence of statement networks is that a change in a value of a particular statement occurring in the network may induce changes in values of other statements.

A statement network may be represented as a graph G

$$G = \langle C, V, E, A \rangle \text{ where } E = E_p \cup E_n , \tag{12}$$

with a static subgraph H defining a constant structure of a statement network and its elements

$$H = \langle C, V, E \rangle, \tag{13}$$

and a dynamic directed subgraph D defining interrelations of statements and actual statement values

$$D = \langle V, A \rangle, \tag{14}$$

where:

C is a set of vertices representing statement contents,
V is a set of vertices representing values of degrees of validity,
E is a set of edges connecting vertices $c \in C$ which represent statement contents with vertices $v \in V$ which represent degrees of validity or invalidity,

E_p is a set of p-edges $e_p(s) = \langle c(s), p(s)\rangle$ connecting contents $c(s)$ of statements s with values of degrees of their validity $p(s)$,
E_n is a set of n-edges $e_n(s) = \langle c(s), n(s)\rangle$ connecting contents $c(s)$ of statements s with values of validity degrees of their contradictions, i.e. with values of degrees of their invalidity $n(s)$,

A is a set of directed edges (arcs) defining relations between values of degrees of validity (or invalidity).

The vertex $v \in V$ representing the value of degree of validity or invalidity may be connected by means of edges $e \in E$ with numerous vertices $c \in C$ representing statement contents, which allows for, inter alia, concurrent considering of equivalent statement contents. Each vertex $c \in C$ may be connected with no more than one p-edge and one n-edge.

Each edge $a = \langle s_1, s_2\rangle \in A$ of the graph G defines a true implicational relationship $s_1 \rightarrow s_2$ between statements s_1 and s_2

$$(a = \langle s_1, s_2 \rangle \in A) \quad \Rightarrow \quad (s_1 \rightarrow s_2) \tag{15}$$

This relationship proves that the antecedent, i.e. statement s_1, is a sufficient condition for the consequent, i.e. statement s_2. The sufficiency condition denotes that recognition of the validity of statement s_1 is always accompanied by recognition of validity of statement s_2.

The necessary and sufficient conditions can be modelled by means of inequalities [3]. For considered intuitionistic values of a pair of statements s_1 and s_2 connected by an edge $a = \langle s_1, s_2\rangle$ and for the selected value $b(s)$ the following sufficiency condition is obtained:

$$(s_1 \rightarrow s_2,\, b(s_1) = \langle p(s_1), n(s_1)\rangle) \quad \Rightarrow \quad (p(s_2) \geq p(s_1)) \tag{16}$$

Furthermore, for the given value $b(s_2)$, the following necessity condition is obtained:

$$(\, s_1 \rightarrow s_2,\, b(s_2) = \langle p(s_2), n(s_2)\rangle \,) \quad \Rightarrow \quad (\, n(s_1) \geq n(s_2)\,) \tag{17}$$

One considers, in the similar fashion, a sufficient condition including one consequent s and a disjunction of a number of antecedents $s_1, s_2, \cdots$

$$(\,(\, \vee_i\, s_i \,) \rightarrow s) \quad \Rightarrow \quad p(s) \geq \max_i p(s_i) \tag{18}$$

as well as conjunctive of a number of antecedents $s_1, s_2, \cdots$

$$(\,(\, \wedge_i\, s_i \,) \rightarrow s) \quad \Rightarrow \quad p(s) \geq \min_i p(s_i) \tag{19}$$

In a set of vertices representing values of degrees of validity and invalidity one can distinguish between three subsets:

$$V = V_{or} \cup V_{and} \cup V_{fix}\,, \tag{20}$$

where:

V_{or} is a set of vertices whose values are defined according to (18), when they appear as consequents of sufficient conditions,

V_{and} is a set of vertices whose values are defined according to (19), when they appear as consequents of sufficient conditions,

V_{fix} is a set of vertices whose values are arbitrarily acknowledged or are defined by means of measurement sets or other systems.

The introduced concept of a statement network is illustrated in Fig.1 which depicts a very simple network, dealing with a pocket flashlight. Statements s_3 and s_4 present the manner in which different statement contents may be allocated to the same vertices representing the degrees of validity.

5 Solving of Dynamic Statement Network

A statement network represents knowledge in a form of graph G (12). While applying the statement network, statement values serve as variable elements of the graph. The remaining elements, i.e. statement contents, undirected edges, as well as directed edges, are not subject to change.

A statement network which is studied as a model of a dynamic object or process should adapt to changing operation environment due to which values $v \in V_{fix}$ of selected statements may be subject to changes. It is essential to emphasize that updates of values $v \in V_{fix}$ may be performed in an asynchronous manner.

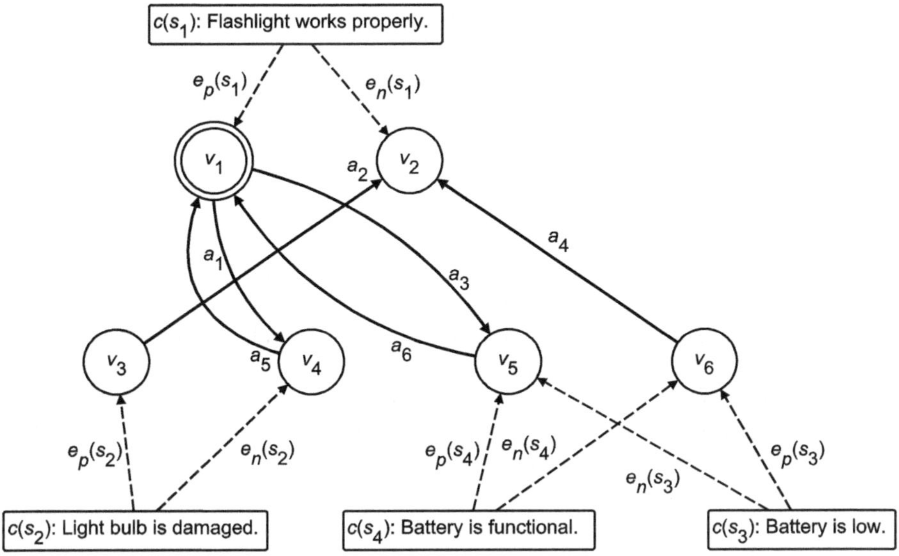

Fig. 1 An example of a statement network, where $v_1 \in V_{and}$, $v_2 \cdots v_6 \in V_{or}$ (see text for details)

Solving of a statement network consists in application of known values of vertices $v \in V_{fix}$ as well as relation $a \in A$ according to (14) in order to determine unknown values of vertices $v \in V \backslash V_{fix}$. A simple algorithm consists in assuming initial zero values of degrees of validity for $v \in V \backslash V_{fix}$ and, subsequently, in iterative increasing selected degrees of validity that fulfils relations $a \in A$. The implemented process of inference is a monotonic one.

In order to limit the monotonic properties of inference, a process of gradual forgetting of the defined degrees of validity may be introduced. This applies to all values $v \in V$, i.e. to both values of primary statements $v \in V_{fix}$ and values of secondary statements $v \in V \backslash V_{fix}$. The values of these degrees are studied as values dependent on a discrete time t, i.e.

$$v_i(t),\ v_i(t+1),\ \cdots . \tag{21}$$

The process of forgetting may be considered in the following form:

$$v_i(t+1) := q_i\ v_i(t)\ ; \quad 0 < q_i \leq 1\,, \tag{22}$$

where the introduced constants q_i assigned separately to all vertices $v_i \in V$ allow for tuning of dynamic properties of the network.

The statement network integrity should be subject to verification during the system development phase as well as application phase. It is assumed that potentially contradicting elements of the network should be detected during network operation, basing on the disagreement level (9)

$$d(s) > 0\,, \tag{23}$$

as elements that are conditionally contradicting, i.e. contradicting for a specific case.

6 Conclusions

A specific class of statement networks has been introduced in which statements values are based on a concept used for intuitionistic fuzzy sets. Statement values include independently determined degree of validity as well as degree of invalidity of a statement. The domain knowledge is represented in a form of a set of necessary and sufficient conditions occurring between the statements. While the statement network is being solved, these conditions are considered as corresponding inequalities between statement values.

The proposed networks with implemented processes of gradual forgetting of statement values may be used as models of dynamic objects or processes. They may be applied, among others, in diagnostic expert systems. The main advantages of these networks include:

- a possibility of independent development of a set of subnetworks which, subsequently, may be connected to form a general network,
- a possibility of evaluation of the disagreement level between values of studied statements enabling identification of erroneous data as well as errors in network design.

Acknowledgements Described herein are selected results of study, supported partly from the budget of Research Task No. 4 implemented under The National Centre for Research and Development in Poland and ENERGA SA Strategic Program of Scientific Research and Development entitled *Advanced Technologies of Generating Energy*.

References

1. Atanassov, K.T.: Intuitionistic Fuzzy Sets: Theory and Applications. Springer, Berlin Heidelberg (1999)
2. Bargiela, A., Pedrycz, W.: Granular Computing. An Introduction. Kluwer Academic Publishers, Boston (2003)
3. Cholewa, W.: Expert systems in technical diagnostics. In: Korbicz, J., Kościelny, J., Kowalczuk, Z., Cholewa, W. (eds.) Fault Diagnosis: Models, Artificial Intelligence, Applications, pp. 591–674. Springer, Berlin, Heidelberg (2004)
4. Cholewa, W.: Mechanical analogy of statement networks. Int. J. Appl. Math. Comput. Sci. **18**(4), 477–486 (2008) DOI 10.2478/v10006-008-0042-7

5. Cholewa, W., Amarowicz, M., Chrzanowski, P., Rogala, T.: Development environment for diagnostic multimodal statement networks. Key Eng. Mater. **588**, 74–83 (2014) DOI 10.4028/www.scientific.net/KEM.588.74
6. Jensen, V.J.: Bayesian Networks and Decision Graphs. Springer, New York (2002)
7. Pawlak, Z.: Information systems: Theoretical foundations. Informat. Syst. **6**(3), 205–218 (1981)
8. Watanabe, S.: Pattern Recognition: Human and Mechanical. John Wiley & Sons, New York (1985)
9. Witten, I.H., Frank, E.: Data Mining: Practical Machine Learning Tools and Techniques, 2nd edn. Morgan Kaufman, San Francisco (2005)

The Shooting Method for Non-standard Boundary Value Problem

Rafał Palej

Abstract The paper deals with a method for solving the non-standard two-point boundary value problem in which the number of boundary conditions is greater than the number of first-order ODEs containing a certain number of unknown parameters. The total number of the unknown parameters depends on the number of boundary conditions and the way they are distributed at both ends of the domain. The presented method consists in the replacement of the boundary value problem by the initial value problem formulated with the use of sensitivity functions. The general discussion has been illustrated by numerical example consisting of two first-order differential equations and four boundary conditions. The difference between the standard and non-standard boundary value problems consists in the size of the initial value problem that needs to be solved.

1 Introduction

The standard boundary value problem is characterized by the same number of first-order ODEs and boundary conditions imposed at both ends of the domain. There are several methods for solving such problems: finite difference method, shooting method, monotype iterative method, quasilinearization method and others [2, 5, 1, 3]. Sometimes the number of boundary conditions is greater than the number of differential equations. Finding a solution to such problems is conditioned by the occurrence of a certain number of unknown parameters in differential equations. In order to find the solution satisfying all boundary conditions it is necessary to determine the values of the unknown parameters. Some computer software offers procedures which can solve simple problems of such type. Unfortunately, more sophisticated problems cannot be solved with the assistance of a ready-made procedure. This paper presents a general concept for solving the non-standard boundary value problem in which the number of boundary conditions may be twice as high as the number of differential equations. The discussed method consists in the

R. Palej (✉)
Department of Applied Computer Science, Cracow University of Technology,
Al. Jana Pawła II 37, 31-864 Cracow, Poland
e-mail: palej@mech.pk.edu.pl

J. Awrejcewicz (ed.), *Applied Non-Linear Dynamical Systems*, Springer Proceedings in Mathematics & Statistics 93, DOI 10.1007/978-3-319-08266-0_26

replacement of the boundary value problem by the initial value problem formulated with the use of sensitivity functions [6, 7]. The general approach to solving the non-standard boundary value problem has been illustrated by the problem of plane curve which should go through two given points and be inclined at given angles at these points [4].

2 Formulation of the Problem

We shall seek the solution to the non-standard two-point boundary value problem consisting of a set of n coupled first-order ODEs containing r unknown parameters $(1 \leq r \leq n)$. The desired solution to this problem is supposed to satisfy q boundary conditions at the starting point and $n+r-q$ boundary conditions at the final point. We assume that the unique solution to such problem exists. The set of n first-order ODEs may be written in the following form:

$$\frac{dy_i}{dx} = f_i\left(x, y_1, \ldots, y_n, p_1, \ldots p_r\right), \; i = 1, \ldots, n, \; a \leq x \leq b \tag{1}$$

At the starting point $x = a$ the solution is supposed to satisfy

$$y_i(a) = y_{ia}, \; i = 1, 2, \ldots, q \; (q \leq n) \tag{2}$$

while at the final point $x = b$ it is supposed to satisfy

$$y_i(b) = y_{ib}, \; i = 1, 2, \ldots, n + r - q \tag{3}$$

where y_{ia} and y_{ib} denote some constants. Adjusting appropriately the values of the guessed parameters we should find the solution to the set of Eqs. (1), satisfying $n+r$ boundary conditions (2) and (3).

3 General Procedure

The method is based on the concept of the shooting method which consists in the replacement of the boundary value problem by the initial value problem. In order to do that we need to complete values of the desired functions y_i at one end of the domain. If, for example, $q > (n+r)/2$, then $n-q$ new parameters $p_i, i = r+1, \ldots, n+r-q$ should be added to complete the boundary conditions (2) at the starting point

$$y_i(a) = p_{i+r-q}, \; i = q + 1, \ldots, n \tag{4}$$

If $q = n$ there is no need to enlarge the number of the unknown parameters. The desired functions y_i depend on variable x as well as on $n + r - q$ unknown parameters

$$y_i = y_i\left(x, p_1, \ldots, p_{n+r-q}\right), \quad i = 1, 2, \ldots, n \tag{5}$$

For the trial values of the unknown parameters $p_i^{(1)}$, $i = 1, 2, \ldots, n + r - q$ the boundary conditions (3) are not usually fulfilled, i.e.

$$y_i\left(b, p_1^{(1)}, \ldots, p_{n+r-q}^{(1)}\right) - y_{ib} \neq 0, \quad i = 1, 2, \ldots, n + r - q \tag{6}$$

In order to decrease the discrepancies in (6) for the subsequent approximation of the guess parameters, we expand all functions with imposed conditions on the final point into Taylor's series

$$y_i\left(b, p_1^{(2)}, \ldots, p_{n+r-q}^{(2)}\right) = y_i\left(b, p_1^{(1)}, \ldots, p_{n+r-q}^{(1)}\right)$$
$$+ \sum_{j=1}^{n+r-q} \left.\frac{\partial y_i}{\partial p_j}\right|_{\substack{x = b \\ \mathbf{p} = \mathbf{p}^{(1)}}} \left(p_j^{(2)} - p_j^{(1)}\right) + \cdots = y_{ib}, \quad i = 1, 2, \ldots, n + r - q \tag{7}$$

The set of Eqs. (7) may be written in the matrix form

$$\mathbf{y}\left(b, \mathbf{p}^{(1)}\right) + \mathbf{A}\left(b, \mathbf{p}^{(1)}\right) . \left(\mathbf{p}^{(2)} - \mathbf{p}^{(1)}\right) = \mathbf{y}_b \tag{8}$$

where

$$\mathbf{y}\left(b, \mathbf{p}^{(1)}\right) = \begin{bmatrix} y_1\left(b, p_1^{(1)}, \ldots, p_{n+r-q}^{(1)}\right) \\ y_2\left(b, p_1^{(1)}, \ldots, p_{n+r-q}^{(1)}\right) \\ \vdots \\ y_{n+r-q}\left(b, p_1^{(1)}, \ldots, p_{n+r-q}^{(1)}\right) \end{bmatrix}, \quad \mathbf{p}^{(2)} = \begin{bmatrix} p_1^{(2)} \\ p_2^{(2)} \\ \vdots \\ p_{n+r-q}^{(2)} \end{bmatrix},$$

$$\mathbf{p}^{(1)} = \begin{bmatrix} p_1^{(1)} \\ p_2^{(1)} \\ \vdots \\ p_{n+r-q}^{(1)} \end{bmatrix}$$

$$\mathbf{A}\left(b,\mathbf{p}^{(1)}\right)=\begin{bmatrix}\frac{\partial y_1}{\partial p_1} & \frac{\partial y_1}{\partial p_2} & \cdots & \frac{\partial y_1}{\partial p_{n+r-q}}\\ \frac{\partial y_2}{\partial p_1} & \frac{\partial y_2}{\partial p_2} & \cdots & \frac{\partial y_2}{\partial p_{n+r-q}}\\ \vdots & \vdots & \ddots & \vdots\\ \frac{\partial y_{n+r-q}}{\partial p_1} & \frac{\partial y_{n+r-q}}{\partial p_2} & \cdots & \frac{\partial y_{n+r-q}}{\partial p_{n+r-q}}\end{bmatrix}\Bigg|_{\substack{x=b\\ \mathbf{p}=\mathbf{p}^{(1)}}},\ \mathbf{y}_b=\begin{bmatrix}y_{1b}\\ y_{2b}\\ \vdots\\ y_{(n+r-q)b}\end{bmatrix}$$

From Eq. (8) we can derive the iteration formula for the subsequent approximations of the parameter values in the form

$$\mathbf{p}^{(k+1)}=\mathbf{p}^{(k)}-\mathbf{A}^{-1}\left(b,\mathbf{p}^{(k)}\right).\left(\mathbf{y}\left(b,\mathbf{p}^{(k)}\right)-\mathbf{y}_b\right),\ k=1,2,\ldots \tag{9}$$

The Jacobian matrix $\mathbf{A}(b,\mathbf{p}^{(1)})$ is composed of derivatives which must be determined. To this end, we differentiate Eqs. (1) with respect to individual parameters

$$\frac{\partial}{\partial p_j}\left(\frac{dy_i}{dx}\right)=\sum_{k=1}^{n}\frac{\partial f_i}{\partial y_k}\frac{\partial y_k}{\partial p_j}+\frac{\partial f_i}{\partial p_j},\ i=1,..,n,\ j=1,\ldots,n+r-q \tag{10}$$

By defining the sensitivity functions with the following formulas:

$$g_{kj}=\frac{\partial y_k}{\partial p_j},\ k=1,..,n,\ j=1,\ldots,n+r-q, \tag{11}$$

we may express Eqs. (10) in the form

$$\frac{dg_{ij}}{dx}=\sum_{k=1}^{n}\frac{\partial f_i}{\partial y_k}g_{kj}+\frac{\partial f_i}{\partial p_j},\ i=1,..,n,\ j=1,\ldots,n+r-q \tag{12}$$

The initial conditions for sensitivity functions (found from Eqs. (2), (4) and (11)) shall have the form

$$g_{kj}(a)=\frac{\partial y_k(a)}{\partial p_j}=\begin{cases}0, & k=1,\ldots,q,\quad j=1,\ldots,n+r-q\\ \delta_{k,j+q-r}, & k=q+1,\ldots,n,\ j=1,\ldots,n+r-q\end{cases} \tag{13}$$

where δ_{kj+q-r} is the Kronecker delta. Finally, in each iteration we have to solve the initial value problem consisting of the system of $n\,(n+r-q+1)$ first-order ODEs, (1) and (12), and the same number of initial conditions, (2), (4) and (13). After determining the values of all parameters from Eq. (9) with the assumed accuracy, we shall obtain the solution to the non-standard boundary value problem, defined by Eqs. (1)–(3), by solving the initial value problem consisting of Eqs. (1) and initial conditions (2) and (4). The solution to this initial problem shall at the same time constitute the solution to the boundary value problem formulated by Eqs. (1)–(3).

The presented method of determining the values of the unknown parameters, which is in fact the Newton-Raphson method, is characterized by quadratic convergence on condition that the first approximation is sufficiently close to the desired solution.

4 Numerical Example

To illustrate the presented method we shall solve a problem consisting of second-order differential equation and four boundary conditions. The differential equation relates the normal reaction force acting on a particle during its slide down the curve to the function $y(x)$ and its derivatives. If we assume that the normal reaction force N is a linear function of x ($N = p_1 - p_2\, x$), then the differential equation takes the following form [4]:

$$\frac{d^2y}{dx^2} = \frac{1}{2}\frac{(p_1 - p_2 x)\left(1 + (y')^2\right)^{\frac{3}{2}} - (y')^2 - 1}{h + h_0 - y} \tag{14}$$

where h denotes the height of the starting point while h_0 denotes the height above starting point to provide the prescribed initial velocity $\left(v_0 = \sqrt{2gh_0}\right)$.

Boundary conditions determine the coordinates of the two points the curve should go through and the angles of tangents at these points. They should therefore adopt the following form:

$$\begin{cases} y(0) = 0 \\ y'(0) = \tan(\alpha) \end{cases} \tag{15}$$

$$\begin{cases} y(1) = h \\ y'(1) = \tan(\beta) \end{cases} \tag{16}$$

To apply the presented method, it is necessary to replace the second-order differential equation (14) with two first-order equations:

$$\begin{aligned} \frac{dy_1}{dx} &= y_2 \\ \frac{dy_2}{dx} &= \frac{1}{2}\frac{(p_1 - p_2 x)\ \left(1 + y_2^2\right)^{\frac{3}{2}} - y_2^2 - 1}{h + h_0 + y_1} \end{aligned} \tag{17}$$

In this example we have $n = r = q = 2$. Now, the differential equations (12) will take the following form:

$$\begin{aligned} \frac{dg_{1,1}}{dx} &= g_{2,1}, \frac{dg_{2,1}}{dx} = h_1(x)g_{1,1} + h_2(x)g_{2,1} + h_3(x) \\ \frac{dg_{1,2}}{dx} &= g_{2,2}, \frac{dg_{2,2}}{dx} = h_1(x)g_{1,2} + h_2(x)g_{2,2} - x\,h_3(x) \end{aligned} \tag{18}$$

where

$$h_1(x) = \frac{1}{2}\frac{(p_1 - p_2 x)\ \left(1 + y_2^2\right)^{\frac{3}{2}} - y_2^2 - 1}{(h + h_0 - y_1)^2}$$

$$h_2(x) = \frac{1}{2}\frac{3\ (p_1 - p_2 x)\ \left(1 + y_2^2\right)^{\frac{1}{2}} y_2 - 2y_2}{h + h_0 - y_1}$$

$$h_3(x) = \frac{1}{2}\frac{\left(1 + y_2^2\right)^{\frac{3}{2}}}{h + h_0 - y_1}$$

The initial conditions (2) and (13) in this case yield

$$\begin{aligned} &y_1(0) = 0, &&g_{1,1}(0) = 0, \quad g_{1,2}(0) = 0 \\ &y_2(0) = \tan(\alpha), &&g_{2,1}(0) = 0, \quad g_{2,2}(0) = 0 \end{aligned} \tag{19}$$

For $h = h_0 = 0.5$, $\tan(\alpha) = 0.2, \tan(\beta) = 1$ and $p_1^{(1)} = 2.0, p_2^{(1)} = 0.5$ we obtain after 4 iterations $p_1^{(5)} = 1.869525017658811$ and $p_2^{(5)} = 0.3544242522592831$. For such values of the parameters, solving the non-standard boundary value problem amounts to solving the initial value problem consisting of Eq. (14) and initial conditions (15). The plot of the desired curve *y*(*x*) is shown in Fig. 1.

One of the important questions arising during the process of solving boundary value problems of this type concerns the region of the solution convergence. Each problem with a unique solution is characterized by a specific region of convergence. The convergence region calculated for simple examples may be unbounded. The region of convergence for parameters p_1 and p_2 is limited and its shape is shown in Fig. 2.

If the starting point is placed inside the region of convergence, then the sequence of subsequent approximations will converge with the solution, marked by a cross in Fig. 2.

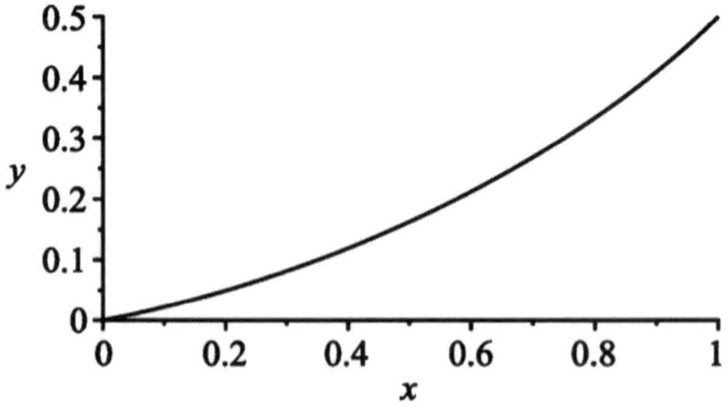

Fig. 1 The plot of the solution of the initial value problem given by (14)–(15)

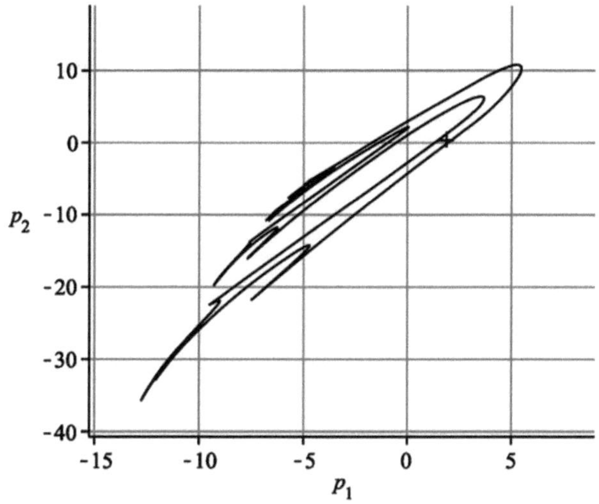

Fig. 2 The region of convergence for parameters p_1 and p_2

5 Conclusions

The method presented in the paper is based on Newton's approach to iterative determination of parameter values p_i. The method allows solving the non-standard boundary value problem in which the number of boundary conditions may exceed the number of first-order differential equations. The difference between the standard and non-standard boundary value problems consists in the size of the initial value problem that needs to be solved in each iteration. When $r = n$ the number of sensitivity functions necessary for determination of the Jacobian matrix equals n^2 which means that the initial value problem will consist of $n\ (n+1)$ first-order ODEs.

References

1. Asher, U.M., Mattheij, R., Russel, R.: Numerical Solution of Boundary Value Problems for Ordinary Differential Equations, SIAM, Philadelphia, 2nd edn. University City Science Center, Philadelphia (1995)
2. Cuomo, S., Marasco, A.: A numerical approach to nonlinear two-point boundary value problems for ODEs. Comput. Math. Appl. **55**, 2476–2489 (2008)
3. Keller, H.B.: Numerical Solutions of Two Point Boundary Value Problems, SIAM, Philadelphia, 4th edn. University City Science Center, Philadelphia (1990)
4. Palej, R., Filipowska, R.: Mathematical modeling of the Inrun profile of a ski jumping hill with the controlled track reaction force. J Theor. Appl. Mech. **47**(1), 229–242 (2009)

5. Press, W.H., Teukolsky, S.A., Vetterling, B.F.: Numerical Recipes. The Art of Scientific Computing, 3rd ed. Cambridge University Press, New York (2007)
6. Rao, S.S.: Applied Numerical Methods for Engineers and Scientists. Prentice Hall, New Jersey (2002)
7. Sung, N.H.: A nonlinear shooting method for two-point boundary value problems. Comput. Math. Appl. **42**, 1411–1420 (2000)

Automatic Sleep Scoring from a Single Electrode Using Delay Differential Equations

Claudia Lainscsek, Valérie Messager, Adriana Portman, Jean-François Muir, Terrence J. Sejnowski, and Christophe Letellier

Abstract Sleep scoring is commonly performed from electroencephalogram (EEG), electrooculogram (EOG), and electromyogram (EMG) to produce a so-called hypnogram. A neurologist thus visually encodes each epoch of 30 s into one of the sleep stages (wake, REM sleep, S_1, S_2, S_3, S_4). To avoid such a long process (about 3–4 hours) a technique for automatic sleep scoring from the signal of a single EEG electrode located in the C_3/A_2 area using nonlinear delay differential equations (DDEs) is presented here. Our approach considers brain activity as resulting from a dynamical system whose parameters should vary according to the sleep stages. It is thus shown that there is at least one coefficient that depends on sleep stages and which can be used to construct a hypnogram. The correlation between manual hypnograms and the coefficient evolution is around 80%, that is, about the inter-rater variability. In order to rank sleep quality from the best to the worst, we introduced a global sleep quality index which is used to compare manual and automatic sleep scorings, thus using our ability to state about sleep quality that is the final goal for physicians.

1 Introduction

Up to 2007, polysomnographic recordings were scored into sleep stages according to the rules introduced by Rechtschaffen and Kales [19] which are mainly based on a spectral analysis. The scoring, accomplished by well-trained neurologist, consists in scoring all 30 s epochs into one of the six stages of vigilance, namely awakeness,

---

C. Lainscsek (✉) • T.J. Sejnowski
Salk Institute for Biological Studies, 10010 North Torrey Pines Road, La Jolla, CA 92037, USA
e-mail: claudia@salk.edu; terry@salk.edu

V. Messager • C. Letellier
Université de Rouen, CORIA, Avenue de l'Université, BP 12, F-76801 Saint-Etienne du Rouvray cedex, France
e-mail: valerie.messager@coria.fr; christophe.letellier@coria.fr

A. Portman • J.-F. Muir
Hôpital de Bois-Guillaume, Bois-Guillaume, France
e-mail: adrianaportmann@yahoo.fr; jean-francois.muir@chu-rouen.fr

J. Awrejcewicz (ed.), *Applied Non-Linear Dynamical Systems*, Springer Proceedings in Mathematics & Statistics 93, DOI 10.1007/978-3-319-08266-0_27

rapid eyes movement sleep (REM), and sleep stages S1, S2, S3, and S4. RK rules were recently modified to overcome the inter-rater variability ([11]). The most important change was that stages 3 and 4 merged into a single stage, named slow-wave sleep or N3. In spite of that, recent studies only showed slight improvements with the new rules ([6]) with an inter-rater agreement slightly greater than 72% ([3]).

Automatic sleep scoring techniques are thus welcome. Most of the computer-assisted scoring techniques stages were based on RK rules ([10, 12, 18]). In fact, most of them try to reproduce what is done by neurologists and which can lead to an overall epoch-by-epoch agreement of 80%, and require a quite complex decisional tree (see Fig. 2 in [2]). With the emergence of "chaos theory," recurrence plots quantifiers, Lyapunov exponents, or correlation dimension were used to obtain hypnograms with an overall agreement which was rarely greater than 60 or 70% ([23]).

Neural networks were also used to distinguish different features exhibited in the spectral domain but were not able to distinguish more than the REM sleep from non-REM sleep ([9]). Another technique was correctly scoring sleep stages but required two EEG channels, one horizontal electrooculogram channel and one chin electromyogram channel ([20]). An automatic sleep classification was able to distinguish wake, slow-wave sleep and rapid eye movements sleep stages ([22]), but a specific sensor, a head accelerometer, was required and must be added to conventional sensors.

Our aim is to develop a reliable automatic technique using a single EEG signal for scoring hypnograms. The subsequent part of this paper is organized as follows. In Sect. 2 the pool of patients which were recorded is described. Section 3 is devoted to our automatic sleep scoring technique and to a new global sleep quality index used to rank a set of hypnograms. In Sect. 4 the results are presented and Sect. 5 gives a conclusion.

2 Patients

This retrospective observational study was conducted at the sleep laboratory at the medical university hospital Intensive Care Unit in Rouen. We selected 38 recordings, but only 35 were associated with a reliable sleep scoring. These patients were long-term ventilated for chronic respiratory failure and grouped into two types. The first type corresponds to an obesity hypoventilation syndrome (OHS) commonly seen in severely overweight people who fail to breathe normally resulting in low blood oxygen levels and high blood carbon dioxide (CO_2) levels. Many of these patients have increased upper airway resistances during sleep (obstructive sleep apnea). This induces a significant amount of wake after sleep onset (WASO) leading to abnormal daytime sleepiness. This disease puts strain on the heart, possibly resulting in heart failure, leg swelling, and various other related symptoms. The second group of respiratory failure, considered here, is associated with chronic obstructive pulmonary disease (COPD). This refers to small airway obstructions

Table 1 Main clinical characteristics of the patients ($n = 34$)

Demographics and respiratory parameters	Mean	(SD)
Age (year)	64.5	(11.7)
Male:female	24:11	
Body mass index ($kg.m^{-2}$)	42.0	(10.5)
PaO_2 (cmH_2O)	9.5	1.1
$PaCO_2$ (cmH_2O)	5.8	(0.9)

Normal values: $(10.7 < PaO_2 < 12.0)$ cmH_2O, $PaCO_2 \approx 5.3\,cmH_2O$, $(18.5 < BMI < 25)\,kg.m^{-2}$ and obesity is defined by BMI$> 30\,kg.m^{-2}$

and emphysema, two commonly coexisting pulmonary diseases in which the airways progressively narrow inducing shortness of breath. In these patients, the airflow limitation is usually nonreversible when treated with bronchodilators and progressively becomes more and more severe. One efficient treatment is to put these patients under noninvasive mechanical ventilatory assistance. In the present case, all patients were ventilated with the bilevel ventilator ResMed VPAP III. All patients included in this study were in stable condition, as assessed by clinical examination and arterial blood gases.

Main characteristics of the thirty-five patients for which the sleep was scored during one night under mechanical ventilation are reported in Table 1. Twenty patients (57%) had OHS and 15 patients (43%) had COPD. Thirteen patients (38%) were diagnosed with obstructive sleep apnea syndrome (defined as more than 10 apneas per hour). Upon study inclusion, the patients were ventilated for a few months. Nineteen patients (56%) were hypercapnic ($PaCo_2 > 5.6\,cmH_2O$).

3 Method

3.1 Automatic Sleep Scoring

A nonlinear delay differential equation has the general form

$$\begin{aligned}\dot{x} = a_1\,x_{\tau_1} + a_2\,x_{\tau_2} + a_3\,x_{\tau_3} + \cdots + a_{i-1}\,x_{\tau_n} + a_i\,x^2_{\tau_1} + a_{i+1}\,x_{\tau_1}x_{\tau_2} \\ + a_{i+2}\,x_{\tau_1}x_{\tau_3} + \cdots + a_{j-1}\,x^2_{\tau_n} + a_j\,x^3_{\tau_1} + a_{j+1}\,x^2_{\tau_1}x_{\tau_2} + \ldots\cdots + a_l\,x^m_{\tau_n}\end{aligned} \tag{1}$$

where $x = x(t)$ and $x_{\tau_j} = x(t - \tau_j)$. In this general form, the DDE has n delays, l monomials with their corresponding coefficients a_i, and a degree of nonlinearity equal to m. In the subsequent part of this paper, we will define a k-term DDE as an equation with only $k < l$ monomials selected from the right-hand side of the general form (1). As for any global modeling technique, there is a significant improvement of capturing main characteristics of the underlying dynamics from observed data by carefully selecting the structure of the DDE model ([1, 14–16]). The minimal mean squared error is used for this process. By structure selection, we mean retaining only monomials in the DDE that have the most significant contribution to classify the data. An equally important task is to select the right time-delays, since linear terms are directly related to the fundamental timescales and nonlinear terms to the nonlinear couplings between them ([16]). This can be performed by using a genetic algorithm ([15]) or by an exhaustive search for the best model among the general form with $n = 2$ and $m = 3$ resulting in $l = 9$ monomials as performed in [16].

Here only models with up to three terms were considered (see Table 2 in [16]). The variable x corresponds to the signal provided by the electrode located in the C_3/A_2 area of the scalp. We ran a genetic algorithm to minimize the least square error of 30 s data windows to select the best models and delays for each 30 s window ([8, 15]). For 95% of the data windows (corresponding to the 35 patients), the four models

$$\dot{x} = a_1\, x_{\tau_1} + a_2\, x_{\tau_2} + a_3\, x_{\tau_1}^2\,; \tag{2}$$

$$\dot{x} = a_1\, x_{\tau_1} + a_2\, x_{\tau_2} + a_4\, x_{\tau_1}\, x_{\tau_2}\,; \tag{3}$$

$$\dot{x} = a_1\, x_{\tau_1} + a_2\, x_{\tau_2} + a_6\, x_{\tau_1}^3\,; \tag{4}$$

$$\dot{x} = a_1\, x_{\tau_1} + a_2\, x_{\tau_2} + a_7\, x_{\tau_1}^2\, x_{\tau_2}\,; \tag{5}$$

were selected as well as delays between $1\,\delta t$ and $4\,\delta t$ with $\delta t = \frac{1}{64}$ s. Among these four models, model (5) is the best to distinguish wake, REM, and S1 from the sleep stages S2, S3, and S4 (see left panel from Fig. 1). Delay $\tau_1 = 1$ is useful to distinguish wake, S2, S3, and S4 from REM and S1 (right panel from Fig. 1). Delay $\tau_2 = 3$ allows to distinguish wake from sleep stages. Thus, combining model (5) with delays $\tau_1 = 1$ and $\tau_2 = 3$ provides the model with the most discriminative ability. Among the three coefficients of model (5), parameter a_2 was found to be the most correlated ($r = 0.95$) to the manually scored hypnogram, as exemplified in Fig. 2 in the case of patient 15. We then used this model and this coefficient to score the sleep for our 35 patients.

It was then necessary to convert the a_2-time series which corresponds to the time evolution of a real number sampled at 0.1 Hz (one point per 10 s) into a sequence of integers from 1 (stage S1) to 6 (awake). This is the tricky part of our technique. In the case of patient 15, we got an automatically scored hypnogram which was quite close to the manually scored one (Fig. 2b).

3.2 Assessing the Sleep Quality

Since patients with chronic respiratory failures are ventilated during their sleep, it is important to assess whether the ventilation improves the sleep quality or,

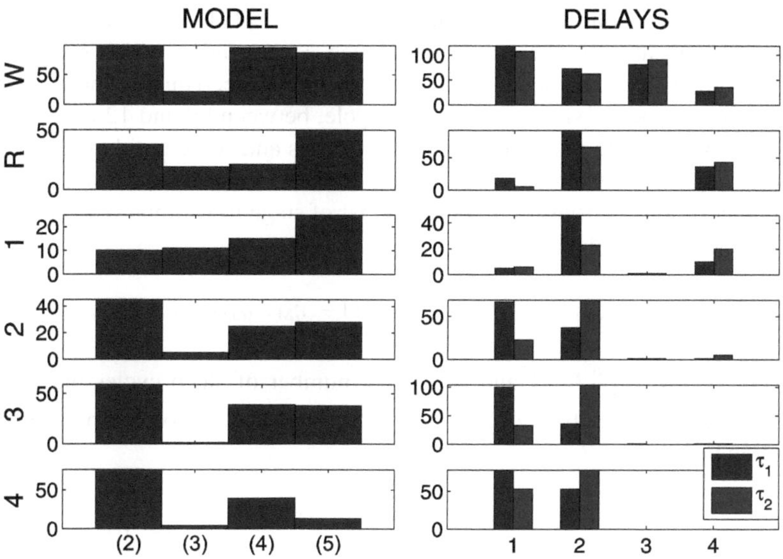

Fig. 1 Histograms of the number of time each of the four selective DDEs (*left*) and each delays (*right*) were selected with minimum error for each sleep stage

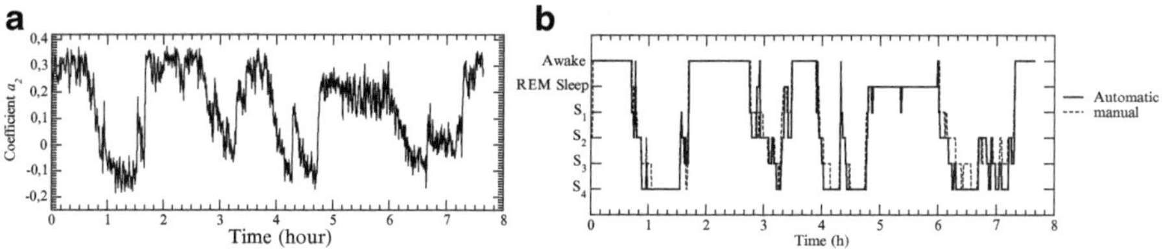

Fig. 2 Time series of coefficient a_2 of the delay differential equation (5) and the corresponding hypnogram. Case of patient 15 (male, 76 years, BMI=50 kg.m^{-2}). The manually scored hypnogram (*green*) is also reported for comparison. (**a**) Raw a_2 time series (**b**) Sequence of integers

at least, that it does not degrade it. In order to do that, it is necessary to be able to rank hypnograms according to sleep quality. From a subjective point of view, sleep quality refers to patient feelings about the refreshing effect of sleep which can be assessed using some sleep diary or the Pittsburgh Quality Index ([4]). The characteristics commonly taken into account in such evaluation are sleep latency, sleep duration, regular sleep efficiency, sleep disturbances (including sleep disruptive events such as snoring, apnea, or pains), use of sleeping medication, and daytime dysfunction ([4]).

Up-to-now, the objective evaluation of sleep quality was based on the same characteristics but directly measured from hypnograms ([11]). Also considered are the arousal index (number of arousals per hour) and the number of various respiratory events. To assess the evolution of sleep quality, all these quantities are then subjectively combined and compared since none of them can alone allow to rank hypnograms according to sleep quality (see [17] for details). In order to avoid this last subjective step, we introduced a new index which combines the most important sleep characteristics. Thus, our global sleep quality index takes into account the number of sleep cycles (each cycle, between 90 and 120 min, contains some slow-wave sleep restoring physical functions and some rapid eye movements restoring cognitive functions), the fraction of WASO, the fraction of stable sleep, the number of micro-arousals, and the number of stage transitions. The global sleep quality index η_{GSQ} is defined as

$$\eta_{\text{GSQ}} = \eta_{\text{cycle}} \cdot \eta_{\text{restoring}} \cdot \eta_{\text{stability}} \cdot (1 - \eta_{\text{M}-\text{frag}}) \cdot (1 - \eta_{\mu-\text{frag}}) \tag{6}$$

where $\eta_{\text{cy}} = \text{Max}\left(\frac{N_{cy}}{6}, 1\right)$ and N_{cy} is the number of sleep cycles that saturates to one when it exceeds 6 cycles; the restoring capacity of sleep is evaluated according to

$$\eta_{\text{restoring}} = \text{Min}\left(\frac{5}{2}\,\frac{\tau_{S3} + \tau_{S4} + \tau_R}{\tau_{S1} + \tau_{S2} + \tau_{S3} + \tau_{S4} + \tau_R}, 1\right) \tag{7}$$

with τ_i being the time duration spent in the ith sleep stage (i = S1, S2, S3, S4, and R) and saturates to 1 when the restorative sleep (S3, S4, and R) exceeds $\frac{2}{5}$ of the effective sleep; the sleep stability is evaluated according to

$$\eta_{\text{stability}} = \frac{\tau'_{S1} + \tau'_{S2} + \tau'_{S3} + \tau'_{S4} + \tau'_R}{\tau_{\text{effective sleep}}} \tag{8}$$

with τ'_i being the time spent in the ith sleep stage without any micro-arousal and not corresponding to an epoch connexe to a stage transition, and $\tau_{\text{effective sleep}}$ being the time duration of sleep stages ($\tau_{S1} + \tau_{S2} + \tau_{S3} + \tau_{S4} + \tau_R$); the sleep macro-fragmentation is evaluated according to

$$\eta_{\text{M}-\text{frag}} = \frac{\tau_{\text{waso}}}{\tau_{\text{waso}} + \tau_{\text{effective sleep}}}; \tag{9}$$

the sleep micro-fragmentation is evaluated according to

$$\eta_{\mu-\text{frag}} = \frac{(\tau_{S1} - \tau'_{S1}) + (\tau_{S2} - \tau'_{S2}) + (\tau_{S3} - \tau'_{S3}) + (\tau_{S4} - \tau'_{S4}) + (\tau_R - \tau'_R)}{\tau_{\text{effective sleep}}} \tag{10}$$

with $\tau_i - \tau'_i$ being the time spent in an epoch of the ith sleep stage with a micro-arousal or connection to a stage transition.

4 Results

The time series of coefficient a_2 were found quite well correlated to the corresponding hypnograms ($\overline{r} = 0.86 \pm 0.1$). To assess the quality of our sleep scorings using the coefficient a_2 we computed the confusion matrix ([13]) which is a specific table layout used to assess performance of classifier. Each column of the matrix represents the instances in a predicted class, while each row represents the instances in an actual class. The confusion matrix for all epochs of all patients is reported in Fig. 3. To get a graphical representation the numbers were also converted to a percentage. A dark diagonal from the upper-left corner to the lower-right corner with all other squares in white would indicate perfect scoring of each data window into the correct sleep stage.

As additional measure of performance we used Cohen's kappa κ [5,7,21] which can be computed directly from the confusion matrix as [13]. $\kappa = \frac{p_a - p_e}{1 - p_e}$, where $p_a = \sum_{k=1}^{q} p_{kk}$, and $p_e = \sum_{k=1}^{q} p_{k+}p_{+k}$ where $q = 6$ for the 6 classes, p_a is the observed percentage of agreement, p_e is the expected percentage of agreement, p_{k+} is the percentage of actual classification, and p_{+k} is the percentage of predicted classification. We got $\overline{\kappa} = 0.51 \pm 0.1$ when comparing automatically scored hypnograms with the manually scored ones. Detailed results are reported in Table 2.

The global sleep quality index η_{GSQ} was first computed from the hypnograms scored by the neurologist. Patients were then ranked according to a decreasing η_{GSQ} (Fig. 4). The hypnogram of the patient with the largest η_{GSQ} (35.4 %) is shown in Fig. 5a: it presents 3 sleep cycles quite well structured. Contrary to this, the hypnogram of patient 22 with the smallest η_{GSQ} (0.1 %) is shown in Fig. 5b: it does not present a single well-structured sleep cycle and the effective sleep time duration is small ($\tau_{\text{effective sleep}} = 146\,\text{min}$).

The rates of each sleep stage was computed for each hypnograms which were ranked according to decreasing η_{GSQ} (Fig. 6). The best hypnogram (patient 34, $\eta_{\text{GSQ}} = 35.4$) presents a good proportion of restorative sleep. Contrary to this,

actual \ predicted	W	R	1	2	3	4
W	3026	1642	147	29	1	4
R	225	1288	762	62	5	11
1	61	733	3841	2388	607	164
2	7	13	191	703	524	528
3	1	34	394	790	1669	97
4	27	54	166	557	1050	8615

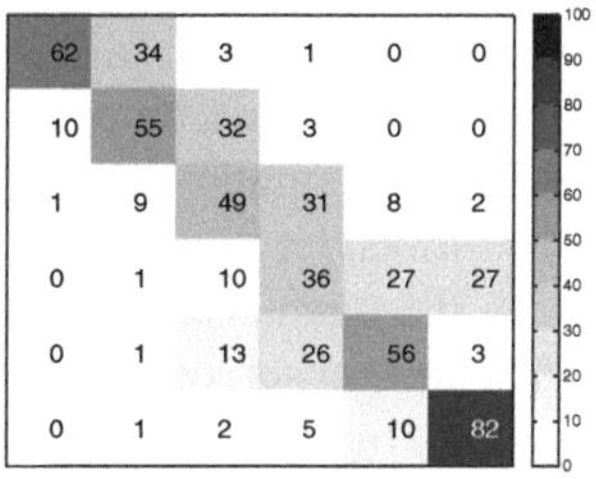

Fig. 3 Confusion matrix for all subjects: the table on the *left side* shows the numbers of predicted and actual sleep stage windows. The plot on the *right side* visualizes the percentage of predicted and actual sleep stage windows

Table 2 Correlation coefficient r and Cohen's κ between the manually scored hypnograms and the time series of coefficient a_2 of model (5) for each subject

#	r	κ	#	r	κ	#	r	κ	#	r	κ	#	r	κ
1	0.82	0.36	9	0.91	0.53	17	0.70	0.28	24	0.95	0.65	32	0.91	0.55
2	0.95	0.61	11	0.80	0.36	18	0.81	0.44	25	0.78	0.41	33	0.84	0.41
3	0.89	0.59	12	0.90	0.64	19	0.87	0.53	26	0.82	0.50	34	0.93	0.66
5	0.91	0.63	13	0.91	0.57	20	0.78	0.59	27	0.92	0.64	35	0.82	0.37
6	0.92	0.51	14	0.76	0.36	21	0.79	0.39	29	0.94	0.61	36	0.89	0.55
7	0.92	0.68	15	0.95	0.67	22	0.80	0.40	30	0.87	0.51	37	0.91	0.59
8	0.79	0.43	16	0.90	0.54	23	0.80	0.41	31	0.79	0.43	38	0.91	0.51

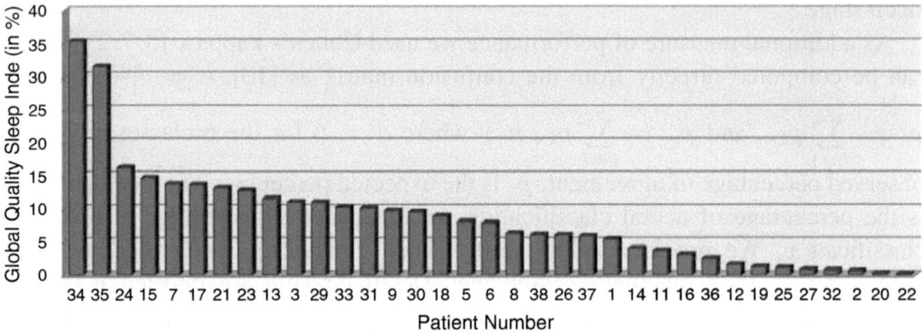

Fig. 4 Global sleep quality index computed from the manually scored hypnograms for the 35 patients of our protocol

the worst hypnogram (patient 22, $\eta_{\text{GSQ}} = 0.1$) associated with a very small fraction of restorative sleep and a large one of WASO. Hypnograms are rather well ranked since the rate of WASO and sleep micro-fragmentation are anticorrelated to η_{GSQ} ($r = -0.65$, $p < 0.0001$ and $r = -0.75$, $p < 0.0001$, respectively). The rate of slow-wave sleep (S3 and S4) and the rate of REM sleep are correlated to η_{GSQ} ($r = 0.83$, $=< 0.0001$ and $r = 0.59$, $p < 0.0001$, respectively). These features and others that are outside the scope of this paper correspond to an increase of the sleep quality with η_{GSQ}.

We now computed the global sleep quality index from the automatically scored hypnograms with our technique (Fig. 7). They were ordered in a slightly different order than the manual hypnograms. In order to quantify this disagreement between these two orders, let us designate by n the rank (n') the rank obtained by computing η_{GSQ} from the manual (automatic) hypnograms. Thus $\Delta n = |n - n'|$ corresponds to the rank shift observed between these two orders. We thus have $\overline{\Delta n} = 4.6 \pm 5.4$, meaning that, in average, the good (bad) hypnograms remain the good (bad) ones. There are four notable exceptions with the hypnograms for patients 11, 15, 24, and 35 for which Δn equals to -23, $+15$, $+20$, and $+11$, respectively.

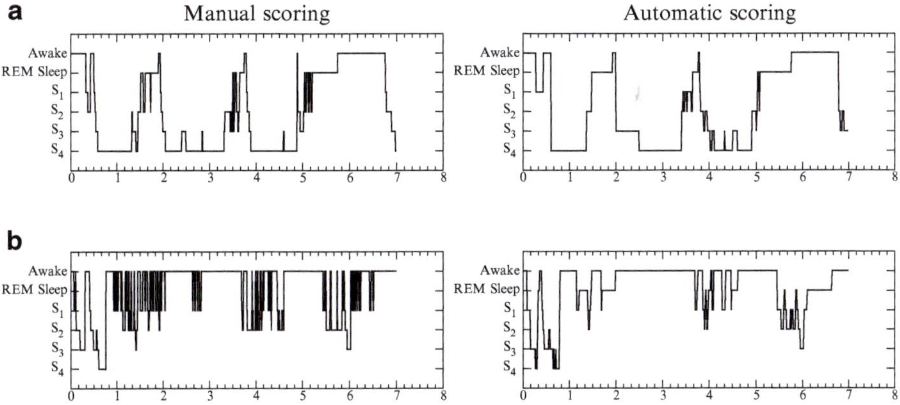

Fig. 5 Hypnograms for two of the 35 patients corresponding to the largest and the smallest global sleep quality index. The gender, age, body mass index, and the rate of synchronous breathing cycles are also reported. (**a**) Patient 34 : male, 82 years, BMI=44.1, 2.1% of asynchronous cycles, and $\eta_{GSQ} = 35.4\%$. (**b**) Patient 22 : male, 83 years, BMI=36.3, 8.0% of asynchronous cycles, and $\eta_{GSQ} = 0.1\%$

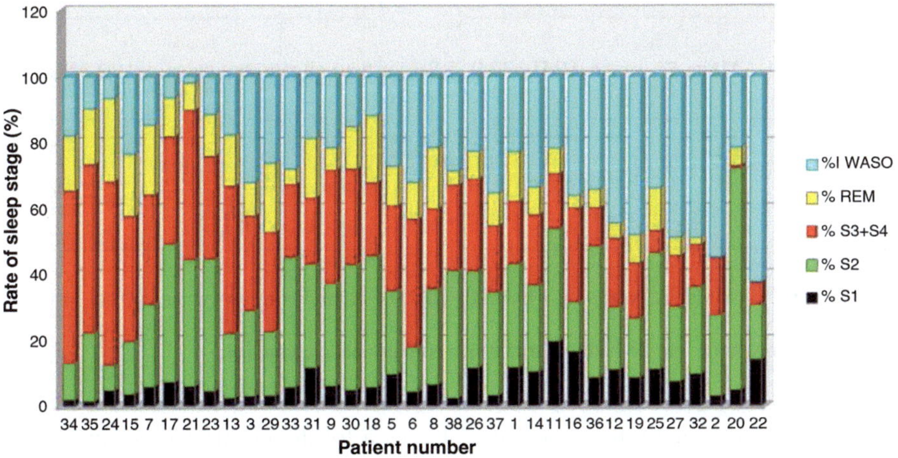

Fig. 6 Fraction of time duration of each sleep stage. Patients are ranked according to the global sleep quality index η_{GSQ}

The manually scored hypnogram of patient 11 (Fig. 8a) presents many fluctuations between wake and stage S1 and a very few epochs in stages S3 or S4 and REM sleep, thus associated with a small global quality sleep index ($\eta_{GSQ} = 3.7\%$). The evolution of the coefficient of the DDE fluctuates a lot between the values corresponding to wake and S1 stages. Consequently, since REM sleep is between these two stages from EEG, our technique returns too often REM sleep (and not WASO). This is significantly increasing the global sleep quality index to 24.9.

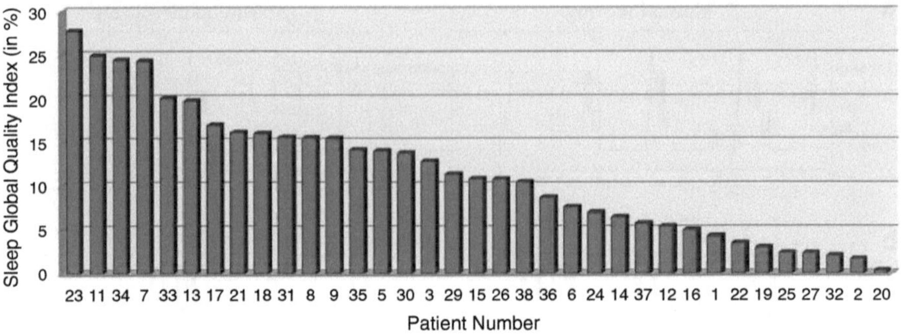

Fig. 7 Global sleep quality index computed from the automatically scored hypnograms for the 35 patients of our protocol

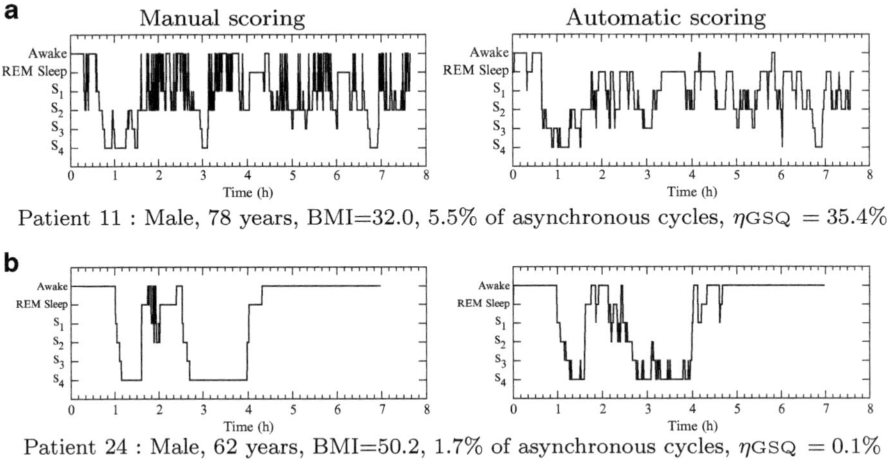

Fig. 8 Hypnograms for two badly scored using our automatic technique

It is important to note that a neurologist uses a lot the electrooculogram and the electromyogram to distinguish REM sleep from awake and S1, two signals which are not considered by our technique.

Contrary to this, the automatically scored hypnograms for patient 24 is characterized by a global sleep quality index $\eta_{GSQ} = 7.0\%$ is significantly smaller than the value (16.2%) obtained from the manual hypnograms (Fig. 8b). There are few reasons explaining such a large departure between these two η_{GSQ}-values. The global sleep duration (between the first and the last sleep epoch) is larger than the one obtained from the automatic scoring (221.5 min and 198 min, respectively), but the number of sleep cycles is 2 in both cases. The rate of WASO in the automatic hypnogram is about three times the rate obtained from the manually scored hypnogram (19.9 and 6.6, respectively). The rate of micro-fragmentation obtained with our technique is about three times the rate returned by the neurologist

(31.8 and 11.1, respectively). The stability is smaller in the hypnogram provided by our technique than in the one scored by the neurologist (38.2 and 58.3, respectively). All these modifications tend to increase the global sleep quality index.

5 Conclusions

In 88% of subjects the overall sleep quality index computed from the DDE hypnograms are in agreement with the sleep quality index computed from the visually scored hypnograms. The difference in 12% of all patients results from converting the real number outputs of the DDE to the integers used for indexing sleep stages (S1, S2, S3, S4, R, and wake). This is the weakest part of the present version of our technique. In spite of this, our hypnograms are already sufficiently close to the manual hypnograms that are used to assess the sleep quality. Importantly, this first study has led to the identification of possible improvements that are currently being developed.

Our automatic scoring technique using DDEs is well correlated to the corresponding visually scored hypnograms ($\overline{r} = 0.86 \pm 0.1$). This excellent agreement becomes even more impressive when considering the use of only one scalp electrode for the DDE method. Indeed, the most promising aspect of our technique is that only one scalp electrode is sufficient to accurately score sleep stages.

References

1. Aguirre, L.A., Billings, S.A.: Improved structure selection for nonlinear models based on term clustering. Int. J. Contr. **62**(3), 569–587 (1995)
2. Anderer, P., Gruber, G., Parapatics, S., Woertz, M., Miazhynskaia, T., Klösch, G., Saletu, B.,Zeitlhofer, J., Barbanoj, M.J., Danker-Hopfe, H., Himanen, S.-L., Kemp, B., Penzel, T., Grözinger, M., Kunz, D., Rappelsberger, P., Schlögl, A., Dorffner, G.: An E-health solution for automatic sleep classification according to Rechtschaffen and Kales: validation study of the somnolyzer 24x7 utilizing the SIESTA database. Neuropsychobiology **51**(3), 115–133 (2005)
3. Basner, M., Griefahn, B., Penzel, T.: Inter-rater agreement in sleep stage classification between centers with different backgrounds. Somnologie **12**,75–84 (2008)
4. Buysse, D.J., Reynolds, C.F., Monk, T.H., Berman, S.R., Kupfer, D.J.: The Pittsburgh sleep quality index: a new instrument for psychiatric practice and research. Psychiatric Res. **28**, 193–213 (1989)
5. Cohen, J.: A coefficient of agreement for nominal scales. Educ. Psychol. Meas. **20**(1), 37–46 (1960)
6. Danker-Hopfe, H., Anderer, P., Zeitlhofer, J., Boeck, M., Dorn, H., gruber, G., Heller, E., Loretz, E., Moser, D., Parapatics, S., Saletu, B., Schmidt, A. Dorfner, G.: Interrater reliability for sleep scoring according to the Rechtschaffen and Kales and the new AASM standard. J.Sleep Res. **18**(1), 74–84 (2009)
7. Fleiss, J.L., Cohen, J.: The equivalence of weighted kappa and the intraclass correlation coefficient as measures of reliability. Educ. Psychol. Meas. **33**, 613–619 (1973)

8. Goldberg, D.E.: Genetic Algorithms in Search, Optimization and Machine Learning. Addison-Wesley, Wokingham (1998)
9. Grötzinger, M., Wolf, C., Uhl, T., Schäffner, C., Röschke, J.: Online detection of REM sleep based on the comprehensive evaluation of short adjacent EEG segments by artificial neural networks. Progr. Neuro-Psychopharmacol. Biol. Psychiatr. **21**(6), 951–963 (1997)
10. Harper, R.M., Schechtman, V.L., Kluge, K.A.: Machine classification of infant sleep state using cardiorespiratory measures. Electroencephalogr. Clin. Neurophysiol. **67**(4), 379–387 (1987)
11. Iber, C., Ancoli-Israel, S., Chesson, A., Quan, S.F. (eds.): The AASM Manual for the Scoring of Sleep and Associated Events: Rules, Terminology, and Technical Specification. American Academy of Sleep Medicine, Westchester (2007)
12. Jansen, B.H., Dawant, B.M.: Knowledge-based approach to sleep EEG analysis-a feasibility study. IEEE Trans. Biomed. Eng. **36**(5), 510–518, (1989)
13. Kohavi, R., Provost, F.: Glossary of terms. Mach. Learn. **30**(2/3), 271–274 (1998)
14. Lainscsek, C., Letellier, C.,Gorodnitsky, I.: Global modeling of the Rössler system from the z-variable. Phys. Lett. A, **314** 409–127 (2003)
15. Lainscsek, C., Rowat, P., Schettino, L., Lee, D., Song, D., Letellier, C., Poizner, H.: Finger tapping movements of Parkinson's disease patients automatically rated using nonlinear delay differential equations. Chaos, **22**, 013119 (2012)
16. Lainscsek, C., Sejnowski, T.J.: Electrocardiogram classification using delay differential equations. Chaos **23**(2), 023132 (2013)
17. Messager, V., Portmann, A., Muir, J.-F., Letellier, C.: A global sleep quality index for ranking hypnograms. in preparation (2013)
18. Principe, J.C., Gala, S.K., Chang, T.G.: Sleep staging automaton based on the theory of evidence. IEEE Trans. Biomed. Eng. **36**(5), 503–509 (1989)
19. Rechtschaffen, A., Kales, A. (eds.): A Manual of Standardized Terminology, Techniques and Scoring System for Sleep Stages of Human Subject. US Government Printing Office, National Institute of Health Publication, Washington (1968)
20. Schaltenbrand, N., Lengelle, R., Toussaint, M., Luthringer, R., Carelli, G. Jacqmin, A., Lainey, E., Muzet, A., Macher, J.P.: Sleep stage scoring using the neural network model: comparison between visual and automatic analysis in normal subjects and patients. Sleep **1**, 26–35 (1996)
21. Scott, W.: Reliability of content analysis: The case of nominal scale coding. Publ. Opin. Q. **19**(3), 321–325 (1955)
22. Sunderam, S., Chernyy, N., Peixoto, N., Mason, J.P., Weinstein, S.L., Schiff, S.J., Gluckman, B.J.: Improved sleep-wake and behavior discrimination using MEMS accelerometers. J.Neurosci. Meth. **163**(2), 373–383 (2007)
23. Šušmáková, K., Krakovská, A.: Discrimination ability of individual measures used in sleep stages classification. Artif. Intell. Med. **44**(3), 261–277 (2008)

Methods for the Quick Analysis of Micro-chaos

Gergely Gyebrószki and Gábor Csernák

Abstract Micro-chaos is a phenomenon when sampling, round-off and processing delay (shortly, digital effects) lead to chaotic oscillations with small amplitude. In previous works [1], the so-called micro-chaos maps of various digitally controlled unstable linear mechanical systems were derived and the possibility of the coexistence of several disconnected attractors was highlighted. The typical size of these attractors is usually negligible from the practical point of view, but the distance of the farthest attractor from the desired state can be rather large. This is why the phenomenon of micro-chaos can be the source of significant control error. In this paper, a set of numerical methods (e.g. cell mapping techniques for the exploration of the phase-space structure) is assembled in order to create a toolkit for the quick analysis of micro-chaotic behaviour. The elaborated methods are tested on models of PD-controlled unstable systems and the practically important characteristics of chaotic behaviour are determined.

1 Introduction

Chaotic vibrations arising due to the digital effects in control are well known for the last 20 years [3, 5]. Sampling and delay are commonly taken into account in control problems; however, rounding is usually neglected. It has been shown [4, 5] that rounding leads to small amplitude chaotic oscillations—referred to as *micro-chaos* because of the small amplitude—where several disconnected attractors may coexist. In a couple of cases (inverted pendulum stabilized with D control without delay [5], D control with delay [1] and PD control without delay [2]), it has been rigorously proven that the vibrations are indeed chaotic. In [2], the coexistence of

G. Gyebrószki (✉)
Budapest University of Technology and Economics, Department of Applied Mechanics, H-1111 Budapest, Műegyetem rkp 5., Hungary
e-mail: gyebro@mm.bme.hu

G. Csernák
HAS-BUTE Research Group on Dynamics of Machines and Vehicles, H-1111 Budapest, Műegyetem rkp 5., Hungary
e-mail: csernak@mm.bme.hu

J. Awrejcewicz (ed.), *Applied Non-Linear Dynamical Systems*, Springer Proceedings in Mathematics & Statistics 93, DOI 10.1007/978-3-319-08266-0_28

chaotic attractors was already shown. Since chaotic behaviour is generally found in theoretical models of digitally controlled systems, in this paper we are focusing on numerical methods to characterize and examine chaotic behaviour. Section 2 introduces the digitally controlled inverted pendulum—as the subject of chaos investigation—while in Section 3 we give a short overview of methods used for characterizing chaotic behaviour. Section 4 shows different scenarios to select the region of interest for numerical methods. Section 5 is devoted to describe the simple cell mapping (SCM) method and other supporting methods (such as fractal dimension calculation) for investigating chaotic behaviour and shows the SCM results for various parameters.

2 System Under Examination

The *pendulum-on-a-cart* is one of the simplest devices used for demonstrating control problems. Neglecting the cart part of this device leads to a simple inverted pendulum with control. In this section, the equation of motion of the digitally controlled inverted pendulum is presented and the derivation of the so-called *micro-chaos map* is shown. The map corresponding to the inverted pendulum is used to illustrate the application of numerical methods described in the next section.

2.1 Inverted Pendulum with Damping and PD Control

Consider an inverted pendulum with damping and digitally implemented PD control with zero-order hold (i.e. the control torque is kept constant between two successive sampling instants) as shown in Fig. 1. The actuator (the neglected cart) is considered to be ideal, i.e. produces the desired control effort immediately. The equation of motion of this system is

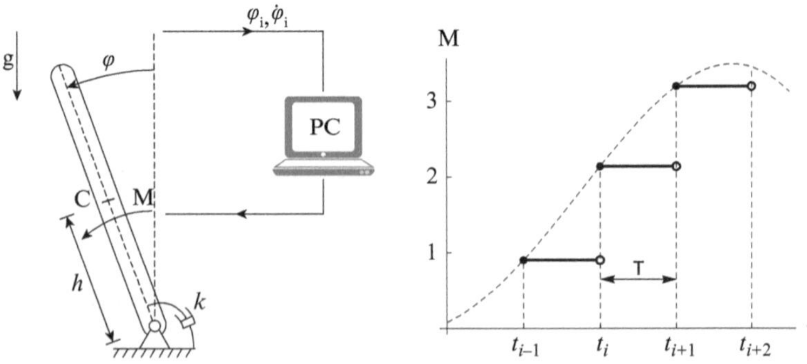

Fig. 1 The digitally controlled inverted pendulum and the control torque with respect to time

$$J\,\ddot{\varphi}(t) = mgh\sin(\varphi(t)) - k\,\dot{\varphi}(t) - p\,\varphi_i - d\,\dot{\varphi}_i, \qquad t \in [i\tau, (i+1)\tau), \tag{1}$$

where m is the mass of the pendulum, J is the mass moment of inertia about the axis of rotation, h is the distance between the centre of mass and the axis of rotation, p and d are control parameters, k is the linear damping coefficient, g is the gravitational acceleration and τ is the sampling time, while $\varphi_i = \varphi(i\tau)$ and $\dot{\varphi}_i = \dot{\varphi}(i\tau)$ are sampled values of the angular position and angular velocity, respectively (at the beginning of the ith time interval). Rearranging, linearizing and dividing (1) by J yields

$$\ddot{\varphi}(t) + 2\beta\dot{\varphi}(t) - \alpha^2\varphi(t) = -P\,\varphi_i - D\,\dot{\varphi}_i, \tag{2}$$

where $\alpha^2 = \frac{mgh}{J}, 2\beta = \frac{k}{J}, P = \frac{p}{J}, D = \frac{d}{J}$.

One can rewrite (2) as a system of first-order differential equations:

$$\dot{\omega}(t) = \alpha^2\varphi(t) - 2\beta\dot{\varphi}(t) - P\,\varphi_i - D\,\omega_i, \qquad t \in [i\tau, (i+1)\tau), \tag{3}$$
$$\dot{\varphi}(t) = \omega(t),$$

with initial conditions $\omega(i\tau) = \omega_i, \varphi(i\tau) = \varphi_i$.

Using the notation $\mathbf{y}_i = \begin{bmatrix} \varphi(i\tau) & \omega(i\tau) \end{bmatrix}^{\mathrm{T}}$ the solution of this equation formulates a 2D map

$$\mathbf{y}_{i+1} = (\mathbf{A} + \mathbf{B}\mathbf{K})\,\mathbf{y}_i, \qquad i \in \mathbb{N}, \tag{4}$$

where

$$\mathbf{A} = \frac{e^{-\beta\tau}}{\gamma}\begin{bmatrix} \gamma\cosh(\gamma\tau) + \beta\sinh(\gamma\tau) & \sinh(\gamma\tau) \\ \alpha^2\sinh(\gamma\tau) & \gamma\cosh(\gamma\tau) - \beta\sinh(\gamma\tau) \end{bmatrix}, \tag{5}$$

$$\mathbf{B} = \frac{e^{-\beta\tau}}{\gamma}\begin{bmatrix} \frac{\beta\sinh(\gamma\tau) + \gamma\left(\cosh(\gamma\tau) - e^{\beta\tau}\right)}{\alpha^2} \\ \sinh(\gamma\tau) \end{bmatrix}, \tag{6}$$

$$\mathbf{K} = -\begin{bmatrix} P & D \end{bmatrix}, \tag{7}$$

where $\gamma = \sqrt{\alpha^2 + \beta^2}$.

Without considering rounding, one can acquire the *dimensionless form of the map*, using the notations $\hat{\alpha} = \alpha\tau$, $\hat{\beta} = \beta\tau$, $\hat{p} = P\tau^2$, $\hat{d} = D\tau$, $x = \varphi/\varphi_{\mathrm{ref}}$, $v = \omega/\omega_{\mathrm{ref}}$, $\hat{\mathbf{y}} = \begin{bmatrix} x & v \end{bmatrix}^T$

$$\hat{\mathbf{y}}_{i+1} = (\hat{\mathbf{A}} + \hat{\mathbf{B}}\,\hat{\mathbf{K}})\,\hat{\mathbf{y}}_i, \tag{8}$$

where

$$\hat{\mathbf{A}} = \frac{e^{-\hat{\beta}}}{\hat{\gamma}} \begin{bmatrix} \hat{\gamma}\cosh(\hat{\gamma}) + \hat{\beta}\sinh(\hat{\gamma}) & \sinh(\hat{\gamma}) \\ \hat{\alpha}^2 \sinh(\hat{\gamma}) & \hat{\gamma}\cosh(\hat{\gamma}) - \hat{\beta}\sinh(\hat{\gamma}) \end{bmatrix}, \tag{9}$$

$$\hat{\mathbf{B}} = \frac{e^{-\hat{\beta}}}{\hat{\gamma}} \begin{bmatrix} \frac{\hat{\beta}\sinh(\hat{\gamma}) + \hat{\gamma}\left(\cosh(\hat{\gamma}) - e^{\hat{\beta}}\right)}{\alpha^2} \\ \sinh(\hat{\gamma}) \end{bmatrix}, \tag{10}$$

$$\hat{\mathbf{K}} = -\begin{bmatrix} \hat{p} \; \hat{d} \end{bmatrix}. \tag{11}$$

Here $\hat{\gamma} = \sqrt{\hat{\alpha}^2 + \hat{\beta}^2}$.

2.2 *Rounding at the Output*

Consider the case when rounding is applied to the calculated control effort (which is the output of the control system). Introducing the rounding to Equation (8), the so-called micro-chaos map (or μ-chaos map) is obtained:

$$\hat{\mathbf{y}}_{i+1} = \hat{\mathbf{A}}\,\hat{\mathbf{y}}_i + \hat{\mathbf{B}}\,\mathrm{Int}(\hat{\mathbf{K}}\,\hat{\mathbf{y}}_{\mathrm{i}}). \tag{12}$$

Here the reference angle in $\hat{\mathbf{y}}$ is $\varphi_{\mathrm{ref}} = r_{\mathrm{out}}\tau^2$, and the reference angular velocity is $\omega_{\mathrm{ref}} = r_{\mathrm{out}}\tau$, while $r_{\mathrm{out}} \left[\frac{\mathrm{rad}}{\mathrm{s}^2}\right]$ is the resolution of the actuated control effort and $\mathrm{Int}(n)$ denotes rounding towards zero (or *truncating* or taking the integer part of n).

2.3 *Rounding at the Input*

When rounding is applied to the measured position and velocity the corresponding form of the *micro-chaos map* is

$$\hat{\mathbf{y}}_{i+1} = \hat{\mathbf{A}}\,\hat{\mathbf{y}}_i + \hat{\mathbf{B}}\,\hat{\mathbf{K}}\,\mathrm{Int}(\hat{\mathbf{y}}_{\mathrm{i}}), \tag{13}$$

where the reference angle is $\varphi_{\mathrm{ref}} = r_{\mathrm{in}}$ (i.e. the resolution of the position measurement), while the reference angular velocity is $\omega_{\mathrm{ref}} = r_{\mathrm{in}}/\tau$ since we assume that the velocity is calculated from the measured position. Function Int() calculates the integer part of every element in the vector, returning a vector of the same dimension.

3 Numerical Methods for Examining Chaotic Behaviour

In this section the general overview of numerical methods which are useful for the quick characterization of control-related chaotic phenomena is presented. Methods are divided into two groups according to the size of the region in which the chaotic behaviour is examined. The first group shows methods which are useful to characterize a single attractor, while the second group enumerates methods for examining an arbitrarily large region in the state space.

3.1 Methods for Examining Local Chaotic Behaviour

3.1.1 Iteration of the Map

One can start repeatedly applying (12)—the *micro-chaos map* for the case when rounding is at the output—to a chosen initial state and will eventually find that the iteration will not converge to a specific stable equilibrium but after N iterations it will arrive at one of the several coexisting chaotic attractors instead (Fig. 2). The following figures are obtained for a set of realistic parameters ($\hat{\alpha} = 6.85 \times 10^{-3}$, $\hat{\beta} = 0$, $\hat{p} = 5.5 \times 10^{-5}$, $\hat{d} = 2.5 \times 10^{-3}$) [4]. Since the number of necessary iterations N (i.e. duration of transients) is unknown, a criterion is needed to determine whether the solution has reached an attractor. Having the exact Lyapunov exponents in hand, an appropriate condition could be derived based on the time necessary for the synchronization of coupled maps used in the Lyapunov exponent estimation method [7].

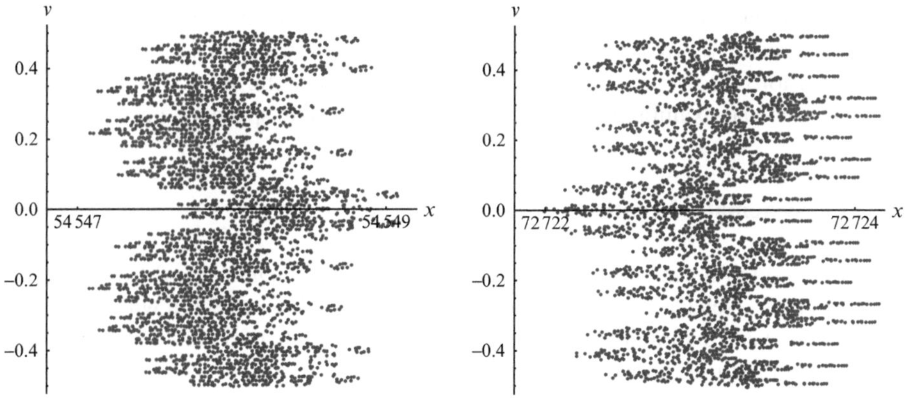

Fig. 2 Chaotic attractors when rounding is applied at the output

3.1.2 Searching for Periodic Points

One can search for periodic points of the map as shown in [2]. A point $\hat{\mathbf{y}}_0$ is p-periodic if $\hat{\mathbf{y}}_p = \hat{\mathbf{y}}_0$. Groups of periodic points (periodic orbits) with high period usually correspond to the skeleton of a chaotic attractor. Consider the case when rounding is applied to the output. Since

$$\hat{\mathbf{y}}_1 \equiv \hat{\mathbf{A}}\,\hat{\mathbf{y}}_0 + \hat{\mathbf{B}}\,\underbrace{\mathrm{Int}(\hat{\mathbf{K}}\,\hat{\mathbf{y}}_0)}_{:=m_1}, \tag{14}$$

$$\hat{\mathbf{y}}_2 \equiv \hat{\mathbf{A}}\,\hat{\mathbf{y}}_1 + \hat{\mathbf{B}}\,\underbrace{\mathrm{Int}(\hat{\mathbf{K}}\,\hat{\mathbf{y}}_1)}_{:=m_2} = \hat{\mathbf{A}}(\hat{\mathbf{A}}\,\hat{\mathbf{y}}_0 + \hat{\mathbf{B}}\,m_1) + \hat{\mathbf{B}}\,m_2, \tag{15}$$

$$\hat{\mathbf{y}}_p \equiv \hat{\mathbf{A}}^p\,\hat{\mathbf{y}}_0 + \hat{\mathbf{B}}\sum_{i=1}^{p}\hat{\mathbf{A}}^{p-i}\,m_i = \hat{\mathbf{y}}_0, \tag{16}$$

$$\hat{\mathbf{y}}_0 = (\mathbf{I} - \hat{\mathbf{A}}^p)^{-1}\,\hat{\mathbf{B}}\sum_{i=1}^{p}\hat{\mathbf{A}}^{p-i}\,m_i, \tag{17}$$

where m_i are integers corresponding to constant control effort. For a p-periodic orbit, checking all m_i integer combinations requires very high computational capacity. Moreover, prior to searching for periodic points, a global analysis should be performed to find specific locations of interest.

3.2 *Methods for Examining Global Chaotic Behaviour*

Methods mentioned in the previous subsection generally give information about a single attractor (corresponding to a specific set of initial states—i.e. the domain of attraction). However, in the case when several chaotic attractors coexist, the exploration and examination of a large region of the state space are necessary.

3.2.1 Systematic Iterations

Starting a set of iterations (repeatedly applying the *micro-chaos map*) from different initial states—selected either randomly or systematically in a region within the state space—one can explore the questioned region and find all attractors within it. Doing so however requires knowledge about the duration of transients or criteria for testing whether the solution has reached an attractor. Moreover executing a large number of simulations requires a large computational capacity and no information on unstable equilibria can be found.

3.2.2 Cell Mapping Methods

Cell mapping (CM) methods are tools for the global investigation of the long-term behaviour of nonlinear dynamical systems [6]. Using CM methods, periodic and chaotic solutions of the equations of motion can be found; moreover, the basin of attraction can also be determined.

4 Selecting the Region of Interest

To obtain a global *image* of the state space, one needs to estimate the region of chaotic behaviour. This can be done using the formula for the maximal possible norm ($\mathbf{y}_\infty$) presented in [2] or the location of the attractors can be calculated based on topological assumptions.

4.1 Topological Pattern

Consider the case when the rounding is applied to the output. A vector plot of (12) in case of parameters $\hat{\alpha} = 6.53 \times 10^{-3}$, $\hat{\beta} = 0$, $\hat{p} = 5.5 \times 10^{-5}$, $\hat{d} = 2.5 \times 10^{-3}$ can be seen in Fig. 3. The location of unstable fixed points can be derived from the following equation:

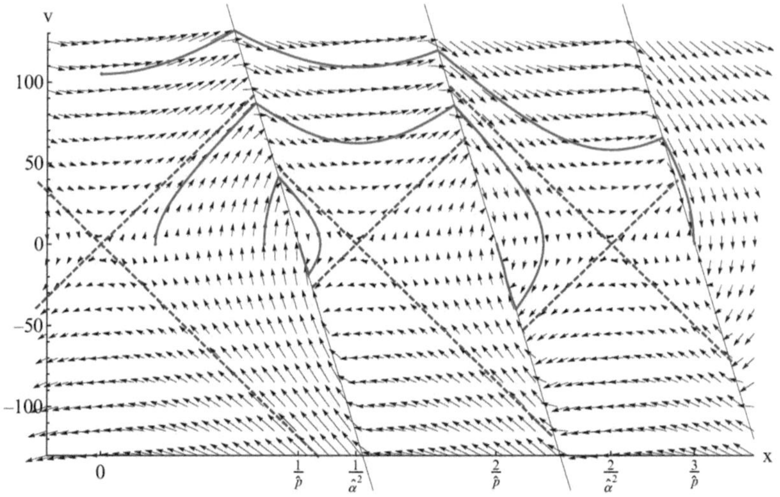

Fig. 3 Vector plot with switching lines (*black lines*), unstable fixed points (intersection of *dashed lines*) and some example trajectories leading to different attractors (*curves*)

$$\begin{bmatrix} x_u \\ 0 \end{bmatrix} = \hat{\mathbf{A}} \begin{bmatrix} x_u \\ 0 \end{bmatrix} + \hat{\mathbf{B}} \,\mathrm{Int}\left(\hat{\mathbf{K}} \begin{bmatrix} x_u \\ 0 \end{bmatrix}\right), \tag{18}$$

$$0 = \frac{e^{-\hat{\beta}} \sinh(\hat{\gamma}) \left(\mathrm{Int}(\hat{p}\, x_u) - \hat{\alpha}^2\, x_u\right)}{\hat{\gamma}}. \tag{19}$$

Since $e^{-\hat{\beta}} \sinh(\hat{\gamma})\, \hat{\gamma}^{-1} \neq 0$,

$$\mathrm{Int}(\hat{p}\, x_u) = \hat{\alpha}^2\, x_u \rightarrow x_u = \frac{k}{\hat{\alpha}^2}, \qquad k \in \mathbb{Z}. \tag{20}$$

The equation of switching lines

$$\mathrm{Int}(\hat{p}\, x + \hat{d}\, v) \rightarrow v = \frac{l - \hat{p}\, x}{\hat{d}}, \qquad l \in \mathbb{Z} \backslash \{0\}. \tag{21}$$

Since the dynamics of the system between two switching lines is unstable, the *stable* equilibria (i.e. attractors) are expected to be on the switching lines. The intersections between the switching lines and the x-axis are

$$x_s = \frac{l}{\hat{p}}, \qquad l \in \mathbb{Z} \backslash \{0\}. \tag{22}$$

Based on topological assumptions, the unstable fixed points and attractors should occur alternately. Since we restrict control parameters to the stable domain, $\hat{p} > \hat{\alpha}^2$, therefore the alternating pattern of unstable fixed points and attractors will end with a virtual unstable fixed point (at $l_{\max}/\hat{\alpha}^2$); thus

$$\frac{l_{\max}}{\hat{p}} < \frac{l_{\max} + 1}{\hat{p}} < \frac{l_{\max}}{\hat{\alpha}^2}. \tag{23}$$

Here the first term corresponds to the location of the last attractor. See Fig. 4. The index of the last attractor is therefore

$$l_{\max} = \mathrm{Int}\left(\frac{1}{\frac{\hat{p}}{\hat{\alpha}^2} - 1}\right) + 1 = \mathrm{Floor}\left(\frac{1}{\frac{\hat{p}}{\hat{\alpha}^2} - 1}\right) \tag{24}$$

So the region of interest along the x-axis is $x \in [-l_{\max}/\hat{p}, +l_{\max}/\hat{p}]$. The region of interest along the y-axis was chosen to include the intersection of the stable manifold of the unstable fixed point in the origin and the neighbouring switching lines. One can also observe the location of unstable fixed points and *stable* attractors by plotting the left- and right-hand side of $\mathrm{Int}(\hat{p}\, x_u)/\hat{\alpha}^2 = x$ (see Fig. 4).

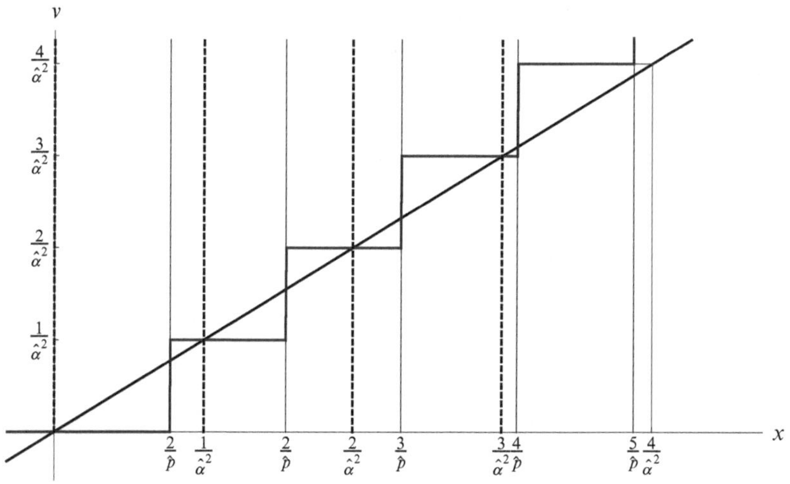

Fig. 4 Solutions of $\mathrm{Int}(\hat{p}\, x_u)/\hat{\alpha}^2 = x$ (in case of parameters $\hat{\alpha} = 6.53 \times 10^{-3}$, $\hat{p} = 5.5 \times 10^{-5}$) yielding the locations of unstable fixed points (*dashed lines*) and attractors (*thin lines*). Here $l_{\max} = 4$, the last attractor is at $l_{\max}/\hat{p}$ and the virtual unstable fixed point is at $l_{\max}/\hat{\alpha}^2$

5 Cell Mapping Method and Results

5.1 Simple Cell Mapping

In simple cell mapping (SCM) [6], the Euclidean state space $\mathbb{R}^N$ of a dynamical system is restricted to a bounded region denoted by Ω, which is divided into M (generally rectangular) cells, indexed by $j \in \{1, .., M\}$. The region $\mathbb{R}^N \backslash \Omega$ is called *sink-cell* (indexed with $j = 0$). The general idea behind SCM is that the state of the examined system is no longer described by its state vector but with the index of the cell corresponding to that state. For each cell one or more *image cells* can be determined where the dynamics of the system leads to. In SCM only one image cell is determined for each cell using the centre point of the cell (i.e. for flows the set of ODEs describing the system are integrated for a fixed time period, or for maps the map is applied to the state corresponding to the centre of the cell). The image cell corresponding to cell z is denoted by $C(z)$. The mapping $z(n + 1) = C(z(n))$ $C : \mathbb{N} \to \mathbb{N}$ is called an SCM and generally means that the next state of the system is completely described by its current state (and explicitly independent of the mapping step n). In SCM, two kinds of cells are distinguished:

- *Periodic cells*: for which $z = C^m(z)$ is true, for $m \in \mathbb{N}$. In this case z is an m-periodic cell. If a cell z is m-periodic, cells $C(z), C^2(z), \ldots, C^{m-1}(z)$ are also m-periodic cells, and such a group of periodic cells is called an m-periodic group. By definition the sink cell is a 1-periodic cell (once the system enters into it, it stays there forever).

- *Transient cells*: which are not periodic cells. Transient cells are mapped to a periodic cell (or the sink-cell) in finite number of steps, thus representing the basin of attraction of periodic groups.

The main procedure of the SCM method is determining the type and properties of every cell. This is done by generating the sequence $C(z), C^2(z), \ldots$ (while marking cells as *visited*) for every cell z until the sequence returns to a previously visited cell. If that cell is the part of the current sequence, a new periodic group and possibly some transient cells leading to that group are found. If the cell is a previously found transient cell, the current sequence contains only transient cells leading to the same periodic group. Similarly in case the sequence leads to a previously found periodic group, all cells in the sequence are marked as transient cells leading to that group. (See the SCM algorithm in [6].) It is clear that in the context of SCM only periodic motions occur, yet cell mapping methods are applicable for chaotic systems by taking the following assumptions stated in [6]:

- Chaotic behaviour is represented by periodic groups with relatively high period.
- A chaotic attractor is represented by a set of periodic cells covering a part of the attractor in the state space.

5.2 *SCM Results*

Some modifications were made to the SCM algorithm [6] to suit the *micro-chaos map*. Adjacent periodic groups are considered to be the part of the same attractor; therefore after the main procedure of SCM periodic groups close to each other were joined. Moreover in order to increase the precision of the method around unstable fixed points, the image cells were determined using multiple steps of the *micro-chaos map* (i.e. if the image cell of a cell was itself, the map was applied once again—until a fixed number of repetitions—to find the image cells of cells where the dynamics of the system is slow). This way false 1-periodic cells were eliminated.

5.2.1 Rounding at the Output

When rounding is applied at the output (Section 2.2), the SCM method shows the expected topological pattern. The chaotic attractors are indeed found on the switching lines and all unstable fixed points are discovered as a set of 1-periodic cells. The area of the basin of attraction of an attractor is larger, the larger the distance of the closest neighbouring unstable fixed point is (i.e. the attractor featuring the largest basin of attraction is the one equally far away from the two neighbouring unstable fixed points). See Fig. 5. One can observe *gateways* on switching lines, which are leading to a specific chaotic attractor (See also Fig. 4.)

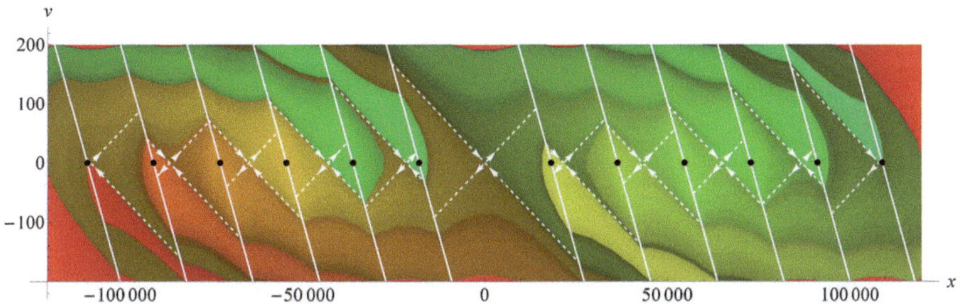

Fig. 5 SCM results with switching lines (*white*), stable and unstable manifolds of unstable fixed points (*white, dashed*) and chaotic attractors (*black dots*) for parameters $\hat{\alpha} = 6.8 \times 10^{-3}$, $\hat{\beta} = 0$, $\hat{p} = 5.5 \times 10^{-5}$, $\hat{d} = 2.5 \times 10^{-3}$. Here $l_{\max} = 6$, and the 3rd attractor has the largest basin of attractions

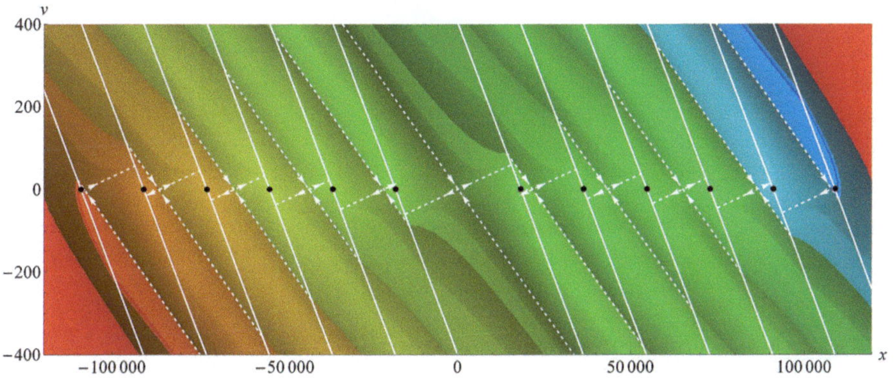

Fig. 6 SCM results for $\hat{\alpha} = 6.8 \times 10^{-3}$, $\hat{\beta} = 4.0 \times 10^{-3}$, $\hat{p} = 5.5 \times 10^{-5}$, $\hat{d} = 2.5 \times 10^{-3}$

5.2.2 Effect of Damping

If the damping ratio is non-zero, the eigenvectors of (9) are *rotating in clockwise direction*; therefore the previously mentioned *gateways* are moving and stretching in the direction of the y-axis (Fig. 6).

5.2.3 Rounding at the Input

In case of realistic parameters rounding at the input leads to a solution where small amplitude chaotic motion is superposed on a periodic orbit (Fig. 7).

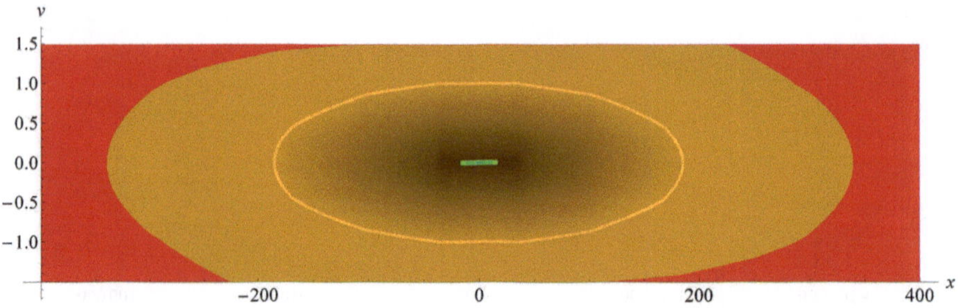

Fig. 7 SCM results for $\hat{\alpha} = 5.0 \times 10^{-3}$, $\hat{\beta} = 0$, $\hat{p} = 5.5 \times 10^{-5}$, $\hat{d} = 2.5 \times 10^{-3}$

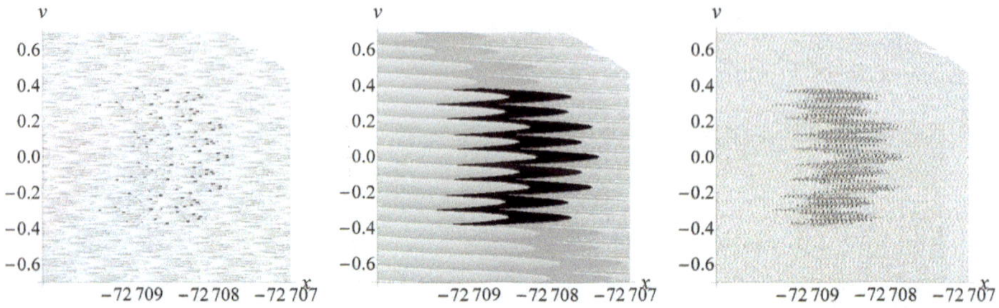

Fig. 8 SCM images of the same chaotic attractor using different cell sizes (from *left* to *right* 940, 1100 and 1220 cells along each axis): $\hat{\alpha} = 6.85 \times 10^{-3}$, $\hat{\beta} = 0$, $\hat{p} = 5.5 \times 10^{-5}$, $\hat{d} = 2.5 \times 10^{-3}$

5.2.4 Fractal Dimension Calculation with SCM

Since SCM utilizes rectangular cells, it could be suitable to determine the fractal dimension (*box counting* or Minkowski dimension) of chaotic attractors. SCMs with different cell sizes (ε) were applied to the close bounding region of chaotic attractors, while the total number of periodic cells $N(\varepsilon)$ was stored. Then, the box-counting dimension by definition is

$$D_{\text{box}} = \lim_{\varepsilon \to 0} \frac{\log N(\varepsilon)}{\log 1/\varepsilon}. \tag{25}$$

The box-counting dimension was calculated by using least squares linear regression to fit a line on $(\log 1/\varepsilon, \log N(\varepsilon))$ points. The slope of the fitted line is D_{box}. However, SCM provides greatly different images even when the cell size varies a little (Fig. 8). Because of this phenomenon, the calculated box-counting dimension will be inaccurate.

6 Conclusion

It has been shown that simple cell mapping (SCM) is a suitable tool to examine systems with complex chaotic behaviour, while it is inaccurate for box-counting dimension calculation. For the case when rounding is applied to the output (Sect. 2.2), the expected topological pattern was obtained by SCM and further examination of *gateways* on switching lines (defined by the stable and unstable manifolds of unstable fixed points) could help reveal unknown properties of the 2D *micro-chaos map*. On the other hand, rounding at the input (Sect. 2.3) shows a completely different behaviour where chaotic motion is superposed on a periodic orbit. In both cases, further investigation with more advanced CM methods is planned.

Acknowledgements This research was supported by the Hungarian National Science Foundation under grant no. OTKA K 83890.

References

1. Csernák, G., Stépán, G.: Digital control as source of chaotic behavior. Int. J. Bifur. Chaos **20**(5), 1365–1378 (2010)
2. Csernák, G., Stépán, G.: Sampling and round-off, as sources of chaos in PD-controlled systems. Proceedings of the 19th Mediterranean Conference on Control and Automation (2011)
3. Delchamps, F.D.: Stabilizing a linear system with quantized state feedback. IEEE Trans. Automat. Contr. **35**, 916–924 (1990)
4. Enikov, E., Stépán, G.: Micro-chaotic motion of digitally controlled machines. J. Vib. Contr. **4**, 427–443 (1998)
5. Haller, G., Stépán, G.: Micro-chaos in digital control. J. Nonlinear Sci. **6**, 415–448 (1996)
6. van der Spek, J.A.W.: Cell Mapping Methods: Modifications and Extensions. Eindhoven University of Technology, Eindhoven (1994)
7. Stefański, A., Kapitaniak, T.: Estimation of the dominant Lyapunov exponent of non-smooth systems on the basis of maps synchronization. Chaos Sol. Fract. **15**, 233–244 (2003)

Bouncing Ball Dynamics: Simple Motion of the Table Approximating the Sinusoidal Motion

Andrzej Okniński and Bogusław Radziszewski

Abstract We study nonlinear dynamics of a small ball moving vertically in a gravitational field and colliding with a moving table. Displacement of the table is approximated in one period of its motion by four cubic polynomials (we shall refer to this model as $\mathcal{M}_C$). The Poincaré map, mapping a state of the ball at an impact to a new state at the next impact, is used to investigate the ball dynamics. Results obtained for the model $\mathcal{M}_C$ are used to elucidate dynamics of the standard model of bouncing ball with sinusoidal motion of the limiter (model $\mathcal{M}_S$). Bifurcation diagrams are computed for the models $\mathcal{M}_C$ and $\mathcal{M}_S$. Then fixed points and their stability are determined for the $\mathcal{M}_C$ model and compared with our earlier results for the model $\mathcal{M}_S$. Finally, the birth of the low-velocity 2-cycle is investigated analytically for the $\mathcal{M}_C$ model.

1 Introduction

In the present paper we investigate dynamics of a small ball moving vertically in a gravitational field and impacting with a periodically moving limiter (a table). This simple but important model has been extensively studied within the framework of nonsmooth and nonlinear dynamical systems which have many applications in technology [1, 2, 5, 11].

In dynamics with impacts it is usually difficult or even impossible to solve nonlinear equation for an instant of the next impact. For example, in the bouncing ball models, the table's motion has been usually assumed in sinusoidal form. This choice of the limiter's motion leads to a complicated nonlinear equation for time of the next impact. To address this problem we proposed a sequence of models in which periodic motion of the table is assumed (in one period of limiter's motion) as a low-order polynomial of time; see [6] and references therein.

A. Okniński (✉)
Kielce University of Technology, 25-314 Kielce, Poland
e-mail: fizao@tu.kielce.pl

B. Radziszewski
Collegium Mazovia Innovative University, 08-110 Siedlce, Poland
e-mail: b@nazu.org

J. Awrejcewicz (ed.), *Applied Non-Linear Dynamical Systems*, Springer Proceedings in Mathematics & Statistics 93, DOI 10.1007/978-3-319-08266-0_29

In the present work we conduct analytical and numerical investigations of the model in which sinusoidal displacement of the table is approximated in one period by four cubic polynomials. We shall refer to this model as $\mathscr{M}_C$. We hope that rigorous results obtained for the model $\mathscr{M}_C$ may cast light on dynamics with sinusoidal motion of the limiter. The latter model is referred to as $\mathscr{M}_S$.

The paper is organized as follows. In Sect. 2 a one-dimensional dynamics of a ball moving in a gravitational field and colliding with a table is reviewed and the corresponding Poincaré map is constructed and models of the limiter's motion $\mathscr{M}_C$ and $\mathscr{M}_S$ are defined. Bifurcation diagrams are computed for $\mathscr{M}_C$ and $\mathscr{M}_S$. In Sect. 3 fixed points and their stability are investigated for the $\mathscr{M}_C$ model and compared with our earlier results for the $\mathscr{M}_S$ model. In Sect. 4 birth of the low-velocity 2-cycle is investigated analytically for the $\mathscr{M}_C$ model. We summarize our results in the last section.

2 Bouncing Ball: A Simple Motion of the Table

We assume that a ball moves vertically in a constant gravitational field and collides with a periodically moving table. The ball is treated as a material point and it is assumed that the limiter's mass is so large that its motion is not affected at impacts. Dynamics of the ball is studied via a Poincaré map which takes a state of the ball at an impact to a new state at the next impact. The corresponding Poincaré map in nondimensional form reads (cf. [7], see also Ref. [4] where analogous map was derived earlier):

$$\gamma Y\left(T_{i+1}\right) = \gamma Y\left(T_i\right) - \Delta_{i+1}^2 + \Delta_{i+1} V_i, \tag{1a}$$

$$V_{i+1} = -RV_i + 2R\Delta_{i+1} + \gamma\left(1+R\right)\dot{Y}\left(T_{i+1}\right), \tag{1b}$$

where T_i denotes time of the i-th impact and V_i is the corresponding postimpact velocity while $\Delta_{i+1} \equiv T_{i+1} - T_i$. The parameters γ, R are a nondimensional acceleration and the coefficient of restitution, $0 \leq R < 1$ [11], respectively, and the function $Y\left(T\right)$ represents the limiter's motion. The limiter's motion has been usually assumed in sinusoidal form, $Y_S(T) = \sin(2\pi T)$. Equations (1) and $Y = Y_S$ lead to the model $\mathscr{M}_S$. This model of limiter's motion leads to serious difficulties in solving the first of Eqns.(1) for T_{i+1}, making analytical investigations of dynamics virtually impossible. We have thus decided to simplify the limiter's periodic motion to make (1a) solvable. The function $Y_C\left(T\right)$ defined below:

$$Y_C\left(T\right) = \begin{cases} f_1\left(T\right), & 0 \leq \hat{T} < \frac{1}{4} \\ f_2\left(T\right), & \frac{1}{4} \leq \hat{T} < \frac{1}{2} \\ f_3\left(T\right), & \frac{1}{2} \leq \hat{T} < \frac{3}{4} \\ f_4\left(T\right), & \frac{3}{4} \leq \hat{T} \leq 1 \end{cases} \tag{2}$$

$$f_1(T) = (32\pi - 128)\hat{T}^3 + (-16\pi + 48)\hat{T}^2 + 2\pi\hat{T} \quad (3a)$$

$$f_2(T) = (128 - 32\pi)\hat{T}^3 + (-144 + 32\pi)\hat{T}^2 + (48 - 10\pi)\hat{T} - 4 + \pi \quad (3b)$$

$$f_3(T) = (128 - 32\pi)\hat{T}^3 + (-240 + 64\pi)\hat{T}^2 + (144 - 42\pi)\hat{T} - 28 + 9\pi \quad (3c)$$

$$f_4(T) = (32\pi - 128)\hat{T}^3 + (336 - 80\pi)\hat{T}^2 + (-288 + 66\pi)\hat{T} + 80 - 18\pi \quad (3d)$$

provides a reasonably good approximation of the function $Y_S = \sin(2\pi T)$ on all intervals $[k, k+1]$, $k = 0, 1, \ldots,$ with $\hat{T} = T - \lfloor T \rfloor$, where $\lfloor x \rfloor$ is the floor function – the largest integer less than or equal to x.

The model $\mathscr{M}_C$ consists of equations (1), (2), (3) with control parameters R, γ. Velocities of the limiter are computed as $g_i(T) = \frac{d}{dt} f_i(T)$, $i = 1, \ldots, 4$.

In Fig. 1 above the bifurcation diagram with velocities and impact times versus γ computed for growing γ and $R = 0.85$ has been shown. It follows that dynamical system $\mathscr{M}_C$ has several typical attractors which undergo a period doubling scenario: fixed points, with one impact per k periods and $V_i = V_* = k$, and 3-cycles with $V_i \approx k$. There are also several small attractors. We shall investigate some of these attractors in the next section combining analytical and numerical approach.

In Fig. 2 bifurcation diagram for the sinusoidal motion has been shown. Similarity of Figs. 1 and 2 suggests that analytical results obtained for model $\mathscr{M}_C$ may shed light on the problem of sinusoidal motion, $\mathscr{M}_S$.

3 Fixed Points and Their Stability

We shall first study periodic solutions of the model $\mathscr{M}_C$ with one impact per k periods and $T \in \left(0, \frac{1}{4}\right)$ since it is suggested by the bifurcation diagram that they are stable. Such states have to fulfill the following conditions:

$$V_{n+1} = V_n \equiv V_*^{(k/1)}, \; T_{n+1} = T_n + k \equiv T_*^{(k/1)} + k \qquad (k = 1, 2, \ldots), \quad (4)$$

where

$$T_*^{(k/1)} \in \left(0, \tfrac{1}{4}\right), \; V_*^{(k/1)} > \gamma \dot{Y}_{c_1}\left(T_*^{(k/1)}\right). \quad (5)$$

The demanded (stable) solution is given by

$$T_{*(s)}^{(k/1)} = \tfrac{\pi-3}{6(\pi-4)} - \tfrac{1}{24(\pi-4)}\sqrt{4(\pi-6)^2 + 6B(\pi-4)}, \quad \left(B = \tfrac{k}{\gamma}\tfrac{1-R}{1+R}\right) \quad (6a)$$

$$V_*^{(k/1)} = k. \quad (6b)$$

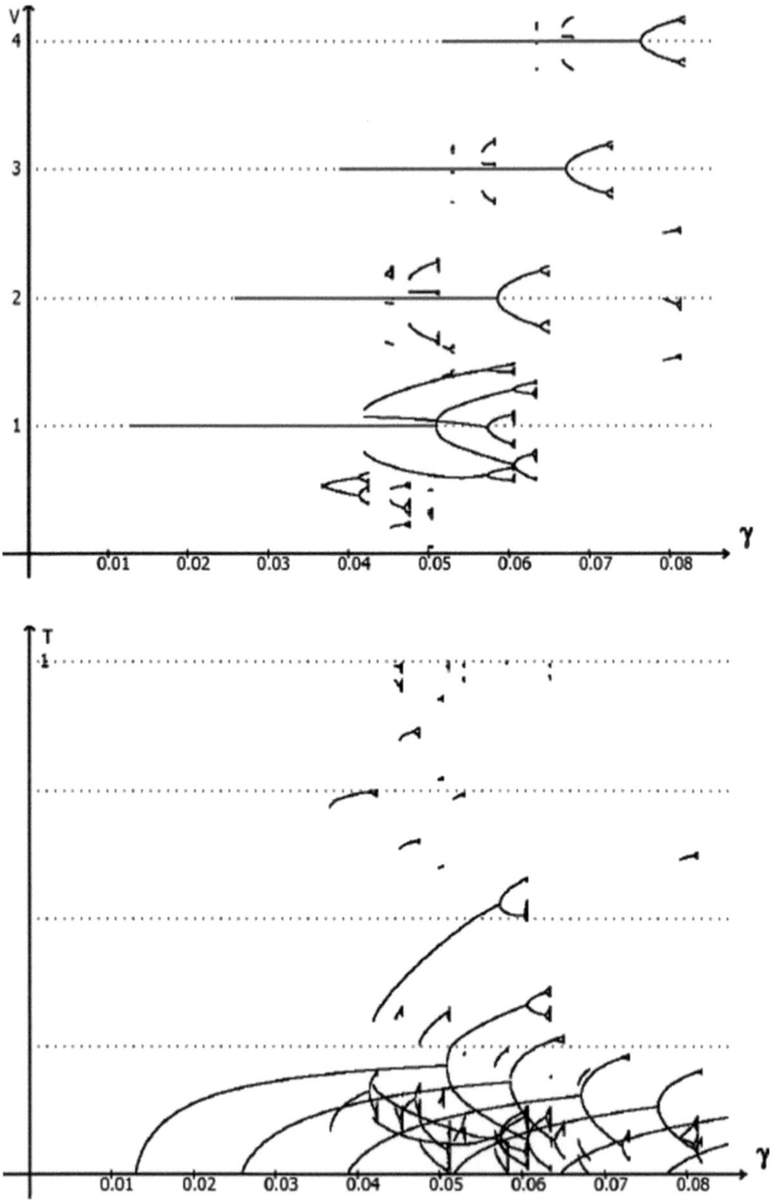

Fig. 1 Bifurcation diagram for the model $\mathcal{M}_C$, $R = 0.85$: velocities after impacts (*top figure*) and times after impacts (*bottom figure*) versus control parameter γ

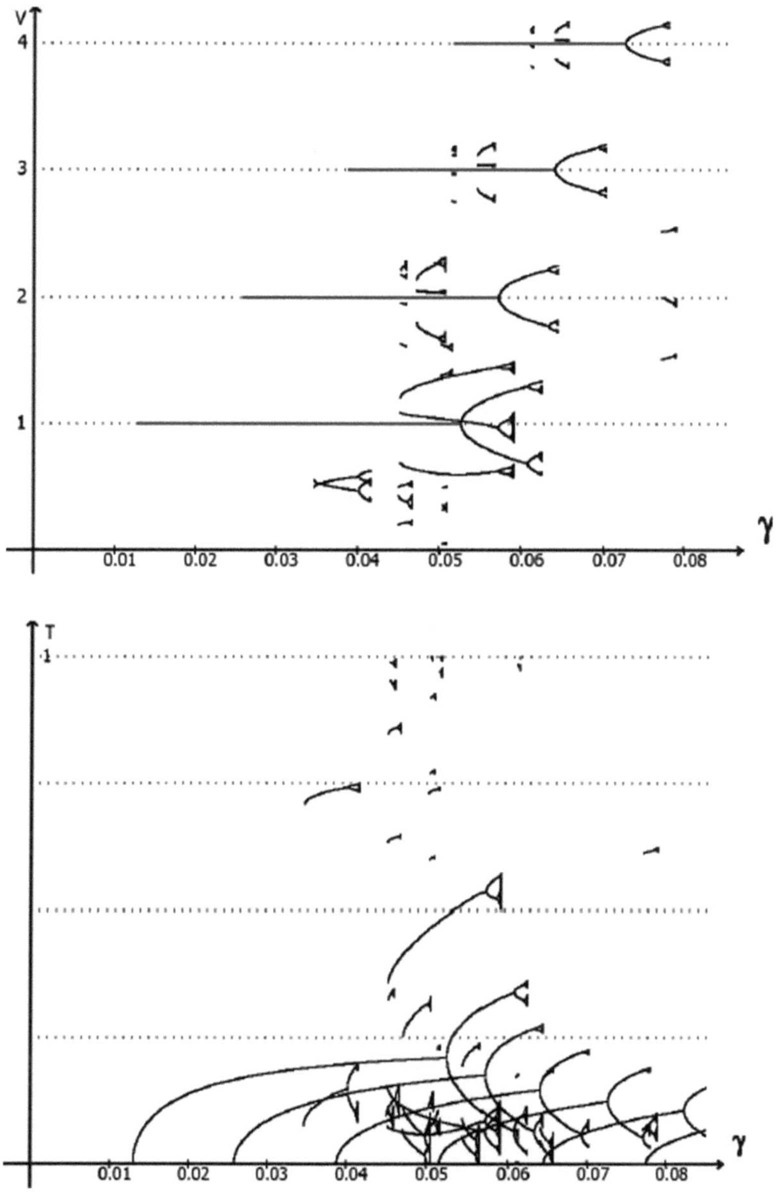

Fig. 2 Bifurcation diagram for the model $\mathscr{M}_S$, $R = 0.85$: velocities after impacts (*top figure*) and times after impacts (*bottom figure*) versus control parameter γ

Since $T_* \in [0,\ 1]$ we demand that $T_* > 0$ and it follows from (6a) that physical solution appears for lower critical value $\gamma > \gamma_{cr_1}^{(k/1)}$ where

$$\gamma_{cr_1,C}^{(k/1)} = \frac{k}{2\pi}\frac{1-R}{1+R}. \tag{7}$$

We have checked by stability analysis that the solution (6a), (6b) is stable for $\gamma > \gamma_{cr_1,C}^{(k/1)}$, i.e. when it is physically acceptable. To determine upper critical value of γ when dynamics loses stability we put into (1):

$$T_i = T_{*(s)}^{(k/1)} + \varepsilon_i, \; T_{i+1} = T_{*(s)}^{(k/1)} + k + \varepsilon_{i+1}, \tag{8}$$

$$V_i = V_* + \mu_i = k + \mu_i, \; V_{i+1} = V_* + \mu_{i+1} = k + \mu_{i+1}, \tag{9}$$

with $Y(T)$ given by (3), and keep only terms linear in perturbations ε_i, ε_{i+1}, μ_i, μ_{i+1} of the fixed point to get

$$\begin{pmatrix} \varepsilon_{i+1} \\ \mu_{i+1} \end{pmatrix} = \begin{pmatrix} 1 & \frac{k}{\gamma f_1(T_*)+k} \\ \gamma (1+R) g_1(T_*) & k\frac{2R+\gamma(1+R)g_1(T_*)}{\gamma f_1(T_*)+k} - R \end{pmatrix} \begin{pmatrix} \varepsilon_i \\ \mu_i \end{pmatrix} \tag{10}$$

where $T_* \equiv T_{*(s)}^{(k/1)}$, $f_1(T)$ is given by (3a), and $g_1(T) = \frac{d}{dT} f_1(T)$.

Since the characteristic polynomial is

$$\begin{cases} X^2 + \alpha X + \beta = 0 \\ \alpha = 4\sqrt{4(\pi-6)^2 + 6k(\pi-4)\frac{1-R}{\gamma(1+R)}}(1+R)^2\gamma - R^2 - 1 \\ \beta = R^2 \end{cases} \tag{11}$$

application of the Schur-Cohn criterion [3]:

$$\beta < 1, \; |\alpha| < \beta + 1 \tag{12}$$

leads finally to the following localization of the fixed points (6): $\gamma_{cr_1,C}^{(k/1)} < \gamma < \gamma_{cr_2,C}^{(k/1)}$, where $\gamma_{cr_1,C}^{(k/1)}$ is given by (7) and

$$\gamma_{cr_2,C}^{(k/1)} = \frac{6k(\pi-4)(R^2-1)+\sqrt{36k^2(\pi-4)^2(1-R^2)^2+4(\pi-6)^2(1+R^2)^2}}{8(\pi-6)^2(1+R)^2}, \quad R < 1. \tag{13}$$

In Fig. 3 stability regions in (R,γ) plane for the $\mathcal{M}_C$ model are shown.

In the case of the model $\mathcal{M}_S$ we have

$$\gamma_{cr_1,S}^{(k/1)} = \gamma_{cr_1,C}^{(k/1)}, \tag{14}$$

$$\gamma_{cr_2,S}^{(k/1)} = \frac{\sqrt{k^2\pi^2(1-R^2)^2+4(1+R^2)^2}}{2\pi^2(1+R)^2}, \quad R < 1, \tag{15}$$

see [7] (note that in [7] we used $Y(T) = \sin(T)$ rather than $Y_S(T) = \sin(2\pi T)$ and it follows that all values of the control parameter λ must be rescaled, $\gamma = \frac{\lambda}{(2\pi)^2}$), and stability regions are very similar to those of model $\mathcal{M}_C$, cf. Fig. 3.

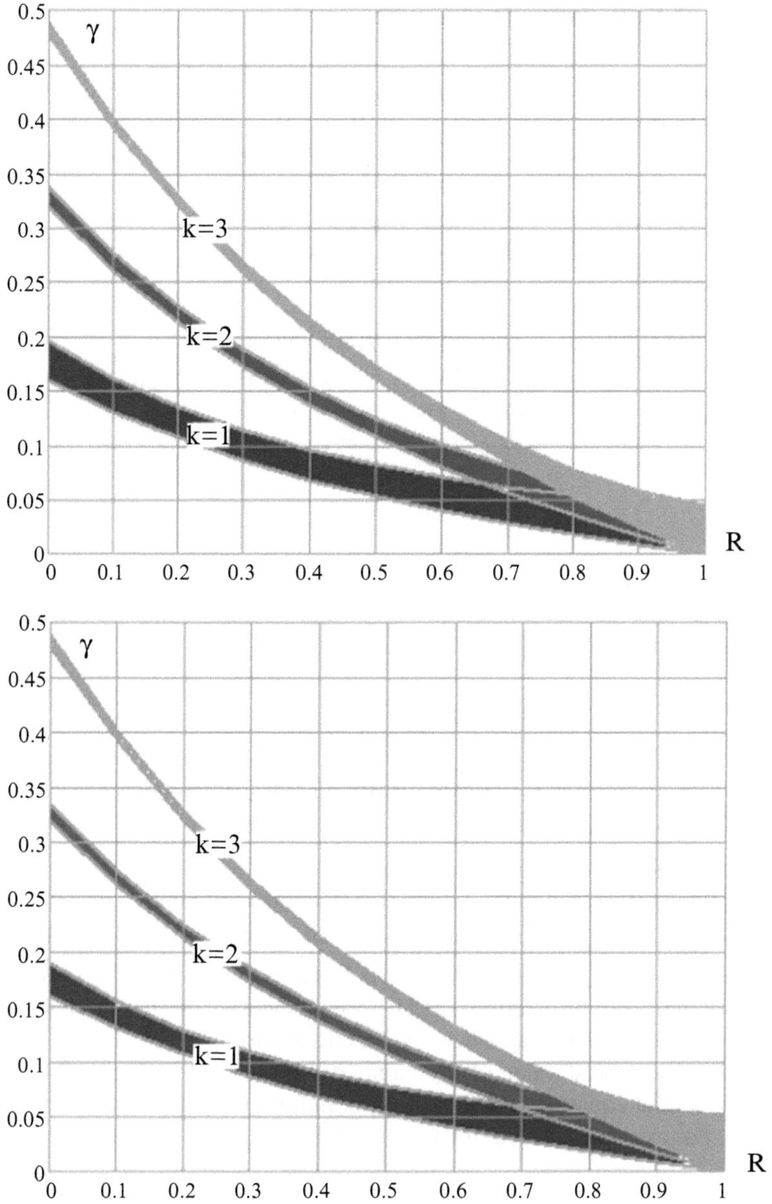

Fig. 3 Stability regions for the model $\mathscr{M}_C$ (*top figure*) and the model $\mathscr{M}_S$ (*bottom figure*)

4 Birth of Low-Velocity 2-Cycle

In this section we shall study birth of low-velocity k-cycles which can be seen in the bifurcation diagram, Figs. 1 for $\gamma > 0.03$ and $V < 1$. In the case of such cycles $T_1, T_2, \ldots, T_k \in (0,1)$ and $T_{k+1} - 1 = T_1$. Numerical tests show that 2-cycle fulfilling conditions $T_1 \in \left(0, \frac{1}{4}\right)$, $T_2 \in \left(\frac{1}{2}, \frac{3}{4}\right)$ and $T_3 = T_1 + 1$ are stable. This 2-cycle can be seen in the bifurcation diagram in Fig. 1 for $\gamma \gtrsim 0.0366$ and $V_1 \cong 0.51$, $V_2 \cong 0.55$ $(R = 0.85)$. Equations to determine T_1, T_2, and V_1, V_2 are shown below:

$$\begin{cases} \gamma f_3(T_2) = \gamma f_1(T_1) - (T_2 - T_1)^2 + (T_2 - T_1)V_1 \\ V_2 = -RV_1 + 2R(T_2 - T_1) + \gamma(1+R)g_3(T_2) \\ \gamma f_1(T_3 - 1) = \gamma f_3(T_2) - (T_3 - T_2)^2 + (T_3 - T_2)V_2 \\ V_3 = -RV_2 + 2R(T_3 - T_2) + \gamma(1+R)g_1(T_3 - 1) \\ T_3 = T_1 + 1 \\ V_3 = V_1 \end{cases} \tag{16}$$

where $f_i(T)$'s and $g_i(T)$'s are defined in Eqn. (3) and the text below.

We were able to simplify Eqns. (16) significantly obtaining equation for $\Delta \equiv T_2 - T_1$ only:

$$F(\Delta) = \sum_{j=0}^{9} d_j \Delta^j = 0, \tag{17}$$

where d_i's are given in the Appendix in [9]. Numerical computations suggest that the 2-cycle appears for $\gamma = \gamma_{cr,C}^{(2)}$ and fixed R, where $\gamma_{cr,C}^{(2)}$ is a critical value, as a double (and stable) solution of Eqns. (16). For $\gamma > \gamma_{cr,C}^{(2)}$ there are two real solutions, one stable (seen in the bifurcation diagram) and another unstable. On the other hand, for $\gamma < \gamma_{cr,C}^{(2)}$, the solutions are complex conjugated and thus unphysical. Moreover, at $\gamma = \gamma_{cr,C}^{(2)}$, the stability matrix has unit eigenvalue. Therefore this is a tangent (saddle-node) bifurcation; see [10] for elementary discussion of the tangent bifurcation in the logistic map when the 3-cycle is born.

To determine critical value of the parameter γ let us note that double solution of the polynomial equation (17) is also the solution of $G(\Delta) = 0$ where $G(\Delta) = \frac{d}{d\Delta}F(\Delta)$. For example, solving for $R = 0.85$ the system of equations:

$$F(\Delta) = \sum_{j=0}^{9} d_j \Delta^j = 0, \tag{18}$$

$$G(\Delta) = \sum_{j=1}^{j} j d_j \Delta^{j-1} = 0, \tag{19}$$

we get $\gamma^{(2)}_{cr,C} = 0.036\,617\,052\,682\,892\,250\,62$, $\Delta_{cr} = 0.634\,279\,960\,677\,747\,355\,95$ (and many other, unphysical solutions) in perfect agreement with numerical computations, see also Fig. 1.

5 Summary

We have studied dynamics of a bouncing ball impacting with a periodically moving limiter within two frameworks of the table motion: $\mathscr{M}_C$ and $\mathscr{M}_S$ defined in Sect. 2. Stability conditions of fixed points have been determined and results for the models $\mathscr{M}_C$ and $\mathscr{M}_S$ have been compared. Then we have found for the model $\mathscr{M}_C$ that the low-velocity 2-cycle is born in tangent bifurcation. Equations for this 2-cycle have been written and the condition for the onset of this attractor was determined and solved numerically. The second part of this study will be devoted to systematic investigation of birth of attractors and chattering in the models $\mathscr{M}_C$ and $\mathscr{M}_S$ [8].

References

1. Awrejcewicz, J., Lamarque, C.-H.: Bifurcation and Chaos in Nonsmooth Mechanical Systems. World Scientific Publishing, Singapore (2003)
2. di Bernardo, M., Budd, C.J., Champneys, A.R., Kowalczyk P.: Piecewise-Smooth Dynamical Systems. Theory and Applications. Springer, Berlin (2008)
3. Jury, E.I.: Inners and Stability of Dynamic Systems. Krieger, New York (1982)
4. Luo, A.C.J., Han, R.P.S.: The dynamics of a bouncing ball with a sinusoidally vibrating table revisited. Nonlinear Dynam. **10**, 1–18 (1996)
5. Mehta, A. (ed.): Granular Matter: An Interdisciplinary Approach. Springer, Berlin (1994)
6. Okniński, A., Radziszewski, B.: Dynamics of impacts with a table moving with piecewise constant velocity. Differ. Equ. Dyn. Syst. **21**(1&2) 165–171 (2013)
7. Okniński, A., Radziszewski, B.: Grazing dynamics and dependence on initial conditions in certain systems with impacts, arXiv: 0706.0257v2, 17 pages (2007)
8. Okniński, A., Radziszewski, B.: Int. J. Nonlinear Mech. **65**, 226–235 (2014)
9. Okniński, A., Radziszewski, B.: Bouncing ball dynamics: simple model of motion of the table and sinusoidal motion, arXiv: 1302.0369v1, 2013, 23 pages.
10. Peitgen, H.O., Jürgens, H., Saupe D.: Fractals for the Classroom. Part Two: Complex Systems and Mandelbrot Set. Springer, New York (1992)
11. Stronge, W.J.: Impact Mechanics. Cambridge University Press, Cambridge (2000)

Periodic Motions of Coupled Oscillators Excited by Dry Friction and Harmonic Force

Madeleine Pascal and Sergey Stepanov

Abstract Vibrating systems excited by dry friction are frequently encountered in technical applications. These systems are strongly nonlinear, and they are usually modeled as spring-mass oscillators. One of the most popular models of stick-slip oscillators consists of several masses connected by linear springs; one (or more) of the masses is in contact with a driving belt moving at a constant velocity. In the past, several authors investigated the behavior of this system, with different friction laws and with or without external actions and damping. In this work, we consider a system composed of two masses connected by linear springs. One of the mass is in contact with a driving belt moving at a constant velocity. Friction force, with Coulomb's characteristics, acts between the mass and the belt. Moreover, it is assumed that the mass is also subjected to a harmonic external force. Several periodic orbits including stick phases and slip phases are obtained in closed form. In particular, the existence of periodic orbits including an overshooting part is proved. In the case of a nonmoving belt, a set of nonsticking periodic solutions is obtained, and we prove that these orbits are symmetrical in space and in time.

1 Introduction

This paper is a continuation of several investigations [5–7, 10] related to vibrating systems excited by dry friction. One of the most popular models of stick-slip oscillators consists of several masses connected by linear springs; one (or more) of the masses is in contact with a driving belt moving at a constant velocity. In the past, several authors investigated the behavior of this system, with different friction laws and with or without external actions and damping [1, 2, 4, 11], mainly via the numerical approach. However, assuming Coulomb's laws of dry friction,

---

M. Pascal (✉)
Universite d'Evry, IBISC, 40 rue du Pelvoux, 91020, Evry, France
e-mail: madeleine.pascal3@wanadoo.fr

S. Stepanov
Department of Mechanics, Computing Center of the RAS,
40 Vavilova street, 119991, Moscow, Russia
e-mail: stepsj@ya.ru

J. Awrejcewicz (ed.), *Applied Non-Linear Dynamical Systems*, Springer Proceedings in Mathematics & Statistics 93, DOI 10.1007/978-3-319-08266-0_30

the corresponding dynamical model is a piecewise linear system, and even for multi-degree-of-freedom cases, some analytical results about the existence and the stability of periodic orbits including stick-slip phases have been obtained [5–7]. One interesting phenomenon is the existence, inside periodic orbits with stick and slip parts, of an "overshooting" slip phase. During this part of the orbit, the mass in contact with the belt moves in the same direction at a higher speed than the belt itself. In this work, we consider the same model of dry friction oscillator subjected to a harmonic external force. Several periodic orbits including stick phases and slip phases are obtained in closed form. In particular, the existence of periodic orbits including an overshooting part is proved. In the case of a nonmoving belt, a set of nonsticking periodic solutions is obtained, and we prove that these orbits are symmetrical in space and in time.

2 Problem Formulation

The system (Fig.1) is composed of two masses m_1, m_2 connected by two linear springs of stiffness k_1, k_2. The second mass is in contact with a belt moving at a constant velocity v_0. A friction force $\tilde{F}$ acts between the mass and the belt. Moreover, the second mass is also subjected to a driving force $\tilde{R} = \tilde{p}\cos(\omega t + \varphi)$ ($\tilde{p}, \omega, \varphi$ are constant parameters). The motion of this system are given by equations

$$x_1'' + x_1 - \chi x_2 = 0, \quad x_2'' + \chi\eta(x_2 - x_1) = \eta u + p\cos(\omega t + \varphi), \tag{1}$$

$$\eta = \frac{m_1}{m_2}, \quad \chi = \frac{k_2}{k_1 + k_2}, \quad u = \frac{\tilde{F}}{k_1 + k_2}, \quad p = \frac{\tilde{p}}{k_1 + k_2},$$

$$t = \Omega\, t', \quad \Omega = \sqrt{\frac{k_1 + k_2}{m_1}}, \quad (O)' = \frac{d(O)}{dt},$$

x_1, x_2 are the displacements of the masses.

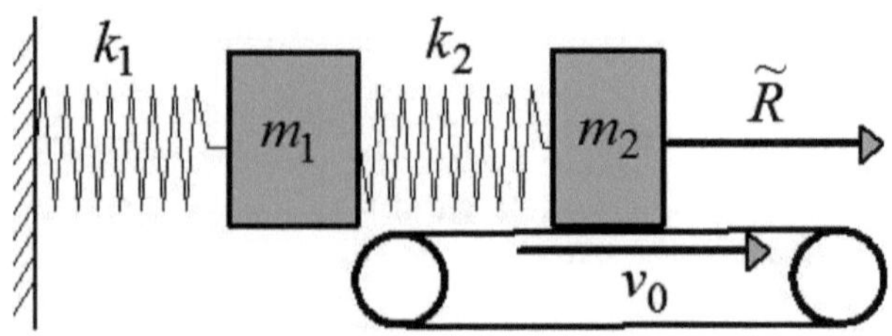

Fig. 1 Dry friction oscillator

The dry friction force u is obtained from Coulomb's laws:

$$\begin{aligned}
&u = u_s \operatorname{sign}\left(V - x_2'\right), && \text{if } V - x_2' \neq 0, \\
&u = \chi\left(x_2 - x_1\right) - r(t), && \text{if } V - x_2' = 0, \quad \left|\chi\left(x_2 - x_1\right) - r(t)\right| < u_r, \\
&u = u_s, && \text{if } V - x_2' = 0, \quad \chi\left(x_2 - x_1\right) - r(t) > u_r, \\
&u = -u_s, && \text{if } V - x_2' = 0, \quad \chi\left(x_2 - x_1\right) - r(t) < -u_r,
\end{aligned}$$

$$0 < u_s < u_r, \quad V = \frac{v_0}{\Omega}, \quad r(t) = \frac{p}{\eta}\cos\left(\omega t + \varphi\right),$$

u_r is the static friction force, and u_s is the dynamic friction force.

3 Three Modes of Motion Exhibited by the System

The dynamical behavior of this oscillator includes several phases of slip and stick motion of m_2. For each kind of motion, the closed form solution is available.

3.1 Slip Motion of m_2 *with* $x_2' < V$

The solution is obtained from a modal analysis of (1) where $u = u_s$

$$Z\left(t\right) = H\left(t\right)\left(Z_0 - F_0\right) + F\left(t\right), \quad F\left(t\right) = \begin{pmatrix} R\left(t\right) \\ R'\left(t\right) \end{pmatrix},$$

$$R(t) = Q\cos(\omega t + \varphi), \quad F_0 = F(0),$$

$$Z = \begin{pmatrix} z \\ z' \end{pmatrix}, \quad Z_0 = Z(0), \quad Z = X - d_0, \quad X = \begin{pmatrix} x_1 \\ x_2 \end{pmatrix},$$

$$H\left(t\right) = \begin{pmatrix} H_1\left(t\right) & H_2\left(t\right) \\ H_3\left(t\right) & H_1\left(t\right) \end{pmatrix}, \quad d_0 = \begin{pmatrix} d_{01} \\ d_{02} \end{pmatrix}, \quad d_{01} = \frac{u_s}{1-\chi}, \quad d_{02} = \frac{d_{01}}{\chi},$$

$$Q = \begin{pmatrix} q_1 \\ q_2 \end{pmatrix}, \quad q_1 = \frac{p\chi}{(\omega^2 - \omega_1^2)(\omega^2 - \omega_2^2)}, \quad q_2 = q_1 \frac{p(1-\omega^2)}{\chi}.$$

The two-by-two matrices $H_i\left(t\right)$ $(i = 1, 2, 3)$ and the natural frequencies ω_1, ω_2 are obtained in analytical form [5].

3.2 Slip Motion of m_2 with $x_2' > V$ (Overshooting)

The solution is obtained from (1) where $u = -u_s$

$$Z(t) = H(t)(Z_0 - F_0) + F(t) + 2L(t)\,d_0 \quad L(t) = \begin{pmatrix} H_1(t) - I \\ H_3(t) \end{pmatrix}, \quad I = \begin{pmatrix} 1 & 0 \\ 0 & 1 \end{pmatrix}.$$

3.3 Stick Motion of m_2 ($x_2' = V$)

This motion is related to the dynamical system

$$x_1'' + x_1 - \chi x_2 = 0, \qquad x_2'' = 0. \tag{2}$$

The solution [5] is given by

$$Z(t) = \Gamma(t)\,Z_0, \quad \Gamma(t) = \begin{pmatrix} \Gamma_1(t) & \Gamma_2(t) \\ \Gamma_3(t) & \Gamma_1(t) \end{pmatrix}.$$

$\Gamma_i(t)$ $(i = 1, 2, 3)$ are obtained from a modal analysis of the system (2). Moreover, during all this kind of motion, the following constraint holds:

$$|\chi\eta(x_2 - x_1) - p\cos(\omega t + \varphi)| < \eta u_r.$$

4 Periodic Orbits Including an Overshooting Part

Several periodic orbits including stick phases and slip phases have been obtained [7]. Among these orbits, a set of periodic orbits including an overshooting part is found. For each period $(0 < t < 2\pi/\omega)$, the motion is composed of three parts. The first one $(0 < t < \tau)$ is a slip motion of m_2 with $x_2' < V$, the next part $(0 < t - \tau < \tau_1)$ is an overshooting slip motion of the mass $(x_2' > V)$, and the last part $(0 < t - \tau - \tau_1 < T,\ T = 2\pi/\omega - \tau - \tau_1)$ is a stick motion of m_2. At the beginning of the motion for $t = 0$, we assume that

$$x_2'(0) = V, \quad \chi\eta(x_2(0) - x_1(0)) = p\cos\varphi + \eta u_r$$

and at $t = \tau$, we assume the conditions

$$x_2'(\tau) = V, \quad \chi\eta(x_2(\tau) - x_1(\tau)) - p\cos(\omega\tau + \varphi) + \eta u_r < 0. \tag{3}$$

The last condition leads to an overshooting motion for $t > \tau$.

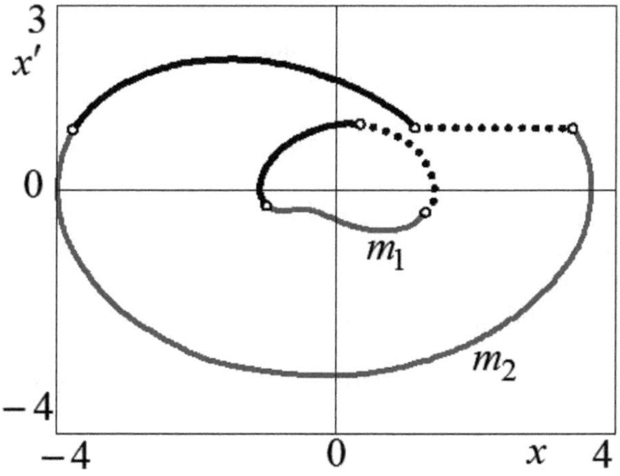

Fig. 2 Phase portrait of the overshooting periodic solution

An example of a periodic orbit with an overshooting part is obtained for the set of data

$$\chi = 0.2,\ \eta = 3.8,\ u_s = 0.059,\ u_r = 0.5347,\ V = 1,\ \omega = 0.6\ ,p = 0.1,\ \varphi = 0.$$

The other parameters are computed:

$$\tau = 4.775,\ \tau_1 = 2.65,\ T = 3.047,\ z_1\,(0) = 1.1471,\ z_2\,(0) = 3.6572,\ z_1'\,(0) = -0.7281.$$

The phase portraits of the system are shown in Fig. 2. The solid lines show slip motion (thick black lines correspond to the overshooting motion); the dotted lines are related to the stick motion.

5 Nonsticking Periodic Solutions

In industrial applications, avoiding sticking phases of motion is sometimes necessary. In the past, several authors [3, 8] investigated the existence of periodic nonsticking solutions of a one-degree-of-freedom oscillator subjected to simple harmonic loading. The mass is in contact with a fixed surface and a dry friction force acts between the mass and the surface. The aim of these works is to obtain some estimates of the minimum external force amplitude needed to prevent this sticking motion. The nonsticking orbit involves for each period a slip motion with a negative mass velocity and a slip motion with a positive mass velocity (overshooting motion). Moreover, the authors assumed that the motion is symmetrical in space and time [3, 8].

In the following, this problem is revisited for the two-degree-of-freedom oscillator considered in this work. Let us consider the system described in Fig.1 with initial conditions

$$x'_{20} = V, \quad \chi\eta\,(x_{20} - x_{10}) > p\cos\varphi + \eta u_r. \tag{4}$$

The nonsticking periodic orbit is composed of two parts: for $0 < t < \tau$ the system undergoes a slip motion ($x'_2 < V$); the second part for $\tau < t < 2\pi/\omega$ is an overshooting motion ($x'_2 > V$). At $t = \tau$ the conditions (3) are fulfilled. We prove [7] that this kind of periodic orbits exists only if

$$V = 0,\ \tau = \pi/\omega,\ x(0) = -x(\tau),\ x'(0) = -x'(\tau).$$

The nonsticking periodic orbits are symmetrical in space and time.

A numerical computation is performed for the parameters

$$\chi = 0.3,\ \eta = 4,\ \omega = .6,\ p = 1,\ u_s = 0.1,\ u_r = 0.2996\ .$$

The corresponding values of the initial conditions and of the time lag φ are obtained:

$$x_{10} = 1.5608,\ x_{20} = 3.3295,\ x'_{10} = 0.1523, \varphi = 0.3925\ .$$

The phase portraits (x_i, x'_i), $i = (1,2)$ of the two masses are shown in Fig. 3 (the thick black parts of the curves are related to the overshooting motion). These curves are symmetrical with respect to the origin.

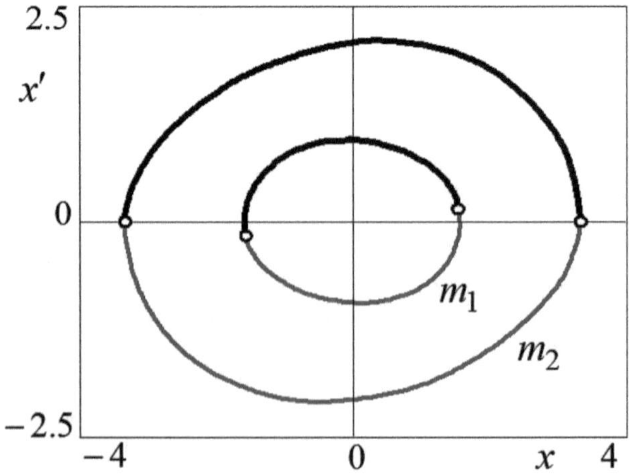

Fig. 3 Phase portrait of the nonsticking orbit

6 Periodic Orbits with Abnormal Stops

The nonsticking orbit investigated in the last paragraph involves for each period two normal stops

$$x_2'(0) = x_2'(\tau) = 0.$$

These stops occur when the displacement of the second mass reaches a local extremum and the mass reverses its direction of motion at the turning point. In [9], for a one-degree-of-freedom oscillator, with dry friction and harmonic load, a set of periodic orbits including abnormal stops have been obtained. Abnormal stops occur when at the turning point the mass moves in the same direction as its motion prior to the stop. The same phenomenon can be observed for the two-degree-of-freedom oscillator investigated in this paper. At $t = 0$ the initial conditions (4) are assumed and for $0 < t < \tau$, the system undergoes a slip motion ($x_2' < 0$). Let us assume that at $t = \tau$

$$x_2'(\tau) = 0, \quad \chi\eta\left(x_2(\tau) - x_1(\tau)\right) - p\cos(\omega\tau + \varphi) - \eta u_r > 0.$$

A new phase of slip motion occurs. This motion ends at $t = \tau + \tau_1$ if at this time

$$x_2'(\tau + \tau_1) = 0, \quad \chi\eta\left(x_2(\tau + \tau_1) - x_1(\tau + \tau_1)\right) - p\cos\left(\omega(\tau + \tau_1) + \varphi\right) + \eta u_r < 0.$$

For $t > \tau + \tau_1$ the system undergoes an overshooting motion ($x_2' > 0$). A periodic orbit of period $2\pi/\omega$ is obtained, but due to the symmetry in space and in time of this nonsticking orbit, the overshooting motion ($\tau + \tau_1 < t < 2\pi/\omega$) involves another abnormal stop for $t = \tau + \tau_1 + T$, with the conditions

$$x_2'(\tau + \tau_1 + T) = 0,$$

$$\chi\eta\left(x_2(\tau + \tau_1 + T) - x_1(\tau + \tau_1 + T)\right) - p\cos\left(\omega(\tau + \tau_1 + T) + \varphi\right) + \eta u_r < 0.$$

Due to the symmetrical property we obtain

$$\tau + \tau_1 = \pi/\omega, \quad T = \tau, \quad x(\tau + \tau_1) = -x(0), \quad x'(\tau + \tau_1) = -x'(0).$$

This kind of orbit exists only if $u_s = u_r$.

An example of periodic orbits with abnormal stops is obtained for the data

$$\chi = 0.7, \ \eta = 3.8, \ \omega = 2\pi/11, \ p = 1, \ u_r = u_s = 0.1717, \ \varphi = 2.6455.$$

The other parameters are computed:

$$\tau = T = 4, \ \tau_1 = 1.5, \ x_1(0) = 2.1249, \ x_2(0) = 2.0451, \ x_1'(0) = -0.7267.$$

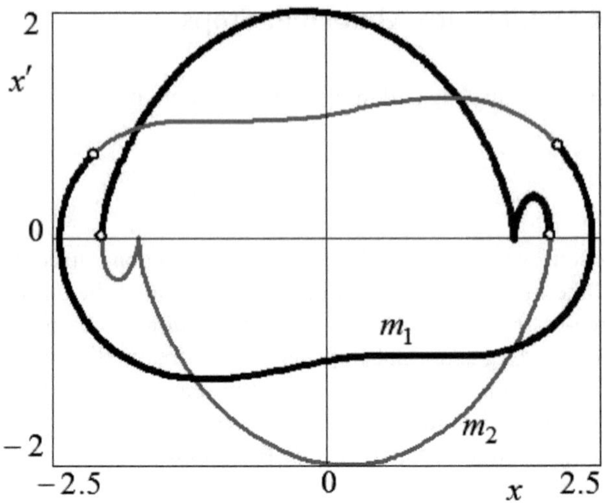

Fig. 4 Phase portrait of the orbit with abnormal stops

The phase portrait of the system is shown in Fig. 4 (the thick black parts of the curves are related to the overshooting motion).

References

1. Andreaus, U., Casini, P.: Dynamics of friction oscillators excited by a moving base and/or driving force. J. Sound Vib. **245**, 685–699 (2001)
2. Awrejcewicz, J., Olejnik, P.: Stick-slip dynamics of a two-degree-of-freedom system. Int. J. Bifur. Chaos. **13**(4), 843–861 (2003)
3. Csernak, G., Stepan, G.: On the periodic response of a harmonically excited dry friction oscillator. J. Sound Vib. **295**, 649–658 (2006)
4. Olejnik, P., Awrejcewicz, J.: Application of Henon method in numerical estimation of the stick-slip transitions existing in Filippov-type discontinuous dynamical systems with dry friction. Nonlinear Dyn. **73**(1–2), 723–736 (2013)
5. Pascal, M.: Dynamics of coupled oscillators excited by dry friction. ASME J. Comput. Nonlinear Dyn. **3**, 20–26 (2008)
6. Pascal M.: New events in stick slip oscillators behavior. J. Appl. Math. Mech. (PMM). **75**, 10–17 (2011)
7. Pascal, M.: New limit cycles of dry friction oscillators under harmonic load. Nonlinear Dyn. **70**, 1435–1443 (2012)
8. Hong, H.K., Liu, C.S.: Non-sticking oscillation formulae for Coulomb friction under harmonic loading. J. Sound Vibr. **244**(5), 883–898 (2001)
9. Hong, H.K., Liu, C.S.: Coulomb friction oscillator: Modeling and responses to harmonic loads and base excitations. J. Sound Vib. **229**(5), 1171–1192 (2000)
10. Stepanov, S.: Limit cycles of a double oscillator excited by dry friction. Int. J. Bifur. Chaos. **21**(10), 3043–3046 (2011)
11. Van de Vrande, B.L., Van Campen, D.H., De Kraker, A.: An approximate analysis of dry-friction-induced stick-slip vibrations by a smoothing procedure. Nonlinear Dyn. **19**, 157–169 (1999)

Limit Cycle Oscillations of an Aerodynamic Pendulum

Yury D. Selyutskiy

Abstract Dynamics of an aerodynamic pendulum in low-speed airflow is studied using the phenomenological model, where the internal dynamics of the flow is simulated using an oscillator attached to the pendulum. Limit cycle oscillations occurring in the vicinity of the "along the flow" equilibrium at a certain range of wind speeds are studied. Their evolution under changing system parameters (wind speed, holder length, etc.) is analyzed. Experimental study of such oscillations of the aerodynamic pendulum is performed in the subsonic wind tunnel of the Institute of Mechanics of Lomonosov Moscow State University for different wind speeds and different structural parameters of the pendulum. It is shown that results of experiments are in good enough agreement with simulation results.

1 Introduction

Aerodynamic pendulum, that is, a wing mounted a flow so that it can rotate around a fixed axis (like a weathercock), belongs to the waste family of aeroelastic systems. Study of different systems of this class began as early as in 30th of the last century in different countries and was driven by problems of quickly developing aviation, and since then, an extensive literature appeared dedicated to research of various issues related with such objects. It is well known that the "trivial" equilibrium of such objects (namely, situation when the wing is stretched along the flow) becomes unstable in a certain range of parameters, and a limit cycle appears. Prediction and control of such cycles is important, as for some technical objects, they are undesired or even dangerous, while for others they can represent working regimes (for instance, for systems for converting flow energy into electric power).

Y.D. Selyutskiy (✉)
Institute of Mechanics of Lomonosov Moscow State University, 119192, Michurinsky pr. 1, Moscow, Russia
e-mail: seliutski@imec.msu.ru

J. Awrejcewicz (ed.), *Applied Non-Linear Dynamical Systems*, Springer Proceedings in Mathematics & Statistics 93, DOI 10.1007/978-3-319-08266-0_31

This effect is described for systems with two degrees of freedom (e.g., [6, 14, 15]) allowing both pitching and plunging motion of the wing. The work [10] shows existence of this phenomenon also for systems with one degree of freedom (only rotational). Papers [8, 9] show experimentally and numerically that such cycles with small amplitude appear in a certain range of wind speeds for the aerodynamic pendulum with one degree of freedom.

It should be noted that these results pertain to the case of elastically mounted pendulum with elastic axes located within the chord of the wing, and hence, the characteristics of limit cycles are influenced by combined action of aerodynamic load and elastic forces (as a rule, assumed nonlinear). However, from the point of view of both theory and applications, it is interesting to examine the pure effect of aerodynamic forces and to study the evolution of limit cycles when the position of elastic axis changes in wide enough range. Works [5, 11] considered behavior of one- and two-degree-of-freedom aerodynamic pendulum without spring mounting under the assumption of quasi-steady aerodynamics, which did not allow describing limit cycles in the domain of small angles of attack.

Another important issue concerning aeroelastic (and not only) systems is developing a plausible and simple enough (that is, containing small number of additional degrees of freedom and relatively few parameters) phenomenological mathematical model of unsteady interaction between the wing and the flow. The actuality of this task is related to the fact that engineering practice requires in many cases to determine values of parameters that are in a certain sense optimal for the given problem. In order to fulfill this requirement, it is necessary either to set up multiple experiments or perform numerous numerical simulation runs. However, making real experiments for studying the dynamical behavior of bodies in flow is very cumbersome, which makes mass testing impracticable. Performing numerical simulation using different CFD-type packages (or vortical methods like those described in [2]) in such kind of problems requires considerable computational resources. This makes it troublesome (if ever possible) to perform quick and efficient parametric analysis in a wide enough range of parameter values.

Therefore, different types of such phenomenological approaches are developed by various groups of researchers (e.g., [3, 4, 7]). All these methods are based on introducing additional empirical variables, so that original dynamic system is supplemented with additional ODEs for aerodynamic forces and/or moments.

The present work develops another approach proposed in [12] where the internal dynamics of the flow is simulated with an additional ("dummy") degree of freedom ("attached oscillator"). Stability of the "along the flow" equilibrium of this object is studied, and stability domains are determined for different combinations of system parameters. Evolution of limit cycle is studied in wide range of wind speeds and positions of rotation axis. Results of mathematical simulation are verified against experiments made in the wind tunnel of the Institute of Mechanics of Lomonosov Moscow State University.

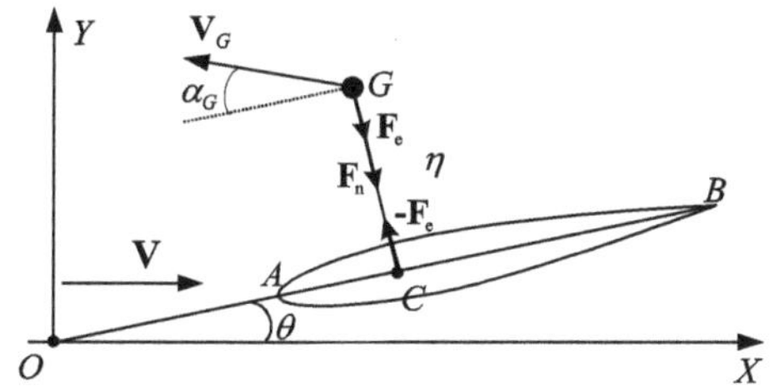

Fig. 1 Aerodynamic pendulum with the attached oscillator

2 Problem Statement and Motion Equations

Consider an aerodynamic pendulum, that is, a rigid body consisting of a weightless holder OA and a wing with symmetric airfoil (Fig. 1) that can rotate about a fixed axis O. The system is placed in airflow having constant speed V at infinity. Denote the angle of rotation of the pendulum about the axis with θ.

We neglect the aerodynamic load upon the holder and suppose that the flow is two-dimensional. In order to take into account unsteady effects arising due to the nonuniform motion of the object under consideration, it is necessary to describe the internal dynamics of the flow. For that, we use the approach proposed in [12], where the internal dynamics of the flow is simulated using an oscillator G attached to the wing. In [13] it was shown that this method allows to provide good enough agreement with results of numerical calculations according to the modified method of discrete vortices [1].

Within the framework of this approach it is assumed that the oscillator is attached to the chord in the center of pressure C of the airfoil and that point G can move along the normal to the chord. Let η be the oscillator length CG, and l be the distance OC. Assume also that the aerodynamic force is applied to the point G, and its normal and tangential components have the following structure:

$$F_n = \frac{\rho S}{2} V_G^2 C_n(\alpha_G), \quad F_\tau = \frac{\rho S}{2} V_G^2 C_\tau(\alpha_G) \tag{1}$$

Here ρ is the air density, S is the wing area, V_G is the effective flow speed, that is, the speed of point G with respect to the flow at infinity, α_G is effective angle of attack, and C_n, C_τ are nondimensional coefficients of normal and tangential force, correspondingly. Their dependence on the effective angle of attack is the same as in steady case and can be determined from static experiments.

Evidently, the effective angle of attack should contain information about orientation and the angular speed of the airfoil, as well as about the attached oscillator speed. Taking this into account, we use the following formula for α_G:

$$\alpha_G = -\theta - \frac{(l + L)\dot{\theta}}{V} - \frac{\dot{\eta}}{V} \tag{2}$$

Here we suppose that $L > 0$.

From steady experiments it is known that for thin enough airfoils and small angles of attack the tangential force coefficient is much smaller than the normal force coefficient. So, we neglect this component of the aerodynamic force. Then the motion equations of the aerodynamic pendulum with the attached oscillator look as follows:

$$J\ddot{\theta} + 2m\eta\dot{\theta}\dot{\eta} + m\eta^2\ddot{\theta} + lm\eta\dot{\theta}^2 = -D\dot{\theta} - lF_e \tag{3}$$

$$m\ddot{\eta} + ml\ddot{\theta} - m\eta\dot{\theta}^2 = F_n + F_e \tag{4}$$

Here J is the moment of inertia of the pendulum with respect to the point O, D is the structural damping coefficient, F_e is elastic force, and m is the mass of the oscillator.

In order to describe nonlinear effects arising during the unsteady interaction between the pendulum and the flow, we introduce nonlinearity into the elastic force F_e:

$$F_e = -k\eta - h\dot{\eta} - H\dot{\eta}\eta^2 \tag{5}$$

Equations (3)–(4) along with relations (1), (2), and (5) make a closed system of equations.

As for symmetric airfoil we have $C_n(0) = 0$, it is evident that the "along the flow" position ($\theta = 0$, $\eta = 0$) is equilibrium. We restrict our subsequent analysis with the domain of small angles of attack. For such values of α_G, the following relation holds:

$$C_n(\alpha_G) = C_n^{\alpha}\alpha_G \tag{6}$$

where C_n^{α} is a positive constant.

Introduce nondimensional time $\tau = b/V_0$ where b is the airfoil chord and V_0 is characteristic flow speed. Then the following nondimensional coordinates and parameters appear:

$$u = \frac{V}{V_0}, \quad \bar{l} = \frac{l}{b}, \quad \bar{D} = \frac{2D}{\rho S V_0 b^2}, \quad \bar{J} = \frac{2J}{\rho S b^3}, \quad \bar{L} = \frac{L}{b},$$

$$\bar{\eta} = \frac{\eta}{b}, \quad \bar{m} = \frac{2m}{\rho S b}, \quad \bar{k} = \frac{2kb}{\rho S V^2}, \quad \bar{h} = \frac{2h}{\rho S V}, \quad \bar{H} = \frac{2Hb^2}{\rho S V}$$

In order to simplify the notation, denote the derivative with respect to τ with dot and omit bars over nondimensional values. Then, taking into account (6), the motion equations look as follows:

$$J\ddot{\theta} + 2m\eta\dot{\theta}\dot{\eta} + m\eta^2\ddot{\theta} + lm\eta\dot{\theta}^2 = -D\dot{\theta} + l\left(ku^2\eta + hu\dot{\eta} + Hu\dot{\eta}\eta^2\right) \quad (7)$$

$$m\ddot{\eta} + ml\ddot{\theta} - m\eta\dot{\theta}^2 = -ku^2\eta - hu\dot{\eta} - Hu\dot{\eta}\eta^2 - C_n^\alpha\left(u^2\theta + u(l+L)\dot{\theta} + u\dot{\eta}\right) \quad (8)$$

In what follows we suppose that the moment of inertia J of the pendulum is large.

3 Stability of the Along the Flow Equilibrium

Motion equations linearized in the vicinity of the abovementioned equilibrium are as follows:

$$J\ddot{\theta} = -D\dot{\theta} + l\left(ku^2\eta + hu\dot{\eta}\right) \quad (9)$$

$$m\ddot{\eta} + ml\ddot{\theta} = -ku^2\eta - hu\dot{\eta} - C_n^\alpha\left(u^2\theta + uL\dot{\theta} + u\dot{\eta}\right) \quad (10)$$

The Hurwitz criterion yields a necessary and sufficient condition of the asymptotic stability of the "along the flow" equilibrium that can be written in the following structural form:

$$Q = A_2k^2 + A_1k + A_0 > 0 \quad (11)$$

$$or$$

$$Q = \sum_{i=0}^{6} B_i l^i > 0 \quad (12)$$

where $A_{0,1,2}$, $B_{0..6}$ are some combinations of system parameters. In particular,

$$A_2 = u^4(ul(l+L)C_n^\alpha + D) * ((J - mlL)JuC_n^\alpha + hu(J+ml^2)^2 + Dm^2l^2),$$
$$B_6 = mhC_n^\alpha ku^6(C_n^\alpha h + km) > 0,$$
$$B_0 = D(h + C_n^\alpha)u^3k\left(J^2ku^2 + DJu(h + C_n^\alpha) + D^2m\right) > 0 \quad (13)$$

Note that $A_2 > 0$ for large enough J.

In order to understand the behavior of the system at different values of parameters, consider first some limit cases (taking into account that $J \gg 1$).

For small values of k we obtain the following relations for the roots of the characteristic polynomial:

$$\lambda_{1,2} = -\frac{D}{2J} - C_n^\alpha lhu\frac{(l+L)(C_n^\alpha + h) - m}{2J\left(C_n^\alpha + h\right)^2} \pm iu\sqrt{\frac{C_n^\alpha lh}{J(C_n^\alpha + h)}} + O(J^{-3/2})$$

$$\lambda_3 = -\frac{uk}{h} + o(k), \quad \lambda_4 = -u\frac{C_n^\alpha + h}{m} + o(1)$$

Under the assumption that $m < L(C_n^\alpha + h)$ real parts of these roots are negative, hence, the equilibrium is asymptotically stable.

For larger values of k the following relations hold:

$$\lambda_{1,2} = -\frac{D}{2J} - \frac{C_n^\alpha ul((l+L)k - C_n^\alpha)}{2kJ} \pm iu\sqrt{\frac{C_n^\alpha l}{J}} + O(J^{-3/2})$$

$$\lambda_{3,4} = u\frac{-(h+C_n^\alpha) + \sqrt{(h+C_n^\alpha)^2 - 4km}}{2m} + o(1)$$

From the previous formulae for $\lambda_{1,2}$ one can readily show that the studied equilibrium is unstable, if parameter l satisfies the following inequalities:

$$\frac{C_n^\alpha - kL}{2k} - \sqrt{\left(\frac{C_n^\alpha - kL}{2k}\right)^2 - \frac{D}{uC_n^\alpha}} < l < \frac{C_n^\alpha - kL}{2k} + \sqrt{\left(\frac{C_n^\alpha - kL}{2k}\right)^2 - \frac{D}{uC_n^\alpha}}$$

Note, in particular, that if $C_n^\alpha < kL$, then for all $l > 0$ the equilibrium is asymptotically stable. However, if $C_n^\alpha > kL$, then there exists a range of positive l where the stability is lost, at least, for large enough J.

For $h \gg 1$ roots of the characteristic polynomial look as follows:

$$\lambda_{1,2} = -\frac{D}{2J} - \frac{C_n^\alpha ul(l+L)}{2J} \pm iu\sqrt{\frac{C_n^\alpha l}{J}} + O(J^{-3/2})$$

$$\lambda_3 = -\frac{hu}{m} + O(1), \lambda_4 = -\frac{uk}{h} + o(h^{-1})$$

Apparently, in this case the "along the flow" equilibrium is asymptotically stable.

Note that from Hurwitz criterion one can readily obtain that for $h = 0$ the equilibrium is asymptotically stable, if

$$k > \frac{uC_n^{\alpha 2} l}{uC_n^\alpha l(l+L) + D} + O(J^{-1}) \quad (14)$$

Note also that from (13) it follows that for $l \ll 1/J \ll 1$ and for $l \gg 1$ the equilibrium in question is asymptotically stable.

Basing on the above results, as well as on representations (11–12) and formula (14), the evolution of the asymptotic stability domain in the plane of parameters (k, h) with the change of l is qualitatively shown in Fig. 2. Here solid lines mean boundaries of stability domains at different l (areas located below each of these lines are domains of instability for the corresponding value of l).

Consider now the situation when the equilibrium position "along the flow" is unstable in more detail.

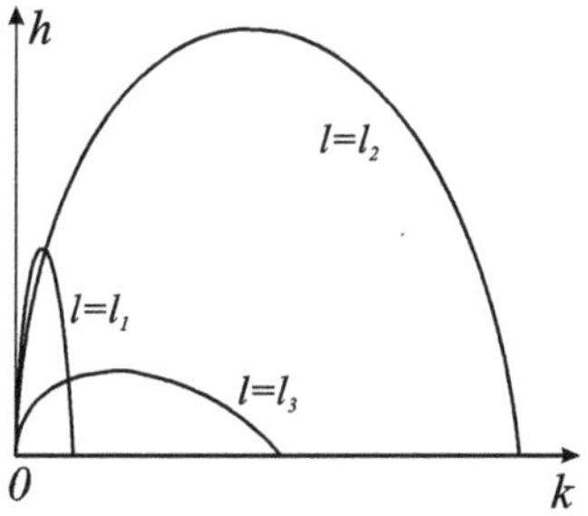

Fig. 2 Evolution of stability domains when the holder length changes ($l_1 < l_2 < l_3$)

4 Limit Cycles

Let's search for limit cycles arising in the system in case of unstable position "along the flow" in the following form:

$$\theta(t) = \theta_c \sin(\omega_c t), \quad \eta(t) = \eta_c \sin(\omega_c t + \Phi) \tag{15}$$

In order to determine the characteristics of the cycle, use the conventional harmonic balance method. Suppose that $\omega_c \ll 1$ and $J\omega_c^2 = O(1)$. Then, substituting (15) into (7–8) and leaving only terms up to the first order of smallness yield the following equations:

$$\begin{gathered}
-J\theta_c\omega_c^2 + C_n^\alpha l\theta_c u^2 - C_n^\alpha l u\eta_c\omega_c \sin\Phi = 0 \\
C_n^\alpha l u\eta_c\omega_c \cos\Phi + C_n^\alpha (l+L)u\theta_c\omega_c + D\theta_c\omega_c = 0 \\
C_n^\alpha u^2\theta_c + \eta_c k u^2 \cos\Phi - uh\eta_c\omega_c \sin\Phi - \\
uC_n^\alpha \eta_c\omega_c \sin\Phi - \frac{1}{4}u\eta_c^3\omega_c H \sin\Phi = 0 \\
C_n^\alpha (L+l)u\theta_c\omega_c + \eta_c k u^2 \sin\Phi + u\eta_c\omega_c h \cos\Phi + \\
C_n^\alpha u\eta_c\omega_c \cos\Phi + \frac{1}{4}u\eta_c H\omega_c \cos\Phi = 0
\end{gathered} \tag{16}$$

Note that from the second equation of (16) it follows that $\cos\Phi < 0$.

One can readily show that amplitudes of $\theta(t)$ and $\eta(t)$ and the phase difference can be expressed in terms of ω_c with the following relations:

$$\theta_c = -\frac{C_n^\alpha l u \eta_c \cos \Phi}{C_n^\alpha l(l+L)u + D},$$

$$\eta_c = 2\sqrt{\frac{-(\omega_c D(h + C_n^\alpha) + \omega_c h C_n^\alpha l(l+L)u + (C_n^\alpha ul(l+L) + D)ku \tan \Phi)}{H\omega_c (C_n^\alpha l(L+l)u + D)}},$$

$$\tan \Phi = \frac{J\omega_c^2 - C_n^\alpha l u^2}{(C_n^\alpha l(l+L)u + D)\omega_c} \tag{17}$$

Solving the system (16) yields the following formula for the frequency of the limit cycle:

$$\omega_c = \sqrt{\frac{2kJ^2\omega_0^2 - kX_1^2 + DulC_n^{\alpha 2} - \sqrt{4X_1X_2kJ^2\omega_0^2 + \left(kX_1^2 - DulC_n^{\alpha 2}\right)^2}}{2J\left(kJ - C_n^{\alpha 2}(L+l)l\right)}} \tag{18}$$

where

$$\omega_0 = u\sqrt{\frac{C_n^\alpha l}{J}}, \quad X_1 = D + uC_n^\alpha l(l+L), \quad X_2 = ulC_n^{\alpha 2} - kX_1$$

In order to verify the accuracy of the obtained approximation, numerical calculations were performed using the following values of parameters:

$$J = 500, \quad C_n^\alpha = 5, \quad D = 1, \quad u = 0.5$$

$$m = 1.5, \quad k = 1, \quad h = 5, \quad L = 1, \quad H = 200$$

Parameter l varied from 0 to 2.

Comparison between exact solution of the motion equations and approximations (17)–(18) is shown in Fig. 3 a,b (boxes denote exact solution, and solid lines denote approximations).

Apparently, the quality of approximation of the cycle frequency can be deemed quite reasonable (Fig. 3a) in the whole range of values of l, for which the limit cycle exists. The quality of approximation of the cycle amplitude (Fig. 3b) is good enough, though it gets somewhat worse near the maximum of the curve and near the ends of the interval where the limit cycle exists.

5 Comparison with Experiments

In order to verify the model, experiments were performed in the subsonic wind tunnel of the Institute of Mechanics of Lomonosov Moscow State University. Characteristics of the wind tunnel are summarized in the following Table 1:

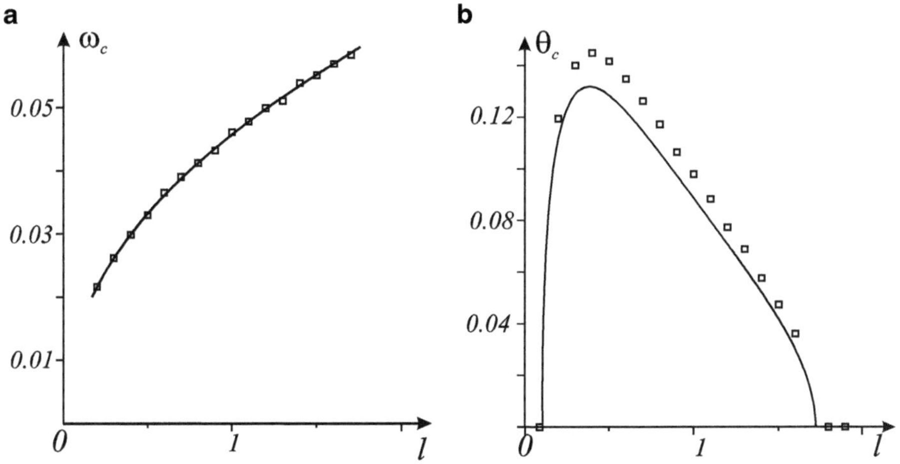

Fig. 3 Cycle frequency (**a**) and cycle amplitude (**b**) as functions of l: comparison between exact solution and approximation formulae

Table 1 Wind tunnel characteristics

Characteristic	Value
Dimensions of the working part	3m x 5m, elliptic
Flow speed range	0–50 m/s
Nonuniformity of the wind speed across the cross section	0.5 %
Flow angularity (both in horizontal and vertical directions)	< 0.25 deg
Gradient of the static pressure	0.002 m^{-1}
Longitudinal component of the turbulence intensity	< 0.2 %

Table 2 Parameters of the experimental setup

Parameter	Case 1	Case 2	Case 3
l	1.12	0.69	0.25
J	419	338	289

The experimental setup represented a pendulum consisting of a frame (made of thin metal rods) and wooden wing installed inside the frame. The structure was mounted in the wind tunnel so that its rotation axis was vertical. The wing parameters were as follows: chord 0.07 m, length 0.32 m, and maximum thickness 0.009 m. Position of the wing with respect to the frame could be changed. Test series were performed for three configurations. Nondimensional parameters of the pendulum for all cases are given in Table 2. The wind speed corresponding to Reynolds number $Re = 10^5$ was taken as the characteristic flow speed V_0.

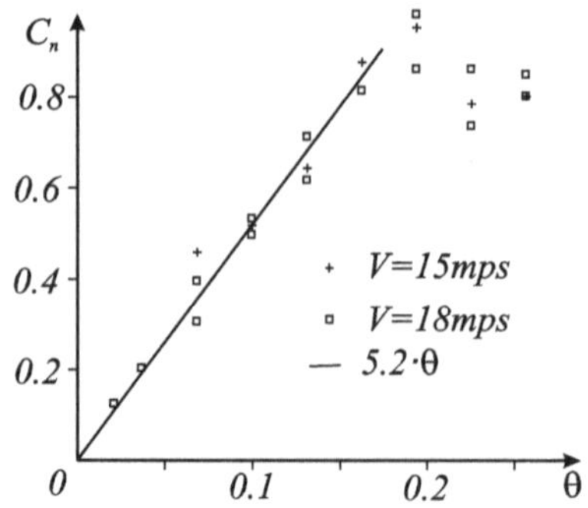

Fig. 4 Normal force coefficient from steady experiments vs steady angle of attack (rad.)

In order to assess the structural damping, a series of experiments was carried out without flow where the rotation axis was horizontal (so that the system behaved like a conventional physical pendulum). In the result, the following estimation was obtained for the nondimensional damping coefficient: $D = 1.6$.

In order to determine static characteristics of the airfoil, a series of experiments was performed, where the pendulum was fixed at different angles, and aerodynamic load was measured acting on the wing. Results are presented in Fig. 4 for two values of the wind speed (boxes denote experimental values of the normal force coefficient C_n, and solid line denotes the approximation curve $C_n(\theta) = C_n^\alpha \theta$, where $C_n^\alpha = 5.2$.

Evidently, agreement with experiments is good for smaller values of θ (up to about 0.15 rad.).

Then dynamic experiments were performed at different flow speeds ranging from 8mps to 25mps (which corresponds to Reynolds numbers from $3.7 \cdot 10^4$ to $1.2 \cdot 10^5$). In the course of testing, the pendulum was deflected from the "along the flow" equilibrium for a certain angle, after which it was released and moved freely. The pendulum rotation angle θ was registered as a function of time using a conventional encoder.

It was found that for case 1 the "along the flow" equilibrium remained asymptotically stable in the entire range of wind speeds. In the same time, for cases 2 and 3 in a certain range of wind speeds (approximately from 10mps to 17mps), this equilibrium became unstable, and a limit cycle appeared. This fact agrees with [8].

Evidently, this phenomenon is due to flow rearrangements taking place in this interval of Reynolds numbers. Such rearrangements having considerable influence upon the interaction between the body and the flow were registered and discussed, in particular, in [9]. In terms of the proposed model, this situation can be interpreted as attached oscillator parameters' dependence on the Reynolds number (or on the normalized flow speed u).

Series of numerical simulations were performed, and parameters of the attached oscillator were chosen (if necessary, as functions of u) in such a way as to provide possibly good agreement with experimental data. In the result, the following values and dependences were selected:

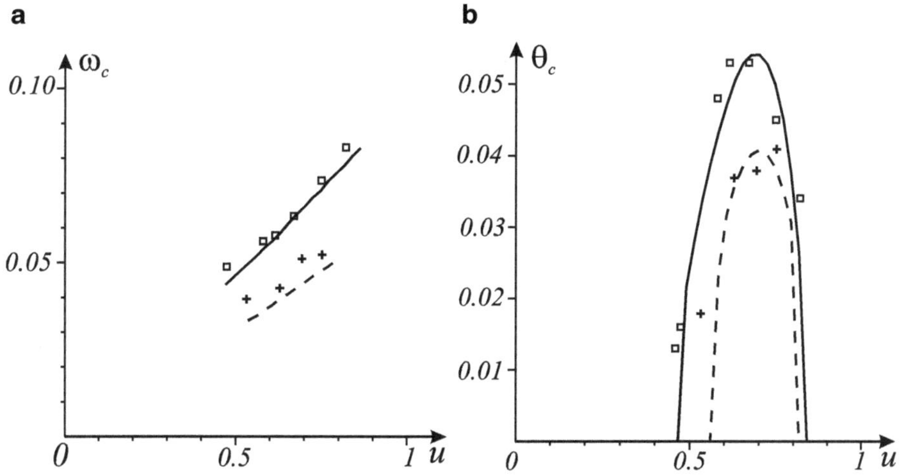

Fig. 5 Cycle frequency (**a**) and cycle amplitude (**b**): comparison between experimental data and numerical simulation

$$m = 1.5, \quad C_n^{\alpha} = 5.2, \quad H = 200, \quad L = 1$$

$$k = \begin{cases} 1.0, & u \leq 0.61 \\ 2u - 0.19, & u < 0.99, \\ 2.4, & u \geq 0.99 \end{cases} \qquad h = \begin{cases} 7.5, & u \leq 0.61 \\ 31.16(u - 0.61)^2 + 7.5, & u < 0.99 \\ 12.0, & u \geq 0.99 \end{cases}$$

Simulation and test data (limit cycle amplitudes and frequencies) are shown in Fig. 5 a,b. Experimental values are represented with boxes (for the case 2) and crosses (for the case 3), and the corresponding calculated curves are shown with solid (case 2) and dashed (case 3) lines. In case 1, the studied equilibrium is asymptotically stable both in simulation and in wind tunnel experiments (as mentioned before) in the whole range of test wind speeds.

Evidently, calculation results qualitatively agree with experimental data. However, the difference between measured and calculated amplitudes becomes larger near the ends of the range of u where the limit cycle exists. This is quite natural, as the cycle amplitude depends rather strongly on u in these areas, and even small error in the determination of the mentioned end points causes relatively large errors in the amplitude. Besides, calculated frequencies in the second case are slightly lower than measured ones. These discrepancies can be due to the presence of additional dissipation factors in the system and to neglection of aerodynamic load upon the frame of the pendulum.

6 Conclusion

This work proposes a novel nonlinear phenomenological model for simulating unsteady interaction between wing-type body and flow. This model is used for the examination of dynamics of the aerodynamic pendulum. Basing on it, stability of the "along the flow" equilibrium of this object is studied, and stability domains are determined for different combinations of system parameters. Formulae are obtained for amplitude and frequency of limit cycles arising upon stability loss of this equilibrium. Comparison with results of wind tunnel test performed in the Institute of Mechanics of Lomonosov Moscow State University shows that the model allows giving adequate qualitative description of experimentally registered features of behavior of the aerodynamic pendulum. This gives ground for applying the model for performing qualitative analysis of dynamics of different aeroelastic systems.

Acknowledgements The work is supported by Russian Foundation for Basic Research (grant No. 12-01-00364).

References

1. Andronov, P.R., Guvernyuk, S.V., Dynnikova, G.Ya.: Vortex Method of Calculation of Unsteady Hydrodynamic Loads. Moscow University Press, Moscow (2006)
2. Belotserkovsky, S.M., Lifanov, I.K.: Method of Discrete Vortices. CRC Press, Boca Raton, Florida (1993)
3. Goman, M.G., Khrabrov, A.N.: State-space representation of aerodynamic characteristics of an aircraft at high angles of attack. J. Aircraft **31**, 1109–1115 (1994)
4. Hansen, M.H., Gaunaa, M., Madsen, H.A.: A Beddoes-Leishman type dynamic stall model in state-space and indicial formulations. Risø-R-1354(EN) (2004)
5. Lokshin, B.Ya., Samsonov, V.A.: On heuristic model of aerodynamical pendulum. Fundam. Prikl. Mat. **4** (3), 1047–1061 (1998)
6. Marsden, C.C., Price, S.J.: The aeroelastic response of a wing section with a structural freeplay nonlinearity: An experimental investigation. J. Fluids Struct. **21**, 257–276 (2005)
7. Meyer, M., Matthies, H.G.: State-space representation of instationary two-dimensional airfoil aerodynamics. J. Wind Eng. Ind. Aerod. **92**, 263–274 (2004)
8. Poirel, D., Harris, Y., Benaissa, A.: Self-sustained aeroelastic oscillations of a NACA0012 airfoil at low-to-moderate Reynolds numbers. J. Fluids Struct. **24**, 700–719 (2008)
9. Poirel, D., Price, S.J.: Random binary (coalescence) flutter of a two-dimensional linear airfoil. J. Fluids Struc. **18**, 23–42 (2003)
10. Runyan, H.R.: Single-degree-of-freedom-flutter calculations for a wing in subsonic potential flow and comparison with an experiment. NACA Reports **1089**, 1–8 (1952)
11. Samsonov, V.A., Sapunkov, G.I.: Stability research of double pendulum construction in a flow. In: Proceedings of 5th International Conference Of Cranes and Textile Mach. EUROCRANE **96**, 250–256 (1996)
12. Samsonov, V.A., Selyutskiy, Yu.D.: Phenomenological model of interaction of a plate with a flow. J. Math. Sci. **146**(3), 5826–5839 (2007)
13. Selyutskiy, Yu.D., Samsonov, V.A., Andronov, P.A.: On oscillations of aerodynamic pendulum. Int. J. Struct. Stability Dynam. (2013), doi: 10.1142/S0219455413400105

14. Shao, S., Zhu, Q., Zhang, C., Ni, X.: Airfoil aeroelastic flutter analysis based on modified Leishman-Beddoes model at low mach number. Chinese J. Aeronaut. **24**, 550–557 (2011)
15. Strganac, T.W., Ko, J., Thompson, D.E., Kurdila, A.J.: Identification and control of limit cycle oscillations in aeroelastic systems. J. Guid. Contr. Dynam. **23**(6), 1127–1133 (2000)

Vibratory Energy Localization by Non-smooth Energy Sink with Time-Varying Mass

Alireza Ture Savadkoohi and Claude-Henri Lamarque

Abstract We study nonlinear interactions of two coupled oscillators at different time scales. The main oscillator which is linear is coupled to a nonlinear energy sink with non-smooth (piecewise linear) potential and time-dependent mass. The overall time is embedded to fast and slow time scales and the behavior of the system at each one of them is revealed.

The invariant of the system at fast time scale is detected. Then we try to have further information about the overall system behavior at the first slow time scale. Finally, analytical developments are compared with numerical results and the possibility of the passive control of the main system by means of the time-dependent NES is commented upon.

1 Introduction

A special kind of light oscillators with essentially nonlinear potential (mainly cubic) is called nonlinear energy sink (NES). It can be coupled to some important systems for passively controlling them and localizing their vibratory energy [16, 17]. The efficacy of NES devices has been verified and proved experimentally as well [3, 4, 13]. Some works were carried out to consider and to design other types of nonlinear NES: namely, vibro-impact [1, 10], non-polynomial [2], and non-smooth [5, 14]. Some works have been carried out in order to consider non-smooth main structure to be controlled by smooth or non-smooth NES systems [6, 11, 15]. In all of those systems, masses of the main system and NES were constant. [7] studied the energy pumping phenomenon between a linear system with time-dependent mass and a non-smooth NES. In this paper we try to study the energy pumping phenomenon between a linear oscillator and a non-smooth NES where the mass of the NES is time-dependent. The organization of the paper is as follows: General presentation of the system, its rescaling process, shifting to the complex

A. Ture Savadkoohi (✉) • C.-H. Lamarque
ENTPE, Université de Lyon, LGCB and LTDS UMR CNRS 5513, Rue Maurice Audin, 69518 Vaulx en Velin, France
e-mail: alireza.turesavadkoohi@entpe.fr; lamarque@entpe.fr

J. Awrejcewicz (ed.), *Applied Non-Linear Dynamical Systems*, Springer Proceedings in Mathematics & Statistics 93, DOI 10.1007/978-3-319-08266-0_32

domain, and averaging are explained in Sect. 2. Behavior of the system at fast and slow time scales is described in Sect. 3. A numerical example is given in Sect. 4 and results are compared with analytical predictions. The application of the proposed system in the passive control of main structures is demonstrated in Sect. 5. Finally summary of the work is collected in Sect. 6.

2 The System and Preliminary Treatments

Let us consider an academic model of a 2-dof system: it consists of a main structure with mass, damping and elastic stiffness as M, C, and k_1 which is coupled to a non-smooth NES with the time-dependent mass $\tilde{m}(t)$. The mass $\tilde{m}$ can move freely in a distance equal to 2δ until it reaches to elastic springs, k_2, at two sides. The overall damping of the NES system is supposed to be $\tilde{\lambda}$. If we assume that the mean velocity at which the mass is "leaving or entering" the mass of the NES in the x direction is zero [18], then governing equations of the system can be summarized as

$$\begin{aligned} &M\ddot{x}_1 + C\dot{x}_1 + k_1 x_1 + \tilde{F}(x_1 - x_2) + \tilde{\lambda}(\dot{x}_1 - \dot{x}_2) = \Gamma \sin(\Omega t) \\ &\tilde{m}(t)\ddot{x}_2 + \tilde{F}(x_2 - x_1) + \tilde{\lambda}(\dot{x}_2 - \dot{x}_1) + \dot{\tilde{m}}(t)\dot{x}_2 = 0 \end{aligned} \tag{1}$$

The non-smooth potential $\tilde{F}$ of the NES is defined as

$$\tilde{F}(z) = -\frac{\partial V(z)}{\partial z} = -\tilde{F}(-z) = \begin{cases} 0 & -\delta \le z \le \delta \\ k_2(z - \delta) & z \ge \delta \\ k_2(z + \delta) & z \le -\delta \end{cases} \tag{2}$$

and the rule of the time dependency of the mass $\tilde{m}(t)$ reads

$$\tilde{m}(t) = m_0(1 + \epsilon m(t)) \tag{3}$$

ϵ is a small parameter which corresponds to the ratio of the initial mass of NES and the mass of the main system, i.e., $0 < \epsilon = \dfrac{m_0}{M} << 1$. We assume that $(1+\epsilon m(t)) \ge 0$ for a time long enough. We would like to rescale the system with respect to the new time domain T where $T = t\sqrt{\dfrac{k_1}{M}} = t\vartheta$. Equation (1) reads ($x_i(t) \rightarrow y_i(T)$):

$$\begin{aligned} &\ddot{y}_1 + \epsilon\zeta\dot{y}_1 + y_1 + \epsilon\hat{F}(y_1 - y_2) + \epsilon\lambda(\dot{y}_1 - \dot{y}_2) = \epsilon f_0 \sin(\omega T) \\ &\epsilon\ddot{y}_2 + \epsilon(1 + \epsilon m(T))^{-1}\lambda(\dot{y}_2 - \dot{y}_1) + \epsilon(1 + \epsilon m(T))^{-1}\hat{F}(y_2 - y_1) \\ &+\epsilon^2(1 + \epsilon m(T))^{-1}\dot{m}(T)\dot{y}_2 = 0 \end{aligned} \tag{4}$$

where $\frac{C}{\sqrt{Mk_1}} = \epsilon\zeta$, $\frac{1}{k_1}\tilde{F} = \epsilon\hat{F}$, $k = \frac{1}{\epsilon}\frac{k_2}{k_1}$, $\frac{\tilde{\lambda}}{\sqrt{Mk_1}} = \epsilon\lambda$, $\frac{1}{k_1}\Gamma = \epsilon f_0$, and $\frac{\Omega}{\vartheta} = \omega$. We assume that $k = o(\epsilon^0)$ and scaled potential of the NES is

$$\hat{F}(z) = \begin{cases} 0 & -\delta \leq z \leq \delta \\ k(z-\delta) & z \geq \delta \\ k(z+\delta) & z \leq -\delta \end{cases} \tag{5}$$

Taking into account that $(1+\epsilon m(T))^{-1} \simeq (1-\epsilon m(T)) + o(\epsilon^2)$, we will have

$$\begin{aligned} &\ddot{y}_1 + \epsilon\zeta\dot{y}_1 + y_1 + \epsilon\hat{F}(y_1 - y_2) + \epsilon\lambda(\dot{y}_1 - \dot{y}_2) = \epsilon f_0 \sin(\omega T) \\ &\epsilon\ddot{y}_2 + \epsilon(1-\epsilon m(T))\lambda(\dot{y}_2 - \dot{y}_1) + \epsilon(1-\epsilon m(T))\hat{F}(y_2 - y_1) \\ &+\epsilon^2(1-\epsilon m(T))\dot{m}(T)\dot{y}_2 = 0 \end{aligned} \tag{6}$$

Let us transfer the system to the coordinates $v = y_1 + \epsilon y_2$ and $w = y_1 - y_2$; equation (6) will be equivalent to

$$\begin{aligned} &\ddot{v} + \frac{1}{1+\epsilon}\epsilon\zeta(\dot{v} + \epsilon\dot{w}) + \frac{1}{1+\epsilon}(v + \epsilon w) + \epsilon^2\lambda m(T)\dot{w} + \epsilon^2 m(T)\hat{F}(w) \\ &+\epsilon^2(1-\epsilon m(T))\dot{m}(T)\frac{1}{1+\epsilon}(\dot{v} - \dot{w}) = \epsilon f_0 \sin(\omega T) \\ &\ddot{w} + \frac{1}{1+\epsilon}\epsilon\zeta(\dot{v} + \epsilon\dot{w}) + \frac{1}{1+\epsilon}(v + \epsilon w) + (1+\epsilon-\epsilon m(T))\lambda\dot{w} \\ &+(1+\epsilon-\epsilon m(T))\hat{F}(w) - \epsilon(1-\epsilon m(T))\dot{m}(T)\frac{1}{1+\epsilon}(\dot{v} - \dot{w}) = \epsilon f_0 \sin(\omega T) \end{aligned} \tag{7}$$

We introduce complex variables of [8] ($\varphi_1 e^{i\omega T} = \dot{v} + i\omega v$ and $\varphi_2 e^{i\omega T} = \dot{w} + i\omega w$ with $i = \sqrt{-1}$) to the system. We can present the functions $\hat{F}(w)$ and $m(T)$ in the form of Fourier series:

$$\hat{F}(w) = \hat{F}\Big(-\frac{i}{2\omega}(\varphi_2 e^{i\omega T} - \varphi_2^* e^{-i\omega T})\Big) = \sum_{j=-\infty}^{+\infty} f_j(\varphi_2, \varphi_2^*) e^{i\omega j T} \tag{8}$$

$$m(T) = \sum_{j=-\infty}^{+\infty} m_j e^{i\omega j T} \tag{9}$$

so,

$$\dot{m}(T) = \sum_{j=-\infty}^{+\infty} i\omega j m_j e^{i\omega j T} \tag{10}$$

where the $.^*$ represents the complex conjugate of the function under consideration. We assume that $\sum_{j=-\infty}^{+\infty} |m_j| < +\infty$ and $\sum_{j=-\infty}^{+\infty} |\omega j m_j| < +\infty$. We would like to analyze the averaged form of equations (7) around the 1 : 1 resonance, i.e., $\omega = 1 + \sigma\epsilon$, by studying its detailed components at different scales of the ϵ. After ignoring some obvious higher-order terms of ϵ in system (7) we will have

$$\begin{aligned} &\dot{\varphi}_1 + \frac{i}{2}\omega\varphi_1 + \frac{\epsilon}{2(1+\epsilon)}\zeta(\varphi_1 + \epsilon\varphi_2) - \frac{i}{2(1+\epsilon)\omega}(\varphi_1 + \epsilon\varphi_2) = \\ &-\frac{i\epsilon f_0}{2} \\ &\dot{\varphi}_2 + \frac{i}{2}\omega\varphi_2 + \frac{\epsilon}{2(1+\epsilon)}\zeta(\varphi_1 + \epsilon\varphi_2) - \frac{i}{2(1+\epsilon)\omega}(\varphi_1 + \epsilon\varphi_2) \\ &-\frac{i}{2}(1+\epsilon)\varphi_2 G_1(|\varphi_2|^2) + \epsilon m_1 G_2(|\varphi_2|^2) + \frac{1+\epsilon}{2}\lambda\varphi_2 \\ &-\frac{\epsilon}{2}\lambda(m_0\varphi_2 + m_2\varphi_2^*) + \frac{\epsilon\omega}{1+\epsilon} i m_2(\varphi_1^* - \varphi_2^*) = -\frac{i\epsilon f_0}{2} \end{aligned} \tag{11}$$

where the functions $G_1(\chi)$ and $G_2(\chi)$ are defined as [14]

$$G_1(\chi) = \begin{cases} 0 & if \quad \frac{\sqrt{\chi}}{\omega} < \delta \\ \frac{k}{\pi\omega}\left(-\frac{2\delta\omega}{\chi}\sqrt{\chi - (\delta\omega)^2} + 2\arccos\left(\frac{\delta\omega}{\sqrt{\chi}}\right)\right) & if \quad \frac{\sqrt{\chi}}{\omega} \geq \delta \end{cases} \tag{12}$$

$$G_2(\chi) = \begin{cases} 0 & if \quad \frac{\sqrt{\chi}}{\omega} < \delta \\ -\frac{2k}{\pi\omega}\sqrt{\chi - (\delta\omega)^2} & if \quad \frac{\sqrt{\chi}}{\omega} \geq \delta \end{cases} \tag{13}$$

We introduce fast ($T = \tau_0$) and slow ($\tau_1 = \epsilon\tau_0, \ldots$) time scales to the system and then we will consider its behavior at different orders of ϵ [9].

3 Time Multi-scale Analysis of the System

ϵ^0 order of the (11) is equivalent to

$$\begin{aligned} &\frac{\partial\varphi_1}{\partial\tau_0} = 0 \Rightarrow \varphi_1 = \varphi_1(\tau_1, \tau_2, \ldots) \\ &\frac{\partial\varphi_2}{\partial\tau_0} + \frac{i\left(1 - G_1(|\varphi_2|^2)\right) + \lambda}{2}\varphi_2 = \frac{i}{2}\varphi_1 \end{aligned} \tag{14}$$

Fixed points of the system, i.e., $\Phi(\tau_1)$ ($\tau_0 \to \infty$, $\dfrac{\partial \varphi_2}{\partial \tau_0} = 0$), can be defined as

$$\frac{i\Big(1 - G_1\big(|\Phi|^2\big)\Big) + \lambda}{2}\Phi = \frac{i}{2}\varphi_1 \tag{15}$$

By considering $\varphi_1 = N_1 e^{i\delta_1}$ and $\Phi = N_2 e^{i\delta_2}$, the following invariant manifold of the system at τ_0 time scale can be obtained:

$$N_1 = N_2\sqrt{\lambda^2 + \big(1 - G_1(N_2^2)\big)^2} \tag{16}$$

Stable areas of the linearized system at the τ_0 time scale can be defined by the following criterion: $(H(N_2^2) = 2N_2G'(N_2^2))$ [14]:

$$\lambda^2 + \big(1 - G_1(N_2^2)\big)\big(1 - H(N_2^2)N_2 - G_1(N_2^2)\big) > 0 \tag{17}$$

Let us study the system at the ϵ^1 order; equations (11) read:

$$\begin{aligned}
&\frac{\partial \varphi_1}{\partial \tau_1} + \frac{1}{2}\Big(\big(i + \zeta + 2i\sigma\big)\varphi_1 - i\varphi_2\Big) = -\frac{i}{2}f_0 \\
&\frac{\partial \varphi_2}{\partial \tau_1} + \frac{i}{2}\Bigg(2G_2\big(|\varphi_2|^2\big)m_1 + \big(\zeta + i(1+\sigma)\big)\varphi_1 + 2im_2\varphi_2^* \\
&+\Big(-iG_1\big(|\varphi_2|^2\big) + \lambda - m_0\lambda + i(-1+\sigma)\Big)\varphi_2 - m_2(2i + \lambda)\varphi_2^*\Bigg) = -\frac{i}{2}f_0
\end{aligned} \tag{18}$$

We would like to analyze the behavior of the structure at the slow time scale τ_1 around obtained invariant manifold at τ_0 time scale. By imposing equation (15) in the second equation of the system (18), assuming $\varphi_1 = N_1 e^{i\delta_1}$, $\varphi_2 = N_2 e^{i\delta_2}$ and taking into account general forms for m_0, m_1 and m_2 as

$$m_j = m_{jr} + im_{ji} \qquad j = 0, 1, 2 \tag{19}$$

the following reduced order forms of system equations can be obtained:

$$\begin{aligned}
\frac{\partial \delta_2}{\partial \tau_1} &= F_1(\delta_2, N_2) \\
\frac{\partial N_2}{\partial \tau_1} &= F_2(\delta_2, N_2)
\end{aligned} \tag{20}$$

where

$$N_2F_1(\delta_2, N_2) = G_1(N_2^2)N_2 + \frac{1}{2}N_2m_{0i}\lambda - \frac{1}{2}N_2\zeta\lambda - N_2\sigma + \frac{1}{2}G_1(N_2^2)N_2\sigma$$
$$-f_0\cos(\delta_2) - G_2(N_2^2)m_{1i}\cos(\delta_2) + G_2(N_2^2)m_{2r}N_2\cos(2\delta_2) + \frac{3}{2}N_2\lambda m_{2i}\cos(2\delta_2)$$
$$+G_2(N_2^2)m_{1r}\sin(\delta_2) + G_2(N_2^2)N_2m_{2i}\sin(2\delta_2) - \frac{3}{2}N_2\lambda m_{2r}\sin(2\delta_2) \tag{21}$$

$$F_2(\delta_2, N_2) = -\frac{1}{2}N_2\zeta + \frac{1}{2}G_1(N_2^2)N_2\zeta + \frac{1}{2}N_2\lambda m_{0r} + \frac{1}{2}N_2\lambda\sigma$$
$$-G_2(N_2^2)m_{1r}\cos(\delta_2) - G_2(N_2^2)N_2m_{2i}\cos(2\delta_2) + \frac{3}{2}N_2\lambda m_{2r}\cos(2\delta_2) - f_0\sin(\delta_2)$$
$$-G_2(N_2^2)m_{1i}\sin(\delta_2) + G_2(N_2^2)N_2m_{2r}\sin(2\delta_2) + \frac{3}{2}N_2\lambda m_{2i}\sin(2\delta_2) \tag{22}$$

Equilibrium points of the system can be obtained by setting $F_1 = F_2 = 0$. Stable areas of the invariant manifold of the system at τ_0 time scale which are given in Eq. (17) are characterized by two border lines, namely fold lines, which can be obtained by the following equation:

$$D = \lambda^2 + \left(1 - G_1(N_2^2)\right)\left(1 - H(N_2^2)N_2 - G_1(N_2^2)\right) = 0 \tag{23}$$

Ture Savadkoohi et al. [10] demonstrated that simultaneous intersections of these two lines with the $F_1 = F_2 = 0$ provide singularities of the system that are called fold singularities.

4 A Numerical Example

We choose a smooth and periodic profile of mass for the NES which reads

$$m(T) = m_{0r} + 2\left(m_{1r}\cos(\omega T) - m_{1i}\sin(\omega T)\right) + 2\left(m_{2r}\cos(2\omega T) - m_{2i}\sin(2\omega T)\right) \tag{24}$$

For our numerical example we assume that $m_{0r} = 1.57$, $m_{0i} = 0$, $m_{1r} = 0$, $m_{1i} = 0.5$, $m_{2r} = 0$, and $m_{2i} = 0.25$. Assumed profile of the mass of the NES is depicted in Fig. 1. Let us consider the following system parameters: $\epsilon = 10^{-3}$, $k = 1.5$, $\delta = 1$, $\lambda = 0.2$, $\zeta = 0.1$, $\sigma = 1$. Moreover, we suppose that the system is under external forcing term $f_0 = 1.176$. Dynamics of the system until infinity of the τ_1 time scale is depicted in Fig. 2.

We can see that the system possesses one fold singular point ($no.$ 1) and an equilibrium point ($no.$ 2). A zoomed area which covers these points is depicted at the same figure. Phase portraits of the system around mentioned points are depicted in Figs. 3 and 4. The singular point $no.$ 1 is saddle and unstable and the equilibrium point $no.$ 2 is unstable as well.

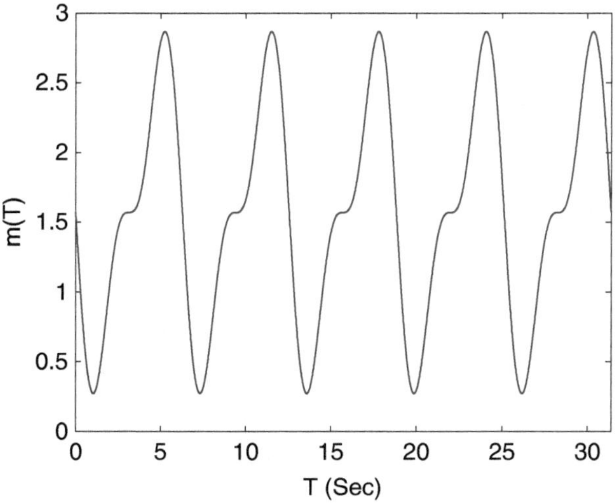

Fig. 1 The profile of periodically varying mass of the NES: $m_{0r} = 1.57$, $m_{1i} = 0.5$, $m_{1r} = 0$, $m_{2i} = 0.25$, and $m_{2r} = 0$

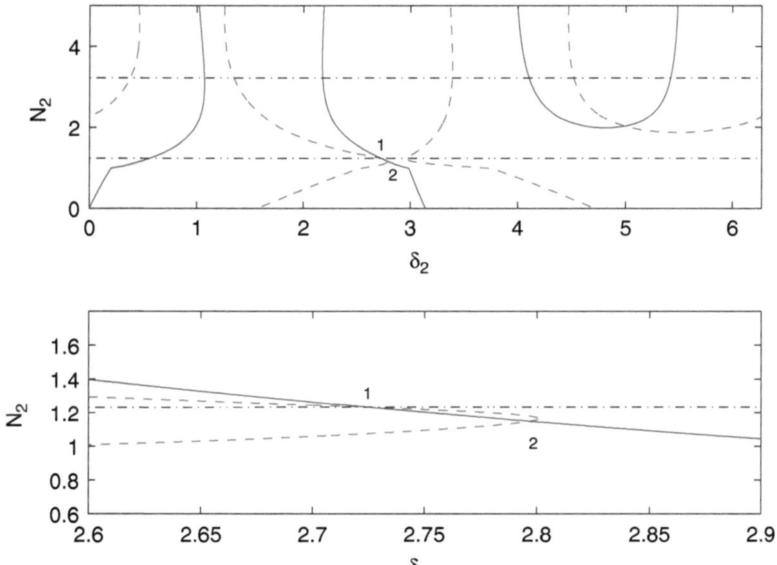

Fig. 2 Positions of equilibrium and fold singularities of the system under $f_0 = 1.176$: $F_1 = 0$ (— — —), $F_2 = 0$ (—), $D = 0$ or fold lines (—· —· —)

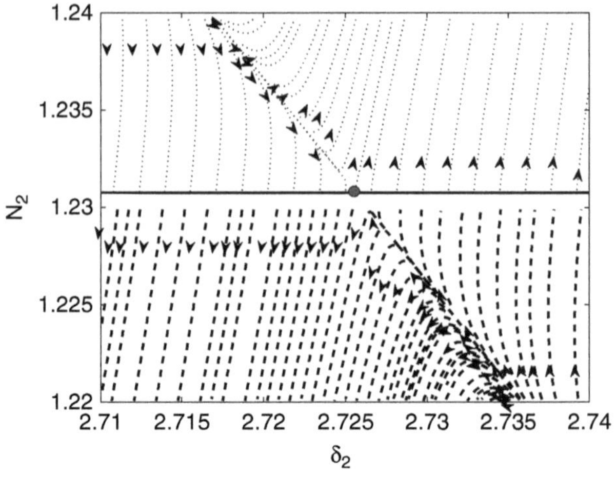

Fig. 3 Phase portraits of the system around fold singular point *no*. 1: $f_0 = 1.176, \sigma = 1$

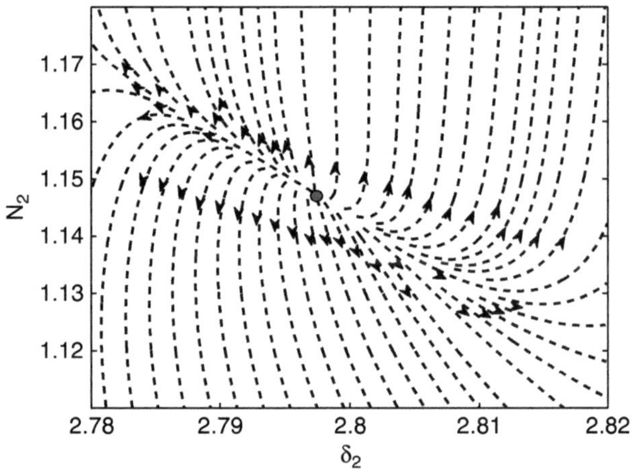

Fig. 4 Phase portraits of the system around fixed point *no*. 2: $f_0 = 1.176, \sigma = 1$

In order to have better envision about the behavior of the system, we present numerical results which are obtained by direct integration of the system (4) by Matlab package. Assumed initial conditions are $y_{10} = 1.5$ and $y_{20} = \dot{y}_{10} = \dot{y}_{20} = 0$. The invariant manifold of the system and its stability borders which are obtained from equations (16) and (17), respectively, are illustrated in Fig. 5. It can be seen that although the analytical invariant manifold is obtained based on keeping first harmonic and truncating higher ones, the averaged form of the real behavior of the system, which contains all harmonics, follows this invariant. When the system reaches to stability border of the invariant, it faces bifurcation to jump to other stable branch. Histories of the change of system amplitudes are depicted in Figs. 6 and 7.

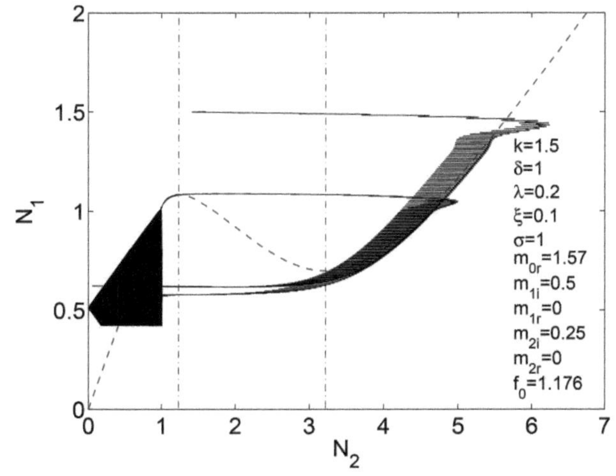

Fig. 5 Invariant manifold (– – –), stability borders (–· –· –), and numerical results (—). Starting values are $(N_2, N_1) = (1.5, 1.5)$: $f_0 = 1.176$, $\sigma = 1$

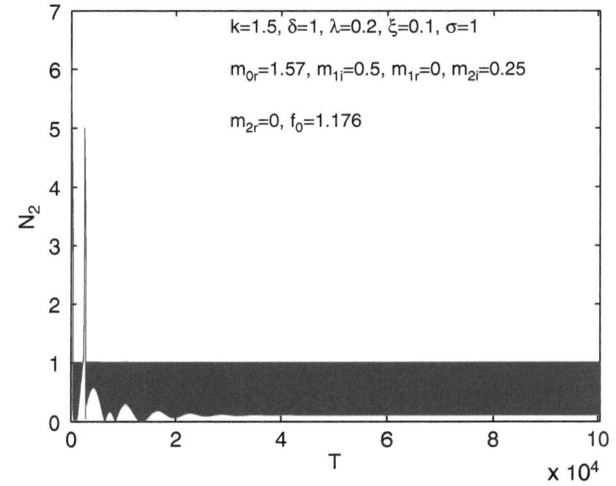

Fig. 6 History of N_2: $f_0 = 1.176$, $\sigma = 1$

These figures show that the system experiences one reverse bifurcation and finally is attracted by the equilibrium point *no*.2. The reason for this behavior is the existence of fold singular point *no*. 1 which is saddle (see Figs. 2 and 3) that makes the system aims to the first fold line to be ready to present a reverse bifurcation (see Figs. 5, 6, and 7). After one complete cycle of bifurcations that forces the system to present beating response, it finally aims to the fixed point *no*. 2 that is unstable during the very long time for τ_1 time scale, i.e., $\tau_1 = \frac{1}{\epsilon^2} = 10^6 Sec$. Although this equilibrium

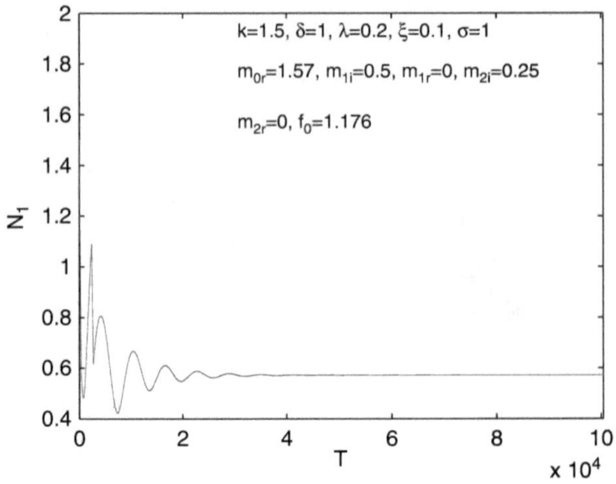

Fig. 7 History of N_1: $f_0 = 1.176, \sigma = 1$

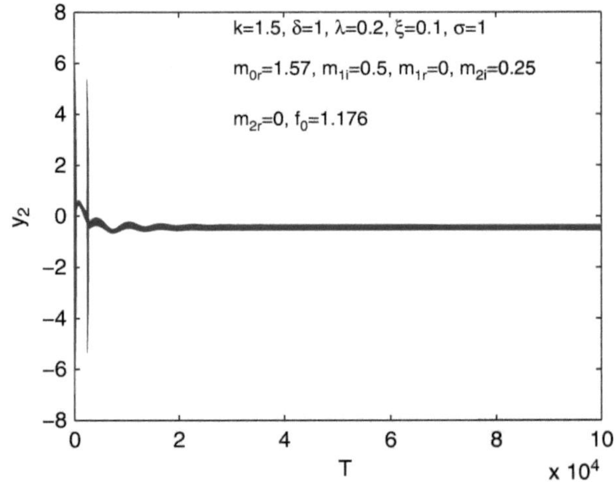

Fig. 8 History of y_2: $f_0 = 1.176, \sigma = 1$

point is unstable after a very long time span, but makes the system oscillates around it for quite long time. This oscillation corresponds to low energy levels for the main system that is interesting from passive control view point.

The displacement histories of two oscillators which are illustrated in Figs. 8 and 9 show that the main structure during the time scale τ_1 behaves in a controlled manner with very low displacement levels. Other behaviors can be observed for the system at other time scales, but we are not going to discuss about them and they are out of the scope of the current work.

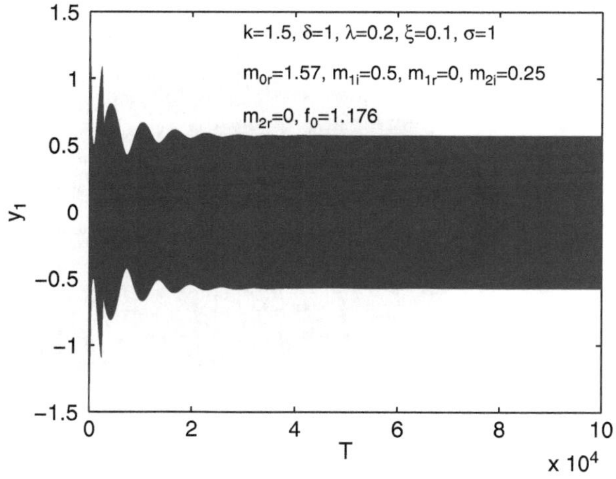

Fig. 9 History of y_1: $f_0 = 1.176, \sigma = 1$

5 Passive Control of Main Systems by NES with Time-Dependent Mass

In order to show the capability of the NES with varying mass in controlling main structures, we present a numerical example by showing the displacement histories of a main structure without and with coupled NES. We consider a worst case for the main system when it is in an exact 1 : 1 resonance with the external forcing terms. This means that we set $\sigma = 0$ so $\omega = 1$. The system in both cases is under external forcing term $f_0 = 1.176$ and assumed initial conditions and also mass profile for the NES are the same as the described example in Sec. 4. The results of the direct integration of system equations for two cases are depicted in Fig. 10. This figure shows that while the single dof main structure (without coupled NES) is presenting a kind of diverging behavior at the exact 1 : 1 resonance, by coupling a NES with varying mass we are able to control the same structure. Its displacement is reduced in a considerable amount and it presents permeant strongly modulated response (SMR) (at least until the maximum time of integration) [5, 12]. This SMR is more distinct by looking at the histories of amplitudes of the main system and the NES which are illustrated in Figs. 11 and 12 showing a beating response of two oscillators due to the SMR.

6 Conclusions

We studied the energy exchange problem between a linear system and a non-smooth energy sink whose mass is time-dependent. Detected invariant manifold of the system at slow scale and its stable zones let us predict the overall surfaces of all

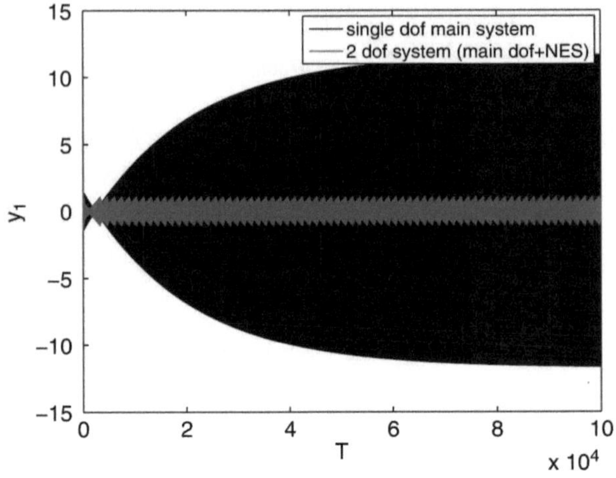

Fig. 10 Histories of y_1 for two cases: the bare structure without coupled NES and 2-dof system. Parameters of the main system are $\xi = 0.1$, $\sigma = 0$, $f_0 = 1.176$. Parameters of the coupled NES are $k = 1.5$, $\delta = 1$ and $\lambda = 0.2$, $m_{0r} = 1.57$, $m_{0i} = 0$, $m_{1r} = 0$, $m_{1i} = 0.5$, $m_{2r} = 0$, and $m_{2i} = 0.25$

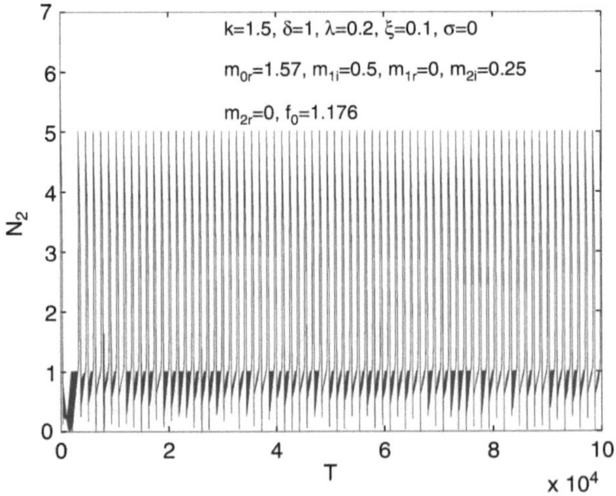

Fig. 11 History of N_2 and beating response of the NES due to the SMR: $f_0 = 1.176$, $\sigma = 0$

fixed and/or pseudo-fixed points of the system by explanations of global bifurcation due to the existence of unstable zone. Enlightened system behavior at first fast time scale let us take further steps for detecting equilibrium points and fold singularities which can explain the behavior of the system during closed-loop bifurcations and/or its attractions to fixed points. By demonstrating numerical examples we showed the capability of the non-smooth nonlinear energy sink in passively controlling a main

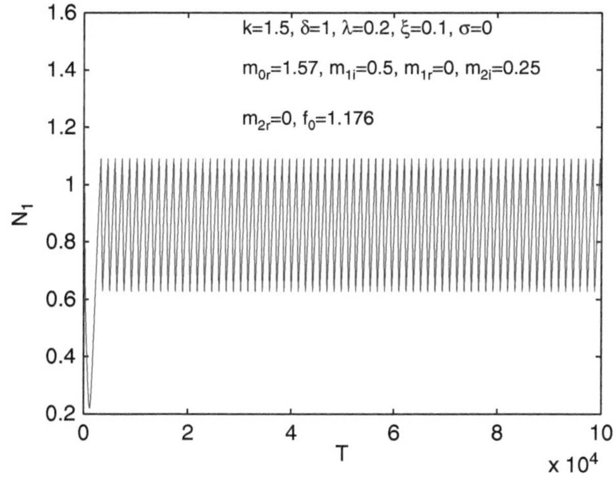

Fig. 12 History of N_1 and beating response of the main oscillator due to the SMR: $f_0 = 1.176$, $\sigma = 0$

structure during a worst case, i.e., exact 1 : 1 resonance. Further studies should be carried out for designing "optimized" nonlinear energy sink devices with varying mass for the aim of passive control of civil/mechanical structures.

Acknowledgements The authors would like to thank the following organization for supporting this research work: LABEX CELYA (ANR-10-LABX-0060) of the "Université de Lyon" within the program "Investissement d'Avenir" (ANR-11-IDEX-0007) operated by the French National Research Agency (ANR).

References

1. Gendelman, O.V.: Analytic treatment of a system with a vibro-impact nonlinear energy sink. J. Sound Vib. **331**, 4599–4608 (2012)
2. Gendelman, O.V.: Targeted energy transfer in systems with non-polynomial nonlinearity. J. Sound Vib. **315**, 732–745 (2008)
3. Gourdon, E., Alexander, N.A., Taylor, C.A., Lamarque, C.-H., Pernot, S.: (2007) Nonlinear energy pumping under transient forcing with strongly nonlinear coupling: theoretical and experimental results. J. Sound Vib. **300**, 522–551 (2007)
4. Kerschen, G., Kowtko, J.J., McFarland, D.M., Lee, Y.S., Bergman, L.A., Vakakis, A.F.: Experimental demonstration of transient resonance capture in a system of two coupled oscillators with essential stiffness nonlinearity. J. Sound Vib. **299**, 822–838 (2007)
5. Lamarque, C.-H., Gendelman, O.V., Ture Savadkoohi, A., Etcheverria, E.: Targeted energy transfer in mechanical systems by means of non-smooth nonlinear energy sink. Acta Mech. **221**, 175–200 (2011)
6. Lamarque, C.-H., Ture Savadkoohi, A., Etcheverria, E., Dimitrijevic, Z.: Multi-scales dynamics of two coupled non-smooth systems. Int J. Bifurcat Chaos. **22**, 1250295:1–18 (2012)

7. Lamarque, C.-H., Ture Savadkoohi, A., Dimitrijevic, Z.: Dynamics of a linear system with time-dependant mass and a coupled light mass with non-smooth potential. Meccanica. (2013) doi: 10.1007/s11012-013-9778-8
8. Manevitch, L.I.: The description of localized normal modes in a chain of nonlinear coupled oscillators using complex variables. Nonlinear Dynam. **25**, 95–109 (2001)
9. Nayfeh, A.H., Mook, D.T.: Nonlinear Oscillations. John Wiley and Sons, New York (1979)
10. Nucera, F., Vakakis, A.F., McFarland, D.M., Bergman, L.A., Kerschen, G.: Targeted energy transfers in vibro-impact oscillators for seismic mitigation. Nonlinear Dynam. **50**, 651–677 (2007)
11. Schmidt, F., Lamarque, C.-H.: Energy pumping for mechanical systems involving non-smooth Saint-Venant terms. Int J. Nonlinear Mech. **45**, 866–875 (2010)
12. Starosvetsky, Y., Gendelman, O.V.: Strongly modulated response in forced 2DOF oscillatory system with essential mass and potential asymmetry. Physica D. **237**, 1719–1733 (2008)
13. Ture Savadkoohi, A., Vaurigaud, B., Lamarque, C.-H., Pernot, S.: Targeted energy transfer with parallel nonlinear energy sinks, part II: theory and experiments. Nonlinear Dynam. **67**, 37–46 (2012a)
14. Ture Savadkoohi, A., Lamarque, C.-H., Dimitrijevic, Z.: Vibratory energy exchange between a linear and a non-smooth system in the presence of the gravity. Nonlinear Dynam. **70**, 1473–1483 (2012b)
15. Ture Savadkoohi, A., Lamarque, C.-H.: Dynamics of coupled Dahl type and non-smooth systems at different scales of time. Int J. Bifurcat Chaos. **23**, 1350114: 1–14 (2013)
16. Vakakis, A.F., Gendelman, O.V., Bergman, L.A., McFarland, D.M., Kerschen, G., Lee, Y.S.: Nonlinear trageted energy transfer in mechanical and structural systems I. Springer, Berlin (2008a)
17. Vakakis, A.F., Gendelman, O.V., Bergman, L.A., McFarland, D.M., Kerschen, G., Lee, Y.S.: Nonlinear Trageted Energy Transfer in Mechanical and Structural Systems II. Springer, Berlin (2008b)
18. Van Horssen, W.T., Pischanskyy, O.V.: On the stability properties of a damped oscillator with a periodically time-varying mass. J. Sound Vib. **330**, 3257–3269 (2011)

Coordinate-Free Formulation of Nonholonomic Constraints for Wheeled Robots

Sanjuan Szklarz Paweł Cesar and Elżbieta Jarzębowska

Abstract A coordinate-free formulation of nonholonomic constraint equations for wheeled vehicles is presented. The no-slip condition for the contact point of the wheel with the ground is translated to a pure geometric formula. Based on this general geometric interpretation, explicit formulas for nonholonomic constraints are derived. The final calculations of the constraint equations are realized in Maxima, a computer algebra system. The general no-slip condition formulation is validated via the direct derivation of nonholonomic constraint equations for well-known robotic systems. The presented approach provides a systematic procedure to calculate the nonholonomic constraint equations for wheel components. The final equations are suitable for symbolic calculations and computer aimed modeling of mechanical systems.

1 Introduction

This paper presents a systematic coordinate-free approach to formulate the no-slip condition for wheeled robot models. The no-slip condition ensures that the contact point between a wheel and a ground has zero velocity. To calculate the velocity of the contact point in a coordinate-free fashion, the wheel displacement and rotations are defined as reference frame isomorphisms over time. This provides constraint equations that can be calculated in any chosen coordinate system in the state manifold.

Based on this general geometric interpretation, explicit formulas for nonholonomic constraints are derived. The final calculations of the constraint equations are realized in Maxima, a computer algebra system. The general no-slip condition formulation is validated via the direct derivation of nonholonomic constraint equations for well-known robotic systems. Presented calculations are symbolically realized in the computer algebra system Maxima [9]. LaTeX formulas are automatically generated.

S.S.P. Cesar (✉) • E. Jarzębowska
Warsaw University of Technology, Institute of Aeronautics and Applied Mechanics, Warsaw, Poland
e-mail: pszklarz@meil.pw.edu.pl; elajarz@meil.pw.edu.pl

J. Awrejcewicz (ed.), *Applied Non-Linear Dynamical Systems*, Springer Proceedings in Mathematics & Statistics 93, DOI 10.1007/978-3-319-08266-0_33

The novelty of the approach relies in the use of the symmetric properties of the wheel geometry. The final nonholonomic equations which are obtained are equivalent to the classical well-known equations [1, 2, 4–8]. However, they provide a considerable improvement in symbolical and numerical calculations for complex system models.

Nonholonomic constraint equations are the basis for the derivation of kinematic control models for wheeled robots and car-like vehicles. Most of these constraints are related to the no-slip condition of the contact point between wheels and the ground. The constraint equations are formulated using selected reference frames and coordinate systems. For multi-wheeled robots or wheeled vehicles of complex designs, the formulation and validation of the resulting constraint equations are not automated and it leads to many mistakes, and it is difficult to verify and debug the final model formulation; additionally the transformations between a global and a wheel reference frames are not shown explicitly; see, e.g., [1, 2, 4–8]. In the case of linearization of the nonholonomic constraints, they require transformations [10]. Also, there are vehicle models derived automatically by computer codes; see, e.g., [11] and references therein. In [11] the code AutoSim was used to generate a variety of linear and nonlinear bicycle models in symbolic forms. Yet, they are not standard in the kinematic model derivations. The validation of models like in [11] requires the verification of the mapping between handmade formulas and computer model descriptions. In the absence of systematic tools for developing the constraint equations, there exist many models derived under special assumptions that make the constraint formulation viable. This approach, in turn, leads to models of the same mechanical systems that are difficult to compare and adopt for other applications.

A coordinate-free formulation of nonholonomic constraints presented in the paper does not require any special assumptions about a mechanical system model and the resulting constraint equations are easily verifiable and repeatable by other users.

2 Notation

It is assumed that the center of the wheel coincides with the origin of the wheel reference frame. The constant r is the wheel radius. Let k be the global reference frame and K the wheel reference frame. The wheel reference frame motion is defined in terms of a translation:

$$T : k \to k, \quad T(a) := a + \varrho \quad \varrho \in \mathbb{R}^3 \tag{1}$$

and a rotation $B \in SO(3)$—the 3D rotation group:

$$B : K \to k. \tag{2}$$

Rotation transformations are defined using the base

$$\Omega_x(a) := \begin{pmatrix} 1 & 0 & 0 \\ 0 & \cos a & -\sin a \\ 0 & \sin a & \cos a \end{pmatrix}, \Omega_y(a) := \begin{pmatrix} \cos a & 0 & -\sin a \\ 0 & 1 & 0 \\ \sin a & 0 & \cos a \end{pmatrix}, \Omega_z(a) := \begin{pmatrix} \cos a & -\sin a & 0 \\ \sin a & \cos a & 0 \\ 0 & 0 & 1 \end{pmatrix}.$$

The translation transformation T is uniquely defined by the displacement vector $\varrho \in \mathbb{R}^3$. We use this relation to identify the space of translations in $\mathbb{R}^3$.

Vectors in the global reference frame k are denoted by lower case letters and the corresponding vectors in the wheel reference frame K are denoted by upper case letters. This means that for $A \in K$ the corresponding $a \in k$ is

$$a = B(A).$$

Displacement $D = T \circ B$ is defined as

$$D : K \to k, \quad D(\cdot) := T(B(\cdot)). \tag{3}$$

For a mechanical system equipped with a wheel, let a state manifold be defined by Φ and a state variable $q \in \Phi$. The wheel reference frame location will be specified with respect to the system state, i.e.,

$$B : \Phi \to SO(3), \quad T : \Phi \to \mathbb{R}^3. \tag{4}$$

For velocity calculation the following notations are used:

$$\dot{q} := \frac{d}{dt}q,$$

$$\dot{B}(q, \dot{q}) := \frac{d}{dt}B(q) = \frac{\partial}{\partial q}B(q)\frac{d}{dt}q,$$

$$\dot{\varrho}(q, \dot{q}) := \frac{d}{dt}\varrho(q) = \frac{\partial}{\partial q}\varrho(q)\frac{d}{dt}q.$$

3 Geometric Formulation of the No-Slip Condition

The no-slip condition states that a contact point between a wheel and a ground has zero velocity. In this section a detailed formulation of the no-slip condition is presented. It is based upon the geometric transformation between the ground and the wheel reference frames.

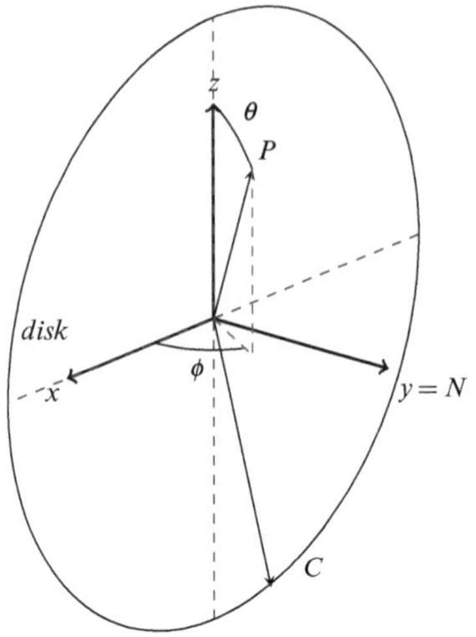

3D projection of the wheel reference frame.

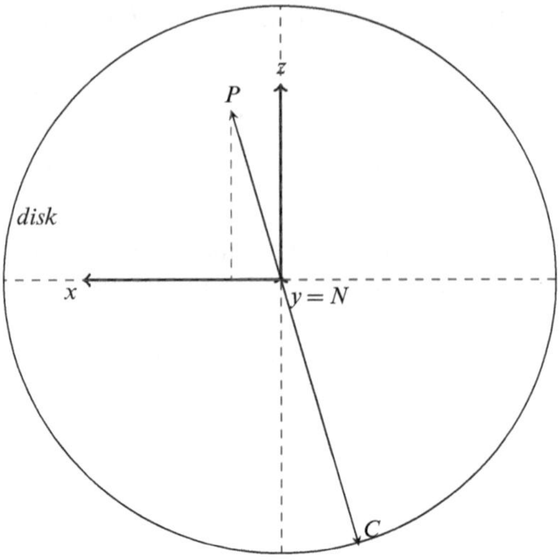

2D projection of the wheel reference frame to the plane xz.

Fig. 1 The ground normal vector P and the wheel normal vector N in wheel reference frame

3.1 Contact Point Location

Define a contact point between the wheel and the ground as the lowest located point with respect to the ground normal direction.

The contact point definition uniquely identifies a point on the wheel, and it is independent from the fact that the wheel center location must match specific conditions to ensure that the contact point in global coordinates is located in the ground plane. Thanks to this property, the contact point for any system state q is well defined and depends only on the rotation $B(q)$.

Following the upper/lower case letter convention described in Sect. 2, define vectors (see Fig. 1):

- c, C is the vector pointing to the contact point position from the wheel center.
- p, P is the normal vector to the ground plane.
- n, N is the normal vector to the wheel plane.

Figure 1 shows the wheel normal vector N and the ground normal vector P in the wheel reference frame. The following properties uniquely identify the contact point vector C:

- It is located in the wheel plane: $C \cdot N = 0$.
- It is aligned with N, P: $C \in span\{P, N\}$.
- It is directed to the ground: $C \cdot P < 0$.
- It is located on the wheel circumference: $\|C\| = r$.

There is a singular case when the wheel lies on the floor, $P = N$. The presented equations are not well defined in this case. In practice, such positions are strongly unfortunate and should be avoided. This is why we do not discuss them.

The first approach to find C may be to project P on the place $\{N\}^{\perp}$ and normalize it using the wheel radius r

$$C = -r\frac{P - (P \cdot N)N}{\|P - (P \cdot N)N\|}.$$

This solution is correct, but the term $\|P - (P \cdot N)N\|$ does not behave well during the equation simplification.

The best approach seems to be the use of the following equation:

$$C = rN \times (\frac{N \times P}{\|N \times P\|}). \tag{5}$$

Equation (5) is a direct consequence of the contact point geometric properties. The intermedium vector $X := \frac{N \times P}{\|N \times P\|}$ is a unitary vector perpendicular to N and P, located in the wheel plane. The contact point is then perpendicular to N and X simultaneously. As N and X are unitary and perpendicular, we obtain equation (5).

Equation (5) exhibits better behavior during its symbolic simplifications. It is due to the fact that in the ground reference frame the vector $p = u_z := (0, 0, 1)^T$ and, in the wheel reference frame, the vector $N = u_y := (0, 1, 0)^T$. Then, the term $\|N \times P\|$ is much simpler than $\|P - (P \cdot N)N\|$. Note that vectors P and N in the wheel reference frame are given by $P = B^{-1}u_z$ and $N = u_y$. Then, equation (5) expands to

$$C = ru_y \times (\frac{u_y \times B^{-1}u_z}{\|u_y \times B^{-1}u_z\|}). \tag{6}$$

Finally, note that for any vector $v := (v_x, v_y, v_z)^T$ the calculation

$$u_y \times v = (v_z, 0, -v_x)^T$$

is just a simple permutation and a sign change. This makes equation (6) easy to use.

Note that equation (6) is independent of any state manifold coordinate selection. It is then the general formulation of the vector C pointing to the contact point position.

3.2 *Velocity of Wheel Fixed Points*

In a given position of a mechanical system, the contact point is a fixed point in the wheel reference frame. It means that the contact point position depends only on the wheel position and not on velocity.

Location of the contact point in the global reference frame is given by

$$p(q) := D(C(q)) = B(q)C(q) + \varrho(q). \tag{7}$$

Then, the velocity in the global frame is

$$\dot{p}(q,\dot{q}) = \dot{\varrho}(q,\dot{q}) + \dot{B}(q,\dot{q})C(q), \tag{8}$$

where the term $B(q)\dot{C}(q,\dot{q})$ is not present as the vector C points to the instantaneous contact point between the wheel and the ground.

Expressing the contact point position in terms of the global reference frame $C = B^{-1}(c)$, the contact point velocity vector can be written as

$$\dot{p}(q,\dot{q}) = \dot{\varrho}(q,\dot{q}) + \dot{B}(q,\dot{q})B^{-1}(c(q)). \tag{9}$$

This leads to the well-known formula

$$\dot{p}(q,\dot{q}) = v_o(q,\dot{q}) + \omega(q,\dot{q}) \times c(q), \tag{10}$$

where $v_o = \dot{\varrho}$ is the linear velocity of the wheel reference frame, operator $\omega(q,\dot{q}) \times (\cdot) := \dot{B}(q,\dot{q})B^{-1}(\cdot)$ is the angular velocity of the wheel reference frame, and $c(q) = B(q)C(q)$ is the position of the contact point with respect to the global reference frame orientation.

Then, the formula (10) and the no-slip condition lead to the nonholonomic constraint formulation as

$$0 = v_o(q,\dot{q}) + \omega(q,\dot{q}) \times c(q). \tag{11}$$

Equation (11) is the desired nonholonomic constraint equation. Note that it is not related to any coordinate selection for the manifold Φ.

3.3 *Final Formulation of the No-Slip Condition*

Using the constraint equation (11) and the contact point location equation (6), we can obtain a general formula for nonholonomic constraint equations for wheeled vehicle models. It is of the form

$$0 = v_o(q,\dot{q}) + \dot{B}(q,\dot{q})B^{-1}(q)\left(rB(q)(u_y \times (\frac{u_y \times B^{-1}(q)u_z}{\|u_y \times B^{-1}(q)u_z\|}))\right). \tag{12}$$

It can be simplified to

$$0 = v_o(q, \dot{q}) + r\dot{B}(q, \dot{q})(u_y \times (\frac{u_y \times B^{-1}(q)u_z}{\|u_y \times B^{-1}(q)u_z\|})). \tag{13}$$

Since $B(q) \in SO(3)$, the calculation of $B^{-1}(q)$ is straightforward; in the matrix representation $B^{-1}(q) = B^T(q)$.

Equation (13) is expressed in terms of the rotation $B(q) \in SO(3)$. It is independent of any coordinate selections, and it does not need any additional assumptions about the model adopted for a mechanical system. Equation (13) presents the no-slip condition and it is the nonholonomic constraint for the wheel.

4 Examples—No-Slip Condition Validation for Robotic Models

Using equation (13) nonholonomic constraints can be calculated for models of well-known wheeled vehicles. The vertical and falling disk constraints calculated using (13) match the equations in [3]. Note that to match the signs in the final constraint equations as in [3], counterclockwise rotations have been adopted. Models presented in [3] do not define the global to the wheel reference frame transformation explicitly.

4.1 Vertical Rolling Disk

Following [3], the configuration space for the vertical rolling disk is $Q = \mathbb{R}^2 \times S^1 \times S^1$ and it is parametrized by the coordinates $q = (x, y, \theta, \phi)$, where:

- x, y is the projection of the center of the wheel in the xy-plane,
- θ is the roll angle of the disk, and
- ϕ is the heading angle of the disk.

The ground to the wheel transformation of reference frames is given by

$$B_v(q) = \Omega_z(\varphi) \cdot \Omega_y(-\theta).$$

Then

$$B_v(q)^{-1}u_z = \begin{pmatrix} -\sin\theta \\ 0 \\ \cos\theta \end{pmatrix}$$

and

$$u_y \times B_v^{-1}(q)u_z = \begin{pmatrix} \cos\theta \\ 0 \\ \sin\theta \end{pmatrix}.$$

Since u_y and $B_v(q)u_z$ are perpendicular, then

$$\|u_y \times B_v(q)u_z\| = 1.$$

Then, we obtain

$$C_v = r u_y \times \left(\frac{u_y \times B_v(q)u_z}{\|u_y \times B_v(q)u_z\|}\right) = r\begin{pmatrix} \sin\theta \\ 0 \\ -\cos\theta \end{pmatrix}. \tag{14}$$

This is correct because in the global coordinates the contact point is always below the center of the wheel and

$$c_v = B_v C_v = B_v \begin{pmatrix} \sin\theta \\ 0 \\ -\cos\theta \end{pmatrix} = \begin{pmatrix} 0 \\ 0 \\ -1 \end{pmatrix}.$$

Nonholonomic constraints (13) for the vertical rolling disk are

$$0 = v_v + \omega_v \times c_v = v_v + \dot{B}_v(q,\dot{q})(C_v) = \begin{pmatrix} \dot{x} \\ \dot{y} \\ \dot{z} \end{pmatrix} + r\begin{pmatrix} -\cos\varphi\,\dot{\theta} \\ -\sin\varphi\,\dot{\theta} \\ 0 \end{pmatrix}. \tag{15}$$

4.2 *The Falling Rolling Disk*

Following [3], the configuration space for the falling rolling disk is $Q = \mathbb{R}^3 \times S^1 \times S^1 \times S^1$, and it is parametrized by the coordinates $q = (x, y, z, \theta, \varphi, \psi)$ where:

- x, y, z are the position coordinates of the center of the disk in the global reference frame,
- θ is the angle between the plane of the disk and the vertical axis,
- φ is the heading angle, and
- ψ is the roll angle.

In [3], the coordinates are that of the contact point. Later we will change the coordinate system to check that the constraints calculated in this paper match the previous result.

The ground to the wheel transformation of reference frames is given by

$$B_f(q) = \Omega_z(\varphi) \cdot \Omega_x(\theta) \cdot \Omega_y(-\psi).$$

Then

$$B_f(q)^{-1} u_z = \begin{pmatrix} -\sin\psi\ \cos\theta \\ \sin\theta \\ \cos\psi\ \cos\theta \end{pmatrix}$$

and

$$u_y \times B_f^{-1}(q) u_z = \begin{pmatrix} \cos\psi\ \cos\theta \\ 0 \\ \sin\psi\ \cos\theta \end{pmatrix}.$$

Note that

$$\|u_y \times B_f^{-1}(q) u_z\| = abs(\cos\theta).$$

Since $\theta \in (-\frac{\pi}{2}, \frac{\pi}{2})$

$$\|u_y \times B_f^{-1}(q) u_z\| = \cos\theta.$$

Then, we obtain that

$$C_f = r u_y \times \left(\frac{u_y \times B_f(q) u_z}{\|u_y \times B_f(q) u_z\|}\right) = r \begin{pmatrix} \sin\psi \\ 0 \\ -\cos\psi \end{pmatrix}.$$

Global coordinates of the contact point are

$$rB_f \begin{pmatrix} \sin\psi \\ 0 \\ -\cos\psi \end{pmatrix} = r \begin{pmatrix} -\sin\varphi\ \sin\theta \\ \cos\varphi\ \sin\theta \\ -\cos\theta \end{pmatrix}. \tag{16}$$

Nonholonomic constraints (13) for the falling rolling disk are

$$0 = v_f + \omega_f \times c_f = v_f + \dot{B}_f(q,\dot{q})(C_f) = \begin{pmatrix} \dot{x} \\ \dot{y} \\ \dot{z} \end{pmatrix} +$$

$$r \begin{pmatrix} -\sin\varphi\ \cos\theta\ \dot{\theta} - \cos\varphi\ \dot{\varphi}\ \sin\theta - \dot{\psi}\ \cos\varphi \\ \cos\varphi\ \cos\theta\ \dot{\theta} - \sin\varphi\ \dot{\varphi}\ \sin\theta - \dot{\psi}\ \sin\varphi \\ \sin\theta\ \dot{\theta} \end{pmatrix}. \tag{17}$$

Equation (17) is not as simple as expected. This is because the constraints are calculated assuming that (x, y, z) are the coordinates of the center of the wheel. Assume now that the coordinates of the contact point are used as provided by (16). They yield

$$\begin{pmatrix} x' \\ y' \\ z' \end{pmatrix} = \begin{pmatrix} x \\ y \\ z \end{pmatrix} + r \begin{pmatrix} -\sin\varphi \sin\theta \\ \cos\varphi \sin\theta \\ -\cos\theta \end{pmatrix}, \tag{18}$$

then

$$\begin{pmatrix} \dot{x}' \\ \dot{y}' \\ \dot{z}' \end{pmatrix} = \begin{pmatrix} \dot{x} \\ \dot{y} \\ \dot{z} \end{pmatrix} + r \begin{pmatrix} -\sin\varphi \cos\theta\, \dot{\theta} - \cos\varphi\, \dot{\varphi} \sin\theta \\ \cos\varphi \cos\theta\, \dot{\theta} - \sin\varphi\, \dot{\varphi} \sin\theta \\ \sin\theta\, \dot{\theta} \end{pmatrix} \tag{19}$$

and equation (17) can be then rewritten as

$$\begin{pmatrix} \dot{x}' \\ \dot{y}' \\ \dot{z}' \end{pmatrix} + r \begin{pmatrix} -\dot{\psi} \cos\varphi \\ -\dot{\psi} \sin\varphi \\ 0 \end{pmatrix} = 0. \tag{20}$$

4.3 Tricycle Robot

A tricycle robot is presented in Fig. 2. The geometry of the tricycle is used to define the state manifold and to illustrate the use of generalized coordinates. Note that calculations of the nonholonomic constraints are realized based on equation (13), i.e., using $T(q)$ and $B(q)$ only. Any change of a coordinate system will only affect the translation $T(q)$ and rotation $B(q)$ mappings. The final calculation of the nonholonomic constraints is automatic so that it can be performed symbolically by any computer algebra system; in this paper Maxima [9] has been used.

The configuration space is parametrized by the coordinates $q = [x, y, \theta, \beta, \phi_l, \phi_r, \phi_f]^T$:

- x, y—position of the rare axis center,
- θ—heading angle of the tricycle body,
- β—heading angle of the front wheel,
- ϕ_l—left wheel roll angle,
- ϕ_r—right wheel roll angle, and
- ϕ_f—front wheel roll angle.

The ground to wheel transformations are denoted by

- T_l, B_l—left wheel translation and rotation,
- T_r, B_r—right wheel translation and rotation, and
- T_f, B_f—front wheel translation and rotation.

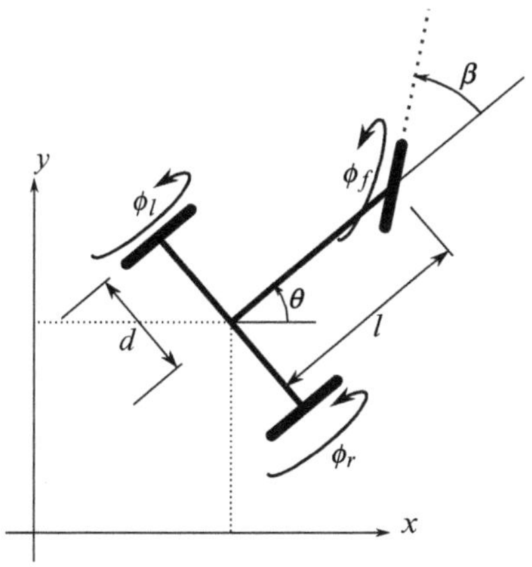

Fig. 2 Tricycle model

Rotation transformations are as

$$\begin{aligned}B_l(q) &= \Omega_z(\theta)\cdot\Omega_y(\phi_l)\,,\\ B_r(q) &= \Omega_z(\theta)\cdot\Omega_y(\phi_r)\,,\\ B_f(q) &= \Omega_z(\theta+\beta)\cdot\Omega_y\left(\phi_f\right),\end{aligned}$$

and translation transformations as

$$\begin{aligned}T_l(q) &= (x,y,0)^T + \Omega_z(\theta)(0,d,0)^T,\\ T_r(q) &= (x,y,0)^T + \Omega_z(\theta)(0,-d,0)^T,\\ T_f(q) &= (x,y,0)^T + \Omega_z(\theta)(l,0,0)^T.\end{aligned}$$

The final result for the nonholonomic constraints is presented. These formulas are calculated symbolically in Maxima, and the LaTeX equations are generated automatically. The z component is omitted, as it is just $0 = 0$.

Nonholonomic constraints (13) for the front wheel are

$$\begin{pmatrix}\dot{x} - l\,\sin\theta\,\dot{\theta} + \dot{\phi}_f\,\cos(\theta+\beta)\\ \dot{y} + l\,\cos\theta\,\dot{\theta} + \dot{\phi}_f\,\sin(\theta+\beta)\end{pmatrix} = \begin{pmatrix}0\\0\end{pmatrix}. \tag{21}$$

Nonholonomic constraints (13) for the left wheel are

$$\begin{pmatrix}\dot{x} - d\,\cos\theta\,\dot{\theta} + \dot{\phi}_l\,\cos\theta\\ \dot{y} - d\,\sin\theta\,\dot{\theta} + \dot{\phi}_l\,\sin\theta\end{pmatrix} = \begin{pmatrix}0\\0\end{pmatrix}. \tag{22}$$

Nonholonomic constraints (13) for the right wheel are

$$\begin{pmatrix} \dot{x} + d\ \cos\theta\,\dot{\theta} + \dot{\phi}_r\ \cos\theta \\ \dot{y} + d\ \sin\theta\,\dot{\theta} + \dot{\phi}_r\ \sin\theta \end{pmatrix} = \begin{pmatrix} 0 \\ 0 \end{pmatrix}. \tag{23}$$

The systematic approach to nonholonomic constraint equation derivation presented herein can be applied to any wheeled robotic system like the one presented in [7, 8]. This approach allows a straightforward derivation of kinematic control models for them.

5 Conclusions

The paper presents a systematic coordinate-free approach for the formulation of the no-slip condition for wheeled robot models. Based upon this one formulation, nonholonomic constraint equations can be derived for many wheeled robotic system models in a unified, verifiable, comparable, and repeatable way. The constraint equations, which are the basis for the derivation of kinematic control models for the robotic systems, help in their derivation in a more systematic and simpler way.

Acknowledgements This work was supported by the Polish National Science Center (NCN) under the Grant No.: 2011/01/N/ST8/01894.

References

1. Awrejcewicz, J.: Classical Mechanics: Dynamics. Advances in Mechanics and Mathematics. Springer, NewYork (2012)
2. Basu-Mandal, P., Chatterjee, A., J.M. Papadopoulos, J.M.:Hands-free circular motions of a benchmark bicycl. Proc. Roy. Soc. A: Math. Phys. Eng. Sci. **463**(2084), 1983–2003 (2007)
3. Bloch, A.M.: Nonholonomic Mechanics and Control, vol. 24. Springer, NewYork (2003)
4. Bonilla, M., Reyes, F., Mendoza, M: Modelling and simulation of a wheeled mobile robot in configuration classical tricycle. In: Proceedings of 5th WSEASA International Conference on Instrumentation, Measurement, Control, Circuits and Systems. (2005)
5. Boussinesq, J.: Aperçu sur la théorie de la bicyclette. J. de Mathématiques Pures et Appliquées. 117–136 (1899)
6. Franke, G., Suhr, W., Rieß, F.: An advanced model of bicycle dynamics. In: Eur. J. Phys. **11**)(2), 116 (2000)
7. Kumar, U., Sukavanam, N.: Dynamic modeling and tracking control of a four wheeled nonholonomic mobile robot. In: Proc. 13th national conference on mechanics and machines (NaCoMM07). 127–134. (2007),
8. Limebeer, D., Robin, S.: Sharp. Bicycles, motorcycles, and models. In: Contr. Syst. IEEE **26** (5), 34–61 (2006)
9. Maxima. A computer algebra system. version 5.31.0 (2012)http://maxima.sourceforge.net/

10. Meijaard, J.P., Papadopoulos, J.M., Ruina, A., Schwab, A.L.: Linearized dynamics equations for the balance and steer of a bicycle: a benchmark and review. In: proc. Roy. Soc. A: Math. Phy. E. Sci. **463**(2084), 1955–1982 (2007)
11. Schwab, A.L., Meijaard, J.P., Papadopoulos, J.M.: Benchmark results on the linearized equations of motion of an uncontrolled bicycle. In: J. Mech. Sci. Techno. **19** 292–304 (2005)
12. Tsai, C., Meng-Bi C., Lin, S.: Dynamic modeling and tracking control of a nonholonomic wheeled mobile manipulator with dual arms. In: J. Intell. Robot. Syst. **47**(4) 317–340 (2006)

Optomechatronic Choppers with Rotating Elements: Design Programs

O. Cira and V.-F. Duma

Abstract Choppers are most used devices for the modulation of light. We have developed a unitary theory of the optical choppers with rotating wheels working with top-hat laser beams (with constant intensity distribution). A novel type of chopper has been introduced, with windows with non-linear edges. The different modulation functions (of the transmitted light flux) were obtained. From this analysis, design programs are developed and presented in this study to tackle with the diverse cases represented by the chopper configurations and by their constructive and functional parameters. Numerical simulations and modeling are carried out based on the mathematical theory we have previously developed. Rules of thumb result to choose the most appropriate type of device and its characteristic parameters with regard to the particular requirements of a specific application. Top-hat (with constant intensity), Gaussian, and Bessel laser beams are considered in this study.

Keywords Optomechatronics • Optical devices • Choppers • Laser beams • Top-hat beams • Gaussian beams • Modulation functions • Analysis Optomechatronics • Design programs

1 Introduction

Optical choppers [1] are devices utilized for a variety of purposes in optical systems, including for the generation of light impulses, the elimination or selection of spectral domains (or of certain wavelengths), or the controlled attenuation of light. The optical setups in which they are used include thermometers and pyrometers [10], photometers and colorimeters [1], lasers [3], telescopes or lidars [11], manufacturing [16], as well as spectral [9, 12] or biomedical systems [13, 14].

Numerous studies, including early ones, were dedicated to the development of both macroscopic and micro-choppers—the latter built as MEMS (micro-electrical-mechanical systems) [15]. Nowadays they are considered an established type of

O. Cira • V.-F. Duma (✉)
Aurel Vlaicu University of Arad, Arad, Romania
e-mail: duma.virgil@osamember.org

J. Awrejcewicz (ed.), *Applied Non-Linear Dynamical Systems*, Springer Proceedings in Mathematics & Statistics 93, DOI 10.1007/978-3-319-08266-0_34

optical modulator, almost as used in laser setups as optical components like lenses, prisms, filters, and mirrors. However, there is not yet a systematic theory for them, which has led us to develop an analysis of "classical" chopper wheels [5]—with rotating wheels with windows with linear edges. We have also introduced [6], to the best of our knowledge, a novel type of device, with rotating wheels with windows with semicircular edges. Due to the way their semicircular edges obscure progressively the section of the laser beam in the plane of the wheel—which resembles a planetary eclipse—we have proposed for these devices the name of "eclipse" choppers [6].

We have presented in previous studies [7, 4] the different types of modulation functions we had developed for these different choppers [5, 6]. Experimental modules were also designed and manufactured, as well as prototype wheels [7] with the different profiles presented above. The next step of this work is to achieve design programs for the two main types of choppers. A first presentation of this program for the classical devices was done in [4], for top-hat (with constant intensity over the entire section) and Gaussian laser beams. In the present chapter this work is further expanded, but for both types of scanners—classical and eclipse—and for three types of the most used laser beams: top-hat, Gaussian, and Bessel.

2 Modulation/Transmission Functions of Classical and Eclipse Choppers

Macroscopic optical choppers (Fig. 1) consist of a rotating wheel with "n" windows and blades. They obscure the laser beam and generate controlled light impulses of frequency $f = n\omega/2\pi$, where $\omega = $ cst. is the rotational speed of the wheel.

The three types of choppers pointed out in the Introduction are presented in Fig. 1. The classical chopper (Fig. 1a) is thus compared with the two types of eclipse choppers: with outward edges (Fig. 1b) and with inward edges (Fig. 1c).

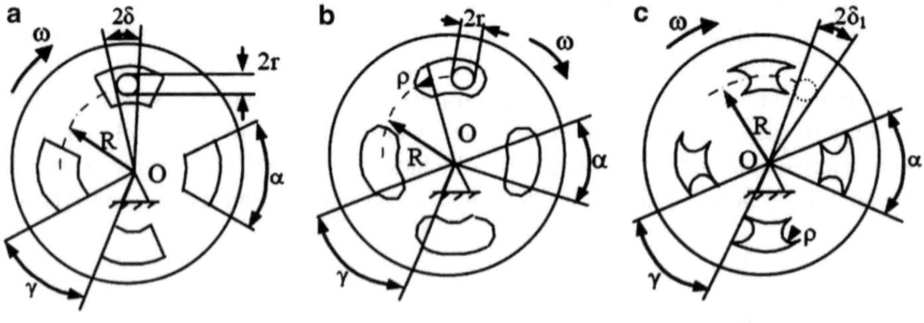

Fig. 1 Optical choppers: (**a**) classical configuration with windows with linear edges, analyzed in [5]; eclipse choppers with windows with outward (**b**) and inward (**c**) semicircular edges, introduced and studied in [6]

One may remark that the classical choppers can be obtained as a particular case of the eclipse ones for the radius of the edges $\rho \to \infty$. The modulation functions we have deduced for the classical devices [5] are therefore a particular case of the modulation functions we have developed [6] for the eclipse choppers.

The other characteristic parameters of the devices are (Fig. 1) α, the window angle; γ, the blade angle; r, the radius of the section of the beam in the plane of the wheel; R, the position of the center of the beam with regard to the pivot O of the wheel; and 2δ, the angle the beam section is seen from the pivot O.

With regard to the relationships between the angle 2δ and the angles α and γ, different types of modulation/transmission functions (radiometric flux of the output beams) were obtained analytically and studied in [5, 6]:

(i) Laser impulses with approximate rectangular profiles are obtained for any type of chopper if the beam is focused in the plane of the wheel.
(ii) Approximate trapezoidal laser impulses were obtained for choppers with blades larger than the beam section ($\gamma \geq 2\delta$) [5]. This case has also been approached experimentally and our theory was confirmed by another group [2].
(iii) Approximate sinusoidal profiles of the output signals can also be obtained for the previous case—if $\alpha = 2\delta = \gamma$. In this case both the full transmission and the full obscuration periods are reduced to zero, and the function of the transmitted flux only consists of transition periods.
(iv) A non-null modulation function results for a single blade in front of the section of the beam (for $\gamma < 2\delta$ and $\alpha > 2\delta$).
(v) Non-null sinusoidal-type profiles are obtained for α, $\gamma < 2\delta$, i.e., when there are always multiple blades placed in front of the beam section.
(vi) Approximate trapezoidal or triangular profiles of the output laser beams can be obtained, as demonstrated in [6], with the eclipse choppers. Thus, while approximate trapezoidal impulses may also be obtained with classical choppers—as pointed out above, at (ii)—for classical devices the transition portions of the graphs of the flux are tangent to the time-axis; this does not happen for the eclipse choppers (except for particular cases). Therefore, if the $\alpha = 2\delta = \gamma$ condition is valid, with eclipse choppers not approximate sinusoidal but approximate triangular impulses will be obtained—as the general case of $\rho \neq \infty$.

3 Design Programs for Classical Choppers

The analysis achieved for the classical choppers (with windows with linear edges) [5] has led to the above rules of thumb to choose the appropriate wheels in order to obtain a desired profile of the laser impulses.

The next steps in our work have thus been (i) to design and manufacture prototype wheels and modules [7] to carry on experimental validations of the theory developed and (ii) to develop a design program to obtain the modulation functions for a given

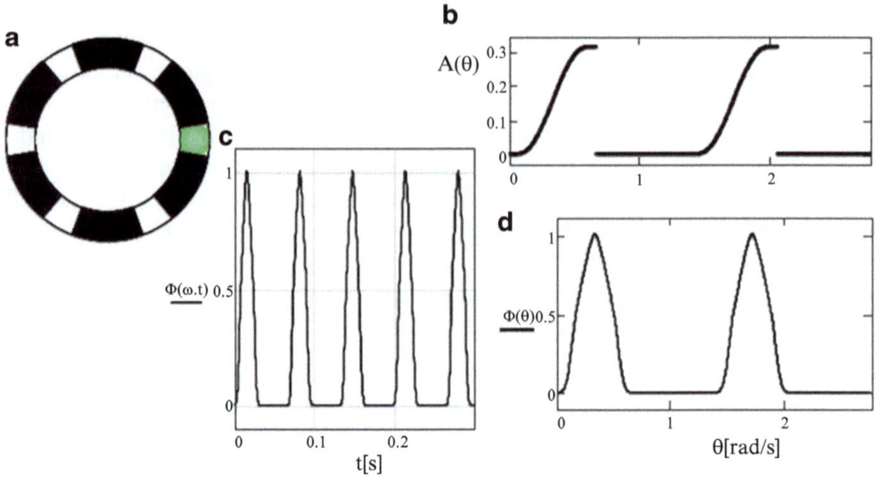

Fig. 2 Results of the design program for the classical chopper for a wheel with $n=6$ wings, for $r=5$ mm, $R=30$ mm, for Gaussian beams, with $m=1$: (**a**) the chopper wheel with the section of the beam in its plane; (**b**) graph of the area covered by the wing with regard to the rotation angle $\theta=\omega t$, where $\omega=$ cst.; (**c, d**) graphs of the transmitted flux (normalized, unitless) with regard to time (**c**) and to the angle θ (**d**)

set of parameters of the chopper. This program also allows the user to adjust the constructive and functional parameters of the setup in order to obtain certain profiles of the laser impulses generated through the chopper wheel.

A brief overview of these possibilities will be done in the following for the types of chopper wheels presented in Fig. 1 and for different types of laser beams.

To characterize the overlapping of the blades and of the beam section, a parameter m was defined [4] as the angular width of the beam section (as seen from the pivot O) and the ratio between the angular width of the window; thus $m=2\delta/\alpha$.

In Fig. 2 an example of the information provided by the program is shown, for a chopper with $n=6$ wings, with $r=5$ mm and $R=30$ mm, working with Gaussian laser beams. We have considered $m=1$ (as an input parameter for the program); thus it generated (Fig. 2a) a wheel with the margins of the window tangent to the circular section of the beam. The program gives the $A(\theta)$ function, which represents the area of the beam section that is covered by the blades—as a function of time or, equivalent, as a function of the rotation angle of the wheel, $\theta=\omega t$ (the latter in Fig. 2b). The final result of the program is the function of the transmitted flux, $\Phi(t)$, Fig. 2c (obtained in this example for $\omega=200$ rad/s), and $\Phi(\theta)$, Fig. 2d.

Another example, this time with fractional values of the parameter m (and with a different number n of blades), is considered in Fig. 3. Only two values of m were considered, from the row of possible values [4], to make a comparison between the modulation functions $\Phi(\theta)$ obtained with a smaller and a larger value of the parameter m.

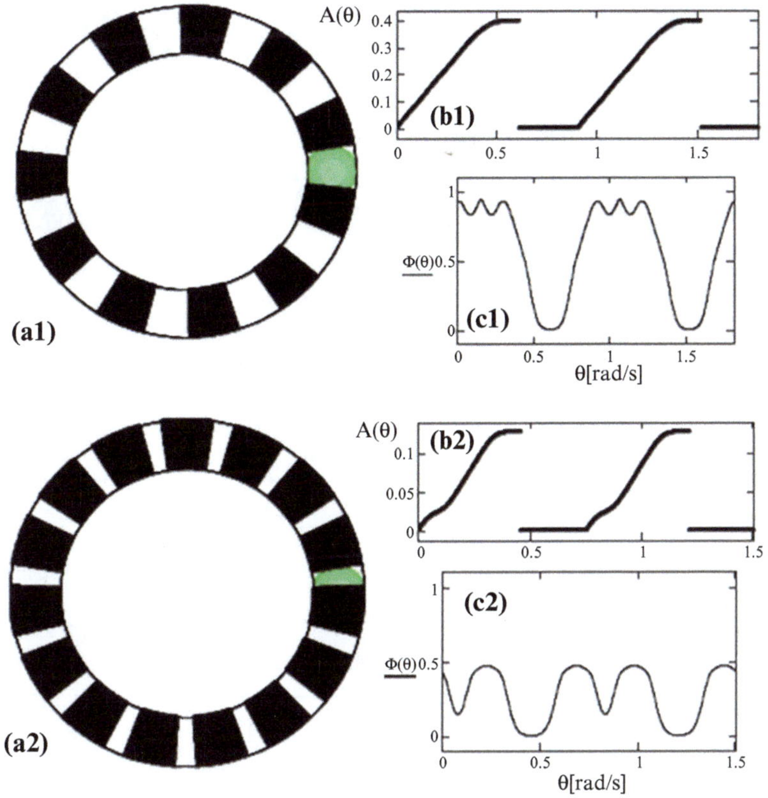

Fig. 3 Classical chopper: applications of the design program developed, for a wheel with (**a**) $n = 11$ and (**b**) $n = 12$, for $r = 5$ mm, $R = 30$ mm, for Gaussian beams, with (1) $m = 11/10$, (2) $m = 15/10$. For each case considered, the program provides: (**a**) the chopper wheel with the section of the beam in its plane; (**b**) graph of the area covered by the wing; (**c**) graph of the transmitted unitless flux—both with regard to the rotation angle $\theta = \omega t$, where $\omega = \text{cst}$

In the examples considered in Fig. 3, these graphs are given in (c1) for $m = 11/10$ and in (c2) for $m = 15/10$. In both cases $m > 1$, which means that the beam section is larger than the window of the wheel. One may remark the irregular shape of the transmitted flux with regard to the regular shape obtained in Fig. 2, in the case of $m = 1$. The irregularities increase, as it can also be seen comparing Fig. 2b1 with Fig. 2b2 with the increase of the difference of m from $m = 1$. Also, comparing the functions in Fig. 3c1, 3c2, the obvious decrease in the overall transmission coefficient of the device with the increase of m can be concluded.

Another comparison is made in Fig. 4, where we have considered other values of the chopper parameters, with three different values of m (one on each column) and three different types of chopper beams (one on each row). The study thus

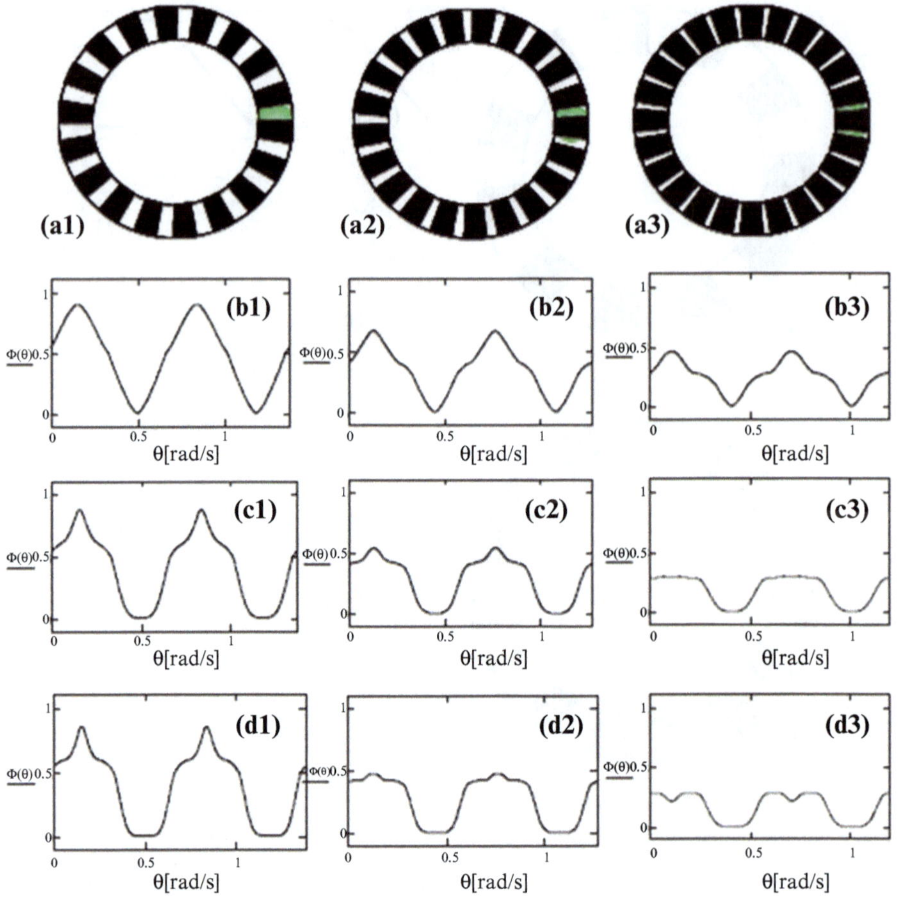

Fig. 4 (**a**) Designs of the wheel generated by the program. Comparison between the output light flux for classical choppers used with different types of laser beams: (**b**) top-hat; (**c**) Gaussian; (**d**) Bessel. The parameters of the setup are: wheels with $n = 12$ wings, for $r = 5$ mm, $R = 30$ mm, with (1) $m = 15/12$, (2) $m = 21/12$; (3) $m = 24/12$

reveals several aspects: (i) regardless of the type of laser beam (top-hat, Gaussian, or Bessel), the transmitted flux decreases with the increase of m—as it has been also concluded from the study in Fig. 3; (ii) the shapes of the laser impulses generated using the same chopper are different for each type of laser beam (for the same *m*, therefore for the same wheel)—although for Gaussian (c) and Bessel (d) beams there is a certain similarity of these shapes; (iii) the decrease in the transmission with the increase of *m* is different for each type of laser beam—it is more severe for Gaussian (c) with regard to top-hat (b) beams, and again more severe for Bessel (c) with regard to Gaussian (c) beams.

4 Design Programs for Eclipse Choppers

The new type of wheels that we have proposed [6] has an interesting property that is also demonstrated in Fig. 5, where an example of applying the program developed for the different types of wheels is shown: more exact trapezoidal profiles can be obtained with these eclipse choppers. While in our previous study [6, 7] we have considered only semicircular edges, outward (Fig. 1b) or inward (Fig. 1c), in order to obtain analytical expressions of the transmitted flux, using the developed program, other types of edges, by example elliptical, can also be tested. Thus, in Fig. 5 the way the program provides for elliptical edges is shown: (a) the shape of a window of the wheel and (b) the graph of the transmitted flux for top-hat and Gaussian beams with regard to the input constant flux $\Sigma = \Phi_S \pi r^2$.

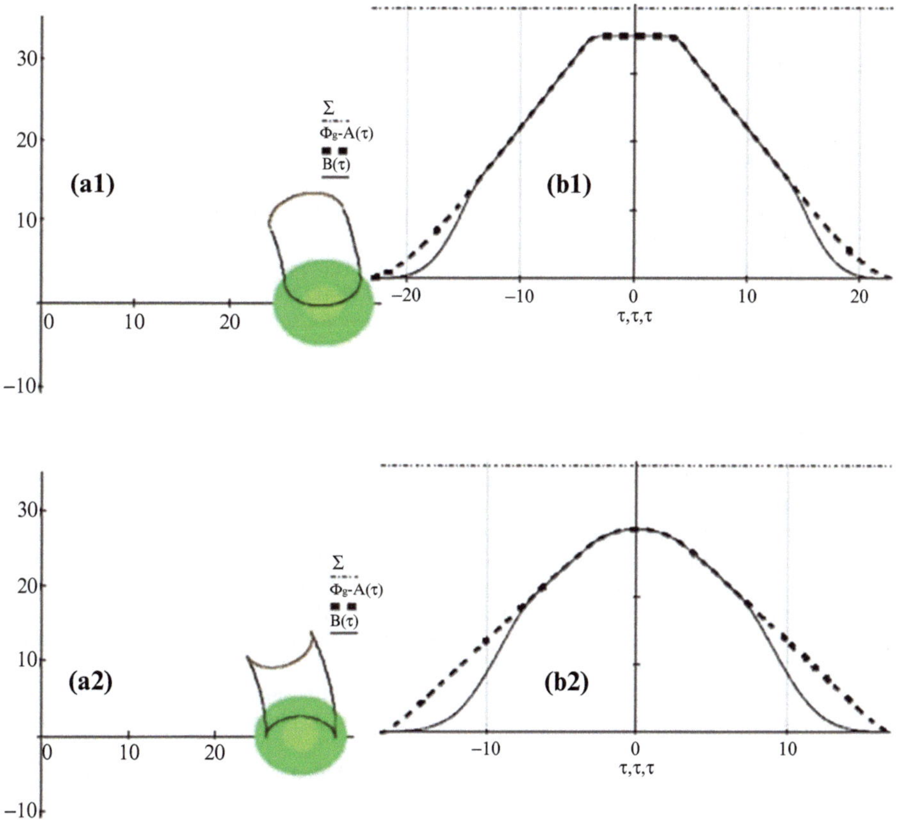

Fig. 5 (**a**) Window with inward edges and section of the laser beam in the section of the beam—for eclipse choppers; (**b**) output flux of the device. Parameters of the chopper: n = 11 windows, $r = 5$ mm, $R = 30$ mm, Gaussian laser beams, with the following semi-axis: (1) a = 4 mm, b = 3 mm (window with outward edges); (2) a = 4 mm, b = −3 mm (window with inward edges)

5 Conclusions

We presented the program we have developed for the design of optical choppers with rotating wheels. We approached classical choppers (with linear edges) and eclipse choppers (i.e., with elliptical edges). The program allows for the design of the chopper wheel for certain parameters of the setup. It also provides the function of the transmitted flux for different types of input laser beams: top-hat, Gaussian, and Bessel. Future work includes other types of choppers and micro-choppers we are currently developing, as well as applications as attenuators [8] or modulators.

Acknowledgments The work reported here was made possible by a Grant of the Romanian National Authority for Scientific Research, CNDI–UEFISCDI project PN-II-PT-PCCA-2011-3.2-1682 (http://3om-group-optomechatronics.ro/).

References

1. Bass, M. (ed.): Handbook of Optics. McGraw-Hill, New York (1995)
2. Benjamin, K., Armitage, A., South, R.: Harmonic errors associated with the use of choppers in optical experiments. Measurement **39**, 764–770 (2006)
3. Benmair, R.M.J., Kagan, J., Kalisky, Y., Noter, Y., Oron, M., Shimony, Y., Yogev, A.: Solar-pumped Er, Tm, Ho: YAG laser. Opt. Lett. **15**, 36–38 (1990)
4. Cira, O., Duma, V.-F.: Transmission functions of optical choppers for Gaussian beam distributions: Modeling and simulations. Proc. SPIE **8789**, 87890E (2013)
5. Duma, V.-F.: Theoretical approach on optical choppers for top-hat light beam distributions. J. Optic. A: Pure Appl. Optic. **10**, 064008 (2008)
6. Duma, V.-F.: Optical choppers with circular-shaped windows: modulation functions. Comm. Nonlinear Sci. Numer. Simulat. **16**, 2218–2224 (2011)
7. Duma, V.-F.: Prototypes and modulation functions of classical and novel configurations of optical chopper wheels. Lat. Am. J. Solids Struct. **10**(1), 5–18 (2013)
8. Duma, V.-F., Nicolov, M.: Neutral density filters with Risley prisms: analysis and design. Appl. Optic. **48**, 2678–2685 (2009)
9. Gondal, M.A., Yamani, Z.H.: Highly sensitive electronically modulated photoacoustic spectrometer for ozone detection. Appl. Optic. **46**, 7083–7090 (2007)
10. He, Y., Jin, W., Liu, G., Gao, Z., Wang, X., Wang, L.: Modulate chopper technique used in pyroelectric uncooled focal plane thermal imager. Proc. SPIE **4919**, 283–288 (2002)
11. McDermid, I.S., Walsh, T.D., Deslis, A., White, M.L.: Optical systems design for a stratospheric Lidar system. Appl. Optic. **34**, 6201–6210 (1995)
12. Nag, A., De, A.K., Goswami, D.: Two-photon cross-section measurements using an optical chopper: z-scan and two-photon fluorescence schemes. J. Phys. B-Atom. Mol. Opt. Phys. **42**(6), 065103 (2009)
13. Olkonnen, H.: Chopper stabilized laser-Doppler skin blood velocimeter. Proc. SPIE **1922**, 219–224 (1993)
14. Podoleanu, A.G., Dobre, G.M., Cucu, R.G.: Sequential optical coherence tomography and confocal imaging. Optic. Lett. **29**, 364–366 (2004)
15. Toshiyoshi, H., Fujita, H., Ueda, T.: A piezoelectrically operated optical chopper by quartz micromachining. J. Microelectromech. Syst. **4**(1), 3–9 (1995)
16. Wan, D.-P., Liu, H.-B., Wang, Y.-M., Hu, D.-J., Gui, Z.-X.: CO_2 laser beam modulating for surface texturing machining. Optic. Laser Technol. **40**, 309–314 (2008)

A Delay Mathematical Model for HIV Dynamics in HIV-Specific Helper Cells

Carla M.A. Pinto and Ana R.M. Carvalho

Abstract In this paper we study a delay mathematical model for the dynamics of HIV in HIV-specific CD4$^+$ T helper cells. We modify the model presented by Roy and Wodarz in 2012, where the HIV dynamics is studied, considering a single CD4$^+$ T cell population. Non-specific helper cells are included as alternative target cell population, to account for macrophages and dendritic cells. In this paper, we include two types of delay: (1) a latent period for the interval of time for cells, with contact with the virus, to be infected by the virions, released by them; (2) a virion production period for the virions to be produced and released to the bloodstream from the infected cells. We compute the reproduction number of the model, R_0, and the stability of the disease-free equilibrium. We find that for values of $R_0 < 1$, the model approaches asymptotically the disease-free equilibrium. We present numerical simulations of this fact. These results suggest that the model is mathematically and epidemiologically well posed.

1 Introduction

HIV/AIDS has been a major global health problem of our time. There have been some breakthroughs, some small conquests, but millions of dollars are spent every year in the treatment of the disease and no cure is available yet. Approximately, 25 million HIV-infected individuals live in sub-Saharan Africa [1, 12].

HIV is the human immunodeficiency virus that acts in the human body by destroying the immunological system. The disease has three stages: the acute stage, characterized by extensive initial virus growth, the chronic or asymptomatic phase, where virus loads are low and the patient appears healthy, and the AIDS phase where there is a sharp increase of virus load and the destruction of the CD4+

C.M.A. Pinto (✉)
School of Engineering, Polytechnic of Porto, Department of Mathematics, Rua Dr António Bernardino de Almeida, 431, 4200-072 Porto, Portugal
e-mail: cap@isep.ipp.pt

A.R.M. Carvalho
Faculty of Sciences, University of Porto, Rua do Campo Alegre s/n 4440-452 Porto, Portugal
e-mail: up080301038@alunos.fc.up.pt

J. Awrejcewicz (ed.), *Applied Non-Linear Dynamical Systems*, Springer Proceedings in Mathematics & Statistics 93, DOI 10.1007/978-3-319-08266-0_35

T helper cells, leading to the failure of the body. The fact that the HIV debilitates not only the immune system but the HIV-specific responses [8], and its ability to rapidly evolve in vivo, escaping the immune responses of the patient [3], makes the HIV one of the most complex viruses to fight.

Many mathematical models have been proposed for the dynamics of HIV infection [4, 7, 9–11]. Many models include delays. The most common delays in the literature are constant delays and continuously distributed delays. Some authors say that constant delays are not biologically realistic, the continuous delays being the most adequate. Gamma distributions are frequently used to model continuous delays [5, 6].

In this paper we study a delay mathematical model for the dynamics of HIV in HIV-specific $CD4^+$ T helper cells. We modify the model presented by Roy and Wodarz in 2012 [9], where the HIV dynamics is studied, considering a single $CD4^+$ T cell population. Non-specific helper cells are included as alternative target cell population, to account for macrophages and dendritic cells. In this paper, we include two types of delay: (1) a latent period for the interval of time for cells, with contact with the virus, to be infected by the virions, released by them; (2) a virion production period for the virions to be produced and released to the bloodstream from the infected cells. We consider that the delays can be approximated by a gamma distribution [4].

The paper is organized as follows. The model is described in Sect. 2. In Sect. 2.2, we compute the reproduction numbers and the stability of equilibria. Simulation results of the full model are presented in Sect. 3. In Sect. 4, we state the main conclusions of this work and discuss future research directions.

2 Model for HIV/AIDS Transmission

We introduce the model studied in this paper.

2.1 *Description of the Model*

We denote by x the uninfected specific helper cells and by y the infected specific helper cells. Variable S models the uninfected non-specific helper cells and I the infected non-specific helper cells. Variable v is the free virus population.

Disease progression is achieved by cells moving from one class to the other.

The HIV-specific helper cells, x, are produced at a constant rate r after being in contact with the virus v. The T cell proliferation is proportional to the total virus population, at low virus load, and saturates for higher values. Saturation of the production of T cells may also occur for greater values of $(x + y)$. This regulates the immune response against the virus. The degree of saturation is given by the parameters ϵ and η. Non-infected specific helper cells, x, become infected, after contact with the virus, at a rate β and die at a rate d. The infected specific helper cells, y, die at a rate a, are produced at a rate β and produce virus at a rate k. The non-specific non-infected target cells, S, are produced at a rate α, become infected, after contact with the virus, at a rate β_{ns}, and die at a rate d_{ns}. Infected non-specific target cells, I, die at a rate a_{ns}, are produced at a rate β_{ns}, and produce virus at a rate k_{ns}. Finally, the free virus population, v, dies at a rate u.

The system of delay ordinary differential equations for the proposed model is given by

$$\begin{aligned}
\dot{x(t)} &= \frac{rx(t)v(t)(\gamma+\epsilon)(\gamma+\eta)}{(x(t)+y(t)+\epsilon)(v(t)+\eta)} - dx(t) - \beta x(t)v(t)\\
\dot{y(t)} &= \beta \int_0^\infty x(t-\tau)v(t-\tau)g_{n_1,b_1}(\tau)e^{-m_1\tau}d\tau - ay(t)\\
\dot{S(t)} &= \alpha - d_{ns}S(t) - \beta_{ns}S(t)v(t)\\
\dot{I(t)} &= \beta_{ns} \int_0^\infty S(t-\tau)v(t-\tau)g_{n_2,b_2}(\tau)e^{-m_2\tau}d\tau - a_{ns}I(t)\\
\dot{v(t)} &= \kappa \int_0^\infty y(t-\tau)g_{n_3,b_3}(\tau)e^{-m_3\tau}d\tau + \kappa_{ns}\int_0^\infty I(t-\tau)g_{n_4,b_4}(\tau)e^{-m_4\tau}d\tau - uv(t)
\end{aligned} \tag{1}$$

The term $e^{-m_i\tau}$ $(i = 1,2,3,4)$ accounts for the cells that are infected at time τ but die before becoming productively infected τ time units later. The distribution function $g_{n_i,b_i}(\tau)$ $(i = 1,2,3,4)$ is defined in [5] as

$$g_{n_i,b_i}(\tau) = \frac{\tau^{n_i-1}}{(n_i-1)!b_i^{n_i}}e^{-\tau/b_i} \tag{2}$$

where the parameters n_i and b_i define the mean delay, $\bar{\tau}_i = n_i b_i$, the variance, $n_i b_i^2$, and the peak, $(n_i - 1)b_i$ $(i = 1,2,3,4)$, of the distribution. We consider $n_i \geq 1$, $i = 1,2,3,4$. If we substitute $b_i = \bar{\tau}_i/n_i$, $i = 1,2,3,4$ in equation (2), then we obtain

$$g_{n_i,\bar{\tau}_i}(\tau) = \frac{n_i^{n_i}\tau^{n_i-1}}{(n_i-1)!\bar{\tau}_i^{n_i}}e^{-n_i\tau/\bar{\tau}_i} \tag{3}$$

where $\bar{\tau}_i^2/n_i$ is the variance of the distribution, $i = 1,2,3,4$. The width of the distribution is set by parameter n_i, and the mean delay gives a way to set the location of the delay.

The inclusion of a continuous delay implies the normalization of the delay kernel so that $\int_0^\infty f_i(\tau)e^{-m_i\tau}d\tau = 1$, $i = 1,2,3,4$, in order for the equilibria of the non-delay model to remain equilibria of the delay one. In the case of a gamma distribution $f_i(\tau) = g_{n_i,b_i}(\tau)$, we rescale the equations of the model to obtain [4]

$$
\begin{aligned}
\dot{x(t)} &= \frac{rx(t)v(t)(\gamma+\epsilon)(\gamma+\eta)}{(x(t)+y(t)+\epsilon)(v(t)+\eta)} - dx(t) - \beta x(t)v(t) \\
\dot{y(t)} &= \bar{\beta}\int_0^\infty x(t-\tau)v(t-\tau)g_{n_1,\hat{b}_1}(\tau)d\tau - ay(t) \\
\dot{S(t)} &= \alpha - d_{ns}S(t) - \beta_{ns}S(t)v(t) \\
\dot{I(t)} &= \bar{\beta}_{ns}\int_0^\infty S(t-\tau)v(t-\tau)g_{n_2,\hat{b}_2}(\tau)d\tau - a_{ns}I(t) \\
\dot{v(t)} &= \bar{\kappa}\int_0^\infty y(t-\tau)g_{n_3,\hat{b}_3}(\tau)d\tau + \bar{\kappa}_{ns}\int_0^\infty I(t-\tau)g_{n_4,\hat{b}_4}(\tau)d\tau - uv(t)
\end{aligned}
\tag{4}
$$

where

$$
\begin{array}{lll}
\bar{\beta} = \frac{\beta}{(1+m_1b_1)^{n_1}} & \bar{\beta}_{ns} = \frac{\beta_{ns}}{(1+m_2b_2)^{n_2}} & \bar{\kappa} = \frac{\kappa}{(1+m_3b_3)^{n_3}} \\
\bar{\kappa}_{ns} = \frac{\kappa_{ns}}{(1+m_4b_4)^{n_4}} & \hat{b}_i = \frac{b_i}{1+m_ib_i} & \int_0^\infty g_{n_i,\hat{b}_i}(\tau)d\tau = 1
\end{array}
\tag{5}
$$

2.2 *Reproduction Numbers and Stability of Disease-Free Equilibria*

In this subsection, we compute the reproduction number of model (1), R_0, using the next-generation method [2]. The basic reproduction number is defined as the number of secondary infections, due to a single infection in a completely susceptible population. The disease-free equilibrium of model (1) is given by

$$
P^0 = (x_0, y_0, S_0, I_0, v_0) = \left(0, 0, \frac{\alpha}{d_{ns}}, 0, 0\right) \tag{6}
$$

Using the notation in [2] on system (4), matrices for the new infection terms, F, and the other terms, V, are given by

$$
F = \begin{bmatrix} 0 & 0 & 0 \\ 0 & 0 & \bar{\beta}_{ns}S_0 \\ 0 & 0 & 0 \end{bmatrix}; \quad V = \begin{bmatrix} a & 0 & 0 \\ 0 & a_{ns} & 0 \\ -\bar{\kappa} & -\bar{\kappa}_{ns} & u \end{bmatrix} \tag{7}
$$

The associative basic reproduction number is given by

$$
R_0 = \rho(FV^{-1}) = \frac{\bar{\kappa}_{ns}\bar{\beta}_{ns}\alpha}{d_{ns}a_{ns}u} \tag{8}
$$

where ρ indicates the spectral radius of FV^{-1}. By Theorem 2 [2], we obtain the following lemma.

Lemma 1. *The disease-free equilibrium P^0 is locally asymptotically stable if $R_0 < 1$ and unstable if $R_0 > 1$.*

Proof. We perform the study of the stability of the disease-free equilibrium P^0 of model (4). The eigenvalues of the jacobian matrix of system (4), around the disease-free equilibrium P^0, are computed through the determinant below:

$$M_1 = \begin{bmatrix} -d-\lambda & 0 & 0 & 0 & 0 \\ 0 & -a-\lambda & 0 & 0 & 0 \\ 0 & 0 & -d_{ns}-\lambda & 0 & -\beta_{ns} S_0 \\ 0 & 0 & 0 & -a_{ns}-\lambda & \bar{\beta}_{ns} S_0 F_2(\lambda) \\ 0 & \bar{\kappa} F_3(\lambda) & 0 & \bar{\kappa}_{ns} F_4(\lambda) & -u-\lambda \end{bmatrix} \tag{9}$$

where

$$F_i(\lambda) = \int_0^{\infty} g_{n_i,\hat{b}_i}(\tau) e^{-\lambda\tau} d\tau \tag{10}$$

The following eigenvalues are easily obtained:

$$-d, \qquad -a, \qquad -d_{ns} \tag{11}$$

The remaining eigenvalues are the roots of the characteristic equation of a 2×2 matrix, calculated through the following determinant:

$$M_2 = \begin{bmatrix} -a_{ns}-\lambda & \bar{\beta}_{ns} F_2(\lambda) S_0 \\ \bar{\kappa}_{ns} F_4(\lambda) & -u-\lambda \end{bmatrix} \tag{12}$$

The determinant is equivalent to

$$(a_{ns}+\lambda)(u+\lambda) - \bar{\kappa}_{ns}\bar{\beta}_{ns} S_0 F_2(\lambda) F_4(\lambda) = 0 \tag{13}$$

and after some algebra manipulation, we obtain

$$\lambda^2 + G_1\lambda + G_2 = G_2 R_0 F_2(\lambda) F_4(\lambda) \tag{14}$$

where

$$G_1 = a_{ns} + u \qquad G_2 = a_{ns} u \tag{15}$$

In order for the characteristic equation (13) to have a purely imaginary root, the eigenvalues must cross the imaginary axis, to become positive. By way of contradiction, we assume that there is some $\omega > 0$ such that $\lambda = i\omega$ is an eigenvalue of (13), that is,

$$-\omega^2 + G_1 i\omega + G_2 = G_2 R_0 F_2(i\omega) F_4(i\omega) \tag{16}$$

Thus,

$$\omega^4 + G_1^2\omega^2 + G_2^2 = G_2^2 R_0^2 |F_2(i\omega)F_4(i\omega)|^2 \leq G_2^2 R_0^2 \tag{17}$$

Therefore, if $R_0 < 1$, the characteristic equation has no purely imaginary roots. We conclude that the disease-free equilibrium is stable for $R_0 < 1$.

3 Numerical Results

In this section, we present the numerical simulations of model (1). The parameter values used in the simulations are $\beta = 0.01$, $d = 0.5$, $a = 1$, $r = 2$, $\epsilon = 10$, $\eta = 10$, $\gamma = 1$, $\kappa = 1$, $u = 1$, $\beta_{ns} = 0.2$, $\alpha = 0.12$, $d_{ns} = 0.1$, $a_{ns} = 0.2$, $\kappa_{ns} = 1$, $m_1 = 0.01$, $m_2 = 0.01$, $b_1 = 0.0625$, $b_2 = 0.0625$, $n_1 = 4$, $n_2 = 4$, $m_3 = 0.048$, $m_4 = 0.13$, $b_3 = 0.25$, $b_4 = 0.25$, $n_3 = 5$, and $n_4 = 6$.

In Fig. 1, we plot the numbers of uninfected and infected non-specific target cells, the numbers of uninfected and infected specific helper cells, and the population of free virus. We observe that, for the given parameter values and initial conditions, the model approaches the stable disease-free equilibrium asymptotically. Biologically, this is explained as follows. The low initial virus load and the low initial number of specific helper cells lead to the extinction of the specific helper cell response.

4 Conclusions

The model presented in this paper is a delay mathematical model for the dynamics of HIV in HIV-specific $CD4^+$ T helper cells. It is a modification of the model proposed in [9], where the HIV dynamics is studied, considering a single CD4+ T cell population. Non-specific helper cells are included as alternative target cell population, to account for macrophages and dendritic cells. Two types of delay are considered: (1) a latent period for the interval of time for cells, with contact with the virus, to be infected by the virions, released by them; (2) a virion production period for the virions to be produced and released to the bloodstream from the infected cells. We compute the reproduction number of the model, R_0, and the stability of the disease-free equilibrium. We find that for values of $R_0 < 1$, the model approaches asymptotically the disease-free equilibrium. We present some numerical simulations. Future work is needed in order to compute the global stability of the disease-free and of the endemic equilibrium.

Acknowledgements The authors wish to thank Fundação Gulbenkian, through *Prémio Gulbenkian de Apoio à Investigação 2003*, and the Polytechnic of Porto, through the *PAPRE Programa de Apoio à Publicação em Revistas Científicas de Elevada Qualidade*, for financial support. CP was partially funded by the European Regional Development Fund through the program COMPETE

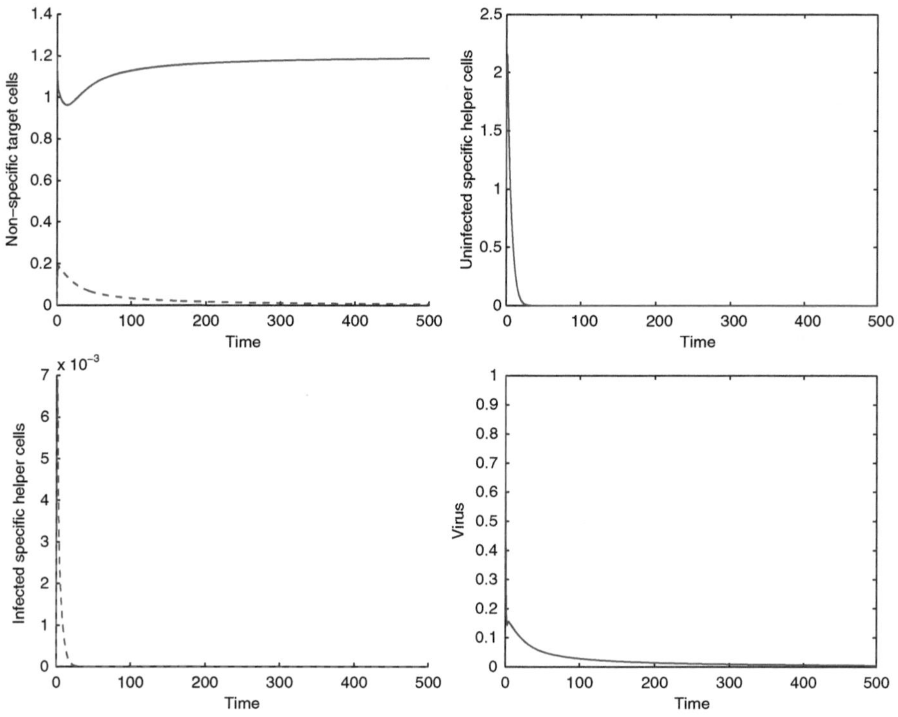

Fig. 1 Uninfected (*solid line, top left*) and infected non-specific target cells (*dashed, top left*), uninfected specific target cells (*top right*), infected specific helper cells (*bottom left*), and free virus population (*bottom right*). Parameter values used in the simulations are as above (see text), and initial conditions are $x(0) = 1$, $y(0) = 0$, $S(0) = 1.2$, $I(0) = 0$, and $v(0) = 1$

and by the Portuguese Government through the FCT - Fundação para a Ciência e a Tecnologia under the project PEst-C/MAT/UI0144/2013. The research of A.C. was supported by a FCT grant with reference SFRH/BD/96816/2013.

References

1. CDC: Center for Disease Control (CDC).http://www.cdc.gov/HIV
2. Driessche, P., Watmough, J.: Reproduction numbers and sub-threshold endemic equilibria for compartmental models of disease transmission. Math. Biosci. **180**(1–2), 29–48 (2002)
3. Goulder, P.J., Walker, B.D.: The great escape – AIDS viruses and immune control. Nat. Med. **5**, 1233–1235 (1999)
4. Lv, C., Yuan, Z.: Stability analysis of delay differential equation models of HIV-1 therapy for fighting a virus with another virus. J. Math. Anal. Appl. **352**, 672–683 (2009)
5. Mittler, J., Sulzer, B., Neumann A., Perelson A.: Influence of delayed viral production on viral dynamics in HIV-1 infected patients. Math. Biosci. **152**, 143–163 (1998).
6. Nelson, P., Perelson, A.S.: Mathematical analysis of delay differential equation models of HIV-1 infection. Math. Biosci. **179**, 73–94 (2002).

7. Perelson, A.S., Kirschner, D.E., Boer, R.: Dynamics of HIV infection of $CD4^{+}$ T cells. Math. Biosci. **114**, 81–125 (1993).
8. Rosenberg, E.S., Altfeld, M., Poon, S.H., Phillips, M.N., Wilkes, B.M., Eldridge, R.L., Robbins, G.K., D′Aquila, R.T., Goulder, P.J., Walker, B.D.: Immunecontrol of HIV-1 after early treatment acute infect. Nature **407**, 523–526 (2000).
9. Roy, S.M., Wodarz, D.: Infection of HIVspecific CD4 T helper cells and the clonal composition of the response. J. Theor. Biol. **304**, 143–151 (2012).
10. Wodarz, D., Hamer, D.H.: Infection dynamics in HIV-specific CD4 T cells: Does a CD4 T cell boost benefit the host or the virus? Math. Biosci. **209**, 14–29 (2007).
11. Wodarz, D., Thomsen, A.R.: Effect of the CTL proliferation program on virus dynamics. Int. Immunol. **17**(9), 1269–1276 (2005).
12. WHO: Malaria and HIV Interaction and Their Implications for Public Health Police. The World Health Organization (WHO) (2004).

The Application of Parametric Excitation in Resonant MEMS Gyroscopes

Barry J. Gallacher, Zhongxu Hu, Kiran M. Harish, Stephen Bowles, and Harry Grigg

Abstract Parametric excitation, via electrostatic stiffness modulation, can be exploited in resonant MEMS gyroscopes. In the case of the rate gyroscope parametric excitation may be used to amplify either the response to angular rate. In the more complex mode of operation, known as "rate integrating," the output of the gyro is angled directly as opposed to angular velocity in the case of rate gyro. In this rate-integrating mode of operation parametric excitation does offer an effective energy control used to initiate and sustain the vibration and minimize damping perturbations. A practical parametric excitation scheme implemented using digital signal processing has been developed for both the rate and rate-integrating gyro. Experimental results are presented demonstrating the benefit of using parametric excitation. By taking advantage of the phase dependence of parametric amplification and the orthogonality of the Coriolis force and quadrature forcing, the response to the applied angular velocity may be parametrically amplified by applying excitation of a particular phase directly to the sensing mode.

1 Introduction

The MEMS gyroscope has received significant attention from MEMS researchers both in industry and academia [1, 14]. Applications in stability, guidance and navigation have driven the technological development of state-of-the-art MEMS gyroscopes to a high level; however, their performance in some key areas does not approach that of other gyroscope technologies. All MEMS vibratory gyroscopes operate by exploiting Coriolis coupling between two modes of vibration of a suspended proof mass [1, 14]. However, due to the small mass of micromachined gyroscopes, this Coriolis coupling is small. Parametric excitation offers several approaches to mitigate against the small Coriolis coupling between the two modes

B.J. Gallacher (✉) • Z. Hu • S. Bowles • H. Grigg
Newcastle University, Newcastle upon Tyne, UK
e-mail: b.j.gallacher@ncl.ac.uk; z.x.hu@ncl.ac.uk; s.r.bowles@ncl.ac.uk; harry.grigg@ncl.ac.uk

K.M. Harish
UTAS, Plymouth, UK
e-mail: kiran.shastri@gmail.com

J. Awrejcewicz (ed.), *Applied Non-Linear Dynamical Systems*, Springer Proceedings in Mathematics & Statistics 93, DOI 10.1007/978-3-319-08266-0_36

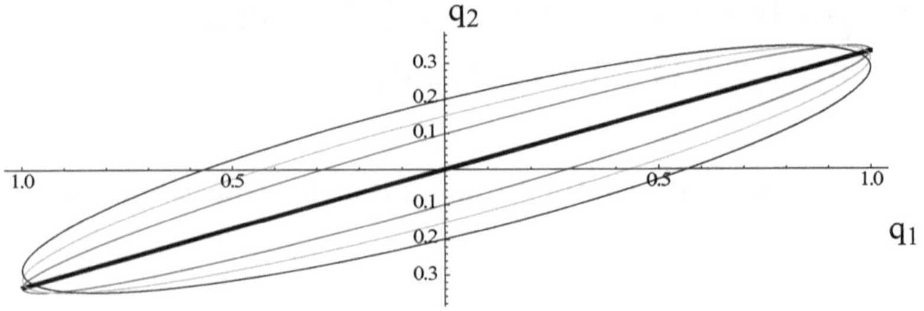

Fig. 1 Fixed elliptical trajectory of the rate gyro

and thus improve performance. The modes of vibration are often referred to as the primary and secondary modes. By far the most common mode of operation for MEMS gyroscopes is where the angular rate is measured. This type of gyro, known as a "rate" gyro, has applications in the automotive, aerospace and consumer sectors. The measurement of angle from a rate gyro requires integration of the angular rate with respect to time. This process has several drawbacks. Noise and bias in the measured angular rate, when integrated, will lead to a diverging error in the angle measurement. In the rate gyro, one of the modes of vibration, known as the primary mode and represented by the generalized coordinate q_1, is forced into resonance and its amplitude is maintained constant. The measurement of the angular rate can be shown to be proportional to the amplitude of a second mode, known as the secondary mode and represented by the generalized coordinate q_2, excited via the Coriolis force. If the vibration is plotted in mode space $[q_1, q_2]$, then the line of oscillation of the total vibration will have its gradient determined by the Coriolis force. The line of oscillation is fixed in the mode space $[q_1, q_2]$ for a given angular velocity applied about the input axis of the gyroscope. Figure 1 shows the fixed elliptical trajectory for the rate gyroscope. Note that the minor axis of the ellipse is related to the structural imperfections and in the ideal case the trajectory becomes a straight line.

A second way of utilizing the inertia of the modes of vibration provides a direct measure of angle and hence avoids the issues associated with numerical integration. This mode of operation known as "rate integrating" has been used successfully since the mid-1980s in macroscopic hemispherical shell gyroscopes [8, 9]. The hemispherical resonator gyroscope (HRG) has been developed to a high degree and is capable of tactical and inertial grade performance. Micromachined "rate-integrating" gyroscope designs are also discussed in the literature; however, these have tended to be suitable for angular rates in the order of deg/s. In the "rate-integrating" gyroscope Coriolis acceleration in a freely vibrating elastic structure couples two modes of vibration such that energy is continuously transferred between the two modes of vibration. The resulting beating modal vibrations have a beat frequency related to the angular rate. If the displacements are plotted in the mode space $[q_1, q_2]$, as shown in Fig. 2, then the line of oscillation of the total vibration will precess with an angular velocity related to the angular velocity applied about

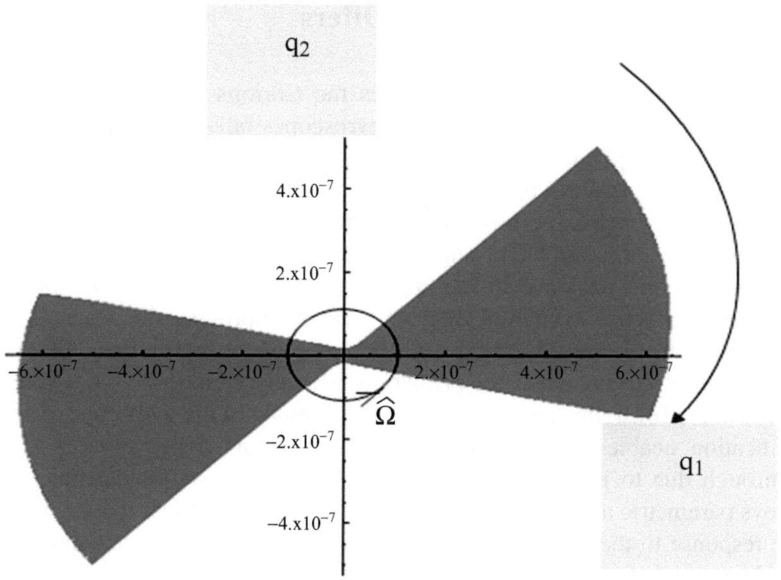

Fig. 2 Precessing elliptical trajectory of the rate-integrating gyro

gyro's input axis. The input axis of a gyroscope is the axis about which maximum Coriolis coupling occurs between the two modes of vibration represented by the generalized coordinates q_1 and q_2. The precession angle will also be related to the applied angle of rotation. This is in contrast to the rate gyro where the orientation of the line of oscillation would remain fixed. For the special case where the two modes share exactly the same natural frequency, the beat frequency and thus the precession will be proportional to the angular rate and the angle of rotation.

2 Axisymmetric Structures and Modal Degeneracy

The property of wave inertia in elastic axisymmetric structures is well known. For axisymmetric elastic structures modes of the same circumferential order share the same natural frequency. This is known as degeneracy. The primary and secondary modes therefore have the same natural frequency; thus excitation of the secondary mode due to Coriolis acceleration will be at its natural frequency. Thus the response is resonant and maximizes the effect of the Coriolis coupling. The theory of structures exhibiting modal degeneracy is also well established [12].

3 What Parametric Excitation Offers

Due to the small mass of MEMS devices the Coriolis force is very small and currently the performance of most micro-gyroscopes falls short of "tactical grade" [14]. This is compounded by the relatively low level of structural precision compared to their macroscopic counterparts and also "electrical feedthrough" of the drive. The principal performance measurands for gyroscopes are bias drift, angle random walk, scale factor and scale-factor stability. Scale factor defines the ratio of the output response to the applied angular rate. In the case of the rate gyroscope parametric excitation offers a way of improving the scale factor and this is linked to the signal-to-noise performance of the gyroscope. This may be achieved by amplifying either the primary or secondary modes of the gyroscope as reported in [7, 16]. When applied to the primary mode of the gyroscope, parametric amplification enables reduced forcing amplitudes and hence reduces electrical feedthrough due to parasitic capacitances. However, the most beneficial method employs parametric amplification of the secondary mode. Recall that the secondary is the response to the Coriolis force. This approach is far more demanding on the control system but it offers the greatest performance enhancements. Importantly, parametric amplification of the Coriolis response occurs before any electrical transduction or amplification. This is particularly advantageous for a range of sensors, not just MEMS gyroscopes, where electronic noise is the major noise contributor. In this case the signal-to-noise ratio and scale factor of the gyro will be amplified. Parametric effects are also reported for other microsystems in [2, 11]. Other resonant devices that could benefit from parametric excitation include the resonant magnetometer.

For the case of the rate-integrating gyro parametric excitation offers a way of negating the effect of damping. By using the appropriate cyclic arrangement of pumping electrodes it is possible to control the vibration energy associated with each mode as well as controlling the total vibrational energy. This is important since differences in the modal damping factors manifest themselves as angle or rate errors. For gyroscopic applications, it is essential that parametric excitation must leave the orientation of the normal modes unchanged. This imposes a particular cyclic symmetric condition on the electrode configuration used to parametrically pump the vibration. A single annular electrode provides a practical way of providing the parametric excitation. However, use of a single annular electrode does not mitigate against differences in the direct or cross damping terms.

4 Structural Form for the Gyro and Its Important Modes of Vibration

Due to its planar structure and relatively simple MEMS fabrication process, the ring structure is an ideal candidate for championing high-performance gyroscopic attributes. Figure 3 shows the ring gyro with its lid removed and also a schematic

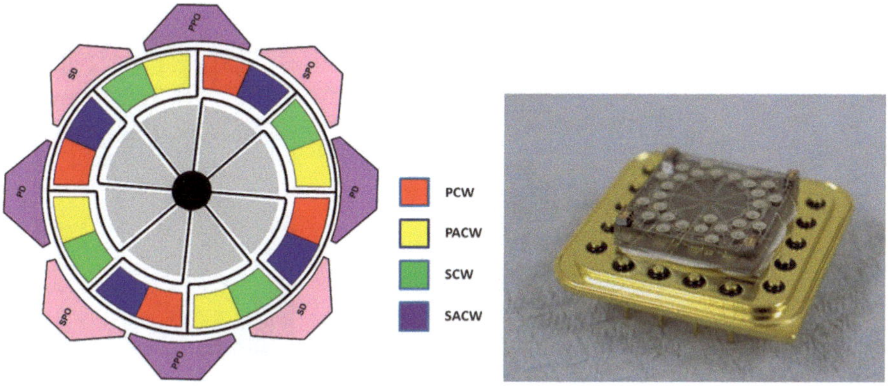

Fig. 3 The MEMS silicon ring gyro [Goodrich and silicon sensing]

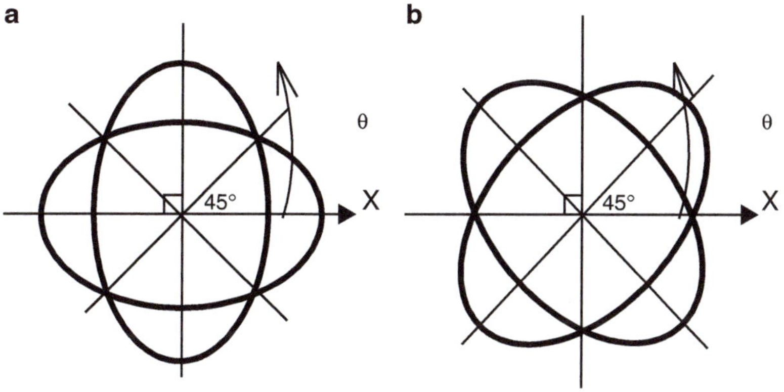

Fig. 4 Radial displacement of the flexural modes of order n = 2. (**a**) Primary mode, (**b**) secondary mode

highlighting the electrode arrangement for electrostatic actuation and sensing. The mode shapes and natural frequencies of a perfect axisymmetric ring are well known and available in the literature [5]. In the single axis ring gyroscope the flexural modes of vibration in the plane of the ring are sensitive to angular rates applied about the polar axis of the ring. The input axis of the ring gyroscope is therefore its polar axis. Of practical interest are the flexural modes of circumferential order $n = 2$, as shown in Fig. 4. The dynamics of the ring will be expressed relative to an axis X passing through the centre of the ring and in its plane, as shown in Fig. 2. With respect to this axis the modes of interest in the rate gyroscope are referred to as the primary and secondary modes. The primary mode is characterized by radial and tangential displacements having spatial variations proportional to cos(nθ) and sin(nθ) and its contribution to the motion is described by the generalized coordinate q_1. The secondary mode is characterized by radial and tangential displacements

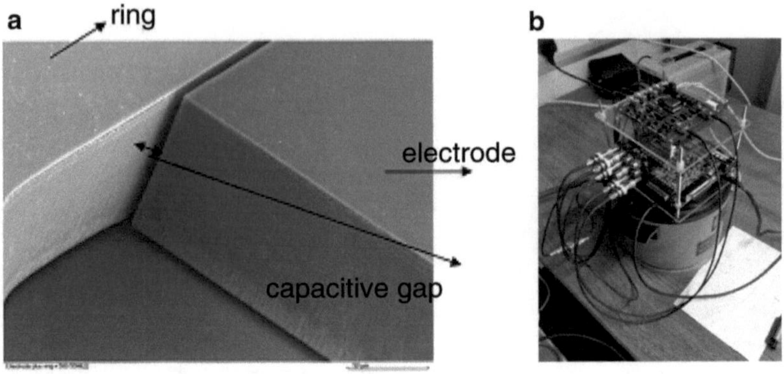

Fig. 5 (**a**) Silicon ring and electrostatic actuation/sensing electrode. (**b**) Experimental setup incorporating the packaged gyro, rate table and DSP control electronics

having spatial variations proportional to sin(nθ) and cos(nθ) and its contribution to the motion is described by the generalized coordinate q_2.

Angular velocities about the polar axis of the ring result in Coriolis coupling between the primary and secondary modes. Figure 4 illustrates the radial displacement of the primary and secondary flexural modes of vibration.

Closed loop parametric amplification of the primary mode and open loop parametric amplification of the secondary mode were demonstrated separately. However, the two procedures may be implemented simultaneously without any added difficulty. Figure 5 shows the experimental setup and includes the DSP control system, the gyro and the rate table. Our work focuses on amplifying the Coriolis response by directly applying a parametric excitation of particular phase to the sense/secondary mode. Direct amplification of the Coriolis response results in significant improvement in the signal-to-noise ratio. A theoretical description of the parametrically amplified gyroscope is presented in order to gain an insight into the principal factors pertinent to its exploitation. The excitation and control scheme have been implemented digitally using a digital signal processing (DSP) development board enabling very high sampling precision and execution speed necessary for gyroscopic applications. Parametric excitation in both the rate and rate-integrating gyro is achieved electrostatically.

5 Basic Gyrodynamics

The homogenous equation of motion of the ring gyroscope in the absence of any parametric excitation is described by Eq. (1) [3]:

$$[M]\,\underset{\sim}{\ddot{q}}(t) + \{[C] + [G]\}\,\underset{\sim}{\dot{q}}(t) + [K]\,\underset{\sim}{q}(t) = \underset{\sim}{0} \tag{1}$$

where

$$\underset{-}{q}(t) = \begin{bmatrix} q_1(t) \\ q_2(t) \end{bmatrix}, G = 2m_o\Omega\,[J], \quad [J] = \begin{bmatrix} 0 & -1 \\ 1 & 0 \end{bmatrix}, \quad \Omega = \left(\frac{2\mathrm{n}}{\mathrm{n}^2+1}\right)\varphi.$$

$$[M] = m_o\overline{M}, \quad [K] = k_o\overline{K}, \quad [C] = 2\xi_o\omega_o m_o\overline{C}$$

and

$$m_o = \pi\rho_0 abd\left(1+\frac{1}{n^2}\right), \quad k_o = \frac{E_o I_{zz}\pi}{a^3}\left(1-n^2\right)^2, \quad \beta_E = \frac{2\pi\varepsilon_o ad}{2{h_o}^3}$$

$$\overline{M} = I + \overline{\delta}, \quad \overline{\mathrm{K}} = I + \overline{\mu}, \quad \overline{\mathrm{C}} = I + \overline{\gamma}.$$

The structural imperfections $\overline{\delta}$, $\overline{\mu}$ may be approximated by assuming that the mass density and elastic modulus vary in a periodic manner circumferentially in the form

$$\rho(\theta) = \rho_0 + \sum_{m=1}^{\infty}\rho_{mc}\cos(m\theta) + \sum_{m=1}^{\infty}\rho_{ms}\sin(m\theta)$$

$$E(\theta) = E_0 + \sum_{h=1}^{\infty}E_{hc}\cos(h\theta) + \sum_{h=1}^{\infty}E_{hs}\sin(h\theta).$$

It may be easily shown that only harmonics with $m = h = 4, 8, 12, 16, etc.$ contribute to the kinetic and strain energies. Single crystal silicon of orientation <111> is used to form the ring structure. This crystal orientation possesses a threefold rotational symmetry which does not perturb the degeneracy of the $n = 2$ in-plane flexural modes.

It has been assumed that the damping imperfections have the same structure as the mass and stiffness imperfections; thus

$$\overline{\delta} = \begin{bmatrix} \overline{\delta}_1 & \overline{\delta}_2 \\ \overline{\delta}_2 & -\overline{\delta}_1 \end{bmatrix}, \overline{\mu} = \begin{bmatrix} \overline{\mu}_1 & \overline{\mu}_2 \\ \overline{\mu}_2 & -\overline{\mu}_1 \end{bmatrix}, \overline{\gamma} = \begin{bmatrix} \overline{\gamma}_1 & \overline{\gamma}_2 \\ \overline{\gamma}_2 & -\overline{\gamma}_1 \end{bmatrix},$$

$$\overline{\delta}_1 = \left(\frac{n^2-1}{2(n^2+1)}\right)\frac{\rho_{4c}}{\rho_o}, \quad \overline{\delta}_2 = \left(\frac{n^2-1}{2(n^2+1)}\right)\frac{\rho_{4s}}{\rho_o},$$

$$\overline{\mu}_1 = \frac{E_{4c}}{2E_o}, \overline{\mu}_2 = \frac{E_{4s}}{4E_o}.$$

The Coriolis coupling between the two modes of vibration is represented by the skew symmetric matrix G. For the case of typical vibratory gyroscopes the imperfections may be adequately represented as perturbations.

6 Rate Gyro Parametric Excitation and Amplification

A brief overview of the linear model of parametric excitation will be presented since a detailed description is available in [4, 6, 16].

6.1 Parametric Amplification of the Primary Mode

Excitation of the primary mode is achieved by applying the drive voltage $V_d(t)$ to the pair of primary drive electrodes (PD). The voltage $V_{PD}(t)$ across the PD electrodes is

$$V_{PD}(t) = V_d(t) - V_{dc}.$$

Harmonic forcing and parametric excitation is made possible by using a drive voltage of the form

$$V_d(t) = V_F(t) + V_P(t),$$

where

$$\begin{aligned} V_P(t) &= V_p \sin\left(2\omega_f t + \varphi_p\right), \\ V_F(t) &= V_f \sin \omega_f t, \end{aligned}$$

and φ_p represents the phase advance of the parametric excitation applied to the PD electrodes with respect to the forcing.

In normal operation $|V_{dc}| > V_f, V_p$. An approximate expression for the equation of motion of the primary mode can be obtained by including only the excitation terms which make a first-order contribution to the response at frequency ω_f. The linearized equation of motion of the primary mode under combined harmonic forcing and parametric excitation is derived in detail in [4, 6, 7, 16] and has the form

$$\ddot{q}_1 + \frac{\omega_1}{Q_1}\dot{q} + \left(\omega_1{}^2 - \eta_p + \eta_{1p} \sin\left(2\omega_f t + \varphi_p\right)\right) q_1 = -\overline{\eta}_{0p} \sin\left(\omega_f t\right) = D_1(t) \tag{2}$$

with ω_1 as the natural frequency of the primary mode and

$$\begin{aligned} \eta_p &= \frac{1}{2}\widehat{\beta}_1 \left(V_p{}^2 + V_f{}^2 + V_{dc}{}^2\right)\left(\frac{1}{m_1}\right) \\ \eta_{1p} &= 2\widehat{\beta}_1 V_p V_{dc} \left(\frac{1}{m_1}\right) \\ \overline{\eta}_{0p} &= 2\widehat{\beta}_0 V_{dc} V_f \left(\frac{1}{m_1}\right) \end{aligned}$$

where

$$\widehat{\beta}_0 = \frac{\varepsilon_o bd}{2{h_o}^2}$$

$$\widehat{\beta}_1 = \frac{\varepsilon_o bd}{2{h_o}^3}.$$

The steady-state solution of Eq. (2) may be written in the form

$$q_1(t) = A_1 \sin\left(\omega_f t + \psi_1\right).$$

By applying the harmonic balance method to Eq. (2), the magnitude of the steady-state amplitude A_1 and phase angle ψ_1 of the primary mode with parametric excitation are described by

$$A_1\left(V_p, \widehat{\omega}_1, \varphi_p\right) = \overline{\eta}_{0p} \frac{\sqrt{{\varsigma_1}^2 + \eta_{1p}\varsigma_1 \cos\left(\varphi_p\right) + {\widehat{\omega}_1}^4 + \eta_{1p}{\widehat{\omega}_1}^2 \sin\left(\varphi_p\right) + \frac{1}{4}{\eta_{1p}}^2}}{{\widehat{\omega}_1}^4 + {\varsigma_1}^2 - \frac{1}{4}{\eta_{1p}}^2}$$

$$\psi_1\left(V_p, \widehat{\omega}_1, \varphi_p\right) = \tan^{-1}\left(-\frac{\varsigma + \frac{1}{2}\eta_{1p}\cos\left(\varphi_p\right)}{{\widehat{\omega}_1}^2 + \frac{1}{2}\eta_{1p}\sin\left(\varphi_p\right)}\right) \tag{3}$$

where

$${\widehat{\omega}_1}^2 = {\overline{\omega}_1}^2 - {\omega_f}^2$$

$${\overline{\omega}_1}^2 = {\omega_1}^2 - \eta_p$$

$$\varsigma_1 = \frac{\omega_1\omega_f}{Q_1}. \tag{4}$$

It is clear from Eq. (3) that the linear model predicts that

$$\frac{1}{Gain\left(V_p, \widehat{\omega}_1, \varphi_p\right)} \to 0 \tag{5}$$

when

$${\widehat{\omega}_1}^4 + {\varsigma_1}^2 - \frac{1}{4}{\eta_{1p}}^2 \to 0.$$

Here $Gain\left(V_p, \widehat{\omega}_1, \varphi_p\right)$ denotes the parametric gain, which is defined as the ratio of the mobility magnitude with parametric excitation "on" to the mobility magnitude with parametric excitation "off" [3]. The DC components η_p present in the electrostatic drive lowers the natural frequency of the primary mode to $\overline{\omega}_1$. From Eq. (5) describing the stability boundary of the primary mode it is clear that the minimum value of η_{1p} causing instability occurs when $\widehat{\omega}_1 = 0$. This means that when the primary mode is excited at the softened resonant frequency, the parametric amplitude needed to cause maximum amplification is a minimum. The excitation parameters correspond to the nose or apex of the stability boundary.

6.2 Parametric Amplification of the Secondary Mode

Parametric amplification of the secondary mode is achieved by applying the pumping voltage $V_P(t)$ to the pair of secondary drive electrodes (SD). The voltage bias V_{dc} is still applied to the ring. The voltage $V_{SD}(t)$ across the SD electrodes is

$$V_{SD}(t) = V_P(t) - V_{dc}$$

where

$$V_P(t) = V_p \sin\left(2\omega_f t + \varphi_s\right)$$

and φ_s represents the phase advance of the parametric excitation applied to the SD electrodes with respect to the Coriolis force.

The linearized equation of motion of the secondary mode under an applied angular rate and parametric excitation has the form

$$\ddot{q}_2 + \frac{\omega_2}{Q_2}\dot{q} + \left(\omega_2^{\;2} - \eta_s + \eta_{2s} \sin\left(2\omega_f t + \varphi_s\right)\right) q_2 = F_2 \dot{q}_1 + F_{12} q_1 \qquad (6)$$

with ω_2 as the natural frequency of the secondary mode, and

$$\eta_s = \frac{1}{2}\widehat{\beta}_1 \left(V_p^{\;2} + V_{dc}^{\;2}\right)\left(\frac{1}{m_2}\right)$$

$$\eta_{2s} = 2\widehat{\beta}_1 V_p V_{dc}\left(\frac{1}{m_2}\right).$$

Note that the phase advance φ_s is measured with respect to the Coriolis force $F_2\dot{q}_1$.

The equation of motion of the secondary mode has two orthogonal forcing terms: $F_2\dot{q}_1$ and $F_{12}q_1$ corresponding to the Coriolis force and quadrature force, respectively. With the primary mode excited at its softened natural frequency $\overline{\omega}_1$, the steady-state solution of Eq. (6) may be written in the form

$$q_2(t) = A_2 \sin\left(\overline{\omega}_1 t + \psi_2\right) - A_{12} \cos\left(\overline{\omega}_1 t + \psi_2\right).$$

By applying the harmonic balance method to Eq. (6), the magnitude of the steady-state amplitudes A_2 and A_{12} and phase angle ψ_2 of the secondary mode with parametric excitation are described by

$$A_2\left(V_p, \widehat{\omega}_2, \varphi_s\right) = F_2 \frac{\sqrt{\varsigma_2^{\,2} + \eta_{2s}\varsigma_2 \cos\left(\varphi_s\right) + \widehat{\omega}_2^{\;4} + \eta_{2s}\widehat{\omega}_2^{\;2}\sin\left(\varphi_s\right) + \frac{1}{4}\eta_1^{\,2}}}{\widehat{\omega}_2^{\;4} + \varsigma_2^{\,2} - \frac{1}{4}\eta_{2s}^{\;2}}$$

$$\psi_2\left(V_p, \widehat{\omega}_2, \varphi_s\right) = \arctan\left(-\frac{\varsigma_2 + \frac{1}{2}\eta_{2s}\cos\left(\varphi_s\right)}{\widehat{\omega}_2^{\;2} + \frac{1}{2}\eta_{2s}\sin\left(\varphi_s\right)}\right) \qquad (7)$$

$$A_{12}\left(V_p, \widehat{\omega}_2, \varphi_s\right) = F_{12}\frac{\sqrt{\varsigma_2{}^2 - \eta_{2s}\varsigma_2\cos\left(\varphi_s\right) + \widehat{\omega}_2{}^4 - \eta_{2s}\widehat{\omega}_2{}^2\sin\left(\varphi_s\right) + \frac{1}{4}\eta_{2s}{}^2}}{\widehat{\omega}_2{}^4 + \varsigma_2{}^2 - \frac{1}{4}\eta_{2s}{}^2} \tag{8}$$

where

$$\begin{aligned}
\overline{\omega}_1 &= \omega_f \\
\widehat{\omega}_2{}^2 &= \overline{\omega}_2{}^2 - \overline{\omega}_1{}^2 \\
\overline{\omega}_2{}^2 &= \omega_2{}^2 - \eta_s \\
\varsigma_2 &= \frac{\omega_2\omega_f}{Q_2}.
\end{aligned}$$

The π/2 phase shift of the quadrature force relative to the Coriolis force means that the quadrature force is shifted by π relative to $V_P(t)$. Also, the DC components η_s present in the electrostatic drive lower the natural frequency of the secondary mode to $\overline{\omega}_2$. From Eq. (7) the linear model predicts that

$$\frac{1}{Gain\left(V_p, \widehat{\omega}_2, \varphi_s\right)} \to 0$$

when

$$\widehat{\omega}_2{}^4 + \varsigma_2{}^2 - \frac{1}{4}\eta_{2s}{}^2 \to 0.$$

The minimum value of η_{2s} causing instability in the secondary mode occurs for $\widehat{\omega}_2 = 0$. The phase advance φ_s may be used to adjust the response amplitudes A_2 and A_{12}. Maximum amplification of the response A_2 and maximum deamplification of the response A_{12} occur when $\frac{dA_2{}^2}{d\varphi_s} = \frac{dA_{12}{}^2}{d\varphi_s} = 0$ which gives the expression

$$\varphi_s = \arctan\left(\frac{\widehat{\omega}_2{}^2}{\varsigma_2}\right). \tag{9}$$

For the ideal case where the mistuning $\widehat{\omega}_2 = 0$, Eq. (9) shows that setting $\varphi_s = 0$ will result in maximum amplification of the response to the Coriolis force. The phase advance φ_s for this ideal case of zero mistuning is therefore described by φ_{so} with $\varphi_{so} = 0$. The response to the quadrature forcing will be deamplified maximally. The phase angle for the ideal case is described by $\psi_2 = \psi_{2o} = -\frac{\pi}{2}$. Equation (9) shows that when $\widehat{\omega}_2 \neq 0$, corresponding to mistuning between the primary and secondary modes, the phase advance resulting in maximum amplification (deamplification)

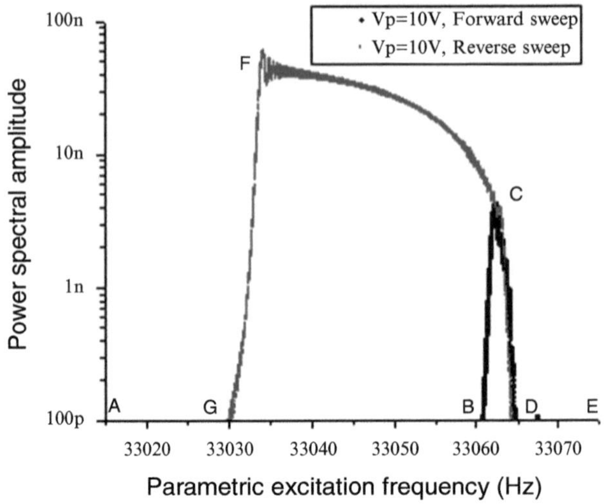

Fig. 6 Forward and reverse frequency sweeps for parametric excitation only

of the response A_2 (A_{12}) will be shifted from that corresponding to the tuned case $\varphi_{so} = 0$. The phase advance will be described by

$$\varphi_s = \varphi_{so} + \arctan\left(\frac{\widehat{\omega}_2{}^2}{\varsigma_2}\right). \tag{10}$$

As demonstrated in [16], the rate of change of parametric amplification with respect to the phase advance is substantial. Therefore, maximum amplification of the response A_2 will require the phase advance φ_s to be calculated and set accurately. Measurements of both the mistuning $\widehat{\omega}_2$ and the damping term ς_2 will be required.

6.3 Determining the Stability Boundary

The stability boundary can be determined experimentally either from the parametrically amplified forced response or the response to parametric excitation only. To determine the shape of the instability boundary the excitation frequency ω was swept, forwards and backwards, over values centred on $2\overline{\omega}$. A typical swept response curve is shown in Fig. 6. For increasing values of frequency, starting from point A, the response is, at first, indistinguishable from the measurement noise. The interpretation of this is that the response corresponds to the trivial solution of the equation of motion. At point B the response suddenly grows and becomes easily seen above the noise. The interpretation now is that the trivial response has become unstable and has developed into a limit cycle, with maximum amplitude occurring at point C. The characteristics of this limit cycle are determined by the

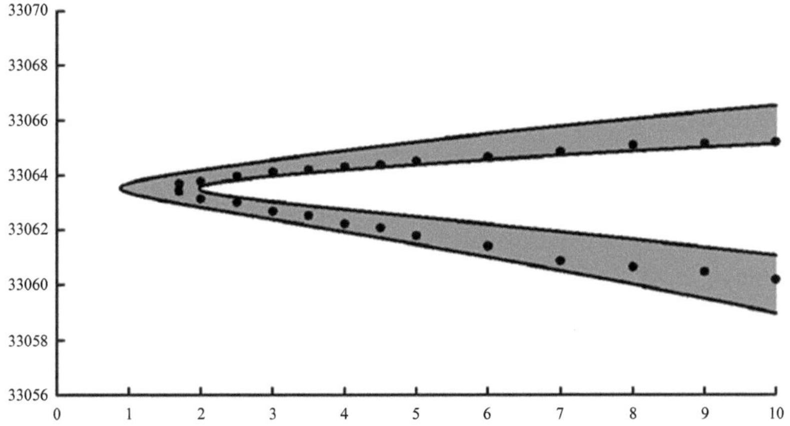

Fig. 7 Measured stability boundary from frequency sweeps with parametric excitation only. Parametric excitation frequency versus parametric excitation voltage

nonlinearity of the electrostatic actuation. Further increases in frequency produce a reduction in the amplitude of the limit cycle until point *D* is reached. Here the response is indistinguishable from the noise and corresponds to the trivial state. This remains the situation as the excitation is swept up to point *E* and back again to point *D*. When the excitation frequency is reduced further the response builds up. The limit cycle is re-established, and its amplitude retraces the previously held values between *C* and *D*. However, when the backward frequency sweep is continued, the amplitude does not reduce along the original path from *C* to *B* but continues to increase until a maximum is reached at point *F*. At *F* the limit cycle state becomes unstable and the response thereafter collapses back to the trivial state at point *G*. This type of behaviour is discussed in [10, 13, 15] and is typical of a Duffing-type system (in this case a system with cubic electrostatic softening nonlinearity) with parametric excitation. This shows that the stability boundary of the linear parametric system is associated with the frequency range *BD*. By conducting a series of such response tests at different values of excitation amplitude V a set of frequency values corresponding to points *B* and *D* were obtained. These are plotted in Fig. 7 as a stability boundary and are compared with the estimated boundary using the measured softened natural frequency and *Q*-factor and the measured geometry of the capacitive actuator.

6.4 Parametric Amplification of the Primary Mode

The parametrically amplified forced response of the primary mode provides an alternative way of determining the stability boundary. Figure 8 confirms the location of the nose of the stability boundary using this method.

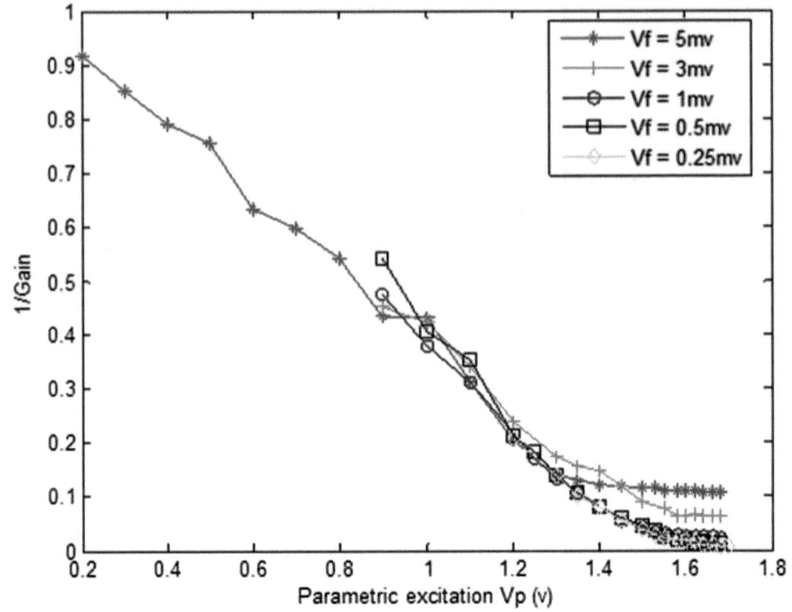

Fig. 8 1/parametric gain to determine the stability boundary

The objective for the amplitude control tests was to ramp down the harmonic forcing voltage whilst sustaining the response amplitude at its set point through using parametric excitation. Figure 9 shows the experimental test results with the parametric gain set to 12. Ramping down of the harmonic forcing voltage commenced at 30 s. At this time the parametric controller was switched on. The response of the parametric controller to step changes in the desired response amplitude was investigated. Figure 9 illustrates that with the gain set to 12, the parametric controller can accommodate for 10 % step changes introduced at the reference amplitude.

6.5 Parametric Amplification of the Coriolis Response

We propose to use parametric excitation to directly amplify the Coriolis response of the sense mode. It is clear from Eq. (7) of the sense mode that if the frequency mistuning is zero or tuned small enough, a parametric excitation with a phase aligned with Coriolis force can be applied to the sense mode to significantly amplify the Coriolis response. The test device already has a frequency mismatch of 1.8 Hz. This frequency split is further reduced by electrostatic tuning to 0.1 Hz. This level of mistuning is small enough to enable a practical demonstration of parametric amplification of the Coriolis response. The phase advance can be set to $\varphi_s = 0$ without any measurable reduction in amplification. Frequency tracking

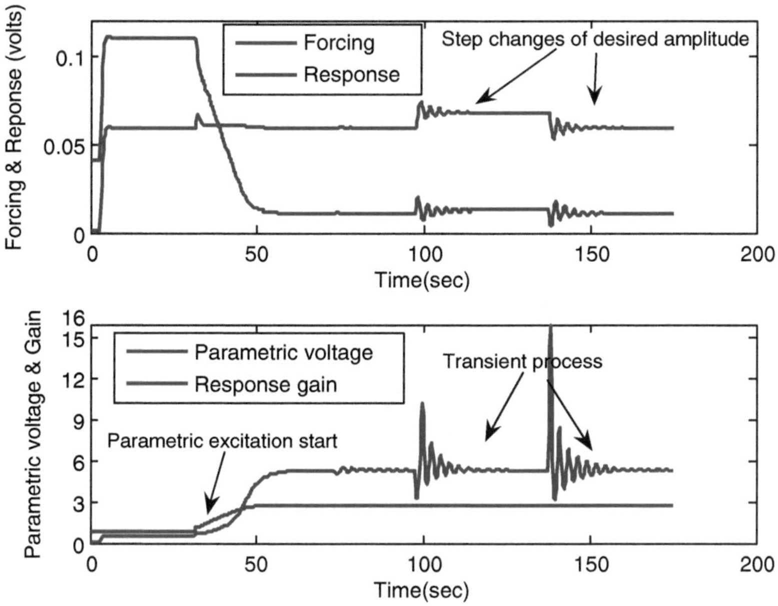

Fig. 9 Parametric gain set to 12. Response to step changes in desired amplitude

of the primary mode was under closed loop control using the PLL. The parametric gain control circuit was not employed on the secondary mode. Figure 10 shows the experimental results for the parametrically amplified Coriolis response and demonstrates a gain by a factor of 11.

6.6 *Signal-to-Noise Performance when Under Parametric Amplification*

For high Q-factor MEMS electronic noise dominates over mechanical noise. As a result parametric amplification of the Coriolis response (the secondary mode of the gyro) will also amplify the signal-to-noise ratio of the gyro, as described by

$$\frac{S}{N} = \frac{S}{N_{mech} + N_{elect}}, \qquad N_{mech} << N_{elect} \text{ and } \frac{N_{mech}}{S} \approx 0$$

$$Gain_{SN} \approx \frac{Gain^{*} S}{N_{elect}}.$$

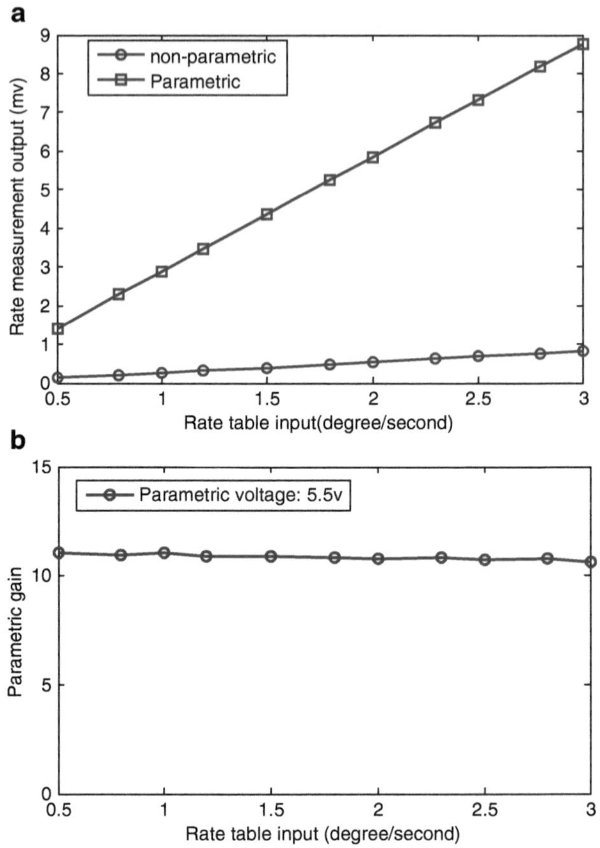

Fig. 10 Open loop rate table tests. (**a**) Amplitude of rate output for sinusoidal rate input. (**b**) Parametric gain

In order to confirm that mechanical noise contributions are small even when parametrically amplified, measurements of the noise floor at the resonant frequency of the gyro were made with parametric excitation and also without parametric excitation for zero rate input.

Figure 11 shows the noise floor of the rate output with and without parametric excitation for zero rate input. The parametric excitation was set to 5.5 V, which corresponds to an amplification of 11 in the Coriolis response. The standard deviation in the noise response increased from 6.35 to 7.35 μV corresponding to an amplification of 1.16. The improvement in the signal-to-noise ratio is therefore given by $Gain_{SN} = Gain/1.16$. Since $Gain = 11$, the signal-to-noise ratio of the gyro has increased by a factor $Gain_{SN} = 9.48$.

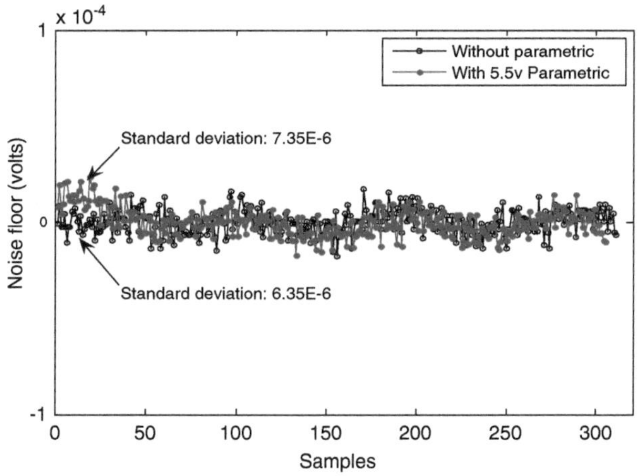

Fig. 11 Measured noise floor with and without parametric amplification applied

7 Parametric Excitation Applied to the Rate-Integrating Gyroscope

The dynamics of the rate-integrating gyro with parametric excitation provided by an annular electrode may be conveniently expressed in terms of the eigenvectors of the dynamical matrix D. Performing the coordinate transformation

$$\underline{q}(t) = [X]\,\underline{x}(t)$$

$$[X] = \left[\underline{X}^{(1)}\underline{X}^{(2)}\right]$$

where

$$[D] = [M]^{-1}\,[K]$$

yields

$$\ddot{\underline{x}}(t) + 2\varepsilon\,(\upsilon_o\omega_o\,[\,[I] + [\gamma]] + \varphi\,[J]\,)\,\dot{\underline{x}}(t)$$
$$+\omega_o{}^2\,[\{[I] + \varepsilon\,[\varsigma]\} - \varepsilon\Phi U(t)\,[I]\,]\,\underline{x}(t) = 0.$$

$\varepsilon << 1$ and may be used as a small parameter.

The structural imperfections are represented by the matrices

$$[\varsigma] = \begin{bmatrix} \varsigma & 0 \\ 0 & -\varsigma \end{bmatrix}, \qquad [\gamma] = \begin{bmatrix} \gamma_1 & \gamma_2 \\ \gamma_2 & -\gamma_1 \end{bmatrix}$$

and

$$\omega_1{}^2 = \omega_o{}^2 (1 + \varepsilon\varsigma) = \omega_o{}^2 (1 + \lambda)$$
$$\omega_2{}^2 = \omega_o{}^2 (1 - \varepsilon\varsigma) = \omega_o{}^2 (1 - \lambda)$$

define the normal mode frequencies.

The method of multiple scales provides an effective method for obtaining a useful solution.

Thus,

$$\underset{-}{x} \left(\bar{t}, \widehat{t}\right) = \underset{-}{x}^{(0)} \left(\bar{t}, \widehat{t}\right) + \varepsilon \underset{-}{x}^{(1)} \left(\bar{t}, \widehat{t}\right) + O\left(\varepsilon^2\right)$$

where $\bar{t}$ and $\widehat{t}$ are the slow and fast timescales described by

$$\bar{t} = \varepsilon t \text{ and } \widehat{t} = t + O\left(\varepsilon^2\right).$$

Assuming the voltage waveform is periodic and represented by the Fourier series

$$U(t) = u_o + \sum_{r=1}^{\infty} u_r e^{ir\omega t} + \sum_{r=1}^{\infty} u_r{}^* e^{-ir\omega t} \quad \text{with} \quad r\omega = 2\widehat{\omega}_o + \varepsilon\delta.$$

gives the zeroth-order and first-order secular equations

$$\underset{-}{x}^{(0)} = \underset{-}{A}_o \left(\bar{t}\right) e^{i\omega_o \widehat{t}} + \underset{-}{B}_o \left(\bar{t}\right) e^{-i\omega_o \widehat{t}}$$

and

$$\frac{\partial \underset{-}{A}_o}{\partial \bar{t}} = -\left\{ \begin{array}{l} \left(\varphi\,[J] + \upsilon_o\omega_o\,[[I] + [\gamma]]\right) \underset{-}{A}_o \\ -i\frac{\omega_o}{2}\,[\varsigma]\,\underset{-}{A}_o + i\frac{\omega_o}{2}\Phi\left(u_o \underset{-}{A}_o + u_r \underset{-}{B}_o e^{i\varepsilon\widehat{\delta t}}\right) \end{array} \right\},$$

$$\frac{\partial \underset{-}{B}_o}{\partial \bar{t}} = -\left\{ \begin{array}{l} \left(\varphi\,[J] + \upsilon_o\omega_o\,[[I] + [\gamma]]\right) \underset{-}{B}_o \\ +i\frac{\omega_o}{2}\,[\varsigma]\,\underset{-}{B}_o - i\frac{\omega_o}{2}\Phi\left(u_o \underset{-}{B}_o + u_r{}^* \underset{-}{A}_o e^{-i\varepsilon\widehat{\delta t}}\right) \end{array} \right\}.$$

Assuming a solution of the form

$$\underset{-}{A}_o \left(\bar{t}\right) = \underset{-}{A}^{(o)} e^{\left(S + i\frac{\delta}{2}\right)\bar{t}}, \underset{-}{B}_o = \underset{-}{B}^{(o)} e^{\left(S - i\frac{\delta}{2}\right)\bar{t}}$$

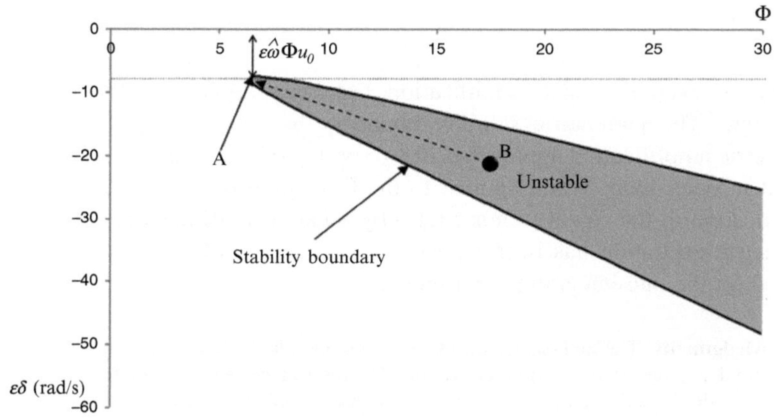

Fig. 12 Stability boundary for the parametrically excited rate-integrating gyroscope

leads to algebraic equation

$$\begin{bmatrix} [n] & c\,[I] \\ c^*\,[I] & [m] \end{bmatrix} \begin{bmatrix} \underline{A}^{(o)} \\ \underline{B}^{(o)} \end{bmatrix} = \underline{0}$$

and characteristic equation

$$a_0 s^4 + a_1 s^3 + a_2 s^2 + a_3 s + a_4 = 0.$$

Purely imaginary solution is guaranteed provided

$$\Delta_3 = a_3\,(a_1 a_2 - a_0 a_3) - {a_1}^2 a_4 = 0.$$

The stability can therefore be plotted and is shown in Fig. 12. The parametric excitation parameters (δ,Φ) resulting in purely imaginary values of s may then be found by solving the equation $\Delta_3 = 0$. This means that the amplitude and frequency of the harmonic voltage necessary to make the net energy dissipation of both normal modes zero are known. The normal mode amplitude envelope can therefore be sustained parametrically without interfering with the precession caused by the Coriolis acceleration. Recall that $\delta/2$ is the detuning of the excitation frequency from the average of the two normal mode frequencies and Φ is the scaling factor for the voltage amplitude. Initialization of the vibration can be achieved by choosing excitation parameters (δ,Φ) that reside within the unstable region. Point (B) in Fig. 12 shows such an arbitrary point with the unstable region. Parametric excitation will cause both modes to be excited. To avoid unbounded growth a control system must adjust the excitation parameters such that (δ,Φ) converge on the stability boundary. Converging on point (A) of Fig. 12 is advantageous since it allows the lowest value of Φ to be used to stabilize the vibration. The square-wave voltage amplitude corresponding to point (A) has the value $\sqrt{6.5}$ V.

8 Conclusions

Parametric excitation and amplification has been applied to the MEMS ring gyroscope. The parametric stability boundary has been mapped; experimental, parametric amplification approaches of two orders of magnitude have been demonstrated in open loop. When applied to the Coriolis response of the rate gyro, an amplification in the signal-to-noise ratio by a factor of 10 has been demonstrated. Parametric excitation has been shown theoretically to offer a suitable excitation method for the rate-integrating gyroscope.

Acknowledgments The authors would like to thank Goodrich Sensors and Integrated Systems (Plymouth UK), the Scottish Microelectronics Centre and the EPSRC for their support in all the projects that contributed to this work. The authors thank Jim Burdess for many useful and interesting discussions in relation to this body of work.

References

1. Acar, C., Shkel, A.: MEMS Vibratory Gyroscopes: Structural Approaches to Improve Robustness. MEMS Reference Shelf Series. Springer, New York (2009)
2. Carr, D.W., Evoy, S., Sekaric, L., Craighead, H.G., Parpia, J.M.: Parametric amplification in a torsional microresonator. Appl. Phys. Lett. **77**, 1545–1547 (2000)
3. Gallacher, B.J.: Principles of a micro-rate integrating ring gyroscope. IEEE Trans. Aero. Electron. Syst. **48**, 658–672 (2012)
4. Gallacher, B.J., Burdess, J.S., Harish, K.M.: A control scheme for a MEMS electrostatic resonant gyroscope excited using combined parametric excitation and harmonic forcing. J. Micromech. Microeng. **16**, 1–12 (2006)
5. Gallacher, B.J., Burdess, J.S., Harris, A.J.: Principles of a three-axis vibrating gyroscope. IEEE Trans. Aero. Electron. Syst. **37**(4), 1333–1343 (2001)
6. Harish, K.M., Gallacher, B.J., Burdess, J.S., Neasham, J.A.: Experimental investigation of parametric and externally forced motion in resonant MEMS sensors. J. Micromech. Microeng. **19**(1), 015021 (2009)
7. Hu, Z.X., Gallacher, B.J., Burdess, J.S., Fell, C.P., Townsend, K.: A parametrically amplified MEMS rate gyroscope. Sensor Actuat. A Phys. **167**(2), 249–260 (2011)
8. Loper, E.J., Lynch, D.D.: The HRG: A new low-noise inertial rotation sensor. In: Sixteenth Joint Services Data Exchange for Inertial Systems, CA, 16–18 Nov 1982
9. Lynch, D.D.: Hemispherical resonator gyroscope in inertial technology for the future. IEEE Trans. Aero. Electron. Syst. **20**(4), 414–444 (1984)
10. Nayfeh, A.H., Mook, D.T.: Nonlinear Oscillations. Wiley, New York (1970)
11. Rugar, D., Grutter, P.: Mechanical parametric amplification and thermomechanical noise squeezing. Phys. Rev. Lett. **67**, 699–702 (1991)
12. Thomas, D.L.: Dynamics of rotationally periodic structures. Int. J. Numer. Meth. Eng. **14**, 81–102 (1979)
13. Thomsen, J.J.: Vibrations and Stability. Springer, Berlin (2003)
14. Yazdi, N., Ayazi, F., Najafi, K.: Micromachined inertial sensors. IEEE Proc. **86**(8), 1640–1659 (1998)
15. Zhang, W., Baskaran, R., Turner, K.L.: Effect of cubic nonlinearity on auto-parametrically amplified resonant MEMS mass sensor. Sensor Actuat. Phys. A **102**(1–2), 139–150 (2002)
16. Zhongxu, H., Gallacher, B.J., Harish, K.M., Burdess, J.S.: An experimental study of high gain parametric amplification in MEMS. Sensor Actuat. A Phys. **162**(2), 145–154 (2010)

Influence of Smart Material on the Dynamical Response of Mechanical Oscillator

Vinícius Piccirillo, Ângelo Marcelo Tusset, José Manoel Balthazar, Davide Bernardini, and Guiseppe Rega

Abstract The dynamical response of systems with shape memory alloy (SMA) or magnetorheological damper (MRD) presents a different behavior due to their nonlinear characteristics. Both systems have a nonlinear behavior due to adaptive dissipation related to their hysteretic behavior. This property is very attractive in engineering field. This paper investigates the nonlinear dynamical behavior of an SMA or MRD oscillator system. The LuGre mathematical model is used to represent the MRD behavior. On the other hand, the SMA model is based on a thermomechanical consistent model with four state variables.

Numerical simulations show different aspects about these two systems.

1 Introduction

In recent years, shape memory alloys (SMA) and magnetorheological dampers (MRD) have attracted considerable attention due to the potential applications as smart structures in innovative research and technological development. In this way, SMA and MRD are being used in various dynamical applications associated with the adaptive dissipation of energy related to their hysteretic behavior. These aspects can be explored both in the adaptive passive and the active control [8, 10, 12, 13, 15].

It is known that SMA exhibit a high damping capacity. A physical reason for that is the energy dissipation during the phase transformation (e.g., friction due to

V. Piccirillo (✉) • Â.M. Tusset
Department of Mathematics, UTFPR - Federal Technological University of Paraná, Curitiba, Brazil
e-mail: piccirillo@utfpr.edu.br; a.m.tusset@gmail.com

J.M. Balthazar
Department of Statistics, Applied Mathematical and Computation, UNESP – São Paulo State University, Rio Claro, Brazil
e-mail: jmbaltha@rc.unesp.br

D. Bernardini • G. Rega
Dipartimento di Ingegneria Strutturale e Geotecnica, Sapienza Università di Roma, Rome, Italy
e-mail: davide.bernardini@uniroma1.it; giuseppe.rega@uniroma1.it

J. Awrejcewicz (ed.), *Applied Non-Linear Dynamical Systems*, Springer Proceedings in Mathematics & Statistics 93, DOI 10.1007/978-3-319-08266-0_37

reorientation). The distinct thermomechanical behavior of SMAs is the result of a transformation from the austenite (parent) phase to martensite (product) phase and vice versa [8].

The asymptotic nonlinear dynamics of SMA oscillators subject to ideal (harmonic) or nonideal (DC motor) mechanical excitation has been studied by various authors employing different constitutive models for the restoring force. For example, [3] used polynomial non-hysteretic models while [2, 5–7] used various types of thermomechanical internal variable hysteretic models.

Magnetorheological (MR) damper is a promising semi-active device in areas of vibration isolation for suspension systems and civil structures. The viscosity of MR fluid is controllable depending on input voltage or current. Since, inherently, the MR damper has uncertain nonlinear hysteresis dynamics, its modeling is an important issue in the semi-active control of vibration isolation control [11].

Some mathematical models have been introduced in recent work [14] to describe MR damper behavior, most of them based on friction phenomena descriptions such as the Bingham and Bouc–Wen models [9]. Although some of these models, e.g., the modified Bouc–Wen model [14], can accurately predict MR damper's dynamics, in most cases, they are complex and difficult to use in control design and parameter adaptation schemes. In [1] an MR damper model based on LuGre model is described. This model expresses the dynamic friction characteristics and the hysteresis effect. Results demonstrate very good model performance and indicate that the proposed model can be readily used in the design of semi-active control algorithms. This paper deals with the nonlinear dynamics of an oscillator studied in two versions where the restoring force is provided by the two kinds of smart materials. The first one incorporates an SMA whose thermomechanical behavior is described by the constitutive model proposed in [5]. The second one incorporates an MRD modeled by means of a modified LuGre dynamical friction model [1]. A comparison of these two oscillators is made in order to gain some information about the influence of the two kinds of nonlinearity on the system dynamics.

2 Equation of Motion of the Physical Model

2.1 SMA Oscillator

A shape memory oscillator (SMO) is characterized by a shape memory device (SMD) that provides a restoring force against the relative displacements of a pair of points of a main structure [2]. An SMD is composed of an arrangement of shape memory materials (SMM) that may be designed to yield various kinds of behavior. In this work the attention is focused on SMDs with pseudoelastic behavior (Fig. 1).

The SMO is considered within a thermomechanical environment characterized by a harmonic forcing $F(t) = \gamma cos(\alpha t)$ and a convective heat exchange $Q(\vartheta) = h(\vartheta_e - \vartheta)$ where γ and α are the excitation amplitude and frequency, respectively, ϑ_e is the environment temperature, and h is the coefficient of convective heat

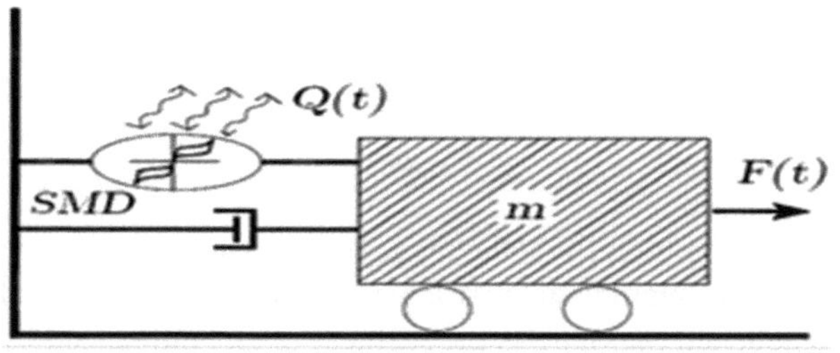

Fig. 1 Schematic representation of an SMD model

exchange [5]. At each time t the state of the SMO is described by displacement $x(t)$, velocity $v(t)$, an internal variable $\xi(t) \in [0, 1]$ that models the internal state of the SMD and by temperature $\vartheta(t)$. To model the complex hysteretic behavior of SMM the state depends not only on the actual value $\xi(t)$ but also on the value $\xi_0(t)$ of ξ at the beginning of the last process of change of ξ that occurred before time t. As discussed in [5] the time evolution of the state takes place according to a system of five-dimensional differential equations

$$\dot{x} = v \tag{1}$$

$$\dot{v} = -x + (sgn(x)\lambda)\,\xi - \zeta v + F \tag{2}$$

$$\dot{\xi} = Z\,[sgn(x)v - JQ] \tag{3}$$

$$\dot{\vartheta} = ZL\left[\frac{\Lambda}{J\lambda} + \vartheta\right][sgn(x)v - JQ] + Q \tag{4}$$

in which Z and Λ are constitutive functions that can take different expressions depending on suitable state-dependent criteria [5]. The system response depends on seven model parameters, $J, \lambda, q_1, q_2, q_3, L, h$, whose physical meaning is discussed later (see also [5]) as well as on the damping coefficient ζ of the viscous damper (Fig. 1).

2.2 *MRD Oscillator*

The friction mechanism is a phenomenon in which two surfaces make contact at a number of asperities at the microscopic level [4]. In the modified LuGre friction model [4] [17], this mechanism is expressed by the average behavior of the bristles. In [1], another MR damper model based on the LuGre model is described as

$$f = \sigma_a z + \sigma_0 z V + \sigma_1 \dot{z} + \sigma_2 \dot{x} + \sigma_b \dot{x} V \tag{5}$$

$$\dot{z} = \dot{x} - \sigma_0 a_0 |\dot{x}| z \tag{6}$$

where f is the damping force, V is the input voltage, and the parameters are defined as follows: $z(t)$ is the internal state variable (m); $\dot{x}$ is the velocity of the oscillator where MRD is attached (m/s); σ_0 is the stiffness of $z(t)$ influenced by V ($N/(m.\ V)$); σ_1 is the damping coefficient of $z(t)$ (Ns/m); σ_2 is the viscous damping coefficient (Ns/m); σ_a is the stiffness of $z(t)$ (N/m); σ_b is the viscous damping coefficient influenced by V ($Ns/(m.\ V)$); and a_0 is a constant value (V/N).

We now consider a one-degree-of-freedom oscillator where the nonlinear restoring damping force is provided by an MRD. We assume that stiffness is linear. Balance of linear momentum is expressed by

$$mx'' + cx' + f + kx = F \tag{7}$$

where m is the mass, c a coefficient of dissipation, and $F(t) = Acos(\omega t)$ an external force. The damping force in MRD can reach approximately 1,500 N when the voltage level in the MRD is 2.5 V [1]. Due to this strong force, the MRD is normalized by the value of 400 corresponding to the input voltage V equal to zero. Thus, the force developed by MRD isn't bigger than the viscous linear damping. It is convenient to work with dimensionless position and time, in such a way that Eq. (7) is rewritten in the following form:

$$\ddot{u} + \xi \dot{u} + F_{MRD} + u = \gamma cos(\alpha t) \tag{8}$$

where

$$\tau = \omega_0 t\ ,\ \xi = \frac{c}{m\omega_0},\ F_{MRD} = \frac{f}{f_{V0} m x_{st} \omega_0^2},\ \gamma = \frac{A}{m x_{st} \omega_0^2},\ \alpha = \frac{\omega}{\omega_0}$$

and x_{st} means a static displacement of the system, τ is the dimensionless time, and f_{V0} is the dimensional MRD damping force when the input voltage $V = 0$ (Fig. 2).

3 Numerical Simulation Results

The evolution of an SMO depends on seven parameters, which may be grouped as follows: mechanical parameters, λ, q_1, q_2, q_3, that reflect the basic features of the device (type and arrangement of the material) and determine the basic shape of the pseudoelastic loop observed in isothermal conditions; thermal parameters, L, h, that reflect the heat production, absorption, and exchange with the environment and

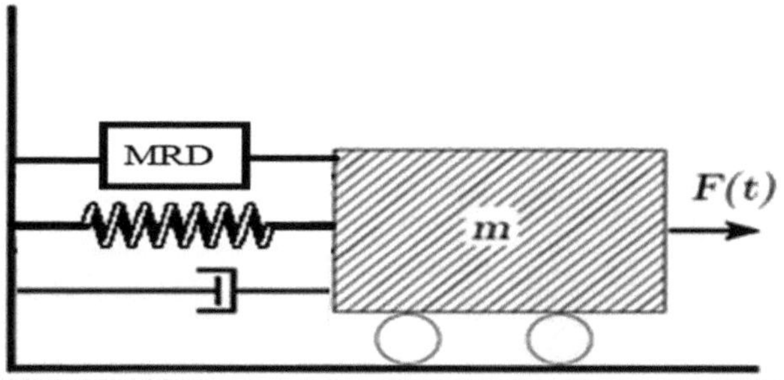

Fig. 2 Schematic representation of an MRD model

Table 1 SMA parameter values used in the simulation

λ	q_1	q_2	L	h	J
8.125	0.98	1.2	1.017	0.12	3.174

Table 2 MRD parameter values used in the simulation

σ_0	σ_1	σ_2	σ_a	σ_b	a_0
8.0×10^5	1.6×10^3	1.5×10^2	4.0×10^5	8.0×10^2	3.0×10^{-3}

therefore determine the temperature variations of the device; and a thermomechanical parameter J that determines the influence of the temperature variations on the transformation forces [5].

Numerical SMA dimensionless parameters are shown in Table 1 which correspond to the values taken from [5] and the parameter values of the MRD model used in the simulation are shown in Table 2 which correspond to the values taken from [1]. The parameters relative to SMA correspond to a typical device with a pseudoelastic loop with medium–high level of hysteresis. On the other hand, the MRD model used in this work has a bilinear form with respect to the input voltage V. The model has the following properties: (a) all parameters are positive because of the characteristic of the friction model; (b) there is no influence of v on the nonlinear dynamics of z; and (c) a viscous damping coefficient is proportional to V [9].

Due to nonlinearities, the SMA and MRD oscillator responses are more complex than the linear oscillator response. This section investigates the numerical dynamics by establishing a comparison between the two different systems. In order to quantify the structure associated with maximal amplitude, we construct the parameter space diagram for α (frequency excitation) versus γ (amplitude excitation), as shown in Figs. 3, 4, 5, and 6. To obtain this diagram, we use a grid of 100×100 cells. For each cell of the grid, we evaluate the largest amplitude response and plot using a color scale (gray, black, red, green, and blue) as shown on the right side of the diagram. Each simulation is carried out at fixed initial conditions, that is, $u = [0, 0]$. It can be observed that the better performance of MRD depends on voltage applied in damper. The parameter space plotted for MRD case is shown in Fig. 3 for the voltage applied

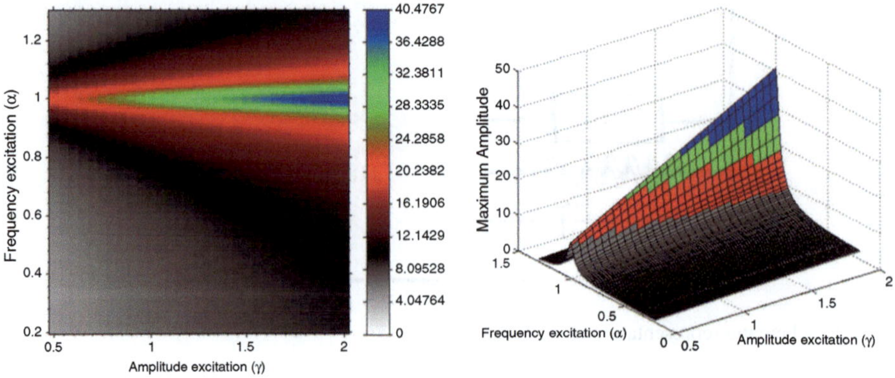

Fig. 3 Parameter space plots of the amplitude versus frequency excitation for the MRD oscillator for $V = 0$

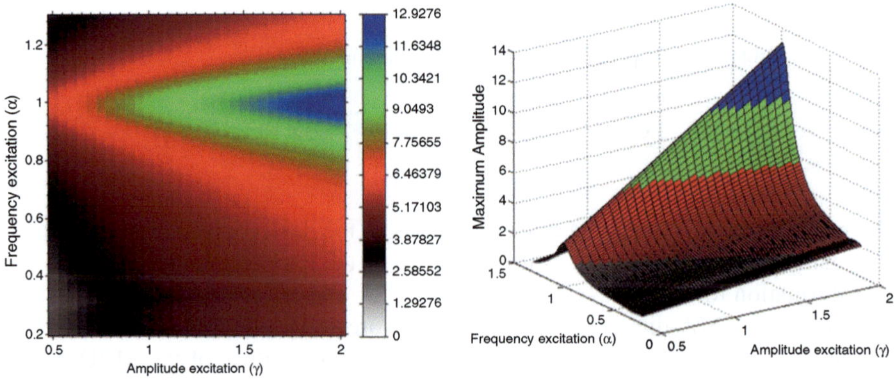

Fig. 4 Parameter space plots of the amplitude versus frequency excitation for the MRD oscillator for $V = 1$

value $V = 0$. It should be noted here that the worst response is for high-frequency excitation $\alpha \approx 1$ and high-amplitude excitation $\gamma \in [1.3, 2]$. Also, in this MRD case, the space parameter region has several small amplitude vibrations, because the damping force is significant to vibration reduction under these parameter conditions (α vs γ). In order to show the influence of the voltage applied in MRD for vibration attenuation of the system, other parameter space plots have been done and shown in Figs. 4 and 5. It should be noted that the damping effect of the MR damper is very significant in oscillator system. Figure 4 shows the response of the MRD oscillator considered $V = 1$. Note that the amplitude of the response is significantly smaller than that for the previous one. However, the critical situation occurs for the same amplitude and frequency. Figure 5 shows that again the maximum amplitude occurs at high-amplitude and high-frequency excitation, but this peak is smaller than the previous cases, thus bettering the damping efficiency.

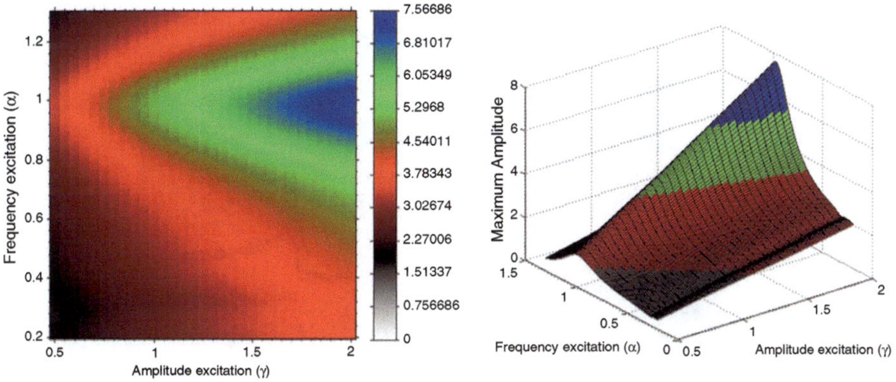

Fig. 5 Parameter space plots of the amplitude versus frequency excitation for the MRD oscillator for $V = 2$

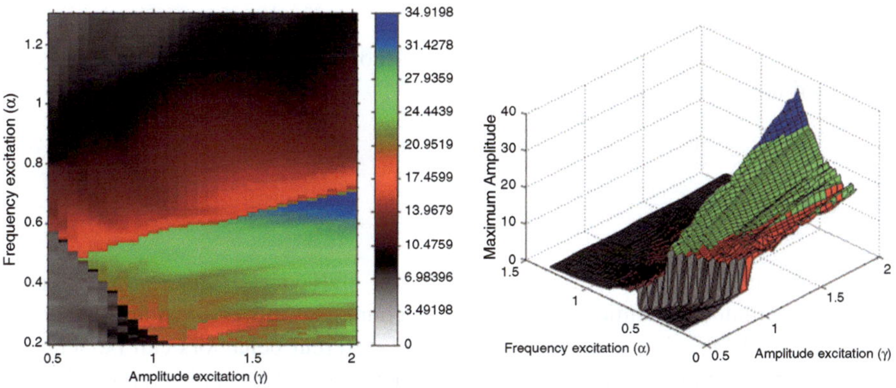

Fig. 6 Parameter space plots of the amplitude versus frequency excitation for the SMO

The parameter space plotted for SMA case is shown in Fig. 6. It is interesting to note that the behavior of the SMA and MRD oscillators is different. It should be observed here that the worst response is for low-frequency excitation $\alpha \approx 0.6$ and high amplitude $\gamma \in [1.5, 2]$. Figure 7 shows the parameter space from Fig. 6 consideringotherperspectives. Note that qualitatively the SMA tends to present better performance for high amplitude and frequency, but the MRD presents smaller amplitude responses for high level of amplitude and low frequency, and for small amplitude and frequency excitation the same response is present.

Initially, numerical simulations were run to investigate the effects of varying two parameters: amplitude and frequency excitation. The maximum force developed during each combination of these parameters is shown in Table 3. The force was evaluated at the amplitude/frequency range and the maximum force was collected considering some amplitude response (see first column in Table 3). All of the results

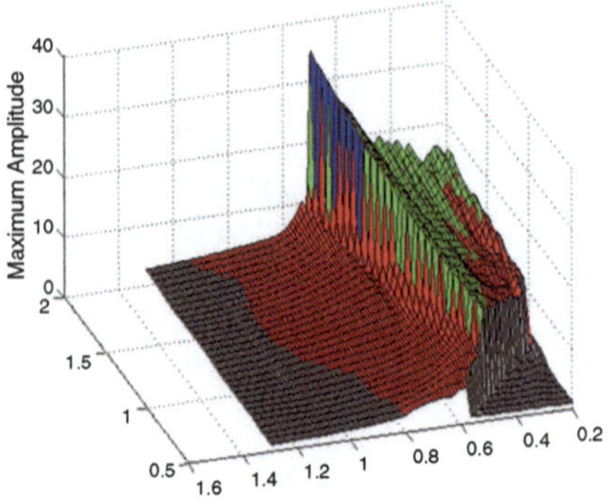

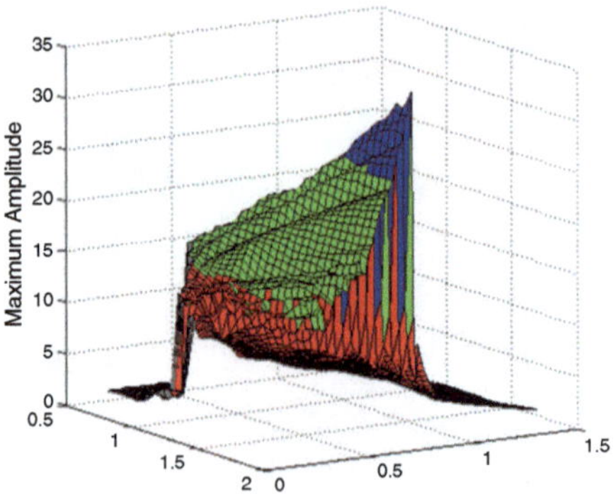

Fig. 7 Three-dimensional representation of the dependence of the maximum amplitude upon frequency excitation and amplitude excitation to SMO

listed are for the MR damper case in which 0, 1, and 2 V was applied in device and SMO. As can be seen, a large variety of nondimensional force responses are observed in both cases.

4 Conclusions

The purpose of this study was to analyze the influence of external parameters on the oscillator dynamics, taking into account two different smart materials: SMA and MRD. It is found that significantly different behaviors can be achieved depending on amplitude and frequency external parameters applied in mechanical oscillator. A nonparametric analysis is carried out considering the effects of system dissipation. It is possible to conclude that the high dissipation capacity of both materials due to the hysteresis loop is capable to improve the dynamical response of the oscillator. However, depending on the external parameter combinations, one material is better than the other. Independent of the voltage applied in MRD, the comparison between MRD and SMA oscillators has been made from a qualitative point of view. In cases of low levels of amplitude and frequency or low amplitude and high frequency the response is similar. In MRD case, the passive damping produces a damping force and therefore is suited for higher level of amplitude and low level of frequency. On the other hand, the SMA dissipation is suited to improve the response when the combination of high levels of amplitude and frequency occurs.

Table 3 Nondimensional force developed during variation of the parameters α and β—MRD and SMA cases

Maximum amplitude	Maximum force for $\mathbf{V} = 0$ (MRD)	Maximum force for $\mathbf{V} = 1$ (MRD)	Maximum force for $\mathbf{V} = 2$ (MRD)	Maximum force—SMA
4	0.13	1.50	0.90	1.16
8	0.26	1.01	1.72	1.30
12	0.27	1.51	–	3.61
16	0.33	–	–	8.13
20	0.44	–	–	11.83
24	0.49	–	–	15.85
28	0.55	–	–	19.79
32	0.63	–	–	23.95
36	0.69	–	–	–
40	0.73	–	–	–

References

1. Awrejcewicz, J., Olejnik, P.: Analysis of dynamic systems with various friction laws. Appl. Mech. Rev. Trans. ASME **58**(6), 389–411 (2005)
2. Bernardini, D., Rega, G.: Thermomechanical modeling, nonlinear dynamics and chaos in shape memory oscillators. Math. Comput. Model. Dyn. Syst. **11**, 291–314 (2005)
3. Bernardini, D., Vestroni, F.: Non-isothermal oscillations of pseudoelastic devices. Int. J. Nonlinear Mech. **38**, 1297–1313 (2003)
4. Canudas, C., Olsson, H., Aström, K.J., Lischinsky, P.: A new model for control of systems with friction. IEEE Trans. Autom. Control **40**, 419–42 (1995)
5. Lacarbonara, W., Bernardini, D., Vestroni, F.: Nonlinear thermomechanical oscillations of shape-memory devices. Int. J. Solids Struct. **41**, 1209–1234 (2004)
6. Lagoudas, D.C., Khan, M.M., Mayes, J.J., Henderson, B.K.: Pseudoelastic SMA spring elements for passive vibration isolation: Part II—simulations and experimental correlations. J. Intell. Mater. Syst. Struct. **15**, 443–470 (2004)
7. Nilkhamhang, I., Mori, T., Sano, A.: Direct and indirect stable adaptive control for suspension systems with MR Damper. In: Proceedings of the 17th World Congress the International Federation of Automatic Control, Seoul, Korea, 2008
8. Piccirillo, V., Balthazar, J.M., Pontes Jr., B.R., Felix, J.L.P.: Chaos control of a nonlinear oscillator with shape memory alloy using an optimal linear control: Part I—ideal energy source. Nonlinear Dyn. **55**, 139–149 (2009)
9. Sakai, C., Ohmori, H., Sano, A.: Modeling of MR damper with hysteresis for adaptive vibration control. In: Proceedings of the 42nd IEEE Conference on Decision and Control, Maui, Hawaii USA, 2003
10. Savi, M.A., Paula, A.S., Lagoudas, D.C.: Numerical investigation of an adaptive vibration absorber using shape memory alloys. J. Intell. Mater. Syst. Struct. **21**, 67–80 (2011)
11. Spencer, B.F., Dyke, S.J., Sain, M.K., Carlson, J.D.: Phenomenological model of a magnetorheological damper. ASCE J. Eng. Mech. **123**(3), 230–238 (1996)
12. Tusset, A.M., Balthazar, J.M.: On the chaotic suppression of both ideal and non-ideal duffing based vibrating systems, using a magneto rheological damper. Differ. Equat. Dyn. Syst. **21**, 105–121 (2013)
13. Tusset, A.M., Rafikov, M., Balthazar, J.M.: An intelligent controller design for magnetorheological damper based on quarter-car model. J. Vib. Control **12**, 1907–1920 (2009)
14. Wen, Y.K.: Method for random vibration of hysteretic systems. ASCE J. Eng. Mech. **102**(EM2), 249–263 (1976)
15. Williams, K., Chiu, G., Bernhard, R.: Dynamic modeling of a shape memory alloy adaptive tuned vibration absorber. J. Sound Vib. **280**, 211–234 (2005)

Dynamical Pendulum-Like Nonconservative Systems

Maxim V. Shamolin

Abstract We have elaborated the methods for the qualitative study of dissipative systems and systems with anti-dissipation that allow us, for example, to obtain conditions for bifurcation of birth of stable and unstable auto-oscillations. We succeeded in generalizing the method for studying plane topographical Poincaré systems to higher dimensions. In three-dimensional rigid body dynamics, we have discovered complete lists of first integrals of dissipative systems and systems with anti-dissipation that are transcendental (in the sense of classification of their singularities) functions that are expressed through elementary functions in a number of cases. We have discovered new qualitative analogs between the properties of motion of free bodies in a resisting medium that is fixed at infinity and bodies in an overrun medium flow.

1 Introduction

We study the nonconservative systems for which the methods for studying, for example, Hamiltonian systems, is not applicable in general. Therefore, for such systems, it is necessary, in some sense, to "directly" integrate the main equation of dynamics. Herewith, we offer more universal interpretation of both obtained cases and new ones of complete integrability in transcendental functions in two-, three-, and four-dimensional rigid body dynamics in a nonconservative force field.

The results of the proposed work are a development of the previous studies, including a certain applied problem from rigid body dynamics [4, 9, 11], where complete lists of transcendental first integrals expressed through a finite combination of elementary functions were obtained. Later on, this circumstance allows us to perform a complete analysis of all phase trajectories and show those their properties which have a roughness and are preserved for systems of a more general form. The complete integrability of such systems is related to symmetries of latent type.

M.V. Shamolin (✉)
Institute of Mechanics, Lomonosov Moscow State University,
1 Michurinskii Ave., 119192, Moscow, Russian Federation
e-mail: shamolin@imec.msu.ru; shamolin@rambler.ru

J. Awrejcewicz (ed.), *Applied Non-Linear Dynamical Systems*, Springer Proceedings in Mathematics & Statistics 93, DOI 10.1007/978-3-319-08266-0_38

As is known, the concept of integrability is sufficiently fuzzy in general. In its construction, it is necessary to take into account the meaning in which it is understood (we mean a certain criterion with respect to which one makes a conclusion that the structure of trajectories of the dynamical system considered is especially simple and "attractive and simple," in which class of functions we seek for first integrals, etc. (see also [8]).

In this work, we accept the approach that as the class of functions for first integrals, takes transcendental functions, and, moreover, elementary ones. Here, the transcendence is understood not in the sense of elementary function theory (for example, trigonometric functions), but in the sense of existence of essentially singular points for them (according to the classification accepted in the theory of function of one complex variable). In this case, we need to formally continue the function considered in the complex domain (see also [13]).

2 Methods for Analyzing Zero Mean Variable Dissipation Dynamical Systems in Spatial Dynamics of a Rigid Body

In this section, we consider the possibilities of extending the results of the plane dynamics of a rigid body interacting with a medium to the spatial case. We analyze the problems of a spherical pendulum placed in the overrunning medium flow and a spatial body motion under the existence of a certain non-integrable constraint and also show the mechanical and topological analogies of the latter two problems.

2.1 Statement of Problem of Spatial Body Motion in a Resisting Medium Under Streamline Flow Around

The conjectures presented in [4], which concern the medium properties, are reflected in constructing the spatial dynamical model of interaction of a medium with a body. In this connection, there arises the possibility of formalizing the model assumptions and obtaining the complete system of equations.

The whole interaction of the medium with the axe-symmetric body is concentrated on the part of the body surface, which has the form of a circular disk (Fig. 1).

Since the interaction is subjected to the streamline flow around laws, the force $\mathbf{S}$ of this interaction is directed along the normal to the disk, and, moreover, the point N of application of this force is determined by at least one parameter, the angle of attack α, which is made by the velocity vector $\mathbf{v} = \mathbf{v}_D$ of the disk point D and the exterior normal at this point (the line CD). The point D is the intersection point of the middle perpendicular dropped from the center of masses C to the plane of disk. Therefore, $DN = R(\alpha, \ldots)$.

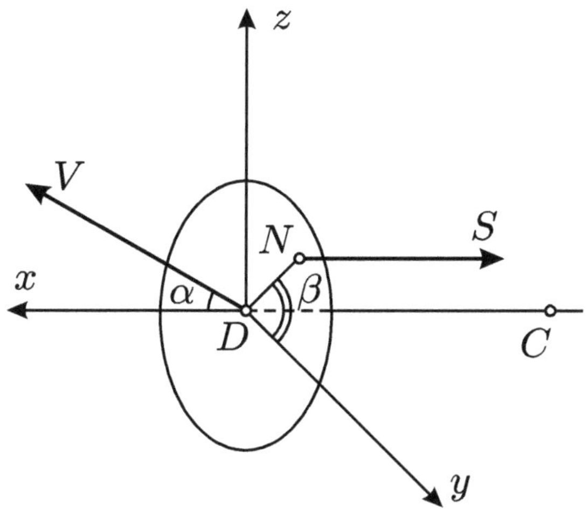

Fig. 1 Spatial model of the dynamical action of the medium on the axe-symmetric body

Assume that the value of the resistance force S has the form $S = s_1 v^2$, where v is the module of the velocity vector of the point D and the resistance coefficient s_1 (as in the plane case; see [9]) is a function of only the angle of attack α: $s_1 = s_1(\alpha)$. Therefore, as previously, we consider such an "extension" of the problem, which was mentioned in [1–4].

As in the plane case, along the line CD, an additional force $\mathbf{T}$ can act on the body; as before, it is called the "following force." The introduction of this force is used for ensuring some given classes of motion (in this case, $\mathbf{T}$ is the reaction of possible imposed constraints). In the case where there is no external force $\mathbf{T}$, the body executes a spatial free drag in the resisting medium.

Denote by $Dxyz$ the coordinate system related to the body (Fig. 1). This coordinate system related to the point D is chosen such that the tensor of inertia in this system has the diagonal form: $\mathrm{diag}\{I_1, I_2, I_3\}$. Assume that the mass distribution is such that the longitudinal principal axis of inertia coincides with the axis CD (this is the axis Dx), whereas the axes Dy and Dz lie in the plane of disk and compose the right coordinate system. Moreover, as it was noted, we consider the case of the dynamically symmetric rigid body, i.e., $I_2 = I_3$.

In this case, to describe the body position in the space, we choose the Cartesian coordinates $(\mathrm{x}_0, \mathrm{y}_0, \mathrm{z}_0)$ of the point D and three angles (θ, ψ, ϕ), which are defined similar to the navigational angles as follows (compare with [5]).

Let us represent the turn from the inertial system $D\mathrm{x}_0\mathrm{y}_0\mathrm{z}_0$ to the system $Dxyz$ as a composition of three turns $T_3(\phi) \circ T_2(\psi) \circ T_1(\theta)$ under which, first, the frame $(\mathbf{e}_{\mathrm{x}_0}, \mathbf{e}_{\mathrm{y}_0}, \mathbf{e}_{\mathrm{z}_0})$ is turned around the vector $\mathbf{e}_{\mathrm{x}_0}$ by the angle θ ($T_1(\theta)$ is executed):

$$(\mathbf{e}_{\mathrm{x}_0}, \mathbf{e}_{\mathrm{y}_0}, \mathbf{e}_{\mathrm{z}_0}) \to^{(T_1(\theta))} \left(\mathbf{e}_{\mathrm{x}_0}, \mathbf{e}^{(1)}_{\mathrm{y}_0}, \mathbf{e}^{(1)}_{\mathrm{z}_0}\right),$$

then the frame $\left(\mathbf{e}_{x_0}, \mathbf{e}_{y_0}^{(1)}, \mathbf{e}_{z_0}^{(1)}\right)$ is turned around the vector $\mathbf{e}_{y_0}^{(1)}$ by the angle ψ ($T_2(\psi)$ is executed):

$$\left(\mathbf{e}_{x_0}, \mathbf{e}_{y_0}^{(1)}, \mathbf{e}_{z_0}^{(1)}\right) \to^{(T_2(\psi))} \left(\mathbf{e}_{x_0}^{(2)}, \mathbf{e}_{y_0}^{(1)}, \mathbf{e}_z\right),$$

and, finally, the frame $\left(\mathbf{e}_{x_0}^{(2)}, \mathbf{e}_{y_0}^{(1)}, \mathbf{e}_z\right)$ is turned around the vector $\mathbf{e}_z$ by the angle ϕ ($T_3(\phi)$ is executed):

$$\left(\mathbf{e}_{x_0}^{(2)}, \mathbf{e}_{y_0}^{(1)}, \mathbf{e}_z\right) \to^{(T_3(\phi))} (\mathbf{e}_x, \mathbf{e}_y, \mathbf{e}_z).$$

In this case, the vectors having the components in the frame $(\mathbf{e}_x, \mathbf{e}_y, \mathbf{e}_z)$ obtain new coordinates in the basis $(\mathbf{e}_{x_0}, \mathbf{e}_{y_0}, \mathbf{e}_{z_0})$. In the basis $(\mathbf{e}_x, \mathbf{e}_y, \mathbf{e}_z)$, such a transformation is given by certain matrix, and then the phase state of the system is characterized by twelve quantities $(\dot{x}_0, \dot{y}_0, \dot{z}_0, \dot{\theta}; \dot{\psi}; \dot{\phi}; x_0, y_0, z_0, \theta, \psi, \phi)$.

Let us consider the spherical coordinates (v, α, β) of the velocity vector $\mathbf{v} = \mathbf{v}_D$ endpoint of the point D in which the angle β is measured from the axis Dy in the disk plane up to the line (DN) which is the intersection of two planes, one of which contains the vector $\mathbf{v}$ and the axis Dx and the other is the disk plane.

The latter spherical coordinates and also the components of the angular velocity are expressed through the phase variables $(\dot{x}_0, \dot{y}_0, \dot{z}_0, \dot{\theta}; \dot{\psi}; \dot{\phi}; x_0, y_0, z_0, \theta, \psi, \phi)$ via non-integrable relations. Therefore, the phase state of the system is determined by the functions $(v, \alpha, \beta, \Omega_x, \Omega_y, \Omega_z, x_0, y_0, z_0, \theta, \psi, \phi)$, and the first six functions are considered as quasi-velocities of the system. Here, the tuple $(\Omega_x, \Omega_y, \Omega_z)$ is defined as

$$\Omega = \Omega_x \mathbf{e}_x + \Omega_y \mathbf{e}_y + \Omega_z \mathbf{e}_z,$$

where Ω is the absolute angular velocity vector of the rigid body.

Since the generalized forces are independent of the body position in the space, the coordinates $(x_0, y_0, z_0, \theta, \psi, \phi)$ are cyclic. This allows us to consider the dynamical part of the equations of motion as an independent subsystem.

Let us introduce the sign-alternating auxiliary function $s(\alpha) = s_1(\alpha)\operatorname{sign}\cos\alpha$.

By the theorem on the motion of the center of masses in the space in projections to the related axes (x, y, z) and the theorem on the change of the kinematic moment with respect to these axes, we obtain the following complete system of differential equations in the dynamical quasi-velocity space $\mathbf{R}_+^1\{v\} \times \mathbf{S}^2\{\alpha, \beta\} \times \mathbf{R}^3\{\Omega_x, \Omega_y, \Omega_z\}$:

$$\dot{v}\cos\alpha - \dot{\alpha}v\sin\alpha + \Omega_y v\sin\alpha\sin\beta - \Omega_z v\sin\alpha\cos\beta + \sigma(\Omega_y^2 + \Omega_z^2) = F_x/m,$$

$$\dot{v}\sin\alpha\cos\beta + \dot{\alpha}v\cos\alpha\cos\beta - \dot{\beta}v\sin\alpha\sin\beta + \Omega_z v\cos\alpha -$$

$$-\Omega_x v\sin\alpha\sin\beta - \sigma\Omega_x\Omega_y - \sigma\dot{\Omega}_z = 0,$$

$$\dot{v}\sin\alpha\sin\beta + \dot{\alpha}v\cos\alpha\sin\beta + \dot{\beta}v\sin\alpha\cos\beta + \Omega_x v\sin\alpha\cos\beta - \tag{1}$$

$$-\Omega_y v\cos\alpha - \sigma\Omega_x\Omega_z + \sigma\dot{\Omega}_y = 0,$$

$$I_1\dot{\Omega}_x = 0,$$

$$I_2\dot{\Omega}_y + (I_1 - I_2)\Omega_x\Omega_z = -z_N\,(\alpha,\beta,\Omega/v)\,s(\alpha)v^2,$$

$$I_2\dot{\Omega}_z + (I_2 - I_1)\Omega_x\Omega_y = y_N\,(\alpha,\beta,\Omega/v)\,s(\alpha)v^2,$$

where $F_x = T - s(\alpha)v^2$.

The coordinates of the point N in the system $(\mathbf{e}_x, \mathbf{e}_y, \mathbf{e}_z)$ take the form

$$\big(0, y_N(\alpha,\beta,\Omega/v), z_N(\alpha,\beta,\Omega/v)\big),$$

$y_N(\alpha,\beta,\Omega/v) = R(\alpha)\cos\beta - h\Omega_z/v$, $z_N(\alpha,\beta,\Omega/v) = R(\alpha)\sin\beta + h\Omega_y/v$.

We can complement the system (1) by the kinematic relations. As before, we use the following notation: σ is the distance DC and m is the mass of the body.

The dynamical system (1) contains the functions $R(\alpha)$ and $s(\alpha)$. To qualitatively describe them, we use the existing experimental information about the properties of the streamline flow around (see [3,5]).

2.2 *Case of Body Motion in a Medium Under Existence of a Certain Non-integrable Constraint and Beginning of Qualitative Analysis*

As in the plane case, we consider the class of motions under which identity

$$v = |\mathbf{v}| \equiv \text{const}, \tag{2}$$

being a non-integrable constraint holds.

2.2.1 On Analytical Integral

In the case of the dynamically symmetric rigid body, system (1) has the analytic first integral

$$\Omega_x \equiv \Omega_{x0} = \text{const}, \tag{3}$$

i.e., the generalized forces admit a body self-rotation around the longitudinal dynamical symmetry axis.

2.2.2 On Appearance of an Independent Subsystem

Let us use the methodological tool of reducing the system order, i.e., from the first equation of (1), let us express the function T so that the total derivative of v by the system (1) vanishes. As a result of this order reduction, system (1) has an independent subsystem of the form

$$\dot{\alpha} v \cos\alpha\cos\beta - \dot{\beta} v \sin\alpha\sin\beta + \Omega_z v\cos\alpha - \Omega_{x0} v \sin\alpha\sin\beta 4-$$

$$-\sigma\Omega_{x0}\Omega_y - \sigma\dot{\Omega}_z = 0,$$

$$\dot{\alpha} v \cos\alpha\sin\beta + \dot{\beta} v \sin\alpha\cos\beta - \Omega_y v\cos\alpha + \Omega_{x0} v \sin\alpha\cos\beta + \tag{4}$$

$$+\sigma\Omega_{x0}\Omega_z + \sigma\dot{\Omega}_y = 0,$$

$$I_2\dot{\Omega}_y = -z_N(\alpha, \beta, \Omega/v)\, s(\alpha) v^2,$$

$$I_2\dot{\Omega}_z = y_N(\alpha, \beta, \Omega/v)\, s(\alpha) v^2,$$

in which the parameter v is added to constant parameters.

2.2.3 Case of Zero Projection of Angular Velocity on Longitudinal Axis and Case of Analytic System

Let us consider the trajectories of motion of system (4) on the level of integral (3) for

$$\Omega_{x0} = 0. \tag{5}$$

In this case, it takes the form

$$\dot{\alpha} v \cos\alpha\cos\beta - \dot{\beta} v \sin\alpha\sin\beta + \Omega_z v\cos\alpha - \sigma\dot{\Omega}_z = 0,$$

$$\dot{\alpha} v \cos\alpha\sin\beta + \dot{\beta} v \sin\alpha\cos\beta - \Omega_y v\cos\alpha + \tag{6}$$

$$+\sigma\dot{\Omega}_y = 0,$$

$$I_2\dot{\Omega}_y = -z_N(\alpha, \beta, \Omega/v)\, s(\alpha) v^2, \; I_2\dot{\Omega}_z = y_N(\alpha, \beta, \Omega/v)\, s(\alpha) v^2. \tag{7}$$

The first two equations (i.e., (6)) reduce to the form

$$\dot{\alpha} v \cos\alpha + v\cos\alpha\left[\Omega_z\cos\beta - \Omega_y\sin\beta\right] +$$

$$+\sigma\left[-\dot{\Omega}_z\cos\beta+\dot{\Omega}_y\sin\beta\right]=0, \tag{8}$$

$$\dot{\beta}v\sin\alpha-v\cos\alpha\left[\Omega_y\cos\beta+\Omega_z\sin\beta\right]+$$

$$+\sigma\left[\dot{\Omega}_y\cos\beta+\dot{\Omega}_z\sin\beta\right]=0, \tag{9}$$

If, for example, S. A. Chaplygin conditions [1]

$$R(\alpha)=A\sin\alpha,\ s(\alpha)=B\cos\alpha;\ A,B>0, \tag{10}$$

hold, then, introducing the notations

$$n_0^2=\frac{AB}{I_2},\ b=\sigma n_0,\ H_1=\frac{Bh}{I_2n_0},\ [b]=[H_1]=1,$$

we have the following analytic system:

$$\dot{\alpha}-bn_0v\sin\alpha+(1+bH_1)\left[\Omega_z\cos\beta-\Omega_y\sin\beta\right]=0, \tag{11}$$

$$\dot{\beta}\sin\alpha-(1+bH_1)\cos\alpha[\Omega_y\cos\beta+\Omega_z\sin\beta]=0. \tag{12}$$

Let us complement it by the equations

$$\dot{\Omega}_y=-n_0^2v^2\sin\alpha\cos\alpha\sin\beta-H_1n_0v\Omega_y\cos\alpha, \tag{13}$$

$$\dot{\Omega}_z=n_0^2v^2\sin\alpha\cos\alpha\cos\beta-H_1n_0v\Omega_z\cos\alpha. \tag{14}$$

The system (11)–(14) is closed. Note that system (11), (12) is equivalent to (8), (9) only outside the manifold

$$\mathcal{O}=\{(\alpha,\beta,\Omega_y,\Omega_z):\ \cos\alpha=0\}. \tag{15}$$

Let us project the angular velocity to the movable coordinate system Dz_1z_2 (turning the system Dyz by the angle $-\beta$) such that

$$z_1=\Omega_y\cos\beta+\Omega_z\sin\beta,\quad z_2=\Omega_z\cos\beta-\Omega_y\sin\beta. \tag{16}$$

In this case, since $\dot{z_1}=\dot{\beta}z_2$, system (6), (7) is equivalent to the following system outside the manifold $\mathcal{O}'$:

$$\dot{\alpha}=-\left(1+b\frac{h}{I_2n_0}\frac{s(\alpha)}{\cos\alpha}\right)z_2+b\frac{v}{I_2n_0}\frac{\mathcal{F}(\alpha)}{\cos\alpha}, \tag{17}$$

$$\dot{z_2}=\frac{v^2}{I_2}\mathcal{F}(\alpha)-\left(1+b\frac{h}{I_2n_0}\frac{s(\alpha)}{\cos\alpha}\right)z_1^2\frac{\cos\alpha}{\sin\alpha}-\frac{hv}{I_2}\frac{s(\alpha)}{\cos\alpha}z_2\cos\alpha, \tag{18}$$

$$\dot{z_1} = \left(1 + b\frac{h}{I_2 n_0}\frac{s(\alpha)}{\cos\alpha}\right) z_1 z_2 \frac{\cos\alpha}{\sin\alpha} - \frac{hv}{I_2}\frac{s(\alpha)}{\cos\alpha} z_1 \cos\alpha, \tag{19}$$

$$\dot{\beta} = \left(1 + b\frac{h}{I_2 n_0}\frac{s(\alpha)}{\cos\alpha}\right) z_1 \frac{\cos\alpha}{\sin\alpha}, \tag{20}$$

where

$$\mathscr{O}' = \{(\alpha, \beta, \Omega_y, \Omega_z) : \ \sin\alpha\cos\alpha = 0\}, \ \mathscr{F}(\alpha) = R(\alpha)s(\alpha). \tag{21}$$

Under S. A. Chaplygin condition (10) and the substitutions

$$< \cdot > \mapsto n_0 v <'>, \ z_k \mapsto n_0 v z_k, \ k = 1, 2,$$

the system (17)–(20) takes the form of the analytic system

$$\dot{\alpha} = -(1 + bH_1) z_2 + b\sin\alpha, \tag{22}$$

$$\dot{z_2} = \sin\alpha\cos\alpha - (1 + bH_1) z_1^2 \frac{\cos\alpha}{\sin\alpha} - H_1 z_2 \cos\alpha, \tag{23}$$

$$\dot{z_1} = (1 + bH_1) z_1 z_2 \frac{\cos\alpha}{\sin\alpha} - H_1 z_1 \cos\alpha, \tag{24}$$

$$\dot{\beta} = (1 + bH_1) z_1 \frac{\cos\alpha}{\sin\alpha}, \tag{25}$$

Outside the manifold $\mathscr{O}'$, from the dynamical system (17)–(20), the independent third-order system (17)–(19) is separated. The system (17)–(20) is considered on the tangle bundle to the two-dimensional sphere (see [13]). Since we have the "separation" of the third-order system, the phase space of our system has a number of bundles.

2.2.4 Symmetries of System Vector Field in Quasi-velocity Phase Space

The vector field of system (17)–(19) has the following three types of symmetries:

1. The central symmetry. Near $(\pi k, 0, 0)$, $k \in \mathbf{Z}$, such a symmetry arises owing to that the field in the coordinates (α, z_2, z_1) alternates its sign under the change

$$\begin{pmatrix} \pi k - \alpha \\ -z_2 \\ -z_1 \end{pmatrix} \longrightarrow \begin{pmatrix} \pi k + \alpha \\ z_2 \\ z_1 \end{pmatrix}$$

2. A certain mirror symmetry (CMS). With respect to the planes Λ_i, $i \in \mathbf{Z}$, where

$$\Lambda_i = \left\{(\alpha, z_2, z_1) : \alpha = \frac{\pi}{2} + \pi i\right\},$$

such a symmetry arises owing to that the α-component of the system vector fields in the coordinates (α, z_2, z_1) is preserved under the change

$$\begin{pmatrix} \pi/2 + \pi i - \alpha \\ z_2 \\ z_1 \end{pmatrix} \longrightarrow \begin{pmatrix} \pi/2 + \pi i + \alpha \\ z_2 \\ z_1 \end{pmatrix}$$

whereas the z_2- and z_1-components alternate their signs.

3. The symmetry with respect to the plane

$$\{(\alpha, z_2, z_1) : z_1 = 0\};$$

precisely, the z_2- and α-components of the system vector field are preserved under the change

$$\begin{pmatrix} \alpha \\ z_2 \\ -z_1 \end{pmatrix} \longrightarrow \begin{pmatrix} \alpha \\ z_2 \\ z_1 \end{pmatrix}$$

whereas the z_1-component alternates its sign.

2.3 *On Transcendental Integrability of System*

This subsection is devoted to studying the possibilities of complete integration of the dynamical system considered. Here, we present first integrals of system (22)–(25) expressed through elementary functions and also discuss the way of integrating the general system (17)–(20).

2.3.1 Complete List of Invariant Relations

At the beginning we compare the third-order system (22)–(24) to the nonautonomous second-order system

$$\frac{dz_2}{d\alpha} = \frac{\sin\alpha\cos\alpha - (1 + bH_1)z_1^2 \cos\alpha/\sin\alpha - H_1 z_2 \cos\alpha}{-(1 + bH_1)z_2 + b\sin\alpha}, \tag{26}$$

$$\frac{dz_1}{d\alpha} = \frac{(1 + bH_1)z_1 z_2 \cos\alpha/\sin\alpha - H_1 z_1 \cos\alpha}{-(1 + bH_1)z_2 + b\sin\alpha}. \tag{27}$$

Let us rewrite the system (26), (26) on algebraic form using the substitution $\tau = \sin\alpha$

$$\frac{dz_2}{d\tau} = \frac{\tau - (1 + bH_1)z_1^2/\tau - H_1 z_2}{-(1 + bH_1)z_2 + b\tau}, \quad \frac{dz_1}{d\tau} = \frac{(1 + bH_1)z_1 z_2/\tau - H_1 z_1}{-(1 + bH_1)z_2 + b\tau}. \tag{28}$$

Later on, if we introduce the uniform variables by the formulas $z_k = u_k\tau,\ k = 1, 2$, we shall reduce the system (28) to the following form:

$$\tau\frac{du_2}{d\tau} = \frac{(1 + bH_1)(u_2^2 - u_1^2) - (b + H_1)u_2 + 1}{-(1 + bH_1)u_2 + b}, \tag{29}$$

$$\tau\frac{du_1}{d\tau} = \frac{2(1 + bH_1)u_1 u_2 - (b + H_1)u_1}{-(1 + bH_1)u_2 + b}. \tag{30}$$

Let us compare the second-order system (29), (30) to the nonautonomous first-order

$$\frac{du_2}{du_1} = \frac{1 - (1 + bH_1)(u_1^2 - u_2^2) - (b + H_1)u_2}{2(1 + bH_1)u_1 u_2 - (b + H_1)u_1}, \tag{31}$$

which is reduced uncomplicated to the complete differential:

$$d\left(\frac{(1 + bH_1)(u_2^2 + u_1^2) - (b + H_1)u_2 + 1}{u_1}\right) = 0. \tag{32}$$

And so, equation (31) has the following first integral:

$$\frac{(1 + bH_1)(u_2^2 + u_1^2) - (b + H_1)u_2 + 1}{u_1} = C_1 = \text{const}, \tag{33}$$

which in former variables is looked like

$$\frac{(1 + bH_1)(z_2^2 + z_1^2) - (b + H_1)z_2 \sin\alpha + \sin^2\alpha}{z_1 \sin\alpha} = C_1 = \text{const}. \tag{34}$$

Later on, let us find the evident form of the additional first integral of the third-order system (22)–(24). At the beginning for this we shall transform the invariant relation (33) for $u_1 \neq 0$ as follows:

$$\left(u_2 - \frac{b + H_1}{2(1 + bH_1)}\right)^2 + \left(u_1 - \frac{C_1}{2(1 + bH_1)}\right)^2 = \frac{(b - H_1)^2 + C_1^2 - 4}{4(1 + bH_1)^2}. \tag{35}$$

As is seen, the parameters of the given invariant relation should satisfy the condition

$$(b - H_1)^2 + C_1^2 - 4 \geq 0, \tag{36}$$

and the phase space of the system (22)–(24) is stratified on the family of the surfaces which is assigned by the equality (35).

Thus, by virtue of the relation (33), equation (29) has the form

$$\tau \frac{du_2}{d\tau} = \frac{2(1 + bH_1)u_2^2 - 2(b + H_1)u_2 + 2 - C_1 U_1(C_1, u_2)}{b - (1 + bH_1)u_2}, \tag{37}$$

where

$$U_1(C_1, u_2) = \frac{1}{2(1 + bH_1)}\{C_1 \pm U_2(C_1, u_2)\}, \tag{38}$$

$$U_2(C_1, u_2) = \sqrt{C_1^2 - 4(1 + bH_1)(1 - (b + H_1)u_2 + (1 + bH_1)u_2^2)},$$

herewith, the constant of the integration C_1 is chosen from the condition (36).

Therefore, the quadrature for the search of the additional first integral of the system (22)–(24) has the form

$$\int \frac{d\tau}{\tau} =$$

$$= \int \frac{(b - (1 + bH_1)u_2)du_2}{2(1 - (b + H_1)u_2 + (1 + bH_1)u_2^2) - C_1\{C_1 \pm U_2(C_1, u_2)\}/(2(1 + bH_1))}. \tag{39}$$

The left-hand side (accurate to the additive constant), obviously, is equal to $\ln|\sin\alpha|$. If

$$u_2 - \frac{b + H_1}{2(1 + bH_1)} = w_1, \; b_1^2 = (b - H_1)^2 + C_1^2 - 4, \tag{40}$$

then the right-hand side of the equality (39) has the form

$$-\frac{1}{4}\int \frac{d(b_1^2 - 4(1 + bH_1)w_1^2)}{(b_1^2 - 4(1 + bH_1)w_1^2) \pm C_1\sqrt{b_1^2 - 4(1 + bH_1)w_1^2}} -$$

$$-(b - H_1)(1 + bH_1)\int \frac{dw_1}{(b_1^2 - 4(1 + bH_1)w_1^2) \pm C_1\sqrt{b_1^2 - 4(1 + bH_1)w_1^2}} =$$

$$= -\frac{1}{2}\ln\left|\frac{\sqrt{b_1^2 - 4(1 + bH_1)w_1^2}}{C_1} \pm 1\right| \pm \frac{b - H_1}{2} I_1, \tag{41}$$

where

$$I_1 = \int \frac{dw_3}{\sqrt{b_1^2 - w_3^2}(w_3 \pm C_1)}, \quad w_3 = \sqrt{b_1^2 - 4(1 + bH_1)w_1^2}. \tag{42}$$

Three cases are possible for the calculation of the integral (42):

I. $|b - H_1| > 2$.

$$I_1 = -\frac{1}{2\sqrt{(b-H_1)^2-4}} \ln\left|\frac{\sqrt{(b-H_1)^2-4} + \sqrt{b_1^2 - w_3^2}}{w_3 \pm C_1} \pm \frac{C_1}{\sqrt{(b-H_1)^2-4}}\right| +$$

$$+\frac{1}{2\sqrt{(b-H_1)^2-4}} \ln\left|\frac{\sqrt{(b-H_1)^2-4} - \sqrt{b_1^2 - w_3^2}}{w_3 \pm C_1} \mp \frac{C_1}{\sqrt{(b-H_1)^2-4}}\right| + \text{const.} \tag{43}$$

II. $|b - H_1| < 2$.

$$I_1 = \frac{1}{\sqrt{4-(b-H_1)^2}} \arcsin \frac{\pm C_1 w_3 + b_1^2}{b_1(w_3 \pm C_1)} + \text{const.} \tag{44}$$

III. $|b - H_1| = 2$.

$$I_1 = \mp \frac{\sqrt{b_1^2 - w_3^2}}{C_1(w_3 \pm C_1)} + \text{const.} \tag{45}$$

When we return to the variable

$$w_1 = \frac{z_2}{\sin\alpha} - \frac{b + H_1}{2(1 + bH_1)}, \tag{46}$$

we shall have the final form for the value I_1:

I. $|b - H_1| > 2$.

$$I_1 = -\frac{1}{2\sqrt{(b-H_1)^2-4}} \ln\left|\frac{\sqrt{(b-H_1)^2-4} \pm 2(1+bH_1)w_1}{\sqrt{b_1^2 - 4(1+bH_1)^2 w_1^2} \pm C_1} \pm \frac{C_1}{\sqrt{(b-H_1)^2-4}}\right| +$$

$$+\frac{1}{2\sqrt{(b-H_1)^2-4}}\ln\left|\frac{\sqrt{(b-H_1)^2-4}\mp 2(1+bH_1)w_1}{\sqrt{b_1^2-4(1+bH_1)^2w_1^2}\pm C_1}\mp\frac{C_1}{\sqrt{(b-H_1)^2-4}}\right|+\text{const.} \tag{47}$$

II. $|b-H_1|<2$.

$$I_1=\frac{1}{\sqrt{4-(b-H_1)^2}}\arcsin\frac{\pm C_1\sqrt{b_1^2-4(1+bH_1)^2w_1^2}+b_1^2}{b_1(\sqrt{b_1^2-4(1+bH_1)^2w_1^2}\pm C_1)}+\text{const.} \tag{48}$$

III. $|b-H_1|=2$.

$$I_1=\mp\frac{2(1+bH_1)w_1}{C_1(\sqrt{b_1^2-4(1+bH_1)^2w_1^2}\pm C_1)}+\text{const.} \tag{49}$$

So, the additional first integral was found right before for the third-order system (22)–(24), i.e., it was presented the complete tuple of the first integrals which are the transcendental functions of its own phase variables.

It is necessary to substitute formally the left-hand side of the first integral (33) instead of C_1 in the expression of the found first integral.

Then the obtained additional first integral has the following structural form (which is similar to the transcendental first integral from the plane-parallel dynamics; see [4]):

$$\ln|\sin\alpha|+G_2\left(\sin\alpha,\frac{z_2}{\sin\alpha},\frac{z_1}{\sin\alpha}\right)=C_2=\text{const.} \tag{50}$$

Thus, there are already found two the independent first integrals for the integration of the fourth-order system (22)–(25). And to complete its integrability it is sufficient to find one more (additional) first integral which "joins" equation (25).

Since

$$\frac{du_1}{d\tau}=\frac{u_1(2(1+bH_1)u_2-(b+H_1))}{(b-(1+bH_1)u_2)\tau},\quad\frac{d\beta_1}{d\tau}=\frac{(1+bH_1)u_1}{(b-(1+bH_1)u_2)\tau}, \tag{51}$$

then

$$\frac{du_1}{d\beta_1}=2u_2-\frac{b+H_1}{1+bH_1}. \tag{52}$$

It is obvious that for $u_1\neq 0$ the following equality is fulfilled:

$$u_2=\frac{1}{2(1+bH_1)}\left((b+H_1)\pm\sqrt{b_1^2-(2(1+bH_1)u_1-C_1)^2}\right), \tag{53}$$

$$b_1^2 = (b - H_1)^2 + C_1^2 - 4,$$

and then the integration of the following quadrature

$$\beta_1 + \mathrm{const} = \pm(1 + bH_1) \int \frac{du_1}{\sqrt{b_1^2 - (2(1 + bH_1)u_1 - C_1)^2}} \tag{54}$$

will bring to the invariant relation

$$2(\beta_1 + C_3) = \pm \arcsin \frac{2(1 + bH_1)u_1 - C_1}{\sqrt{(b - H_1)^2 + C_1^2 - 4}}, \quad C_3 = \mathrm{const}. \tag{55}$$

In other words, the equality

$$\sin[2(\beta_1 + C_3)] = \pm \frac{2(1 + bH_1)u_1 - C_1}{\sqrt{(b - H_1)^2 + C_1^2 - 4}} \tag{56}$$

is fulfilled and under the transition to the old variables

$$\sin[2(\beta_1 + C_3)] = \pm \frac{2(1 + bH_1)z_1 - C_1 \sin\alpha}{\sqrt{(b - H_1)^2 + C_1^2 - 4 \sin\alpha}}. \tag{57}$$

In principle, it makes possible to stop on the latter equality to achieve the additional invariant relation "joining" equation (25), if we add to this equality that it is necessary to substitute formally the left-hand side of the first integral (33) instead of C_1 in the latter expression.

But we shall make the certain transformations which is reduced to obtaining the following evident form of the additional first integral (herewith, the equality (33) is used):

$$\mathrm{tg}^2[2(\beta_1 + C_3)] =$$

$$= \frac{((1 + bH_1)u_1^2 - (1 + bH_1)u_2^2 + (b + H_1)u_2 - 1)^2}{u_1^2(2(1 + bH_1)u_2 - (b + H_1))^2}. \tag{58}$$

Returning to the old coordinates, we shall obtain the additional invariant relation as the form

$$\mathrm{tg}^2[2(\beta_1 + C_3)] =$$

$$= \frac{((1 + bH_1)z_1^2 - (1 + bH_1)z_2^2 + (b + H_1)z_2 \sin\alpha - \sin^2\alpha)^2}{z_1^2(2(1 + bH_1)z_2 - (b + H_1)\sin\alpha)^2}, \tag{59}$$

or finally

$$-\beta_1 \pm \frac{1}{2}\times$$

$$\times \mathrm{arctg}\frac{(1+bH_1)z_1^2-(1+bH_1)z_2^2+(b+H_1)z_2\sin\alpha-\sin^2\alpha}{z_1(2(1+bH_1)z_2-(b+H_1)\sin\alpha)} = C_3 = \mathrm{const}. \tag{60}$$

And so, the system of dynamic equations (1) under S. A. Chaplygin conditions (10) has five invariant relations in considered case: there exist the analytical non-integrable constraint (2), the cyclic first integral (3), (5), and the first integral (34), and also there exists the first integral expressed by the relations (43)–(50) which is the transcendental function of its phase variables (in sense of complex analysis also) and expresses in terms of finite combination of the elementary functions, and finally, there exists the transcendent first integral (60).

Theorem 1. *The system (1) under the conditions (10), (2), (5) possesses five invariant relations (the complete tuple), three of which are the transcendental functions from the complex analysis view of point. Herewith, all the relations express in terms of the finite combination of the elementary functions.*

2.3.2 On Pendulum with Variable Dissipation

The system (17)–(20) is also a pendulum zero mean variable dissipation system. The motion is executed under the action of the following two forces: the potential force

$$\frac{v^2}{I_2}\mathscr{F}(\alpha)$$

and the linear-in-velocity force

$$b\frac{v}{I_2 n_0}\alpha'\frac{d}{d\alpha}\frac{\mathscr{F}(\alpha)}{\cos\alpha}$$

with variable coefficient. In a certain subspace, this coefficient has the strictly positive sign, and, therefore, the energy pumping from the is occurred here. In the other subspace, this coefficient has the strictly negative sign. Therefore the force scatters the energy forcing the body to damp its motion.

The latter remarks show that we deal with dissipative *zero mean variable dissipation system*. The phase space is decomposed into a union of alternating domain, in each of which there is a dissipation of only one sign.

Remark. In the case of the system (17)–(20) the search for integrals reduces to integrating the Riccati equations whose solutions are not expressed in elementary functions in the most general case.

2.4 Problem on Spatial Pendulum in Overrunning Medium Flow

Analogously to the plane case [4], let us consider the problem of the spatial pendulum placed in the overrunning medium flow.

Let a convex plane domain (circular disk, for simplicity) be clamped perpendicularly to the holder OD by a spherical hinge, and let it be in the overrunning medium flow, which moves with a constant velocity $\mathbf{v}_\infty \neq \mathbf{0}$. Assume that the holder does not create any resistance (Fig. 2).

The total force $\mathbf{S}$ of the medium flow action on the body is directed parallel to the holder, and the point N of application of this force is determined by only one parameter, the angle of attack α measured between the velocity vector $\mathbf{v}_D$ of the point D with respect to the flow and the holder. Therefore, the force $\mathbf{S}$ is directed along the normal to the side opposite to the direction of the velocity $\mathbf{v}_D$ and passes through a certain point N of the plane domain displaced from the point D forward with respect to the direction of $\mathbf{v}_D$. Such conditions arise in using the model of streamline flow around spatial bodies [4].

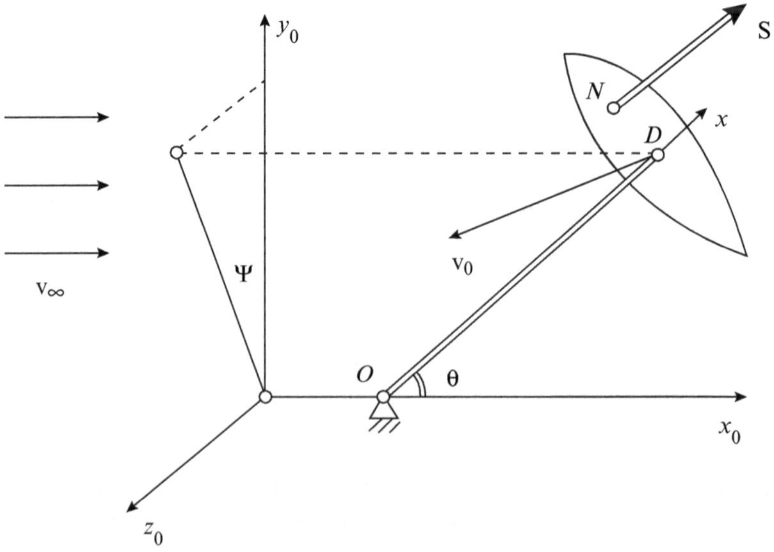

Fig. 2 Spherical pendulum in the homogeneous overrunning medium flow

The vector **e** determines the orientation of the holder. Then $\mathbf{S} = s_1(\alpha)v_D^2\mathbf{e}$, where the resistance coefficient has the form $s_1 = s_1(\alpha) = s(\alpha)\text{sign}\cos\alpha$. Let $O\text{x}_0\text{y}_0\text{z}_0$ be the immovable coordinate system. The direction of the overrunning flow coincides with the direction of the axis x_0. Let us relate the coordinate system $Dxyz$ with the body, where the axis Dx is directed along the holder and axes Dy and Dz are rigidly related to the plane domain.

The coordinates of the point N in the system $Dxyz$ have the form $(0, y_N, z_N)$. Analogously to the problem of the free body motion, we introduce the function $R(\alpha)$ and also the angle β measured in the plane Dyz. In thus case for simplicity, let S. A. Chaplygin properties (10) hold. For any admissible function $R(\alpha)$, the analysis is performed analogously.

If the body is dynamically symmetric (I_1 and $I_2 = I_3$ are principal moments of inertia in the system $Dxyz$) and (Ω_x, Ω_y, and Ω_z) are projections of the angular velocity in the system $Dxyz$, then the equations of motion take the following form, which is analogous to (13), (14) ($H_1 = 0$, for simplicity):

$$\Omega_y' = -n_0^2v_D^2 \sin\alpha\cos\alpha\sin\beta,\ \Omega_z' = n_0^2v_D^2 \sin\alpha\cos\alpha\cos\beta. \tag{61}$$

The resistance force admits the existence of the first integral (3), and, in this case, the condition $\Omega_{x0} = 0$ is taken into account in equations (61).

Let us introduce the angles (θ, ψ) determining the orientation of the pendulum (Fig. 2). The angle θ is measured from the axis x_0 to the holder, whereas ψ is measured from the projection of the holder on the plane $O\text{y}_0\text{z}_0$ to the axis y_0. Then

$$\cos\theta = \cos\psi\cos\phi, \quad \sin\theta\cos\psi = \cos\psi\sin\phi, \quad \sin\theta\sin\psi = \sin\psi. \tag{62}$$

2.4.1 Complete System of Equations

The relations connecting (v_D, α, β) and $(\theta, \psi, \Omega_y, \Omega_z)$ (l is the holder length) have the form

$$\begin{aligned} v_D\cos\alpha &= -v_\infty\cos\theta, \\ v_D\sin\alpha\cos\beta &= l\Omega_z + v_\infty\sin\theta\cos\psi, \\ v_D\sin\alpha\sin\beta &= -l\Omega_y - v_\infty\sin\theta\sin\psi. \end{aligned} \tag{63}$$

Later on, we have

$$\begin{aligned} \dot\theta &= -\Omega_y\frac{\sin\phi}{\cos\psi}, \\ \dot\phi &= \Omega_z + \Omega_y\sin\phi\frac{\sin\phi}{\cos\psi}, \\ \dot\psi &= \Omega_y\cos\phi, \end{aligned} \tag{64}$$

whence we easily deduce that

$$\Omega_y = \frac{\dot{\psi}}{\cos\phi}, \ \Omega_z = \dot{\phi} - \dot{\psi}\frac{\sin\phi}{\cos\phi}\frac{\sin\psi}{\cos\psi}. \tag{65}$$

Using properties (62) and (65), we have the identities

$$\Omega_y = \dot{\theta}\sin\psi + \dot{\psi}\frac{\sin\theta}{\cos\theta}\cos\psi, \ \Omega_z = \dot{\theta}\cos\psi - \dot{\psi}\frac{\sin\theta}{\cos\theta}\sin\psi. \tag{66}$$

Equations from (61), (63), and (66) compose a complete system for determining the pendulum motion on the level of the integral $\Omega_{x0} = 0$.

2.4.2 Systems of Differential Equations and Topological Analogy

Starting from three groups of equations two of which are differential and the third is algebraic, it is easy to prove the following proposition.

Theorem 2. *The complete system of the pendulum motion on the tangent bundle to the two-dimensional sphere has the following form:*

$$\ddot{\theta} + ln_0^2 v_\infty \dot{\theta}\cos\theta + n_0^2 v_\infty^2 \sin\theta\cos\theta - \dot{\psi}^2 \tfrac{\sin\theta}{\cos\theta} = 0, \tag{67}$$

$$\ddot{\psi} + \dot{\theta}\dot{\psi}\left(\tfrac{1+\cos^2\theta}{\cos\theta\sin\theta}\right) + ln_0^2 v_\infty \dot{\psi}\cos\theta = 0. \tag{68}$$

As in the case of a free body, system (67), (68) has symmetries. It also has a complete tuple of first integrals, and the angle ψ is a cyclic coordinate.

Theorem 3. *System (67), (68) is topologically equivalent to the system (17)–(20). Therefore, as in the plane case, there is a mechanical analogy between the pendulum in the medium flow and the free body under the presence of a certain non-integrable constraint.*

Remark. The angle α for a free body is equivalent to the angle θ, whereas the angle β is equivalent to the angle ψ. Moreover, for systems (22)–(25) and (67), (68) to be identical, it is necessary to set $l = -\sigma$ and $v_\infty = v$. The constant velocity of the characteristic point of the circular disk for a free body corresponds to the constant velocity of the overrunning flow on the pendulum. The relation $l = -\sigma$ tells us that for a free body, the stationary motion $\alpha \equiv 0$ is exponentially unstable, and for the pendulum, the stationary motion $\theta \equiv 0$ is exponentially stable.

2.5 *Topological Structure of the Phase Portrait of the System Studied*

In this subsection, we present a scheme of global qualitative analysis of the dynamical system (17)–(19) on the whole phase space $\{\alpha, z_2, z_1\}$. For any function $\mathcal{F}$, the phase portrait of system (17)–(19) has the same topological type.

System (17)–(19) has no trajectories having infinitely distant points of the phase space as its α- and ω-limit sets. Moreover, the system has no simple and complicated limit cycles (see also [10]).

2.5.1 Reducing System to the Form Studied

Let us consider the case of the absence of an additional dependence of the functions y_N, z_N on angular velocity (i.e., for simplicity, $H_1 = 0$). For convenience of drawing the three-dimensional phase portrait and preserving the right-oriented coordinate system $\{\alpha, z1_2, z_1\}$, let us make the formal change $\alpha \to -\alpha$. Moreover, because of the existence of symmetries, we study the domain

$$\{(\alpha, z_2, z_1) : -\pi < \alpha < 0, z_1 > 0\}. \tag{69}$$

In this case, the system takes the form

$$\begin{aligned} \dot{\alpha} &= z_2 + b\frac{v}{I_2 n_0}\frac{\mathcal{F}(\alpha)}{\cos\alpha}, \\ \dot{z_2} &= -\frac{v^2}{I_2}\mathcal{F}(\alpha) + z_1^2\frac{\cos\alpha}{\sin\alpha}, \\ \dot{z_1} &= -z_1 z_2\frac{\cos\alpha}{\sin\alpha}, \end{aligned} \tag{70}$$

and under S. A. Chaplygin condition (10), the analytic system has the form

$$\begin{aligned} \dot{\alpha} &= z_2 + b\sin\alpha, \\ \dot{z_2} &= -\sin\alpha\cos\alpha + z_1^2\frac{\cos\alpha}{\sin\alpha}, \\ \dot{z_1} &= -z_1 z_2\frac{\cos\alpha}{\sin\alpha}. \end{aligned} \tag{71}$$

For simplicity, let us study the system (71), and let us "rectify" the field along the cylinders $\{(\alpha, z_2, z_1) : z_2 + b\sin\alpha = 0\}$; precisely, making the change of the phase variables $u = z_2 + b\sin\alpha$, we pass from system (71) to the system

$$\begin{aligned} \dot{\alpha} &= u, \\ \dot{u} &= -\sin\alpha\cos\alpha + z_1^2\frac{\cos\alpha}{\sin\alpha} + bu\cos\alpha, \\ \dot{z_1} &= -z_1[u - b\sin\alpha]\frac{\cos\alpha}{\sin\alpha}. \end{aligned} \tag{72}$$

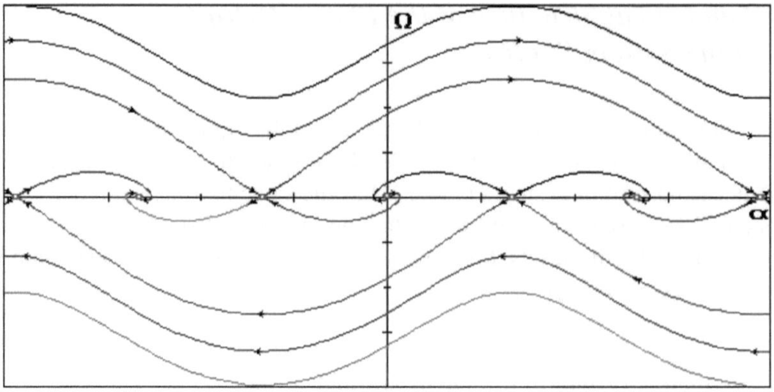

Fig. 3 Phase portrait of the system (72) for $z_1 \equiv 0$ and $\Omega \to u$ formally

For $b = 0$, system (72) (denoted by (72′)) has two analytical integrals. The following but important proposition is obvious.

Theorem 4. *The plane*

$$\{(\alpha, u, z_1) : z_1 = 0\} \tag{73}$$

is integral for the system (70).

Theorem 5. *Plane (73) "contains" the portrait of the system from the plane dynamics (see Fig. 3 if we extend the system field to the lines $\{(\alpha, u) : sin\alpha = 0\}$ by continuity).*

Let us introduce the family of (three-dimensional) layers

$$\Pi_{(\alpha_1,\alpha_2)} = \{(\alpha, x_1, x_2) \in \mathbf{R}^3 : \alpha_1 < \alpha < \alpha_2\}. \tag{74}$$

2.5.2 Conservative Third-Order Comparison System

We have already presented many assertions concerning Poincaré topographical systems (PTS) and more general comparison systems on two-dimensional manifolds [6,7]. To study third-order systems, we need PTS and comparison systems of higher order. We do not dwell on the general theory of PTS and comparison systems of higher dimension and restrict ourselves to its application to the system studied.

Since system (72′) has two analytical integrals

$$\Phi_1 = z_1^2 + u^2 + \sin^2\alpha = C_1^0,\ \Phi_2 = z_1 \sin\alpha = C_2^0,\ C_1^0, C_2^0 = \text{const}; \tag{75}$$

the latter fiber the phase space at each point of which we can draw two surfaces given by relations (75) that intersect along the phase characteristic of system (72′). For each point of the phase space of system (72), let us define two pairs of subspaces in each of which the characteristic of system (72) enters or emanates from. The first integrals (75) "help us" to study the behavior of phase trajectories of system (72) (compare with [14, 15]).

If $g_1 = \{\sin\alpha\cos\alpha, u, z_1\}$, $g_2 = \{z_1\cos\alpha, 0, \sin\alpha\}$ are the gradients of surfaces Φ_1, Φ_2, then the inner products

$$\chi_k = (g_k, \bar{v}),\ k = 1, 2$$

($\bar{v}$ is the vector field of system (72)) have all the properties of the characteristic functions (see also [12]).

Theorem 6. *The characteristic functions χ_k have the form*

$$\chi_1(\alpha, u, z_1) = b\cos\alpha[u^2 + z_1^2], \quad \chi_2(\alpha, z_1) = bz_1\sin\alpha\cos\alpha.$$

2.5.3 Equilibrium Points of System Studied

The system (72) under study has the following equilibrium states:

1. The repelling point $(0, 0, 0)$
2. The attracting point $(-\pi, 0, 0)$
3. The saddles $(-\pi/2, 0, Q_1)$ and $Q_1 \in [0, n_0)$ in each small area parallel to the plane $O\alpha z_2$
4. The centers $(-\pi/2, 0, Q_2)$ and $Q_2 \in (n_0, +\infty)$ in each small area parallel to the plane $O\alpha z_2$

In cases 3 and 4, the singular points are *not isolated*.

Through the whole section, we have performed the qualitative analysis in sufficient detail. The next subsection is a consequence of the previous material.

2.5.4 Phase Portrait Structure

Theorem 6 implies the following assertions:

1. Point 1 is the α-limit set of the separatrices entering points 3 in the layer $\Pi_{(-\pi/2,0)}$.
2. Point 2 is the ω-limit set of the separatrices emanating from points 3 in the strip $\Pi_{(-\pi,-\pi/2)}$.
3. The ω-limit (α-limit) sets of the separatrices emanating from (entering) points 3 in the layer $\Pi_{(-\pi/2,0)}$ (in the layer $\Pi_{(-\pi,-\pi/2)}$) are the same points.
4. The part of the phase space containing points 4 entirely filled with closed trajectories.

2.6 *Trajectories of Spherical Pendulum Motion and Case of Its Nonzero Twist Near Longitudinal Axis*

2.6.1 Pendulum Trajectories on Sphere

In accordance with the properties of the phase space partition into trajectories, the typical trajectories of the point D of the plane domain fall into classes:

1. The trajectories corresponding to the oscillatory domain. Such trajectories are curves on the sphere that unboundedly approach the poles of the sphere (along the flow) as $t \to \pm\infty$.
2. The trajectories corresponding to the rotational domain. Such trajectories are curves on the sphere that almost always fill annulus-like domains on the sphere symmetric with respect to the equator.

2.6.2 Spherical Pendulum Under Nonzero Proper Twist

We immediately present the equations of the pendulum motion under the condition $\Omega_{x0} \neq 0$. These equations have the following form:

$$\ddot{\theta} + ln_0^2 v_\infty \dot{\theta} \cos\theta + n_0^2 v_\infty^2 \sin\theta\cos\theta - \dot{\psi}^2 \frac{\sin\theta}{\cos\theta} - \frac{I_1}{I_2}\Omega_{x0}\dot{\psi}\frac{\sin\theta}{\cos\theta} = 0, \quad (76)$$

$$\ddot{\psi} + \dot{\theta}\dot{\psi}\left(\frac{1+\cos^2\theta}{\cos\theta\sin\theta}\right) + ln_0^2 v_\infty \dot{\psi}\cos\theta + \frac{I_1}{I_2}\Omega_{x0}\dot{\theta}\frac{\cos\theta}{\sin\theta} = 0. \quad (77)$$

Now let us immediately pass to the classification of possible pendulum trajectories on the sphere:

1. Trajectories analogous to trajectories 1 for the case $\Omega_{x0} = 0$. The asymptotics of behavior of such trajectories is the same as above.
2. The trajectories analogous to trajectories 2 for the case $\Omega_{x0} = 0$. Such trajectories are everywhere dense on the whole sphere [15].

Acknowledgements This work was supported by Russian Fund of Basic Research, project 12-01-00020-a.

References

1. Chaplygin, S.A.: On motion of heavy bodies in an incompressible fluid. In: A Complete Collection of Works [in Russian]. Vol. **1**, Izd. Akad. Nauk SSSR, Leningrad 133–135 (1933)
2. Chaplygin, S.A.: Selected Works [in Russian]. Nauka, Moscow (1976)
3. Gurevich, M.I.: Jet Theory of Ideal Fluid [in Russian]. Nauka, Moscow (1979)
4. Samsonov, V.A., Shamolin, M.V.: Body motion in a resisting medium. Moscow Univ. Mech. Bull. **44**(3), 16–20 (1989)

5. Sedov, L.I.: Continuous Medium Mechanics [in Russian]. vol. 1, 2. Nauka, Moscow (1983, 1984)
6. Shamolin, M.V.: Applications of Poincaré topographical system methods and comparison systems in some concrete systems of differential equations. Vestn. MGU, Ser. 1, Mat., Mekh. **2**, 66–70 (1993)
7. Shamolin, M.V.: Spatial Poincaré topographical systems and comparison systems. Usp. Mat. Nauk. **52**(3), 177–178 (1997)
8. Shamolin, M.V.: On integrability in transcendental functions. Usp. Mat.Nauk. **53**(3), 209–210 (1998)
9. Shamolin, M.V.: New Jacobi integrable cases in dynamics of a rigid body interacting with a medium. Dokl. Ross. Akad. Nauk. **364**(5), 627–629 (1999)
10. Shamolin, M.V.: A new family of phase portraits in spatial dynamics of a rigid body interacting with a medium. Dokl. Ross. Akad. Nauk. **371**(4), 480–483 (2000)
11. Shamolin, M.V.: New integrable cases and families of portraits in the plane and spatial dynamics of a rigid body interacting with a medium. J. Math. Sci. **114**(1), 919–975 (2003)
12. Shamolin, M.V.: Comparison of Jacobi integrable cases of plane and spatial body motions in a medium under streamline flow around. Prikl. Mat. Mekh. **69**(6), 1003–1010 (2005)
13. Shamolin, M.V.: Integrability of Some Classes of Dynamic Systems in Terms of Elementary Functions. Vestn MGU, Set. 1, Mat., Mekh. **3**, 43–49 (2008)
14. Shamolin, M.V.: On integrability in elementary functions of certain classes of nonconservative dynamical systems. Contemporary Mathematics and Its Applications. Geometry Mech. **62**, 131–171 (2009)
15. Trofimov, V.V., Shamolin, M.V.: Geometrical and dynamical invariants of integrable Hamiltonian and dissipative systems. Fund. Prikl. Mat. **16**(4), 3–229 (2010)

Index

J. Awrejcewicz (ed.), *Applied Non-Linear Dynamical Systems*, Springer Proceedings in Mathematics & Statistics 93, DOI 10.1007/978-3-319-08266-0

MIX
Papier aus verantwortungsvollen Quellen
Paper from responsible sources
FSC® C105338

If you have any concerns about our products,
you can contact us on
ProductSafety@springernature.com

In case Publisher is established outside the EU,
the EU authorized representative is:
Springer Nature Customer Service Center GmbH
Europaplatz 3, 69115 Heidelberg, Germany

Printed by Libri Plureos GmbH
in Hamburg, Germany